Rethinking Building Skins

Woodhead Publishing Series in Civil and Structural Engineering

Rethinking Building Skins

Transformative Technologies and Research Trajectories

Edited by

Eugenia Gasparri

Arianna Brambilla

Gabriele Lobaccaro

Francesco Goia

Annalisa Andaloro

Alberto Sangiorgio

Woodhead Publishing is an imprint of Elsevier
The Officers' Mess Business Centre, Royston Road, Duxford, CB22 4QH, United Kingdom
50 Hampshire Street, 5th Floor, Cambridge, MA 02139, United States
The Boulevard, Langford Lane, Kidlington, OX5 1GB, United Kingdom

Copyright © 2022 Elsevier Inc. All rights reserved.

No part of this publication may be reproduced or transmitted in any form or by any means, electronic or mechanical, including photocopying, recording, or any information storage and retrieval system, without permission in writing from the publisher. Details on how to seek permission, further information about the Publisher's permissions policies and our arrangements with organizations such as the Copyright Clearance Center and the Copyright Licensing Agency, can be found at our website: www.elsevier.com/permissions.

This book and the individual contributions contained in it are protected under copyright by the Publisher (other than as may be noted herein).

Notices

Knowledge and best practice in this field are constantly changing. As new research and experience broaden our understanding, changes in research methods, professional practices, or medical treatment may become necessary.

Practitioners and researchers must always rely on their own experience and knowledge in evaluating and using any information, methods, compounds, or experiments described herein. In using such information or methods they should be mindful of their own safety and the safety of others, including parties for whom they have a professional responsibility.

To the fullest extent of the law, neither the Publisher nor the authors, contributors, or editors, assume any liability for any injury and/or damage to persons or property as a matter of products liability, negligence or otherwise, or from any use or operation of any methods, products, instructions, or ideas contained in the material herein.

British Library Cataloguing-in-Publication Data
A catalogue record for this book is available from the British Library

Library of Congress Cataloging-in-Publication Data
A catalog record for this book is available from the Library of Congress

ISBN: 978-0-12-822477-9 (print)

ISBN: 978-0-12-822491-5 (online)

For information on all Woodhead Publishing publications visit our website at https://www.elsevier.com/books-and-journals

Publisher: Matthew Deans
Acquisitions Editor: Glyn Jones
Editorial Project Manager: Rachel Pomery
Production Project Manager: Anitha Sivaraj
Cover Designer: Matthew Limbert

Typeset by MPS Limited, Chennai, India

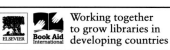

Contents

List of contributors	ix
Foreword	xv
Preface	xvii
Acknowledgements	xxi

1 Façade innovation: between 'product' and 'process' 1
Eugenia Gasparri, Arianna Brambilla, Gabriele Lobaccaro, Francesco Goia, Annalisa Andaloro and Alberto Sangiorgio

2 Façade innovation: an industry perspective 15
Alberto Sangiorgio

Part A Product Innovation 59

3 Urban overheating mitigation through facades: the role of new and innovative cool coatings 61
Mattia Manni, Ioannis Kousis, Gabriele Lobaccaro, Francesco Fiorito, Alessandro Cannavale and Mattheos Santamouris

4 The pursuit of transparency 89
Lisa Rammig

5 Advanced fenestration—technologies, performance and building integration 117
Fabio Favoino, Roel C.G.M. Loonen, Michalis Michael, Giuseppe De Michele and Stefano Avesani

6 Embedding intelligence to control adaptive building envelopes 155
Fabio Favoino, Manuela Baracani, Luigi Giovannini, Giovanni Gennaro and Francesco Goia

7 Biomimetic adaptive building skins: design and performance 181
Aysu Kuru, Philip Oldfield, Stephen Bonser and Francesco Fiorito

8	Building integrated photovoltaic facades: challenges, opportunities and innovations	201
	Francesco Frontini, Pierluigi Bonomo, David Moser and Laura Maturi	
9	Unitized Timber Envelopes: the future generation of sustainable, high-performance, industrialized facades for construction decarbonization	231
	Eugenia Gasparri	
10	Industrialized renovation of the building envelope: realizing the potential to decarbonize the European building stock	257
	Thaleia Konstantinou and Charlotte Heesbeen	
11	Vertical farming on facades: transforming building skins for urban food security	285
	Abel Tablada and Vesna Kosorić	
12	Interactive media facades—research prototypes, application areas and future directions	313
	Martin Tomitsch	

Part B Process Innovation 339

13	The building envelope: failing to understand complexity in tall building design	341
	José L. Torero	
14	Resilience by design: building facades for tomorrow	359
	Mic Patterson	
15	Inverse design for advanced building envelope materials, systems and operation	377
	Roel C.G.M. Loonen, Samuel de Vries and Francesco Goia	
16	Towards automated design: knowledge-based engineering in facades	403
	Jacopo Montali, Michele Sauchelli and Mauro Overend	
17	Additive manufacturing in skin systems: trends and future perspectives	425
	Roberto Naboni and Nebojša Jakica	
18	Mass customization as the convergent vision for the digital transformation of the manufacturing and the building industry	453
	Gabriele Pasetti Monizza and Domink T. Matt	

19	Automation and robotic technologies in the construction context: research experiences in prefabricated façade modules *Kepa Iturralde, Wen Pan, Thomas Linner and Thomas Bock*	475
20	Life cycle assessment in façade design *Linda Hildebrand, Kim Tran, Ina Zirwes and Alina Kretschmer*	495
21	Circular economy in facades *Mikkel K. Kragh and Nebojša Jakica*	519
22	Facades-as-a-Service: a business and supply-chain model for the implementation of a circular façade economy *Juan F. Azcarate-Aguerre, Annalisa Andaloro and Tillmann Klein*	541

Afterword by Ulrich Knaack 559
Index 561

List of contributors

Ajla Aksamija Department of Architecture, University of Massachusetts Amherst, Amherst, United States

Annalisa Andaloro Energy Efficient Buildings Group, Institute for Renewable Energy, Eurac Research, Bolzano, Italy

Stefano Avesani Energy Efficient Buildings Group, Institute for Renewable Energy, Eurac Research, Bolzano, Italy

Juan F. Azcarate-Aguerre Department of Architectural Engineering+Technology, Chair Building Product Innovation, Faculty of Architecture and the Built Environment, Delft University of Technology, Delft, The Netherlands

Manuela Baracani Department of Energy, Polytechnic University of Turin, Torino, Italy

Thomas Bock Chair of Building Realization and Robotics, Technical University of Munich, Munich, Germany

Pierluigi Bonomo University of Applied Sciences and Arts of Southern Switzerland (SUPSI), Manno, Switzerland

Stephen Bonser School of Biological, Earth and Environmental Sciences, University of New South Wales, Sydney, NSW, Australia

Keith Boswell Skidmore, Owings & Merrill LLP, San Francisco, CA, United States

Arianna Brambilla School of Architecture, Design and Planning, The University of Sydney, Sydney, NSW, Australia

Alessandro Cannavale Polytechnic University of Bari, Bari, Italy

Andrew Cortese Grimshaw Architects, Sydney, NSW, Australia

Giuseppe De Michele Energy Efficient Buildings Group, Institute for Renewable Energy, Eurac Research, Bolzano, Italy

Samuel de Vries Department of the Built Environment, Eindhoven University of Technology, TU/e, Eindhoven, The Netherlands

John Downes Lendlease, London, United Kingdom

Fabio Favoino Department of Energy, Polytechnic University of Turin, Torino, Italy

Francesco Fiorito School of Built Environment, University of New South Wales, Sydney, NSW, Australia; Department of Civil, Environmental, Land, Building Engineering and Chemistry, Polytechnic University of Bari, Bari, Italy

Francesco Frontini University of Applied Sciences and Arts of Southern Switzerland (SUPSI), Manno, Switzerland

Eugenia Gasparri School of Architecture, Design and Planning, The University of Sydney, Sydney, NSW, Australia

Giovanni Gennaro Department of Energy, Polytechnic University of Turin, Torino, Italy

Luigi Giovannini Department of Energy, Polytechnic University of Turin, Torino, Italy

Francesco Goia Department of Architecture and Technology, NTNU Norwegian University of Science and Technology, Trondheim, Norway

Charlotte Heesbeen Department of Architectural Engineering+Technology, Chair Building Product Innovation, Faculty of Architecture and the Built Environment, Delft University of Technology, Delft, The Netherlands

Linda Hildebrand Chair for Reuse in Architecture, Faculty of Architecture, Rheinisch-Westfälische Technische Hochschule (RWTH) Aachen, Aachen, Germany

Kepa Iturralde Chair of Building Realization and Robotics, Technical University of Munich, Munich, Germany

Nebojša Jakica Department of Technology and Innovation, University of Southern Denmark (SDU), Odense, Denmark; Section for Civil and Architectural Engineering, University of Southern Denmark (SDU), Odense, Denmark; FACETS lab, Odense, Denmark

Tillmann Klein Department of Architectural Engineering+Technology, Chair Building Product Innovation, Faculty of Architecture and the Built Environment, Delft University of Technology, Delft, The Netherlands

Ulrich Knaack Institute of Structural Mechanics and Design, Chair of Façade Structures, Department of Civil and Environmental Engineering, Technical University of Darmstadt, Darmstadt, Germany; Department of Architectural Engineering+Technology, Chair Design of Construction, Faculty of Architecture and the Built Environment, Delft University of Technology, Delft, The Netherlands

Thaleia Konstantinou Department of Architectural Engineering+Technology, Chair Building Product Innovation, Faculty of Architecture and the Built Environment, Delft University of Technology, Delft, The Netherlands

Vesna Kosorić Balkan Energy AG, Starrkirch-Wil, Switzerland; Daniel Hammer Architekt FH AG, Olten, Switzerland

Ioannis Kousis Department of Engineering, CIRIAF—Interuniversity Research Center on Pollution and Environment 'Mauro Felli', University of Perugia, Perugia, Italy

Mikkel K. Kragh Department of Civil and Architectural Engineering, Aarhus University, Aarhus, Denmark

Alina Kretschmer Chair for Reuse in Architecture, Faculty of Architecture, Rheinisch-Westfälische Technische Hochschule (RWTH) Aachen, Aachen, Germany

Aysu Kuru School of Built Environment, University of New South Wales, Sydney, NSW, Australia; School of Architecture, Design and Planning, University of Sydney, Sydney, NSW, Australia

Thomas Linner Chair of Building Realization and Robotics, Technical University of Munich, Munich, Germany

Gabriele Lobaccaro Department of Civil and Environmental Engineering, Faculty of Engineering, NTNU Norwegian University of Science and Technology, Trondheim, Norway

Roel C.G.M. Loonen Department of the Built Environment, Eindhoven University of Technology, TU/e, Eindhoven, The Netherlands

Mattia Manni Department of Engineering, CIRIAF—Interuniversity Research Center on Pollution and Environment 'Mauro Felli', University of Perugia, Perugia, Italy

Domink T. Matt Faculty of Science and Technology, Free University of Bozen-Bolzano, Bolzano, Italy

Laura Maturi Energy Efficient Buildings Group, Institute for Renewable Energy, Eurac Research, Bolzano, Italy

Michalis Michael Engineering Department, Glass and Façade Technology Research Group, University of Cambridge, Cambridge, United Kingdom

Jacopo Montali Algorixon Srl, Parma, Italy

David Moser Energy Efficient Buildings Group, Institute for Renewable Energy, Eurac Research, Bolzano, Italy

Roberto Naboni CREATE, Odense, Denmark; Section for Civil and Architectural Engineering, University of Southern Denmark (SDU), Odense, Denmark

Philip Oldfield School of Built Environment, University of New South Wales, Sydney, NSW, Australia

Mauro Overend Department of Architectural Engineering+Technology, Chair of Structural Design & Mechanics, Delft University of Technology, Delft, The Netherlands

Wen Pan Chair of Building Realization and Robotics, Technical University of Munich, Munich, Germany

Gabriele Pasetti Monizza Process Engineering in Constructions, Fraunhofer Italia Research, Bolzano, Italy

Saverio Pasetto Skanska, London, United Kingdom

Mic Patterson Façade Tectonics Institute, Newington, CT, United States; University of Southern California, Los Angeles, CA, United States

Sophie Pennetier Enclos, Los Angeles, CA, United States

Lisa Rammig Eckersley O'Callaghan, Los Angeles, CA, United States; Delft University of Technology, Delft, The Netherlands

Kieran Rice Aecom, Melbourne, VIC, Australia

Damian Rogan Eckersley O'Callaghan, London, United Kingdom

Davina Rooney Green Building Council of Australia, Sydney, NSW, Australia

Alberto Sangiorgio Grimshaw Architects, Sydney, NSW, Australia

Mattheos Santamouris School of Built Environment, University of New South Wales, Sydney, NSW, Australia

Michele Sauchelli WSP UK Ltd, London, United Kingdom

Haico Schepers ARUP, Sydney, NSW, Australia

Dominic Shillington Laing O'Rourke, Sydney, NSW, Australia

Abel Tablada Faculty of Architecture, Technological University of Havana J.A. Echeverría, Havana, Cuba

Martin Tomitsch School of Architecture, Design and Planning, The University of Sydney, Sydney, NSW, Australia; Central Academy of Fine Arts, Beijing Visual Art Innovation Institute, Beijing, P.R. China

José L. Torero Civil, Environmental and Geomatic Engineering, University College London, London, United Kingdom

Kim Tran Chair for Reuse in Architecture, Faculty of Architecture, Rheinisch-Westfälische Technische Hochschule (RWTH) Aachen, Aachen, Germany

Ina Zirwes Chair for Reuse in Architecture, Faculty of Architecture, Rheinisch-Westfälische Technische Hochschule (RWTH) Aachen, Aachen, Germany

Marc Zobec Permasteelisa, Sydney, NSW, Australia

Foreword

Building skin is one of the most significant contributors to the energy budget and the comfort parameters of any building. As energy and other natural resources continue to be depleted, it has become clear that technologies and strategies that allow us to maintain our satisfaction with interior environment while consuming fewer of these resources are major objectives for the contemporary facade designs. Moreover, life cycle considerations for facade design, engineering, production, installation, maintenance and disassembly are integral for the future of building skins. Therefore systematic and integrated approach for design, engineering, construction and operation is essential, and collaboration between different stakeholders is paramount for this endeavor.

Building facades perform two primary functions: first, they are the barriers that separate a building's interior from the external environment; second, more than any other component, they create the image of the building. High-performance facades can be defined as exterior enclosures that use the least possible amount of energy to maintain a comfortable interior environment, which promotes the health and well-being of buildings' occupants. This means that high-performing facades are not simply barriers between interior and exterior; rather, they are building systems that create comfortable spaces by actively responding to the building's external environment and significantly reduce buildings' energy consumption. In high-performance facade design, it is crucial to consider climate and environmental factors, materials, components, building function, methods for improving occupants' comfort, fabrication, construction, impacts on other building systems, operation, and ultimately deconstruction and/or refurbishment. Therefore emerging building technologies, systems, tools and production methods are essential in our quest to improve building skin performance and consequently performance of buildings and the built environment.

This book presents strategies, technologies and research trajectories associated with contemporary building facades. The primary goal of the book is to discuss trends that can transform facade industry, including design, engineering, fabrication, construction, operation, maintenance and end-of-life considerations. The book is organized into introductory section that outlines the role of building facades in the built environment and current challenges, followed by an overview of state-of-the-art technologies and methods. Then, two major parts present different global perspectives on product innovation (Part A) and process innovation (Part B). Product innovation refers to new materials, technologies and systems, new concepts and forms, and new needs and performance requirements. Process innovation considers new design methods, fabrication and construction techniques, and operational modes.

Chapters in *Part A: Product Innovation* discuss relationships between facades and urban design, coatings and transparency, advanced fenestration systems, adaptive facades and intelligent systems, biomimicry and building skin design, energy-producing facades, modular timber envelopes, retrofitting strategies for existing buildings, vegetated systems and interactive media facades.

Chapters in *Part B: Process Innovation* present complexities in tall building facade design, resiliency, inverse design methods, automated design and knowledge-based engineering, additive manufacturing, mass customization and digital transformations, robotic technologies for prefabrication, life cycle assessments and circular economy as it relates to building skin and supply-chain model.

The remarkable aspect of this book is that it provides different viewpoints and perspectives from faculty members, researchers and industry professionals around the world. It sets the course of action for transforming the design, engineering and construction of facade systems. Innovation requires intense research and development, and the current challenges facing our industry can only be solved by dedicated research, technological developments and solutions that improve the performance of building skin and ultimately our buildings and the built environment. Strategies and methods that are presented in this work can certainly help us in our quest to create better performing, environmentally sensitive enclosures of the future.

I hope that you will enjoy reading this important book and that you will find inspiration for further innovations in facade systems. Ultimately, we are designing our buildings for people — to be inhabited, used for learning, working and all other aspects of our lives. Future innovative building skins can create beautiful, well-performing, inspirational buildings that improve our environment and our well-being and perhaps even impact our happiness. Perhaps in the future, we might start measuring our buildings according to the level of happiness that they provide to their occupants — until then, let us focus on improving their performance through innovative design, production and collaboration.

Ajla Aksamija
Department of Architecture, University of Massachusetts Amherst,
Amherst, United States

Preface

The world we live in today is facing unprecedented challenges which are shaking our society and communities at their core. The health and economic crises caused by the global pandemic together with the incremental demand for climate action have been urging many industries to reconsider the way they operate and function towards more sustainable models, efficient approaches and equitable and safe practices.

Building construction is the largest industry in the global economy and one of the great accomplishments and incremental sustained innovation. Over the last few decades, building construction has been significantly underperforming showing on average very modest productivity growth. The industry is highly fragmented and risk-averse, which make it less responsive to adaptation and transformational change. It is less digitised than nearly any other industry and historically reliant on manual labour, both factors contributing to amplify its vulnerability to present and future disruptions, such as, for instance, the recent COVID-19 global disruption pandemic. The building construction sector is also responsible for considerable resource depletion and greatly contributes global energy consumption and carbon emissions.

The way buildings are designed, assembled, used and maintained and ultimately disassembled needs to undergo a profound transformation in the very near future. Recognising the complexity of the sector ecosystem is central to promote better practices of knowledge sharing and cross-disciplinary collaboration. Systemic innovation and sectoral digitalisation are key factors in driving the change and reframing the construction paradigm for the industry to meet global sustainability goals towards decarbonisation, keep up with the ever-changing needs of our society and grow prosperously.

Building skins play a pivotal role in architecture. They determine the aesthetic appearance of a building, its relationship with the surrounding context, and they are the interface between the built and human dimensions of our cities. They mediate between the indoor and outdoor environments, largely contributing to the overall building energy performance, and to occupants' comfort and well-being. Ultimately, facades dramatically influence the construction process efficiency and safety of a building, particularly in high-rises that have established a relevant presence in our highly urbanised cities. Therefore façade technologies have enormous potential to drive change in the construction sector and represent a perfect incubator to explore opportunities for transformative innovation.

Rethinking Building Skins: Transformative Technologies and Research Trajectories builds up on these premises by showcasing major research trends in the field of façade technology throughout its entire life cycle, including design, fabrication, construction, operation and maintenance and, ultimately, end of life. The book opens with the editors' critical analysis on the development of façade technologies through the 'fil-rouge' of innovation, identifying main stepping-stones for advancement over the last century and future trajectories to rethink the facades of tomorrow. The second chapter gives voice to a rich array of highly influential experts from the different sectors of the industry, who share their views on the future of façades and discuss major challenges and opportunities to innovation. The main body of the book presents a compelling and highly diverse collection of the most forward-looking research projects in façade design and applied technologies from top-tier international research groups. The research themes are organised into two main domains, one dealing with *product* innovation and the other with *process* innovation. The dichotomous structure is deemed functional to the narrative, but the two proposed domains are not to be intended as separate entities, rather as bodies of knowledge meant to inform and complement one another.

Part A: Product Innovation focuses on the interpretation of façade as a tangible asset, providing an overview of the most advanced products, components, technologies and systems, being them close to market or experimental rising star innovations.

Part B: Process Innovation focuses on the façade project life cycle and value-chain development, providing insights into different phases of a project and showcasing novel cross-industry approaches to current and perspective challenges in the façade world.

The look at innovation within each domain is bifocal. It includes studies which thoughtfully address current challenges to guarantee a more sustainable, resilient, and environmentally conscious future, as well as studies exploring new opportunities brought about by the digital era to achieve a more efficient, transparent, collaborative and connected future.

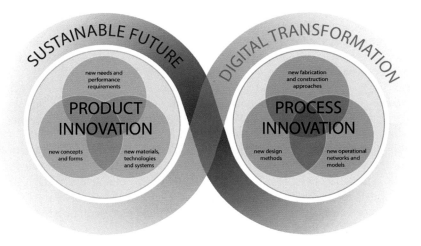

The book is not a theoretical or technical manual nor a specialist guide for the practice. It rather aims to be a platform for innovation, creating a fertile ground for a multilevel, multidisciplinary discussion and contributing to the definition of an international research agenda in this sector for the years to come. Our intent is to reach a wide audience, encompassing academics, practitioners and students from different discipline background and expertise, as well as innovation enthusiasts.

To date, this book is unique in its genre. It brings together the expertise, creativity and critical thinking of more than 60 global innovators and leading experts in the field of façades, both from academia and the industry.

We hope this is only the beginning of a participatory discussion on the topic that opens wide the doors of a future where the principles of 'collaboration' and 'knowledge sharing' are central in catalysing innovation in construction. Let us start 'Rethinking Building Skins'!

Eugenia Gasparri, Arianna Brambilla, Gabriele Lobaccaro, Francesco Goia, Annalisa Andaloro and *Alberto Sangiorgio*

Acknowledgements

We would like to express our gratitude to all authors who contributed with their time, commitment and energy by delivering high-quality chapters on their most forward-looking research works and to the industry experts who have enthusiastically shared their views about future trends in façade innovation.

We also acknowledge those industries and public entities that contributed, either technically or financially, to the development of the research featured in the book's chapters.

A heartfelt thank you to Ajla Akšamija and Ulrich Knaack who have endorsed the value and relevance of this book, respectively, through their foreword and afterword.

Eugenia Gasparri, Arianna Brambilla, Gabriele Lobaccaro, Francesco Goia, Annalisa Andaloro and *Alberto Sangiorgio*

Façade innovation: between 'product' and 'process'

Eugenia Gasparri[1], Arianna Brambilla[1], Gabriele Lobaccaro[2], Francesco Goia[3], Annalisa Andaloro[4] and Alberto Sangiorgio[5]

[1]School of Architecture, Design and Planning, The University of Sydney, Sydney, NSW, Australia, [2]Department of Civil and Environmental Engineering, Faculty of Engineering, NTNU Norwegian University of Science and Technology, Trondheim, Norway, [3]Department of Architecture and Technology, NTNU Norwegian University of Science and Technology, Trondheim, Norway, [4]Energy Efficient Buildings Group, Institute for Renewable Energy, EURAC Research, Bolzano, Italy, [5]Grimshaw Architects, Sydney, NSW, Australia

Abbreviations

AEC	Architecture, Engineering and Construction
AR	augmented reality
BIPV	Building Integrated Photovoltaics
BIST	Building Integrated Solar Thermal
GDP	Gross domestic product
LCA	life-cycle approach
LCC	Life-cycle cost
OLEDs	Organic Light Emitting Diodes
R&D	Research and Development
RES	Renewable Energy Sources
SME	Small and Medium Enterprises
VR	virtual reality

1.1 Introduction

Building construction is the largest industry in the global economy, accounting for 13% of the world's gross domestic product and employing around 7% of the global workforce. However, its annual productivity has only increased by 1% on average over the last 20 years, highlighting a sectoral resistance to change and adaptation (Barbosa et al., 2017).

Innovation in construction has been a highly debated topic for decades. However, nowadays more than ever, it is becoming pivotal worldwide as current social and environmental challenges request rapid and far-reaching measures. Pressing concerns around climate change and resource depletion have put the building construction sector under the spotlight, as responsible for more than 35% of global energy consumption, nearly 40% of global carbon emissions, as well as

Rethinking Building Skins. DOI: https://doi.org/10.1016/B978-0-12-822477-9.00025-5
© 2022 Elsevier Inc. All rights reserved.

ranking amongst the most wasteful industries (IEA, 2019). The magnitude of the problem is further exacerbated by the steep growth of global population, expected to increase by over 2 billion in the next 30 years, together with the fast-paced urbanization that contributes to further congest our densely populated cities (United Nations, 2019). Finally, the present COVID-19 pandemic has been shaking the world population and is forcing to reconsider the way people live and 'inhabit' spaces, toward more sustainable, equitable and safer built environments.

These unprecedented and multifaceted challenges come at a time where advances in digital technologies are opening up a whole new world of possibilities toward automation of manufacturing and industrial processes, the so-called Fourth Industrial Revolution (Industry 4.0). New materials and technologies are rising at the horizon, with the fields of nanotechnologies and robotics gaining momentum. New strategies and models for value creation through circularity have started to be implemented across the market. Global digital trends uptake and cross-sectoral knowledge contamination offer enormous potential to revitalize the construction and building sector, creating favourable conditions for transformational change (Ribeirinho et al., 2020).

In this vibrant panorama, where new vast potentials meet traditional resistance toward change, it becomes natural to question how the façade industry will contribute to shaping the future of the architecture, engineering and construction sector. Indeed, the building envelope, other than the aesthetical, is key to the climate performance, occupant's health and well-being, construction process efficiency and overall environmental impact. Therefore facades represent the ideal playground to explore innovation in construction and play a pivotal role in scaling up the change.

Building on this premises, this chapter—*Façade Innovation: between 'product' and 'process'* —presents a critical analysis of the façade sector through the 'fil-rouge' of innovation. It is organized in two main sections:

- 'Facades today' presents an overview of the main technological advancements in the field of facades to date, provides a picture of the façade industry status quo and examines present challenges and barriers to sectoral innovation.
- 'Facades tomorrow' discusses new opportunities and drivers for the sector to embrace transformational change and defines main research trajectories and pathways for innovation to rethink the facades of tomorrow.

1.2 Facades today

Facades, as we intend them today, have dramatically changed over the history of building design and construction, acquiring stand-alone prominent architectural roles with respect to the structural part of the building fabric. Over the last century, façade design has become a central and independent technical discipline in the construction industry, attracting the interest of many and initiating a series of new roles in the field where façade experts have different backgrounds and specializations, being them architects or designers, building technologist and engineers and sometimes also modern critics as 'web-influencers' (Fig. 1.1). Nowadays, the façade industry is a fully formed, mature sector with dedicated market products,

Façade innovation: between 'product' and 'process' 3

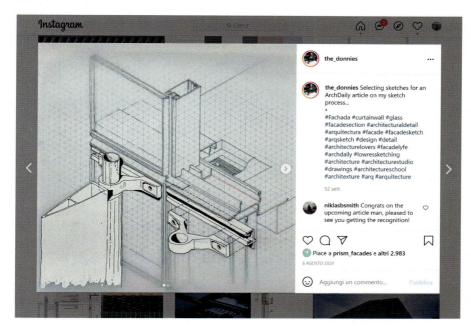

Figure 1.1 Façade sketches from @the_donnies. Credit: Troy Donovan. Troy is Principal at Prism Facades (Sydney, Australia) and has more than 300k followers on Instagram.

technologies and systems as well as an ecosystem of established professionals and entities eager to contribute to sectoral innovation and development.

This evolutionary climax has taken place in a relatively short time frame, thanks to the advancements in material science and production techniques brought about by the industrial revolution. At the same time, the sudden market growth and sector specialization has resulted in an extremely competitive environment, characterized by process scattering and fragmentation that hinder innovation momentum. This section draws upon these considerations in two steps:

- It navigates through major technological advancements occurred during the last centuries that contributed to redefine and transform the façade as a 'product'. It discusses and reflects upon the major drivers to innovation in relationship with their enabling factors.
- It presents the 'process' of façade design and construction today, by identifying the relationship among main actors and activities. It discusses and reflects upon major barriers to innovation and opens the floor to new perspectives and ideas for the sector to prosper and thrive moving forward.

1.2.1 Façade technological advancements: a 'product' innovation narrative

Façade systems classification is today vast and multifaceted and terminology encompasses a wide range of definitions. The juxtaposition between opaque multilayer and transparent envelopes is acknowledged by the market and is employed within this section for the purpose of the narration. Both groups have been evolving through history pursuing improved energy efficiency and comfort performance, enabled by technical advancements in the field of material science, design and industrialization.

The façade was initially a solid structural wall supporting the roof, optionally equipped with holes for letting air or light in. Made of a single material, product, or product combination, such as stone or bricks and mortar, it provided for structural resistance and shelter. Rain and wind were kept out from apertures by moveable elements, such as windows, doors, or shutters. The need for optimized use of resources brought in a transition from stacked to infill walls, which set the basis for the disassociation of the façade from the sole load-bearing function. Along with novel indoor environment comfort requirements and energy consumption concerns, the façade system developed into a more complex system where multiple layers ensured the achievement of better performance and climate control levels (Knaack, 2014).

The uptake of new structural concepts favouring open floor plan configurations and higher flexibility of spaces allowed the façade to develop further as an independent element, the architectural and technological identity of which spans beyond any load-bearing duty. This brought to a progressive increase of envelopes' glazed surfaces, from window to wall size, which soon became a global trend in high rises as a way to optimize daylight penetration, therefore maximizing floor plates' depth and increasing buildings' value.

Transparency in façade evolved continuously starting from the 19th century in America, where mass production techniques, such as the invention of the Bessemer process, introduced the use of timber for window frames and some use of wrought iron frames for larger industrial buildings where buildings' own weight have been carried by structural steel frames in place of masonry walls. Furthermore, the Industrial Revolution witnessed the exponential growth of new factories with larger glazing area (Condit, 1975). However, stone, bricks and iron, the structural resistance of which is counterbalanced by heavy weight, limiting building height, were still being used to form the spandrels or solid sections of wall, yet no longer load bearing. During the skyscraper boom of the 1930s, steel engineering unleashed the potential of high-storied buildings characterized by lighter weight and this form of construction has been widely employed and developed until present times. In parallel, the advent of modernist architecture brought a gradual elimination of glazed decoration areas and moved the focus on their functionality. In that regard, architects began to explore how the façade could be disassociated form its structural function to achieve higher transparency.

From the post—Second World War, architectural styles pulled together several new materials facilitating three major stages of the modern curtain wall developments. The first stage (1970—80) sees simple designs, mainly stick systems and on-site assembly and visible aluminium frame. Aluminium became increasingly common in construction due to its robustness and ductility as well as the advantage to be extruded into fine and precise shapes and then tempered for higher strength, making it an ideal material to form window sections. The following decade, 1980—90, witnessed an increased application of systems designed and built according to standards and manufacturer recommendations, with stick systems still dominating on others. Only from 1990, given the challenge of more and more high-rises construction, American manufacturers developed stick curtain wall systems by employing the third new component, widely available gaskets and seals, to form complete curtain wall systems. This, coupled with new glass processing techniques on float glass (eg, glass fins, point support glazing), has enabled a further push for transparency and lighter construction wall systems. Furthermore, unitized systems started to take over in larger buildings due to their economic benefits and ease of installation, as well as improved energy performance. The economic driver is the main difference in development drivers against opaque facades, where cost logics have not been so prominent in driving façade system evolution.

Unitized glazed systems have recently been evolving into advanced active or adaptive façade systems that enable increased comfort control in indoor environments. As a matter of fact, the integration of shading systems within or outside the cavity, coupled with energy systems or distribution ducts installation in the spandrel part of the unitized façade cell, enabled a brand-new construction concept, saving room in the enclosed space thanks to effective allocation of active components above and under floor slabs, leveraging through-floor connections located in the façade space. The same space-saving driver is also being developed in opaque envelopes, especially in the retrofit sector.

Alongside the postwar developments, façade systems continued to evolve in complexity. The current century is characterized by multifunctional and multiperformance building envelopes, designed and developed to optimize building occupants' comfort together with operation costs. The façade system is now asked to work harder than ever before: it should be able to stand up, bear horizontal loads (ie, wind and any earthquakes), adapt to building movements, guarantee air and water tightness, keep air contaminants out, guarantee thermal insulation and avoid overheating, let natural ventilation through (ie, warming sun and cooling breezes), offer views out and daylight, protect from outside noise, fire spreading through the building, intrusion and blasts. The complex integration of all these performance requirements, coupled with aesthetic ambitions, calls for the use of a wide range of materials, technologies and systems, as well as the need for cross-disciplinary, integrated design approaches.

1.2.2 Façade design and construction: a 'process' innovation barrier

Architecture and building construction are complex and dynamic systems. The process of designing, assembling and managing a building involves a consistent amount of different interdependent activities, several actors and technical skills. The façade discipline reflects perfectly the complex and dynamic nature of the building construction sector. As a matter of fact, façade design has become a highly specialized, multifaceted technical discipline characterized by fast progress and evolution. New regulations, norms and standards in the field are continuously established, new technical competences and expertise required and an increasing number of professionals from different backgrounds and knowledge domains involved across the whole process of façade design and construction.

The role of the façade expert (or façade engineer) has become prominent in building construction projects, as the one who is aware of each product's technical specifications and individual component's performance requirements and, at the same time, accountable to deploy optimal façade technological solutions through function integration and holistic design approaches. The façade engineer is also involved in the process of façade design and construction, overviewing all project stages and engaging with the different actors across the value chain, including architects and other design specialists (Fig. 1.2), as well as suppliers, fabricators, installation teams and contractors (Fig. 1.3).

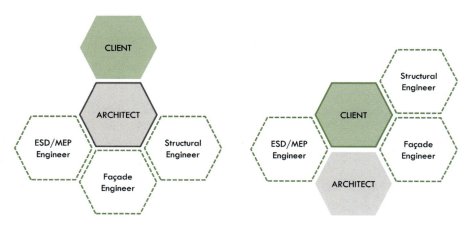

Figure 1.2 Typical contract organization chart—design phase (on the left: engineers employed by the architects; on the right: engineers employed by the client).

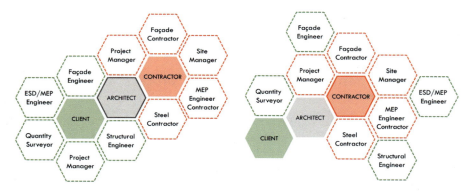

Figure 1.3 Typical contract organization chart—design and construction phase (on the left: tender; on the right: novation).

Façade innovation: between 'product' and 'process' 7

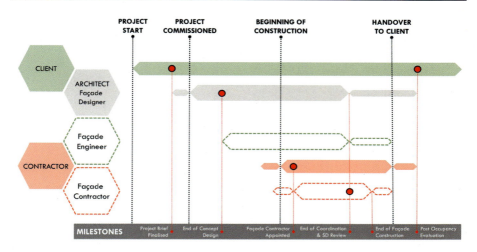

Figure 1.4 Typical project timeline with actors' early-stage involvement in the design and construction phase.

In this context, it seems evident how the organization and management of interactions among the various experts at the different stages of a project is highly challenging and requires substantial efforts in coordination. However, at the current stage, the fragmentation of the design scope among the various designers and consultants involved in façade projects, who are often scattered and/or working in isolation, results in significant process inefficiencies and frictions. As shown in Fig. 1.4, the different actors involved in the process of façade design and construction come into play at different stages of a project, often at a point in time when their efficacy in guiding critical decisions is very limited, or design changes would require substantial investments in time and cost. From an environmental perspective, for example, early-stage involvement of façade engineers and technical experts can enable an integrated façade design outcome. This guarantees high energy performances on a building scale, together with reduced environmental impact and positive contribution toward more comfortable indoor spaces and sustainable urban environments.

Today, the endemic fragmentation of the industry and the related lack of transparency, knowledge isolation and productivity stagnation constitute the main barrier to optimization and innovation in facades and building construction more broadly. The 'one-off' nature of construction projects adds a second dimension to fragmentation, where the lack of knowledge transfer not only happens within the different phases of a project, but also among different projects. Another factor hindering innovation can be identified in the risk-averse nature of the business, particularly affecting the large multitude of small and medium enterprises characterized by negative cash-flow patterns and low margins. This necessarily results in general underinvestment in technical skills development and research and development, with nonhomogeneous growth patterns across the sector in favour of few players who

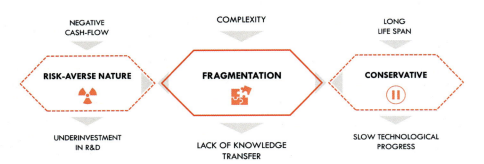

Figure 1.5 Barriers to construction innovation.

can afford large capital investment. Finally, conversely to other technology-based industries, the products of which are oftentimes 'designed to fail', the construction sector is conservative as its assets must serve extended life spans. This logically results in slower technological advancements and more 'cautious' innovation attempts (Fig. 1.5).

In this scenario, it is clear how 'isolated' and 'incremental' innovation widely adopted in construction so far become inadequate to guarantee the sector's sustainable development and economic growth in the years ahead. The construction industry will need to look at the problem holistically and move toward 'systemic-radical' innovation approaches (Blayse & Manley, 2004) where 'business-as-usual' practices are coherently integrated within novel platform models and networks to enable the creation of collective value.

Today, a significant shift in mindset is already permeating the industry at different levels, with several actors embracing sustainable development goals and green practices. Moreover, new digital megatrends introduced by Industry 4.0 seem to offer enormous potential to overcome present barrier to innovation and open the floor to an announced 'construction revolution', also known as 'Construction 4.0'.

1.3 Facades tomorrow

1.3.1 Transformational change: drivers and opportunities

The process of innovation is generally driven by the rise of new needs, being them of 'economical' nature (for instance related to productivity/profitability optimization) or 'sociopolitical' nature (for instance, aiming at the improvement of societal/environmental conditions) (IEA, 2019). In modern capitalistic economies, these two domains are often misaligned or entwined in cost—benefit logics, as the pursuit of the common good can become an obstacle or restrain the pursuit of private profit.

As discussed in the previous sections, facades developments so far have been primarily characterized or enabled by material-related or technological innovations (Knaack, Klein, Bilow, & Auer, 2007), pertaining to the broader domain of 'technical' innovation, either 'product-' or 'process-based' (Blayse & Manley, 2004). Over the last few decades, since the Kyoto protocol agreement was signed in 1997, sustainability has represented the major 'sociopolitical' driver to innovation in construction. However, the idea of sustainability in buildings was mainly bound to the concept of 'operational energy' reduction and occupants' comfort, achieved through better climate control. This contributed to bring a wide range of new façade products, technologies and systems to the market, with the purpose to maximize the performance of the building envelope. However, the increasing demand for more stringent requirements has today reached a point where incremental improvements are either not sufficient or even uneconomical and unsustainable (eg, the thickness of thermal insulation in certain climatic regions). Moreover, the concept of 'embodied energy' across the entire building life cycle has now become prominent, bringing process-based innovation logics at the centre of present and future facades technological developments.

Sustainability in construction has evolved (United Nations Framework Convention on Climate Change UNFCCC, 2015) and acquired different nuances that can lead to the realignment of 'economical' and 'sociopolitical' drivers according to win-win logics, favouring the implementation of 'systemic-radical' innovation pathways that bring entirely new perspectives to the façade (and construction) industry.

> *The view that negative impacts are an inevitable consequence of development has blinded us to the obvious. We could design development to increase the size, health and resilience of natural systems, while improving human health and life quality.*
> **Janis Birkeland**.

Today, the idea of sustainability goes beyond the net-zero energy or carbon paradigms. The reduction of the sector environmental impact is not enough anymore. Instead, the goal of future sustainable design and construction practices is to add or create value through the introduction of innovative platform models where economic, environmental and social goals can thrive simultaneously and harmoniously.

Future sustainability and innovation approaches will not only be 'systemic', thus embracing holistically all the different phases of a project (cradle-to-gate), but they will have to be 'ecosystemic', spanning beyond the traditional life cycle of a project to nourish future developments (cradle-to-cradle). This vision is encompassed in the regenerative and circular economy theories, which aim at decoupling the concept of growth from consumption, rather grounding itself on virtuous models such as resilience and adaptability (UNEP, 2006).

As mentioned in the previous section, Industry 4.0 and digital megatrends uptake represent the key to success and can be major enablers for the construction sector transformation. Despite being at its embryonic stage, industry digitalization has already started and research in the field has been demonstrating huge potential in accelerating innovation processes toward breakthrough, disruptive approaches to the way we design, build and operate facades. Digital platforms will also enable the actualization of new design approaches that will reshape the building construction industry at its core, by leveraging expertise in technology, favouring knowledge sharing and transparent processes, redefining business models for value creation and paving the ground for transformational change through both 'product' and 'process' innovation.

1.3.2 Rethinking building skins: pathways to innovation

In light of what explained throughout this chapter, *Rethinking Building Skins* would imply the exploration of new concepts and forms, advanced materials and technologies, novel approaches to design, more efficient fabrication and construction techniques, innovative operating networks and models. We have envisioned a future that will be profoundly transformed by the digital era, where new innovation models will enable integral economic, environmental and social sustainability to thrive.

With a two-step approach, we aimed at identifying the main research trajectories in façade design and technology, pertaining to the proposed domains of 'product' and 'process' innovation (Fig. 1.6). First, we present a simple and intuitive classification for 'the façade product' of tomorrow, through the identification of four main macro-categories. Currently, there is no clear classification of the different façade systems that is representative of their main functions and characteristics and terminology in the field is rather contextual and often confusing (Carlucci, 2021; Romano, Aelenei, Aelenei, & Mazzucchelli, 2018). As a second step, we illustrate the opportunities for 'the façade process' to uptake digital transformation across each different stages of a project (design, fabrication, construction) and, holistically, by looking at the entire value chain.

Part A: Product Innovation

We have imagined a future where 'the façade product' can be classified in:

A.1 *Eco-active facades*—adopting active strategies to 'give back to' or to 'repair' the ecosystem they work within. This might include conversion and exploitation of energy or biofuel from renewable energy resources, water capture or food harvesting, CO_2 capture and air purification. Among the technologies that could serve these functions are building-integrated photovoltaics, building-integrated solar thermal, wind turbines, algae, hydroponic systems for vertical farming, nanocoatings and/or bioinspired systems for water capture.

A.2 *Re-active facades*—reacting and adapting to changing environmental conditions—such as changes in temperature, wind patterns, atmospheric moisture levels and sunlight, or even extreme weather events—to minimize buildings' energy consumption, optimize occupants' comfort and well-being and guarantee safety and security. This façade systems mimic biological systems adaptation strategies and involve different levels of user intervention, from manual to fully automated control strategies. Technologies that could serve these functions might include the use of smart materials and nanocoatings, photo- or electrochromic devices, dynamic shadings, built-in sensors, artificial intelligence and machine learning for improved building−environment interaction.

A.3 *Inter-active facades*—interacting, informing and communicating via interior or exterior environments with humans, being them, respectively, building's occupants or citizens. This offers new and exciting opportunities for collaboration across the fields of architecture and neuroscience, so far mostly unexplored. Among the technologies that could serve these functions might include the use of media screens for placemaking, data feeds and sensors, organic light-emitting diodes for whole-surface illumination, augmented reality (AR) and virtual reality (VR) for immersive user experience, IoT technologies, cyber-physical systems, artificial intelligence and machine learning for improved building−human interaction.

A.4 *Circular facades*—reducing consumption of finite resources and designing-out waste. This could be enabled for instance by using renewable or carbon-negative materials (eg, timber technologies), products that can be reused or recycled at the end-of-life, plug-and-play and reversible connection systems, modular facades and flexible components that can be easily assembled and disassembled for programmed maintenance and repair across their life-cycle.

The abovementioned identified categories are not meant to be exclusive. Instead, the integration of different functions, methods and operating models is desirable (or perhaps necessary in certain contexts) to maximize benefits and enable 'systemic-radical' innovation in façade design and technology. The technologies listed for each category are not exhaustive and they might have different levels of maturity. Categories A.1, A.2, A.3 define the 'façade product' throughout a set of functionalities guaranteed during operation, while the A.4 category refers to systems materiality and characteristics of assemblies or technologies throughout their life cycle − being therefore inclusive of the former three.

Part B: Process Innovation

Similarly, we have imagined a future where 'the façade process' will integrate:

B.1 *Design methods*—using strategies for knowledge integration through the use of digital twins and pushing the boundaries of automated design through artificial intelligence and machine learning.

B.2 *Fabrication processes*—adopting advanced industrialized processes, which involve for instance the use of additive manufacturing technologies, advanced prefabrication and mass-customization methods, robotics for high-quality and efficient assembly.

B.3 *Construction practices*—applying advanced digital tools to increase site efficiency and safety, such as smart helmets integrating AR and VR technologies for site activities real-time monitoring and activities forecasting, drones providing real-time on-site control and tracking, Building Information Modelling and robotics for efficient and safe control and installation.

B.4 *Life cycle approaches (LCA)*—designing for systems' resilience, adopting LCA and life cycle cost approaches and implementing service platforms and business models that aim at extending the façade service life over the traditional cradle-to-grave approach, for instance, using the façade as a 'material bank'.

The abovementioned identified categories are not to be considered as independent but rather deeply intertwined, particularly when looking at new digitalization horizons. Here again, the technologies listed for each category are not exhaustive and they might have different levels of maturity. Categories B.1, B.2, B.3 deal with three different stages of the 'façade process', while the B.4 category looks at the whole-life process holistically − therefore aiming to 'oversee' and integrate the former three.

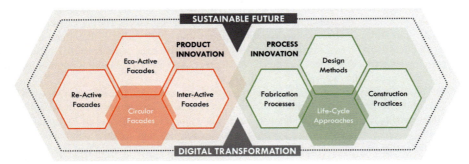

Figure 1.6 Façade Innovation – between 'product' and 'process'.

1.4 Conclusion

This chapter provided an overview of major barriers, drivers and opportunities to innovate the façade industry. It framed the current ecosystem looking at technological advancements throughout recent history which shaped its looks and organization, defining the main challenges ahead for the sector.

Future pathways for innovation are presented according to two domains, the 'product' and the 'process' of facades. As technical and technological innovation has already reached very high momentum, the optimization of processes and industry organizational structure are expected to gain further strength in the years to come, thanks to the disruption brought about by digital transformation and the adoption of Industry 4.0 approaches.

Total sustainability of future façade will be the driver for industry development and economic growth, enabled by radical life cycle approaches to design, construction and fabrication. In this frame, advanced sectoral digitalization will unlock the potential for façade increased integrability, adaptability and climate resilience to meet global targets toward decarbonization, keep up with the ever-changing needs of our society and grow prosperously.

This chapter is a progressive introduction to the book conception, organization and innovation topics, through the narrative of 'product' and 'process' innovation in the field of facades. These two pillars are gradually investigated within the book, benefitting from contributions by a wide group of established experts from academia and the industry toward the definition of an international research agenda for the façade industry to uptake transformational change in future.

References

Barbosa, F., Woetzel, J., Mischke, J., Ribeirinho, M. J., Sridhar, M., Parsons, M., ... Brown, S. (2017). *Reinventing construction: A route to higher productivity*. McKinsey Global Institute.

Blayse, A. M., & Manley, K. (2004). Key influences on construction innovation. *Construction Innovation*, *4*(3), 143−154.

Carlucci, F. (2021). A review of smart and responsive building technologies and their classifications. *Future Cities and Environment*, *7*(1), 10.

Condit, C.W. (1975). *Designing for industry: The architecture of Albert Kahn*.

IEA. (2019). *Global status report for buildings and construction*. Paris: IEA.

Knaack, U. (2014). Potential for innovative massive building envelope systems − Scenario development towards integrated active systems. *Journal of Façade Design and Engineering*, *2*(3−4), 255−268.

Knaack, U., Klein. T., Bilow, M., Auer., T. (2007), *Fassaden. Prinzipien der Konstruktion*, Birkauser: Basel, <https://doi.org/10.1007/978-3-7643-8319-0>.

Ribeirinho, M. J., Mischke, J., Strube, G., Sjödin, E., Blanco, J. L., Palter, R., . . . Andersson, T. (2020). *The next normal in construction: How disruption is reshaping the world's largest ecosystem*. McKinsey & Company.

Romano, R., Aelenei, L., Aelenei, D., & Mazzucchelli, E. S. (2018). What is an adaptive façade? Analysis of recent terms and definitions from an international perspective. *Journal of Façade Design and Engineering*, *6*(3), 65−76. Available from https://doi.org/10.7480/jfde.2018.3.2478.

UNEP. (2006). *Circular economy: An alternative for economic development*. Paris: UNEP DTIE.

United Nations. (2019). *World urbanization prospects: The 2018 revision (ST/ESA/SER.A/420)*. New York, NY: United Nations.

United Nations Framework Convention on Climate Change (UNFCCC). (2015). *The Paris agreement*. New York, NY: UNFCCC.

Façade innovation: an industry perspective

Alberto Sangiorgio
Grimshaw Architects, Sydney, NSW, Australia

Abbreviations

BIM	Building Information Modelling
CWCT	Centre for Window and Cladding Technology
DfMA	Design for manufacturing and assembly
DGU	Double Glazed Units
EC	Embodied carbon
ECI	Early contractor involvement
EPD	Environmental Product Declaration
LCA	Life cycle assessment
LCIs	Life cycle inventories
NZC	Net-zero carbon
PV	photovoltaic
TB	Text boxes
TGU	Triple Glazed Units

2.1 Introduction

Innovation is seldom a unidirectional process. On the contrary, it is usually a convoluted process involving different drivers, players, stakeholders and constraints. The interaction between these factors determines the pace and the potential for innovation to flourish and spread.

Construction projects have always been characterized by a high level of complexity and recent studies indicate that both technical and organizational complexities are gradually increasing (Luo et al., 2017). The effects are already visible within the construction industry, as well as in its relationships with other sectors, from manufacturing to information technology.

Compared to other markets, the construction industry, including the façade sector, comprises a large and growing number of stakeholders: clients, urban planners, architects, engineers, consultants, contractors, subcontractors, fabricators, regulators and certifiers, to name just a few. Stakeholders have different and sometimes competing agendas, targets, risks and contingencies, focus on different aspects of the process (such as design, installation, maintenance and end-of-life), have different expertise and are regulated by a great variety of national and regional building codes. This landscape results in a highly fragmented industry which, as a whole, moves and innovates slowly, often focusing on an incremental and field-specific innovation process in which major opportunities to create radical and cross-field innovation processes are less likely to occur.[1]

The process of 'rethinking building skins' has to start with an understanding of the factors that drive and limit innovation within the façade industry, taking into consideration its relationships with other industries, academia and state government. This chapter presents a wide-ranging overview of the façade industry's perspective on present and future drivers, challenges and opportunities for innovation in building envelopes through contributions from leading experts and companies in the façade design and construction industry. Its purpose is to provide fertile ground for a multilevel discussion across disciplines and fields, to assess the engagement of the façade industry in different research streams and to serve as a catalyst for the development of synergic collaborations.

2.2 Method

Considering the great variety of players and stakeholders involved, the task of rendering an overview of the façade industry in relation to innovation is challenging. The method adopted to meet this challenge was to collect the insights of 11 selected experts affiliated to global tier-one companies, leaders in the façade design and construction industry. These companies, due to their size and revenue, are major players in R&D activities, since big projects and big contracts are necessary to meet the initial costs of development and testing of new ideas and, potentially, generating the benefits of economy of scale. These global companies can be considered the highways on which innovation is developed, tested, applied and, most importantly, spread across the market to other smaller entities and stakeholders and, eventually, across the globe, including developing countries.

This analysis does not claim any statistical relevance, as it interrogates a limited number of purposefully selected experts. However, it creates a multidimensional overview of façade innovation that takes account of the great variety of roles and expertise that characterize the façade industry and includes recognized leaders in different sectors, such as architecture and design, engineering and consultancy, façade contractors and general contractors, as well as policy makers. Further, it acknowledges the importance of differences in geographical context by inviting authors from Australia, the United Kingdom and the United States.

Table 2.1 shows the complete list of industry experts, their affiliated company, the field of expertise and country of operation.

---

[1] Incremental innovation is defined as 'small and based on existing experience and knowledge'; radical or disruptive innovation is defined as 'a breakthrough in science or technology' (Blayse & Manley, 2004).

Table 2.1 List of industry experts.

Ref	Façade expert	Company of affiliation	Field of expertise	Country
TB01	Andrew Cortese	Grimshaw Architects	Architecture-Design	Australia
TB02	Keith Boswell	Skidmore, Owings & Merrill	Architecture-Design	United States
TB03	Kieran Rice	AECOM	Engineering	Australia
TB04	Damian Rogan	Eckersley O'Callaghan—Engineers	Engineering	United Kingdom
TB05	Haico Schepers	ARUP	Engineering	Australia
TB06	Marc Zobec	Permasteelisa	Façade contractor	Australia
TB07	Sophie Pennetier	Enclos	Façade contractor	United States
TB08	Saverio Pasetto	Skanska UK	Contractor	United Kingdom
TB09	John Downes	Lendlease	Contractor	United Kingdom
TB10	Dominic Shillington	Laing O'Rourke	Contractor	Australia
TB11	Davina Rooney	Green Building Council of Australia	Policy maker	Australia

TB, Text boxes.

The experts were asked to provide a written contribution, with an indicative limit of 1000 words, addressing the following two questions:

Question 1: What are the most relevant innovations that you foresee will play a major role in shaping the way the next generations of building envelopes will be designed, procured, built, maintained and dismantled?

Question 2: In your current practice, what are the future innovations you are actively fostering/exploring and how do you expect these innovations to change your role, your approach to design, engineering, fabrication, procurement, construction and the outcome of the design?

The experts were asked to use these questions as guidelines, focusing on their specific field of expertise and personal experience. The structure of the response was intentionally flexible, leaving the contributors free to combine the answers or address them separately.

The following section scaffolds on the industry experts' contributions, summarizing their feedback and providing a snapshot of present and future drivers, challenges and opportunities for innovation in the façade industry. In it, the author provides a commentary that discusses the main points made by the industry experts and contextualizes them in relation to the broader industry and relevant literature to create a mental map of the key themes and topics that are discussed by each expert. The 11 contributions from the experts are included in full in the 'Industry insights' section in the form of 'text boxes' (TB) and are presented in the same order as shown in Table 2.1.

2.3 Façade innovation: drivers, challenges and opportunities

2.3.1 What drives innovation? Why do we innovate?

These questions are apparently simple, but their answers are complex and embedded in the way our global economy and society operate. During his career the Italian design Master, Bruno Munari, championed creativity and radical innovation and proposed an interesting theory: 'Laziness is the engine of progress. It is the stimulus that pushes us to obtain what we desire with the minimum of physical effort; the maximum result with the minimum effort is now a law of economics'.[2] (Munari, 2018).

In more general terms, industrial innovation aims to improve the operations of a specific field with reference to specific targets. In the façade construction industry, these targets are as follows:

Economic/productivity growth: This is the type of innovation to which Munari refers it is driven by what he playfully calls laziness, but which can be related to the more conventional ideas of efficiency, optimization and competition between

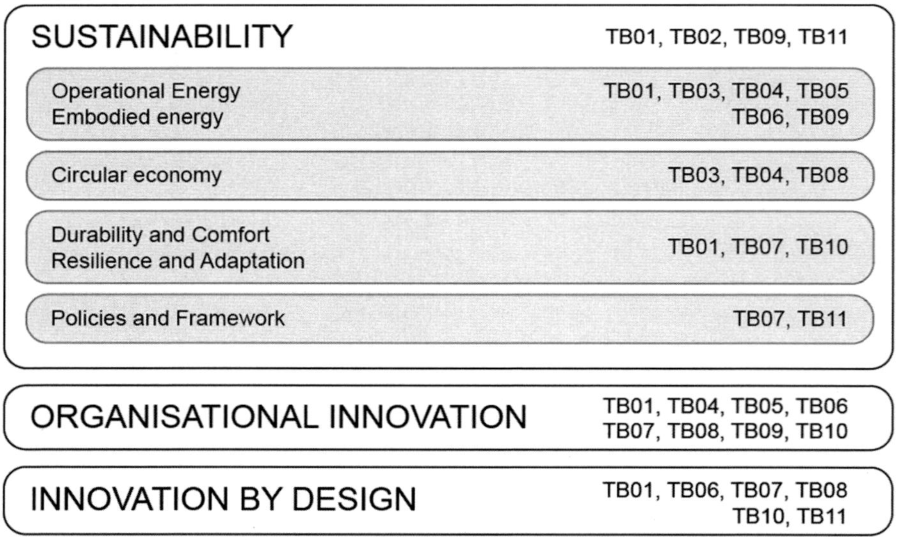

SUSTAINABILITY	TB01, TB02, TB09, TB11
Operational Energy Embodied energy	TB01, TB03, TB04, TB05 TB06, TB09
Circular economy	TB03, TB04, TB08
Durability and Comfort Resilience and Adaptation	TB01, TB07, TB10
Policies and Framework	TB07, TB11

ORGANISATIONAL INNOVATION	TB01, TB04, TB05, TB06 TB07, TB08, TB09, TB10

INNOVATION BY DESIGN	TB01, TB06, TB07, TB08 TB10, TB11

Figure 2.1 Summary of the main topics and contributions from the industry experts cited in the text.

---

[2] 'La pigrizia è il motore del progresso. È lo stimolo che ci spinge a ottenere ciò che desideriamo facendo il minimo di fatica fisica; il massimo risultato col minimo sforzo è ormai una legge di economia'. Translated by the author from Munari (2018).

companies. One example is the innovation that aims to find new fabrication processes to improve a product or reduce its time and cost of production (Slaughter, 1998).

Social/environmental drivers: In this case, innovation aims to improve humans' personal and social conditions with respect to comfort, safety, health and well-being. It is enabled by the adoption of policies or codes. Innovation as a means to achieve higher targets in sustainability and building performance can be included in this category (Slaughter, 1998).

Creativity: There are innovations that do not target growth from an economic perspective, nor from a social/political one. A good example related to façade construction is research that focuses on the achievement of new architectural languages, including free forms and which is justified by aspects of our society that have more in common with fashion and creativity than with the rationality that characterizes the previous two categories.

In the process of rethinking the facades of the future, innovation will find its *raison d'être* in one or more of the three categories listed earlier. The same applies to all the innovations mentioned by the industry experts in their contributions.

The following sections summarize the key topics and concepts mentioned by the industry experts and divide the discussion into themes. It starts from the most popular topic—sustainability—and moves to the other recurrent themes, linking them with one another in a continuous discussion. References are given to the respective TB in which these topics are mentioned (Fig. 2.1).

2.3.2 Sustainability

Today, buildings are responsible for up to 39% of global emissions, a figure that is expected to grow by 50% by 2050, despite the global effort to minimize their impact (D. Rooney TB11). In this context, facades are pivotal in determining the final energy and carbon consumption of a building. This fundamental role is acknowledged by all industry contributions, which indicate that sustainability is a trending topic and driver for future innovation. The major role played by sustainability comes as no surprise, considering that the market has been increasingly focusing on this topic over recent decades, due to the growing impact of environmental policies. This awareness found new strength during 2019 and 2020, when increasing temperatures and the unprecedented spread of bushfires ramped up the discussion on effects, causes and solutions to climate change, while the global pandemic put health and well-being in the built environment under the spotlight.

The aftermath of the Paris Agreement and the Sustainable Development Goals released in 2015 saw an increase in global awareness of sustainability across all fields and especially in construction. In a strong response, most of the signatory countries committed to delivering net-zero carbon (NZC) new constructions by 2030 and to operate within a completely NZC by 2050.

To facilitate this process the private industrial sector and, more specifically, the construction industry reacted with the creation of voluntary organizations such as the World Green Building Council and Architect/Engineers/Contractors Declare, among others. The aim is to raise awareness of climate and biodiversity emergencies, to advocate for faster change in the industry toward sustainable and regenerative design practices and to promote the exchange of knowledge and research. The ultimate goal is to reduce greenhouse gas emissions in the construction industry, in alignment with the international commitments.

The success and relevance of these initiatives across the construction sector and the increasingly tight building codes and certification requirements, demonstrate a growing social responsibility toward the environment, which is reflected in the industry experts' contributions: today and in the next 30 years, innovation in building skins will primarily focus on reducing energy and carbon emissions. 'Creating a pathway to net carbon zero is most likely the biggest challenge that the property industry has ever faced' (J. Downes—TB09), but not enough has been done in designing toward the NZC goals. However, the industry as a whole is slowly moving in the right direction, acknowledging the urgency of the problem and upskilling to reach the set targets.

From a building and façade design perspective, the journey toward NZC is already creating a paradigm shift, gradually moving away from a *design-for-compliancy* approach —where a building's energy and carbon performances are assessed and eventually adjusted only after the completion of the concept design phase—to an *integrated design* approach. In the latter, architects and consultants work together to achieve a holistic balance among performance against all targets, optimizing the consumption of energy and carbon 'without losing focus on the other fundamental qualities of buildings and facades, such as amenity, variety, experience, visual connection, etc'. (A. Cortese—TB01).

In this context, innovations that focus on only one of these aspects, overlooking the others, will be unlikely to succeed (K. Boswell—TB02). Finding a balanced approach to innovation will require novel early-stage data-based design methods and workflows, as well as multidisciplinary environments and procurement routes to allow designers and engineers to work together with contractors. In such a truly integrated design approach, it is possible to make informed decisions and align design and construction performances.

This new arrangement will be necessary to achieve good and holistic zero-energy building design and it will require a reassessment of the role of the architect and the design team in general. These will have to extend their scope of work from concept to completion, expanding their knowledge and expertise to include deeper awareness of performative design, construction processes and advanced methods of research (K. Boswell—TB02). The advent of new high-performing envelopes will also be an opportunity for architects to renovate the façade design language, so that 'the array of architecture and sustainability elements [...] are rationalized and used to build the tectonic and the filigree of the facades' (A. Cortese—TB01).

In any analysis of the relationship between sustainability targets and innovation with a focus on the NZC goals, it is important to identify the main parameters in the façade design and construction process that have an impact on carbon emissions. These are as follows:

- operational energy and embodied energy;
- 'circular economy', that is, the use and disposal of resources;
- durability and comfort, resilience and adaptation; and
- policies and frameworks.

Operational energy and embodied energy

In the last 20 years, one of the primary objectives of building policies and, consequently, of façade design, has been to improve the thermal performance of the building skin as a means of reducing the heating and cooling energy consumption. However, very little focus has been given to a whole-of-life carbon approach (K. Rice—TB03; D. Shillington—TB10).

This has resulted in a focus on the development and diffusion of a new generation of high-performance facades, designed to suit the new challenging targets in operational energy. Different strategies have proved successful in addressing low energy targets, leading to the development of new façade typologies that include, for example, high-performance defensive, reactive and active (A. Cortese—TB01), as well as dynamic facades. Dynamic facades in particular are repeatedly cited in most industry contributions, which acknowledge their great potential in managing and optimizing daylight and solar gains under changing weather conditions. This is due to their responsive behaviour, which works similarly to human skin (M. Zobec—TB06).

In the future, this new generation of high-performance dynamic facades will require sensors, motors, moving components and, ultimately, integration with the building management system and smart home appliances, or the Internet of things (H. Schepers—TB05). These components will allow the façade to meet the needs and comfort requirements of the building occupants while decreasing the building's operational energy usage and carbon emissions. The development of smart solutions to manage the increasing complexity of future static and dynamic facades will be a fundamental step to ensure the success of these technologies, especially when the operational energy benefits are compared to the potential downsides in relation to service life, maintenance costs and increased embodied energy (D. Rogan—TB04).

In recent years, growing interest in embodied energy and carbon has prompted new policies and pathways intended to limit the CO_{2e} embedded in building components, especially facades and structures, which are recognized as the main contributors to the total embodied energy content (Azari & Abbasabadi, 2018). This has allowed new tools for carbon accounting to be implemented and led to the development of new life cycle inventories (LCIs).

Today, the use of life cycle assessment (LCA) software as a design tool is still subject to debate, due to its limitations. Indeed, LCA software and embodied calculators are limited by many factors that can jeopardize their use, especially in the early stages of façade design. The first factor is represented by the market's limited knowledge of embodied energy (J. Downes—TB09). A second limiting factor is the low accuracy of the input and the lack of accurate, consistent, local-specific data and standard methodology. Both factors affect the reliability of the tools: assessing the same design with the different tools that are available can lead to great differences in the embodied energy accounting (Dixit et al., 2012). To respond to these challenges, the next generation of building skins will need more precise and local-specific LCIs, a shared definition of the scope of investigation (scopes 1, 2 and 3 of embodied energy as defined by the Green House Gas protocol) and an agreed method of accounting.

Last but most importantly, one of the great issues associated with the use of embodied energy accounting as a design tool is the number of assumptions required at an early design stage, whereas decisions about materials procurement are finalized only at a later phase, when the contractor is involved (D. Rogan—TB04). This issue touches the foundations of the design process and its resolution will require a deep organizational shift, forcing designers and contractors to work closely to determine façade systems and procurement much earlier in the process.

In the path toward a more informed and truly energy- and carbon-sensible design resolution, another issue that must be addressed during the design phase is the sometime competing nature of embodied and operational impacts. Kieran Rice (TB03) provides a practical example in the form of a project-specific research undertaken by AECOM Victoria (Australia), which aimed to compare the use of triple-glazing unit (TGU) and double-glazing unit (DGU) on a full LCA, measuring the costs/benefits in relation to operational factors and embodied carbon (EC) consumption for both alternatives. The results showed that, for that specific project and climatic context, despite the fact that the TGU seemed to offer operational advantages, the additional glass panel would take 33 years to offset its additional embodied energy, leading to a negligible carbon emission saving over the design life of the building.

The design of future facades will have to address this problem through a whole-of-carbon design approach which relies on multivariable optimization processes necessary to achieve the lowest energy and carbon intensity in the LCA and, ultimately, a zero carbon design.

Circular economy

Another sustainability-related concept that has emerged in recent years will likely be a major driver of future innovation and research in buildings and facades. This concept, the 'circular economy', represents the shift from a linear pattern of consumption that starts with material extraction and ends in disposal to a circular approach, where materials are reemployed at the end of their life (MacArthur, 2013). This shift has the potential to minimize both waste disposal and the demand for new materials (D. Rogan—TB04), with significant benefits for the environment.

The basic principles of circularity were embedded in the economy and construction as long ago as the Middle Ages, or even earlier in the architecture of the Roman Empire. Due to the high cost of materials and craftsmanship, construction materials were a valuable resource, so recycling was most often cheaper than the use of new resources. For this reason, buildings were designed and built to last and their components often reused at the end of the building's life (S. Pasetto—TB08).

The industrial revolution contributed to an increase in the availability of construction materials through the employment of a highly engineered mass production chain, resulting in a reduction in the cost of new materials for the end client. As a consequence, the strategy of 'reuse and recycle' became a less appealing solution, breaking the circularity and opening the door to the current consumerist 'throwaway' model. Of course, this is a crude simplification of a process that took a century to change the way our economy and society work.

The presence of 'loops' in the economy was formally theorized for the first time in an EU report commissioned in 1976, called 'Jobs for Tomorrow: the potential for substituting manpower for energy', which focused on car manufactory and building construction (Stahel & Reday-Mulvey, 1981). Some 30 years later, new ideas in relation to 'circularity' and 'cradle to cradle' were introduced, but very little has been done to implement them in the industry (Braungart & McDonough, 2009).

Some steps have been undertaken on the recyclability front, with increasing the use of recycled material in the fabrication of aluminium, steel and glass. However, most current façade systems employ a variety of materials with unique provenance, making it very difficult to deploy a responsible reuse and recycle process at the end-of-life of the façade elements (D. Rogan—TB04).

A simple example is again provided by DGUs, whereby glass is processed and combined with other components and materials (such as solar control/low-e coatings, sealant, space bar) to improve its thermal performance. Although glass is generally recyclable, its use in a complex product makes the recycling process more complex, expensive and less convenient. A similar point can be made in relation to durability performance: simple float glass has a much longer design life compared to the more complex multiple-glazed products that are the market's standard and which have a lifespan of only 25–30 years (Patterson et al., 2016).

Considering that a more durable and easy-to-recycle product has a higher potential to save energy for fabrication and reduce the amount of material that needs to be disposed of, it can be concluded that the increased complexity necessary to achieve higher targets on operational energy competes with the targets on circular economy and embodied energy.

As a consequence of these intercorrelations between operational energy, embodied energy and circular economy, the façade industry will have to make a 'fundamental reappraisal of what constitutes a carbon efficient building skin', shifting the present focus on operational energy to a more 'nuanced appreciation of what gives the most optimal overall carbon foot print across the building's lifecycle' (K. Rice—TB03).

Durability and comfort, resilience and adaptation

In this fast-changing reality in which the façade systems or 'species' appear to be in competition to increase their success in the market in some kind of Darwinian model, 'threat, adaptation, survival' will be other important drivers for innovation (S. Pennetier—TB07). In the process of 'rethinking building skins', considerations of durability and comfort, resiliency and adaptation of facades will be central in efforts to reduce embodied energy and optimize the use of resources. A durable façade is capable of matching all minimum performance targets established during the design phase, from the beginning to the end of its intended service life.

Today more than ever, we live in an environment characterized by unprecedented rates of change in economic and climatic conditions. In imagining the façade of the future, designers and researchers will have to foresee these factors and take them into account to extend the adequacy and durability of façade solutions, making them resilient and capable of adapting to constant and sudden changes within and outside the building construction industry.

An example of a long-term trend that is already impacting the durability of facades and buildings is the pursuit of higher energy efficiency. Increasingly tighter provisions have confronted indoor environmental health and quality with unprecedented challenges, not always successfully. Evidence can be found in the dramatic increase over the past 50 years in the number of cases of 'sick building syndrome' and mould growth. It has been estimated that the proportion of buildings currently damaged by mould is 45% in Europe, 40% in the United States, 30% in Canada and 50% in Australia (Brambilla & Sangiorgio, 2020). These statistics are the result of multiple factors that include the rapid uptake of new façade systems and materials in the absence of sufficient understanding of their hygrothermal performance, the increase difference in the hygrothermal conditions between internal and external environments due to the broad diffusion of air conditioning systems and more generally poor design practices that fail to address the mid-long-term consequences of these problems.

Over the design life of a building, all these factors have a negative impact on the durability of materials and building components, with consequences for the financial, energy and carbon budgets and a detrimental effect on the building's air quality (Brambilla & Sangiorgio, 2020). Future research will have to address these problems not only to increase the durability of the envelopes but also to improve the health and well-being conditions for the building's users.

An example of sudden change outside the industry that has heavily impacted our lives and the way in which we utilize our buildings is the global pandemic of COVID19, which has escalated discussion around resiliency and adaptation of spaces and facades to suit different functions and occupation patterns. A further source of long-term impact is represented by climate change, which will heavily impact the selection of the new 'species' of building envelopes. In the next 50 years, new buildings and facades will have to respond to restrictions on energy/carbon consumption, while dealing with a range of environmentally related emergencies, such as the increase in temperature and urban heat island effects, the growing level of urbanization and the problem of food security, together with the likely scenario of new global pandemics and economic crises. In this fast-changing environment, design and innovation for 'resilience' will be fundamental for the new generation of facades and, more generally, for the future of façade industry operations (A. Cortese—TB01; D. Shillington—TB10).

Policies and frameworks

All experts agree on the major role that increasingly aspirational sustainability targets will play in reshaping building skins. Nonetheless, high-performance facades, which employ innovative and advanced technologies and materials and sophisticated investigative and data-based design processes, usually imply higher initial costs, which can conflict with the industry's basic principles of competitiveness. Although in most cases this initial investment is balanced if not exceeded by the savings in operational costs—something that is well understood by inspired and well-informed clients—the systematic development, diffusion and success of innovation driven by sustainability/energy performance is still strictly dependent on building codes and policies.

For example, a building developer is less interested in reducing the operational energy and carbon of the building if it is intended for sale or rental after construction. In this kind of scenario, among others, sustainable policies are essential to regulate the construction industry and will play a major role in improving the energy performance of future facades, increasing the minimum standards and then incentivizing the industry to do better (D. Rooney—TB11). A recent example is provided by Sophie Pennetier, who witnessed an instantaneous change in façade glazing from DGU to TGU when New York's Law 97 passed in 2019, placing carbon caps on roughly 50,000 residential and commercial properties across NYC. This case is a perfect example of adoption issues, rather than the lack of technology and knowledge (S. Pennetier—TB07).

Innovation of appropriate technologies driven by sustainability depends on the ability of institutions to respond progressively to climate change (Rodima-Taylor et al., 2012). Regulators and policy makers have a primary role in reestablishing the complex balance between 'economic/productivity' and 'social/environmental' drivers and in the definition of the innovation agenda, including targets and implementation strategies.

In accordance with the recent trends identified by the industry experts, it is certain that future envelopes will have to comply with increasingly stringent regulations and policies, which will not only provide tighter operational energy consumption targets but also thresholds for embodied energy (or carbon emissions) and more specific metrics to measure performance related to the circular economy.

In the future the current so-called green or sustainable design approach, which focuses on the mitigation of a building's environmental impacts, will be implemented with the new regenerative design principles, which interpret buildings as a whole in their social−cultural and environmental context (Cole, 2012). Hence, as part of this trend, we will witness the rise of new regenerative design policies, frameworks and tools aimed at encouraging innovative design solutions capable of shifting the focus from 'green' and 'sustainable' toward regenerative design thinking.

In conclusion, the reappraisal of all sustainability design and construction metrics and parameters has the potential to drive and promote innovation in architectural expressions, materials, fabrications and construction processes. The magnitude of this potential will depend on how well researchers, policy makers and industry work together to define targets, frameworks and pathways to achieve them.

2.3.3 Organizational innovation: collaboration and education

The façade construction sector and the building industry in general innovate at a slow pace (S. Pasetto—TB08). Slow innovation and operational inefficiencies have common causes and these must be identified both within and outside the façade industry. From an internal perspective the fragmented supply chain, the lack of coordination between design, supply chain and construction and the low level of investment in R&D are definitely among the major problems (Dubois & Gadde, 2002; Dulaimi et al., 2002). This fragmentation is the result of multiple factors, including the multidisciplinary design process typical of façade design, the typical one-off nature of the projects and the use of multiple products/materials assembled in complex and highly customizable systems, which involves a highly specialized supply chain with several different stakeholders, such as designers, consultants, contractors and subcontractors, organized in a compartmentalized chain of separated entities. In this panorama, every player in the game has a limited scope of work and, at the same time, has to rely on the work of many other stakeholders; this increases the risks and contingencies while reducing efficiency at every step of the design and construction process.

This environment also reduces the parties' ability and willingness to innovate (Blayse & Manley, 2004), curtailing any opportunity of collaboration along the design/supply chain. The result is a systematic failure of cross-field innovation, which increases the risks and constrains the development of novel facades (M. Zobec—TB06). Risk management will be key to the success of innovation (H. Schepers—TB05) and reducing the inherent risk of innovation will require (1) a reconsolidation of the market toward the development and coexistence of new forms of collaboration and cross-fertilization between different parties; and (2) new standardized digital management tools and workflows. The first is often defined in the literature as 'organizational innovation' (Blayse & Manley, 2004), as it aims to change the organizational structure by introducing advanced management techniques and implementing new corporate strategic orientations. Changing the inner structure of the market to foster a positive climate of change is an essential step to enhance opportunities for exchange and, ultimately, support a paradigm shift in the way the market works (Demircioglu, 2016).

Further research in this field will be fundamental to define the new generation of business models, project management strategies and tools, tendering processes, contracts, scope-of-works and responsibilities, insurance schemes and procurement systems. These will bridge the gap between designers, contractors and subcontractors, minimizing the existing barriers to innovation.

In the future, new contracts and project management schemes will transform the design methodology, paving the way for the achievement of ambitious sustainability goals in relation to EC and circular economy by forcing procurement choices at earlier stages of the design (D. Rogan—TB04). An example is provided by prefabricated mega-panels, which have the benefit of supporting local manufacturing and reducing transport costs, but which cannot be developed without an 'earlier coordination of designers and contractors' and an 'earlier engagement with the client' (D. Shillington —TB10). Hence, it is clear that new strategies for involving all stakeholders form the early design stage will be key to a radical transformation of the market.

A first step in this direction is represented by the increasing use of early contract involvement (ECI) practices, although these are not sufficient on their own to drive change. A more radical shift will require a deeper integrated design approach and will represent, for example, a new opportunity for architects to play a fundamental role in this transition (A. Cortese—TB01).

Another notable direction in policy change is represented by the shift toward the façade-as-a-service model, where the façade is 'leased or offered as a service, rather than bought' (S. Pasetto—TB08). This might also be an interesting way of encouraging the reuse and recycling of façade elements and, ultimately, reduce buildings' environmental impact.

Nevertheless, organizational innovation must be supported by the development of new management tools that can facilitate the change. In the construction sector, this may include a holistic integrated design, fabrication and management tool that can be employed across the field, allowing a more effective and standardized platform for collaboration between different parties. Current R&D projects are working in this direction, aiming to develop 'total façade management systems' (M. Zobec—TB06), 'digital twins' (J. Downes—TB09) and the technical tools necessary to support design for manufacturing and assembly (DfMA).

To support multidimensional and multidisciplinary information flows within the industry and across different sectors, it will be necessary to establish new highways for data and knowledge exchange. Indeed, the fragmentation of the façade market is reflected in a discontinuous knowledge development process, both within and between organizations, which prevents the development of an 'organizational memory' (Blayse & Manley, 2004). In the past, little effort has been devoted to transmitting data, knowledge and experience between projects, disciplines and companies, resulting in a system that penalizes the spread of advantageous mutations to standard products and processes (Dubois & Gadde, 2002).

However, the development and diffusion of innovations resonates with the way 'we develop our skills, our knowledge and our relationships' (H. Schepers—TB05). Today, intellectual capital is becoming as important as financial capital as the basis of future economic growth (Etzkowitz, 2003) and institutions, such as Arup University, are already investing in the creation of innovative internal learning platforms to bring global experts together and enable the creation of an 'organizational memory'.

External learning platforms and opportunities are also key to sharing knowledge and innovation between companies (J. Downes—TB09; S. Pennetier—TB07) and, therefore, enabling innovation. Examples include the involvement of Lendlease UK, Skanska UK and other contractors in the Centre for Window and Cladding Technology (CWCT) and the strategic committee on Sustainability and Embodied Carbon (J. Downes—TB09), or the growing number of industry−academia research programmes that constitute 'a very fertile soil for open source, tax supported collaboration and innovation' (S. Pennetier—TB07). The development of internal knowledge-capture platforms and external knowledge-generation opportunities will increase the already compelling need for the identification of more efficient ways of extracting/distilling codified knowledge from experience, overcoming the problems given by the project-specific nature of the market and making this knowledge easily accessible and transferable.

Today, the main opportunities for cross-field multidisciplinary innovation are found in temporary coalitions of firms enabled by large-scale projects and in the relationships between the construction industry, academia and state governments. While the first can be called the 'Innovation by Design' model, the structure that emerges from the latter involves what has been labelled the 'Triple Helix Model of Innovation' (Etzkowitz, 2003).

This model, formulated at the end of the 20th century, describes the latest trends in innovation. It identifies competition and cooperation between industry, academia and government as the 'sweet spot' where innovation can thrive, bridging the gap between disciplines, reducing risks and contingencies and boosting private resources with public funding and state government grants (Etzkowitz, 2003).

The Triple Helix model includes a great variety of submodels that reflect different possible forms of collaboration between these three entities, characterized by different results and success rates. The model also offers a critical analysis of the benefits and downsides of the different possible approaches. For instance, the *laissez-faire* Triple Helix model (Etzkowitz, 2003) sees state, industry and academia working as separate entities with limited iteration. Despite its extensive assistance with ideas and innovation during the early development phases, the weak relationships between the parties and the lack of agreed and coordinated targets and goals in the research activities may limit its success rate. This scenario often leads to innovative research that struggles to deal with the harsh realities of industry production and commercialization and fails to provide the level of support that is necessary to take the important step from prototype to product (technology transfer) in the case of product innovation, or from theory to practice, in the case of process innovation.

A closer interaction between state, academia and industry can help to create an environment that favors innovation. Examples of successful practices that reflect this idea can be found on either side of the helix, for example, where industry increases its investment in upskilling and embraces new ideas; where academia extends its focus from the resolution of isolated problems to the development and implementation of 'strategies of change' (Braungart & McDonough, 2009); and where governments increase their influence at the 'innovation table', not only to ensure adequate economical and structural support but also to employ the Triple Helix as a tool in the policy-making process.

2.3.4 Innovation by design

The one-off nature of production in the construction industry is often considered deleterious for innovation, a deterrent to the implementation of management techniques successfully used in other industries to improve productivity and an obstacle to the creation and diffusion of structured knowledge in the sector (Dubois & Gadde, 2002). The difference is clear in comparison with, say, the automotive industry, where repetition and mass production allow a much greater level of rationalization and optimization of the design, fabrication and construction process. This, in turn, generates an exponential increase in the economic benefits of more efficient products and processes, making innovation more appealing to industry stakeholders and facilitating the transfer of innovation across the market (Blayse & Manley, 2004).

Today, in pursuit of the beneficial effects of repetition and mass-production, the façade industry demonstrates a higher level of standardization and industrialization compared to other sections of the construction market, which rely more on on-site workmanship. Nonetheless, it is still not immune from the one-off nature of construction, which provides opportunity for project-driven innovations, where 'each project is a new opportunity for new designs or methods' (S. Pennetier—TB07).

In this context, the application of the concept of 'innovation by design' to the façade industry is and will continue to be of great importance in determining innovation potential, as a project-specific design can still encourage innovation by

1. challenging the state of the art in relation to materials, products and processes through the design of buildings that feature complex shapes or unprecedented performances;
2. developing project-specific innovations, exploiting the economy of scale provided by large-scale projects; these innovations can then be applied in other contexts and diffuse across the market; and
3. requiring the creation of temporary coalitions of different high-quality firms/companies, where collaboration is essential to bring the work to completion; these temporary coalitions encourage the exchange of knowledge and the cooperation of a range of experts in multidisciplinary teams.

Exceptional projects in terms of scale or magnitude can indeed support incremental innovation, especially in the fields of technology and manufacturing, as well as enable further diffusion and adoption of innovative processes and products across the market. Many successful recent examples of design-driven innovation can be identified, from the challenging geometry of the Sydney Opera House, which triggered many innovations, including the first use of computational technologies for the design and structural assessment of the envelope, to the titanium cladding of the Guggenheim Museum in Bilbao, which was first to apply aerospace manufacturing techniques to the façade industry, unlocking a cross-fertilization between the two industries (Cardellicchio & Stracchi, 2020).

Another successful example of innovation by design is represented by the increasing dimensions of glass panels (S. Pasetto—TB08) and the increasing use of glass as a structural material. Apple Inc. was able to activate this process of incremental innovation through the development of a number of flagship stores with unprecedented glazed facades and staircases. To achieve this design intent a multidisciplinary design team of engineers, architects and fabricators challenged the standard design process, the material properties and the fabrication processes, ultimately enabling the diffusion of the jumbo-sized glass that we see today in many shopfronts and buildings and which has been transferred to adjacent market sectors, such as in the design of dissipative safety glass.

In these examples, designers, fabricators, project managers and others from companies with different backgrounds worked together in temporary coalitions to complete a project, creating exceptional, innovative design solutions to resolve unprecedented challenges. Innovation by design has proved to be a valuable tool not only to 'push the boundaries of the design targets, but also as a mean to find new paradigms of collaboration' (A. Cortese—TB01).

In more recent years and on a broader level, the necessity of free-forms and customization of the façade elements has been a powerful driver for innovations in advanced manufactory systems (M. Zobec—TB06). These new advanced fabrication technologies, especially when hybridized with standardized façade systems, have demonstrated great potential in giving 'freedom from "customary" construction constrains particularly with regards to modulation and standardization' (M. Zobec—TB06) while enabling local manufacturing opportunities (D. Shillington—TB10). With particularly large and demanding projects, innovation by design has the potential to bring advantages typical of mass production and to bridge the gap between experimental prototypes and marketable products/processes, bypassing the 'technological lock-in' by covering the inherent costs which, in other circumstances, can delay or negate the transition.

Furthermore, the pioneering nature of design-driven innovation often leads to iconic projects that have high resonance on a local and global scale, with great potential in promoting and diffusing technological innovation (D. Rooney—TB11). Examples can be found in the Arthur Phillip High School (APHS) (Sydney) and the Woodside Building for Technology and Design (Melbourne) designed by Grimshaw Architects (Sydney). The first uses a reactive façade able to maintain comfortable indoor conditions without resorting to mechanical ventilation for 60% of the time, while the second employs a high-performance façade, featured on the cover of this book, which has been key to achieving Passivhaus certification. These buildings not only promote innovation but are 'living laboratories', allowing 'the future generation of engineers and designers to experience first-hand the technology employed in structure, façade and services and, hopefully, to inspire them toward the interest in technology and innovation'. (A. Cortese—TB01).

2.4 Industry insights

The previous section has summarized the key topics and concepts mentioned by the industry experts, organizing the contributions into a cohesive snapshot of innovation in the façade industry. Nonetheless, it has inevitably omitted some of the themes discussed in the 11 essays. This section represents the core of the industry perspective, presenting integrally the insights into innovation in the façade industry generously provided by the experts.

TB01

Andrew Cortese
Managing Partner Sydney – Design Director @ Grimshaw Architects
Sydney, NSW, Australia

Unprecedented changes in the social, economic, and climatic conditions are urging the whole construction industry to quickly adapt and set up a new framework to respond to these changes. The next generation of buildings and building skins will have to face great challenges, especially regarding comfort, sustainability, and resilience.

The Australian Building Code (BCA) is increasingly raising the bar on sustainability, asking to reduce the buildings operational energy, while improving the minimum thermal performance requirements for facades. Additionally, all energy performance certificates (EPCs), including the Australian Greenstar, are pushing the bar even further, aiming at Net Zero Carbon targets in both operational and embodied energy and promoting regenerative design approaches.

As architects and designers, we believe that it's important to face these challenges holistically, without losing the focus also on the other fundamental qualities of buildings and facades, such as amenity, variety, experience, visual connection, etc. Instead of "design for compliancy", Grimshaw Architects focuses on a design process that uses sustainability as a paradigm to shape the building. Sustainability, in its broader meaning, is used as a filter to balance competing performances that include, thermal performance, daylight factor, access to view, variety of space etc.

The design process starts with an early-stage data-based integrated-design approach where the main design parameters are defined: the array of architecture and sustainability elements, such as external shading, presence of photovoltaic (PV) panels, and the contrast between solidity and transparency, are rationalized and used to build the tectonic and the filigree of the façades. It is interesting how, in the last couple of years, following this same design method and using similar energy performance targets, Grimshaw Architects (Sydney) delivered 3 buildings that employ completely different approaches defined as Defensive, Reactive and Active.

A. With a "Defensive" envelope the building skin works as a barrier, limiting the air and thermal exchange between the external and internal environments. This approach, which is typical of the Passivhaus certification, put the stress on the services to manage the air exchange rates and the air quality. A good example is the Woodside building for Monash University, which received the Passivhaus certification in 2020 and is, today, one of the biggest and most innovative education building in Australia and beyond.

Considering the 19.000 m^2 of Gross Floor Area (GFA), the extremely short construction time (16 months), and the high performance modular unitized façade, this building it is relevant to the research and innovation on high-performance building in Australia since it's the first PH fully designed and built in Australia, without heavily relying on European products.

Furthermore, the building is designed as a living laboratory, allowing the future generation of engineers and designers to experience firsthand the technology employed in structure, façade, and services and, hopefully, to inspire them towards the interest in technology and innovation.

In this building, the façade and the adjacent internal space work together and act as a filter, the first, and as a buffer zone, the latter, activating the informal teaching spaces located close to the façade with variable environmental conditions and providing a more constant and environmentally controlled space in the center of the building, where the formal teaching spaces are located.

(Continued)

(cont'd)

B. With a "Reactive" envelope the façade works as a filter managing and optimizing the thermal and air exchange between internal and external environment. This approach was employed in the design of the Arthur Phillip High School (APHS), the first Australian high-rise school.

In this building, the façade "breathes" and uses the mild Sydney climate to minimize the use of the mechanical services in being able to work without the use of any mechanical ventilation and air conditioning for 60% of the year.

The façade is divided in the shading system and the weatherproofed window wall, with an external space that works as a buffer zone and that, is used for the external activities and recreation.

Similarly to Monash, also, in this case, the development of the façade is strictly correlated with the development of the floor plan, that maximizes the opportunity of cross-ventilation, and the services, that are designed with flexible temperature setpoints, using the "adaptive comfort" principles.

C. Finally, the use of an "Active" envelope with focus on the sun and water harvesting and the integration of the façade with the building services was employed for the design of the Cockle Bay Tower was to the external shading system was given an active role in contributing to the energy performance of the building.

Although these three projects showcase different technologies and different approaches to façade design, they all contributed to the development and spread of innovation in the market, establishing new benchmarks in terms of design processes, and design outcomes.

Looking forward, we believe in the relevance and the importance of "Innovation by design", not only to push the boundaries of the design targets but also as a mean to find new paradigms of collaboration between ourself, our clients, consultants, contractors and subcontractors.

The main reasons behind the slow-paced innovation in the façade construction market, and the construction market in general, is the diffused fragmentation in many different stakeholders with different agendas and contingencies. In the last few years, we experienced increased use of Early Contractor Involvement (ECI) whose benefits are widely debated. The future of facades and buildings will require a much more profound paradigm shift towards deeper and more meaningful cross collaborations between designers, fabricators, and contractors. As architects, we are in the perfect spot to fill the gaps between all these players and provide fertile ground for innovation.

TB02

Keith Boswell
FAIA Partner @ Skidmore, Owings & Merrill LLP,
San Francisco, CA, United States

To address the 2 questions posed below relating to innovation in Exterior Building Skins (Enclosures), a fundamental and non-negotiable parameter to be included is: The exterior building enclosure must be Good Design. To achieve "good", exterior enclosure design must equally address qualitative (visual) and quantitative (performance) items specifically identified for the design problem - or design opportunity - at hand. Innovation that favors either visual or performance too heavily at the expense of the other will not be successful. This requires the designer(s) and team participants to exercise left- and right-brain thinking resulting in a complete or holistic enclosure design. Exterior building enclosure design is a multi-faceted design opportunity requiring 3-dimensional thinking and exploration resulting in a clearly documented format that can be fabricated, constructed, and maintained. My views are from an architect perspective with hopefully an acknowledgement on the contributions of all participants.

In addition to defining "good", innovation must also be generally defined. A dictionary definition of "innovation" is: *a new method, idea, product, etc*. Dictionary definitions usually utilize the operative word "new". I strongly support an expanded definition of innovation to include: *The clever and appropriate use of materials and systems implemented by clever and knowledgeable people*. To address "knowledgeable" in the definition of innovation motivated project team participants must possess or acquire fundamental understanding of items including but not limited to; design performance principles, materials, composition, high school physics and geometry, a farmer's mentality of climate, capacity for research, and fabrication/assembly/construction basics.

Question 1 — What are the most relevant innovations that you foresee will play a major role in shaping the way the next generations of building envelopes will be designed, procured, built, maintained, and dismantled?

The most relevant topic for architects is designing and documenting zero energy buildings. Therefore the primary objective is to identify innovations in the methods and procedures — with resulting design solutions — to achieve zero energy building designs that are affordable and can be constructed, operated, and maintained for a long period of time with the ability to disassemble and recycle. For exterior building enclosures this translates to very low energy transfer through the enclosure. Additionally, to achieve "good" and "holistic" zero energy building design, low embodied carbon content for exterior materials and systems that are appropriate for its respective climate. In short, an exterior enclosure is different for hot, cold, and everything in between climates. All participants in the delivery of building enclosures understand that goals must be lofty — yet attainable. Is there a more lofty goal than designing the building and exterior enclosure with full respect to our planet, natural resources, and the local environment?

So the question is: how to do this? Architects must possess or acquire a more informed understanding of exterior enclosures from concept through

(Continued)

(cont'd)

completion. This is not a goal—it is a necessity. Exterior enclosure design, documentation, and delivery is a team sport. However, architects — my profession — must focus on more than the visual and composition aspects. Rather, we must be cognizant on all aspects so architects are active participants - not observers - waiting for another project team member or participant to figure it out. Once you figure out what you know and equally important what you don't know, assemble a team of motivated and knowledgeable folks that can fulfill the knowledge gaps. The practice of architecture sees two primary topics where innovation is critically necessary. These are:

A. Zero operational energy buildings with the associated highest performance exterior enclosure.
B. Use of low embodied carbon in the materials, systems, and processes of fabrication, assembly, and construction. (by the way- there is no such thing as zero carbon.)

Innovation occurs where there is a need that cannot be achieved through existing means. These are primary topics confronting the built environment.

Question 2 - In your current practice, what are the future innovations you are actively fostering / exploring and how do you expect these innovations to change your role and the outcome of the design?

In our quest to design higher performing building enclosures with lower embodied carbon in the resulting enclosure system design, we are actively researching materials with a focus on embodied carbon in materials and their manufacture process and developing enclosure designs to determine actual performance and embodied carbon values. We are exploring two enhanced steps in our design process.

The first is a more nimble manner to "test" enclosure designs at a design development level to quickly and accurately determine thermal transfer - on paper. We have established evaluation metrics by climate zone, building type, building use, enclosure orientation, and size for the exterior enclosure and how these contribute to advancing and achieving zero energy building design. The second is a re-evaluation of materials used in our enclosure design work and overhaul of materials utilized in our specifications requiring validation of actual embodied carbon content. This is our "D-Spec" initiative.

Our research over the last several years has broadened our understanding of the carbon performance component of holistic enclosure design to be a design parameter along with other performance metrics. This still must be balanced with the visual component to achieve the earlier stated goal of "good" design. It is not surprising that "natural" or human made materials utilizing natural products in an intelligent and knowledgeable manner consistently rise to the top in performance.

It is our hypothesis that our findings will provide the ability to provide enclosure design that can be objectively evaluated utilizing understandable metrics.

> Our goal is to exceed performance beyond jurisdiction requirements — codes/standards etc. Innovation and innovative enclosure solutions can get a boost by elevating and simplifying jurisdiction requirements so low performing enclosures can be eliminated. Higher and understandable performance standards will mandate innovation since current industry practice is just not good enough. Why not just require: "The completed building must be operational zero energy?" This will require owners, architects, builders, and trades to increase their knowledge and ability to execute designs and buildings that are truly timeless.

TB03

Kieran Rice
Technical Director, Façade Practice Lead — Australia / New Zealand @ Aecom
Melbourne, VIC, Australia

Question 1 — What are the most relevant innovations that you foresee will play a major role in shaping the way the next generations of building envelopes will be designed, procured, built, maintained, and dismantled?

Over the last 20 years, one of the primary objectives of façade design has been to improve the thermal performance of the building skin as a means of reducing the energy required to heat and cool the building's interior spaces. Manufacturers have developed new technology leading to improved glazing and framing suites, allowing modern façade systems to achieve ever lower thermal conductivity. Glass coatings have also advanced so that modern glazing systems provide good levels of protection from solar radiation and heat gain whilst still allowing natural light transmission into interior spaces.

The quest to reduce the energy required to heat and cool buildings is driven by the consideration that mechanical plant typically uses electricity to cool and natural gas to heat. Natural gas is a fossil fuel that generates carbon emissions, and electricity when taken from national grids also has associated carbon emissions (though these vary from country to country and indeed from region to region).

It is becoming apparent however that the technological advances made over the last twenty years are now producing decreasing marginal gains in the energy used during the operating cycle. It is clear to me that increasingly we will focus on the hidden part of the iceberg, the embodied emissions that are associated with the extraction of raw materials, the production of the façade systems and their transport to the building site (cradle to site emissions).

(Continued)

(cont'd)

Furthermore, there are trends in most developed countries to achieve zero carbon grid electricity by 2050 (or even earlier). The reduction in emissions associated with grid energy will further weaken the case for concentrating on minimising buildings' energy use during operations.

We are starting to see more focus on embodied emissions across all sectors of the building industry. Processes and procedures are being developed and improved so that accurately measuring embodied emissions becomes easier. Traditionally this has been the task of the sustainability specialist, however I expect that these skills will become much more mainstream and that all members of the design team will be involved in assessing and optimising embodied emissions during the design of the building.

For the façade industry, I think that when the carbon emissions associated with the building skin are assessed from cradle to disposal, this will lead to a fundamental reappraisal of what constitutes a carbon efficient building skin. The race to improve performance at the expense of embodied emissions will evolve to a more nuanced appreciation of what gives the most optimal overall carbon footprint across the building's life-cycle.

Question 2 - In your current practice, what are the future innovations you are actively fostering / exploring and how do you expect these innovations to change your role and the outcome of the design?

I think that there will be a more pronounced push toward cross-disciplinary collaboration. As we exchange more frequently with our colleagues, this will lead to previously unexplored areas of practice. Two examples are given below of areas of cross-disciplinary practice that we are exploring.

My team and I have recently carried out research, based on a mid-sized commercial development in regional Victoria in order to examine the difference in embodied emissions in a double glazed and triple glazed curtain wall system and how the improved thermal performance of the triple glazed system impacts the heating and cooling energy required during the Use stage.

The results showed that the triple-glazed system would add around 105 tonnes of embodied CO_{2e} emissions, and would save just over 3 tonnes per annum of CO_{2e} emissions during the Use stage. In other words, it could take around 33 years to offset the increased embodied emissions in the triple-glazed system through emissions savings during the Use stage.

Like most developed countries, every state and territory in Australia has committed to achieved zero carbon grid energy by 2050. When this trend is taken into account in our study, it showed that the increased embodied carbon in the triple-glazed system would never be offset by decreases in operating emissions.

It is worth noting that the study building is situated in one of the colder climates in Australia. The case for using higher performance building skin systems is easier to make in colder climates as reducing heating loads (principally derived from natural gas) is the main gain from improving the system's conductivity. Our expectation is that in warmer climates, the gains from triple glazing would be even smaller.

Given the results of our research, it seems evident that we should carefully weigh the carbon costs and benefits of the building skins that we are designing. We are developing tools and processes that allow close collaboration with our sustainability colleagues so that we can undertake accurate assessments of the embodied carbon of any given system as well as its impact on operating emissions.

I expect that these workflows will drive a much more granular appreciation of how to effectively minimise carbon in our building skins, and ultimately will lead to skin systems that are more finely tuned to the building's location and function.

We are also interested in how new developments in modelling and fabricating façade systems will drive more organic forms in our buildings.

Freeform architecture has provided an expression of something fundamental in the human spirit, our connection with nature and natural forms. Through gradual evolution, nature has developed living forms which are inseparable from their function. Perhaps the biggest weakness of freeform architecture is that the connection between form and function has not been fully explored.

Engineering has been slow to catch up — efforts have been directed to the processes which enable to form to be achieved, but the engineering rationale for the form itself is lacking in most cases. This is understandable, the engineering challenges in creating freeform buildings are significant. I do expect however that as our modelling and fabrication tools improve, the engineering focus will shift from simply enabling architectural free forms to driving organic forms that have an engineering justification and allow demonstrable reductions in material and greater levels of performance.

It is likely in my opinion that this drive away from rectilinear to more organic building forms will involve modular construction. This is another area of cross-disciplinary practice that we are actively looking to develop further.

TB04

Damian Rogan
Director of Façade Engineering @ Eckersley O'Callaghan
London, United Kingdom

Question 1 — What are the most relevant innovations that you foresee will play a major role in shaping the way the next generations of building envelopes will be designed, procured, built, maintained, and dismantled?

(Continued)

(cont'd)

In recent years the 'circular economy' has emerged as a key concept in design thinking, representing a shift from a consumptive economy — where raw materials are extracted, processed to high value products, consumed, then disposed of — to an economy where processed materials are used as raw materials for future products, reducing the demand for resource extraction and waste disposal and bringing associated benefits in energy demand.

While circular economy principles can be applied to all manner of industries, they have particular challenges when applied to façades, as each is unique and contains a multitude of materials from different producers, installed by different contractors. There are many different ways to realise a façade of a given appearance, and the external finished material is often separate from the many materials providing the weathering functions, insulation and so forth. Cost, time, and buildability are the typical drivers dictating the form of construction on a particular project, but the ability to apply circular economy principles rarely features in the decision-making of clients, designers, and contractors. This must change if we are to reform the façade industry to the extent needed to meet our net-zero commitments.

We are already seeing an awareness of circular economy principles among suppliers. Some aluminium producers now offer alloys with post-consumer recycled content, and certain glass suppliers are increasing the amount of waste product returned to float glass. While these are positive steps, a more holistic approach is required when considering that modern façades are complex compositions containing potentially dozens of materials, each with a unique provenance. More suppliers must improve the recyclability of their products and establish the means to separate their materials within multi-layered façade build-ups.

Architects and façade engineers must also embrace circular economy principles in their designs; they can be adopted at many different scales of construction. At a product level, this involves using readily separable and recyclable products, avoiding composites and overly complex build-ups. At a building level, it could involve designing façades using fully modular systems that can be removed wholesale and replaced with upgraded systems, then refurbished and passed on to the next building.

Designers and clients will need a much greater awareness of the products and systems that go into their façades. They will need to take responsibility for the potential dismantling and recyclability of their designs and ensure that these issues are given equal importance to buildability, embodied carbon and whole-life performance. This will require a fundamental reassessment of our design approach and the response of industry to a crucial future challenge.

Question 2 - In your current practice, what are the future innovations you are actively fostering / exploring and how do you expect these innovations to change your role and the outcome of the design?

For 20 years or more, the tightening of building regulations around operational energy consumption has led inevitably to a focus on façade design. Façades have a significant role to play in low-energy building design, particularly as relates to lighting, heating, and cooling loads. The demands placed on façades are often competing, the most obvious example being the need to provide good internal daylight at the same time as controlling solar gain. To meet such competing demands, industry has developed dynamic façades that can vary their performance to changing weather conditions, though this has introduced the need for sensors, motors and moving components, increasing the complexity of façades and often reducing their service life. Static façades are likewise becoming more expressive of their performance, incorporating external shading features and more solidity, using a variety of materials such as stone, brick, and terracotta rather than lightweight metal. Thus, with both static and dynamic designs we are seeing a marked increase in the complexity of façade constructions compared to the International Style that dominated past decades.

Too often, though, insufficient attention is given to the complexity of the façade systems in their construction and the embodied carbon (EC) resulting from their manufacture. The question that presents itself is this: does the EC inherent in these complex façades provide a payoff in whole-life carbon, considering the benefits they offer in operational energy? To answer this, we at Eckersley O'Callaghan are currently investing a great deal of time and effort in developing tools which can help us calculate the EC of façades and weigh this up against the operational carbon benefits of alternative options.

EC calculation is a relatively straightforward procedure if you know the exact materials to be used, but in reality many aspects are not known until the contractor is selected and material origin is finalised through the contracted supply chain. The tools we are developing help us to predict at an early stage the likely EC using a database of historic benchmark data and also to determine the variables we can play with to optimise the EC of a given design. This changes the process of façade design in a fairly fundamental way, as it pushes system selection forward to an earlier design stage rather than leaving it to the contractor. It also allows for a conversation about whole-life carbon accounting for EC in initial construction and throughout the building life, including façade replacement and disposal. This enables us to have much more sophisticated conversations about the benefits of complex façades compared to simpler, more durable ones.

It also forces us to have honest conversations about the desired lifespan of different façade components and the underlying construction. Clients and architects typically approach projects believing that their building will be a static object with only internal fit-out refurbishments occurring over time, ignoring the fact that styles and occupant demands change, and that a large degree of façade refresh can be desirable and thus planned for. If we build in

(Continued)

> **(cont'd)**
> the potential to adapt our façades, it could have a profound impact on the prevailing architectural aesthetic and revive a tectonic approach where façade materials are not mere cladding but portray an honesty about their permanence or temporality.

TB05

Haico Schepers
Principal – Building physics @ ARUP
Sydney, NSW, Australia

Question 1 – What are the most relevant innovations that you foresee will play a major role in shaping the way the next generations of building envelopes will be designed, procured, built, maintained, and dismantled?

The next generation of building envelopes will have significantly higher environmental performances. This will manifest in multiple ways. Thermal and operational performance will be improved through optimised designs, probably reduced windows, better ventilations systems, better insulation and shading. Innovations in analysis and automated manufacture will enable envelopes to perform for their specific location in a building creating greater system diversity. Materials used in envelopes will be selected on their life cycle impacts built with more structed materials that can be disassembled, more recycled products, and better product chain of custody. Innovations in digital tracking and modern manufacturing techniques will create envelopes with detailed materials passports, certification, and embodied carbon data. Envelopes will generate more power with an increase in the availability of a variety of photovoltaic cladding systems. In some cases, this power will create autonomous facades that will power themselves to perform better environmental control, provide a security function. Self-powered envelopes may be connected to google home or IoT systems and respond to occupancy patterns and weather forecasts. Innovations in robotic manufacture will drive systems that reduce materials use, minimise component parts, reduce site labour and increase installation safety. The cost effectiveness and ubiquitous nature of robotic sensing will convert building maintenance units, that transport window cleaners, into smart inspection systems that will continuously monitor envelope quality and identify potential risks such as water leaks and crack detection.

Finally, innovations in the industry knowledge, changes to regulations, and ability to manage risk will have the most significant role in the uptake of any of these technical innovations, if any.

Question 2 - In your current practice, what are the future innovations you are actively fostering / exploring and how do you expect these innovations to change your role and the outcome of the design?

There is a difference between fostering and exploring.

Exploring is really about scanning the future and trying to identify future trends, innovations and processes that maybe prevalent in the future work. Typically, we take a now, new and next approach. We try to identify what is current, what is 5 years away and what is 20 years way. We work with our foresight and innovation team to investigate these opportunities and trends and try to use them to inform our design approach. In a way we road test the items to see how future ready they are, how likely they are to have a significant impact in future envelope designs.

For our organisation fostering is about how we develop our skills, our knowledge and our relationships to enable the design outcomes we believe are future ready. We do this in multiple ways. Instilling an ethos that is curious and willing to innovate is critical. Performing research internally or with project partners that builds specific knowledge is a cost-effective way to grow new skills organically. Dissemination of projects, research and learning pathways is rolled out by our façade skills network and Arup University through a digital platform that seeks to bring experts together. A lot of innovations are trial and error so finding a safe place to innovate is essential, what I like in these projects is the way these projects develop skills that can then be transferred to other projects.

More specifically the skills and knowledge we are fostering is in digital skills sets, from automating design process to working on how our design documentation can lead to direct manufacture. We are looking at how our design tools can give quick feedback for environmental decisions such as carbon, material passport and embedded manufacture data. We are developing knowledge in machine learning processes and robotic systems design and specification. We are looking at how to collaborate on more complex value chain systems to instigate change in the industry. Finally, we test smaller prototypes to develop proof of concept and learn the issues associated with the product design. This is focused on new materials and on smart self-powered facades developing the PV, battery, control and system nexus to create smarter façade systems. Also investigating applicability of onsite 3D printing systems for facades and brackets.

TB06

Marc Zobec
Former Group Technical Director — Executive Consultant Special Projects Manager @ Permasteelisa
Permasteelisa, Sydney, NSW, Australia

(Continued)

(cont'd)

Question 1 — What are the most relevant innovations that you foresee will play a major role in shaping the way the next generations of building envelopes will be designed, procured, built, maintained, and dismantled?

Over the past decades, great advancements have been made with regards to 3D and spatial modelling as well as the use and interfacing of these systems with advanced manufacturing systems including computer numerical control (CNC) machining, sheet metal forming technologies, 3D printing and additive manufacturing. These technologies have given architects and developers freedom from "customary' construction constraints particularly with regards to modulation and standardization of façade systems and materials. True design freedom for architects will not lie with rapid and accurate modelling and freeform surface manufacturing but the development of integrated holistic façade management systems that will holistically manage design, procurement, manufacturing, production, and site installation.

The constraint to many new advanced façade innovations will be risk. Innovations in total façade management systems that enable total process integration will be an essential tool to manage such risk. With the use of Building Information Modelling (BIM) in construction and an increasing trend for building contractors to self-procure facades by outsourcing each various stage to non-aligned stakeholders, coordination through total façade management systems will become imperative particularly in order to trace materials and detect any issues that may have occurred through what should be a holistic process. Failure to do so will see buildings incurring problems with no discernable traceability and responsibility resulting in risk to the owner and developer. In addition to this, Total Façade Management Systems will also be able to integrate materials/panel delivery tracking, quality control, non-conformance reporting, traceability and health and safety management. This will become increasingly important for architects and developers in order to manage and minimize risk. In the past, materials used in façade construction had performance bench marking stemming back decades and even centuries. The use of glass, masonry, natural stone to name a few, all had a degree of historical certainty and reliability. With the rapid development of alternate materials including composites, engineered timber, polymers, adhesives technology and materials used in additive surface manufacturing have yet to prove the test of time and as such traceability and "total seamless" quality control will be essential.

With greater integration and alignment of modelling and fabrication technologies with regards to free-form surface design, the issue of module standardization; through the use of total façade management systems, will be replaced with standardization of the total façade procurement process which will enable free-form surfaces, formerly constrained by economies of scale through repetition of standard units to be realized not only economically but with markedly reduced risk.

Question 2 - In your current practice, what are the future innovations you are actively fostering / exploring and how do you expect these innovations to change your role and the outcome of the design?

Building facades have been metaphorically referred to as building "skins" and like the human skin façade design and innovations progressed objectively behave in a responsive manner. Recent research by Permasteelisa has focused on the development of advanced façade systems with the aim of the façade becoming and integrated component of the total building environmental system, rather than a static element. Energy flows both in and out of the building are best filtered through the use of effective, intelligent shading systems which not only control energy flows but whose variability and use of various materials and configurations result in not only a functional element but also provide a new architectural aesthetic.

Recent developments by Permasteelisa have focused on the integration of façade and building energy systems through the use of advanced algorithms in building control systems. It is not uncommon for shading systems to be linked to building weather stations to respond to solar irradiation, but consideration is being given to a multitude of factors in the algorithms that balance the response of integrated shading systems including occupant thermal comfort, acoustics, glare control and external views. These algorithms have been developed not only to provide an optimum thermal balance but also provide occupants the ability to have an override control when conditions permit. Wireless and internet integration also allows the building control systems to be linked to weather forecasting alerts allowing the façade to not simply respond but pre-empt weather events.

Many elements and components within building facades are often specified with differing warranty and life cycle requirements. This may at first to appear as a logical approach but often when one minor component reaches its' life cycle expiry, the entire façade or major upgrade is required. Permasteelisa has been developing systems that not only allow effective replacement of components that reach their "use-by" dates, but also systems address these shortcomings. The advent, development and advancement of the MFreeS close cavity façade as aimed at using multiple glazing layers and shading systems that negates the need to replace sealed IGU's that inevitably fail not due to the glazing itself, but the edge seals that prevent condensation. The simple replace-ability of individual glass layers will not only lengthen the façade life but also allow upgrades to be undertaken simply and economically.

In recent years, buildings in city centres have seen changes in their use from once predominantly commercial design to mixed use and in particular high-rise residential buildings centred around effective transport hubs. Premium land prices and advances in structural engineering systems have seen taller and slender residential buildings with high aspect ratios. The challenge in recent years has been how to ensure occupant comfort due to building wind induced accelerations. Often mass tuned dampers are used but the placement of these systems

(Continued)

(cont'd)

is usually at the upper premium rental levels which also occupy valuable rental space. Permasteelisa has been developing the façade as a multiple integrated mass tuned damper (MiMTD), which uses the distributed façade mass as well as integrated mass dampers within the spandrel areas to not only reduce accelerations but also provide savings in the building structure itself.

It is an unfortunate though ever present fact that protective design is becoming more prevalent as a design parameter in facades; namely blast and ballistic design. The use of large glazed surfaces should not be a limiting factor when design for blast enhancement. Permasteelisa have developed true balanced dissipative façade systems and brackets that enable not only the use of large glass surface areas but also reduce the risk of glass fragmentation injuries.

Finally, the greatest innovation in façade design and engineering will not lie in technology itself, but a new collaborative approach from builders and developers through the commissioning of façade contractors as part of the early stage "integrated" building design allowing true and holistic building design, procurement, fabrication and installation.

TB07

Sophie Pennetier
Associate Director — Special Projects @ Enclos
Los Angeles, CA, United States

Question 1 — What are the most relevant innovations that you foresee will play a major role in shaping the way the next generations of building envelopes will be designed, procured, built, maintained, and dismantled?

Facades are a fascinating field. They represent a holistic challenge of combined performances of shelter to the elements and structural integrity. Systems are optimized for balancing physics objectives, sometimes secondary to project challenges such as budget, schedule, durability, life cycle. These often are in turn outweighed by behavioral, societal, ecological, and political challenges.

I will always remember that design workshop at the NY Centre for Architecture, after hurricane Sandy in 2012. Our city had just washed its first of its kind hurricane, I had witnessed the South half of Manhattan shut off into the dark, had volunteered with SEAoNY to structurally inspect houses, washed and burned, and set green and sad red stickers. At the workshop where we were envisioning (and honestly processing) the rapidly changing design challenges for our city shoreline, I realized for the first time really, that most of us do not truthfully design and build to last. I am not referring to the materials resistance to corrosion, but to the greater lasting effects of our collective efforts. We, designers, contractors, stakeholders, from both my European and US experiences, share this moral responsibility to be still thinking short term.

Most Life Cycle Analyses find that over 2/3 of the carbon footprint of an office building is its HVAC energy use[3]. However how many have snoozed this design conversation since tenants are the ones ultimately paying the energy bill? Why, since New York's Law 97 passed in 2019, are most offices facades suddenly designed triple glazed[4]? The market change was instantaneous because the problem was adoption, not technological innovation.

The relevant innovation in my opinion, is the social and political innovation that we have secluded ourselves from. How can we, not just create systems or products but how can we make them adopted as the new norm?

There is an abyssal disconnect between the areas of knowledge of a project and a dire need for partnership. While the finite scope of user-oriented design trades is coordinating well technically (SMEP etc.), the collateral imprint of the project into the society fabric is still largely overlooked. The challenges lie in the business models and individual pressures: sustaining workforces and business, Intellectual roperty (IP) and return on investment, are some examples of essential immediate needs to be balanced with long arc thinking such as the generational impacts of parity, diversity, employment opportunity, education, health, and limited material resources. These necessary innovations have remained walled out in the policy world.

In Europe in 2010, I joined an Industry-Academia research program funded by the EU. Such partnership is a very fertile soil for open source, tax supported collaboration and innovation. It works when paired with the adequate policies and standards that protect the markets and societies in/for which these ideas have been nurtured. There's much room for innovation in ways we collaborate: In 2020 we witnessed many grassroot initiatives, organic and inspiring responses to the stress test on our society. Grassroot initiatives also exist in our industry: to name a few, the Living Building Challenge, Passive House Accelerator, Cradle to Cradle, Health Product Declaration (HDP) collaborative, and the Facades Tectonics Institute represent excellent pointers for innovation in terms of positive handprint and social co-benefits.

Innovation often comes from threat, adaptation, survival and - in a less Darwinian fashion - serendipity. Innovation also relies on fundamental research. In the wake of the Great Recession, the 1930s turned out to be one

(Continued)

---

[3] Life Cycle Assessment of an Office Building Based on Site-Specific Data, Ylmen, Penaloza, Mjornell, 2019

[4] The Law 97 places carbon caps on most buildings larger than 2,400 m^2 —roughly 50,000 residential and commercial properties across NYC. These caps start in 2024 and will become more stringent over time, eventually reducing emissions by 80% by 2050.

(cont'd)

of the most innovative decades of the twentieth century from a manufacturing standpoint[5]. Over the past decades, the comfort of secured energy resources has slowed the development of manufacturing and construction technologies, while another convenience, production outsourcing, has punted down the road the social and ecological disaster of short-term decisions onto our workers, families, consumers, builders, community members, voters. We are an ecosystem.

Since I transitioned from design consulting to the contractor's world, I realized how much less time there is for any reconsideration, and feel I had spent my energy on the wrong optimization problems. What are daring cantilevers, freeform or transparency worth if my façade is still mostly made of non-recycled content, remains the building's worst energy sink and responsible for the increasing energy needs of our society? I had no idea of the impact and power I held, even limited.

The most pressing need for innovation lays before us when we start early in design, to evaluate components not just for cost, tonnage, or carbon footprint, but also socio-ecological impact, widening our multi-objective optimization scope. Our work from design to manufacturing and construction is getting increasingly automated, freeing us up for tackling these objectives. That is probably the greatest challenge and the most exciting opportunity ahead for the next generations (of building (envelopes)).

Question 2 - In your current practice, what are the future innovations you are actively fostering / exploring and how do you expect these innovations to change your role and the outcome of the design?

Enclos designs, manufactures, and installs high-performance unitized curtainwall in the US. BIM and Big Data had long been a core platform for production engineering, coordination, and digital fabrication when I joined in 2018. Starting in sales with quantity take-offs, BIM is also leveraged for design documentation, manufacturing, operations means & methods strategies. I look forward to bidding packages with standardized pre-existing data schedules, carbon (or social) tracking presets, thoroughly linked shared object families, which we often must repair or restart from scratch during the bidding process.

Production engineering largely leverages BIM for coordinating and staging interfaces such as anchors, support structures, maintenance units, etc. On large scale projects such as our most recent sport facilities, large volumes of custom units are parametrically defined and documented for paperless fabrication. Laser tracking is leveraged in the shop and on site to evaluate, anticipate or

---

[5] The rise and retreat of American manufacturing, Vaclav Smil, 2013.

adjust fabrication and installation tolerances, which is especially crucial in complex geometry projects.

Supply chain management needs innovation: it is mostly budget, and schedule driven. To date, most of the HPD/EPD (Environmental Product Declaration) documentation has been handled in parallel to more pressing (sic) parameters. Hopefully, bidding and procurement become more holistic, and our micro-scale scope data is evaluated within a macro-scale analysis beyond the project itself.

We have hosted various areas of research in-house, such as some of our recent work on cold bent glass and fostered select investigations such as in ultra-thin glass and large-scale 3D printing prototyping at the Autodesk Technology Centers in Boston. 3D printing is frequently leveraged in production for either rapid prototyping of tabletops, Quality Control measurement jigs for complex 3D conditions in the shop or on site, or at times for in-house developed project specific manufacturing tooling.

Most of our innovation is project driven, which means that it is constrained by limited timelines and budgets. Because of its shapeable nature, the design of custom extruded aluminum mullions alone is an incommensurable opportunity for new processes, materials, and design philosophies. This is, I believe, the asset of our industry compared to mass-produced markets: each project is a new opportunity for new designs or methods.

Hopefully, the role of contractors will grow in the earlier stages of design, in a healthy setup for instant feedback. I also hope to see innovations towards cross-training opportunities and pathways between consulting and contracting, to share experiences, visions, and ultimately join forces towards a more sustainable industry, both economically and ecologically.

TB08

Saverio Pasetto
Head of Façades @ Skanska,
London, United Kingdom

The façade construction sector, and in a way the building industry as whole, with its complexity and fragmentation is slow when it comes to innovation, especially when compared with other industries. This also applies to the glass industry, which is an essential part of the façade industry. There is a need to change the status quo to enable the application of novel technologies to make a tangible impact and move the industry forward.

Nevertheless, the façade industry, with its passion for façade and glass engineering disciplines, is rich with new ideas some of which are well presented in this publication.

(Continued)

(cont'd)

Robotic - and drone - applications, often in combination with Artificial Intelligence (A.I.), appear to be very close to become part of our day-to-day lives. They are already integrated in other industries (I.T., automotive, aerospace, logistics, military to name a few), are becoming increasingly intelligent and will inevitably become part of our industry too. These innovative, autonomous technologies may be applied to the façade industry off-site, for example to exploit factory prefabrication and assembly of components including their quality controls, and on-site, for example to carry out surveys, assess progress and for installation of components. The façade industry already makes use of manipulators, e.g. for installation and glazing activities, however human interaction is currently required, and it is fundamental. The "step change" in our industry will happen when current technologies become autonomous and research projects are already on going and with promising results. Personnel safety, e.g. working at hight and lifting activities, quality of product, accuracy and speed of installation can all potentially benefit from these technologies.

One aspect that could facilitate and improve the implementation of innovation in the façade construction sector is the cross-contamination with other industries, such as those mentioned above. We tend to look inward for new ideas when perhaps we need to look outward and see how new technologies are currently being used in other sectors and what their future looks like. A good example of this is the involvement of leading architectural and engineering practices with space architecture research projects, like exploring how to build in space, on Mars or the Moon. I truly believe that our industry has potentially a lot to learn from these research projects especially in the long term, for example on construction methodologies and novel materials with due consideration to life cycle of structures.

The latter is certainly something that our industry has been looking into for some time in relation to circular economy. It is progressing at a slow pace, however an increasing interest by designers and developers has been noticed, often alongside building future proofing strategies.

It is interesting to note how the basic principles of circularity are often perceived as a new concept, as the future. We seem to have forgotten that in a not so distant past, prior to our throwaway society and maybe unintentionally, most items were first built to last and then regularly reused at the end of their life cycle. This included buildings and their key components: with limited or no design input, reuse of stone blocks and bricks, windows, timber beams and even tombstones on other buildings was very common. The effect of this practice is still visible for instance on pre-medieval European buildings like churches, castles, arenas and houses. Nowadays we certainly have the technology to re-use and re-cycle this concept and perhaps what is required is to change our attitude and in turn our approach to the way we design our buildings and their envelopes, as well as the materials we use. This may require clients and architects to accept some compromises to ensure that the value of the materials used in not wasted but is retained for further use. Conversely, one interesting way to deliver this latter principle may involve innovative procurement models where the curtain walling is "leased" or offered as a "service", rather than bought.

Applying innovation to the façade industry appears to be particularly challenging and real changes are rarely seen. Building skins can be complex, they are multidisciplinary matters fragmented into a great variety of products and suppliers and these aspects perhaps contribute to preventing change and progress. However, these aspects are present in other industries too, yet they are innovating and at faster rates.

Sometimes it seems that all we need is the right client, with the right vision, and the right project. An example of this is the increase in glass sizes we have all witnessed in recent years. Still the industry cannot rely solely on this type of opportunity, which are clearly limited in number, and something else must be done. In the automotive industry there are examples of leading manufacturers joining forces to develop new cars and perhaps this could be a model worth exploring.

One extremely positive aspect of the façade construction sector is its network: it is an industry where, despite its large size, many people know each other and there are real opportunities to connect at a global scale, as demonstrated by this book. This existing network, and the various industry bodies therein, should be used to enable innovation.

What changes are required in façade construction to enable innovation, capitalise on opportunities, and help transform the industry so that it can best face future challenges? How can lessons learnt in other industries be applied to the façade?

Perhaps this is an area where research should be focussed in order to unlock this potential.

TB09

John Downes
Head of Technical @ Lendlease Europe
London - UK

Question 1 — What are the most relevant innovations that you foresee will play a major role in shaping the way the next generations of building envelopes will be designed, procured, built, maintained, and dismantled?

If every reader of this article was polled on the most challenging problem facing building envelopes over the next 20 years, what would be the answers?

(Continued)

(cont'd)

It would be very surprising if any would say 'finding a pathway to achieving net zero carbon'.

It is for this reason that 'creating a pathway to net carbon zero' is most likely the biggest challenge that the property industry has ever faced and, specifically, the building envelope, steel and concrete sectors. As an industry, the reduction of embodied carbon is simply not high enough on our collective agendas. Our Industry generally considers aesthetics, cost, safety, buildability and operational design constraints in the first instance but where do we specify or consider embodied carbon?

Lendlease are fortunate to work on some of the most prestigious projects around the world and with the best design teams, supply chain partners and manufacturers. We do see an inconsistency across our industry in the activity and interest in embodied carbon.

We see education being one of the greatest challenges in the reduction of embodied carbon. There is widespread confusion around terminology that is not consistent. Net Zero Carbon (NZC), Carbon Zero, Net Zero with offsets and other terms may seem very simple, but the reality is that there is much confusion with this inconsistent terminology. Scopes 1,2 & 3 (as defined by The Green House Gas Protocol), offsetting and how is it relevant are all terms that supply chain are struggling to understand. As façade designers and engineers, we are very comfortable discussing U-value, g-value, air permeability – all the design parameters in which we have been educated. We are very familiar with operational targets and boundary conditions that are part of our everyday working life, our DNA.

However, if we were to challenge our industry right now and ask this question – what is the embodied carbon (specified or calculated) for the façade on your current project? It would be entirely surprising if there was an understanding of how to start an evaluation or what is the metric.

Our challenge is a daunting one. We need to educate our façade industry across all disciplines to get to a point where we understand the embodied carbon challenges at hand. The pathway to net carbon zero will not be achieved by the few. We need to collectively identify the solutions and start to execute to achieve incremental milestones.

In 2020, a high number of countries and clients have begun to set net zero carbon targets for as early as 2030. The challenge for the façade industry is are we equipped to respond?

Question 2 - In your current practice, what are the future innovations you are actively fostering / exploring and how do you expect these innovations to change your role and the outcome of the design?

The global façade industry has a considerable challenge to achieve net zero carbon and to be 1.5 degree aligned by 2050. The main products produced for the majority of façade realisation are extremely energy intensive and as an industry we are struggling to respond with more sustainable solutions.

There is only a minute section of our industry who are considering a façade in terms of embodied carbon and using $KgCO_{2e}/m^2$ (façade area) as a metric. Industry is quite slow to set project and business milestone targets that are challenging and demonstrating progression to a NZC position?

In the Lendlease business, we are setting our milestone targets for 2025 and focusing on some key issues which will set our business up for our target of achieving absolute carbon zero (with no offsets) by 2040.

We are choosing supply chain partners that have similar sustainability credentials as Lendlease. We are selecting supply chain partners that are aligned with our appetite to increase recycled content in aluminium, glass and architectural precast concrete. These three product types are the areas that we believe will see the most R&D and product development in the very near future.

We are setting EC targets on projects for 2025 in terms of $KgCO_{2e}/m^2$ façade area. It is a daunting task for our Industry to imagine a net zero position without offsetting. Specifying a façade system in terms of embodied carbon is most likely a new concept to most of the supply chain. We currently see disparate levels of knowledge base across supply chain in this area.

There are good levels of understanding in the product manufacturing sector (particularly the European centric organisations) however technological solutions to net zero solutions in the key focus areas are not on the horizon. We do see innovation emerging that is set to deliver significant reductions in embodied carbon, however, we have recognised that we will need to invest and partner with likeminded supply chain to realise innovative and dynamic solutions that show progression to an absolute zero carbon goal.

When we review the façade fabricator and installer knowledge base, we have recognised that there is a disparity across the supply chain. Unfortunately, we foresee a situation emerging where ambitious and challenging targets will be set for embodied carbon and the fabrication / installation supply chain will not have the tools to respond. This will be reflected across computation tools for analysis and demonstration of compliance together with the skilled workforce required. There could be a situation emerging where innovative products, designs and tools are developed but the delivery supply chain will not be equipped to cope with bringing the overall concept together.

We see education being the first step in that journey. Lendlease is a sponsor and supporter of the Centre for Window and Cladding Technology (CWCT). The CWCT in the UK has assembled a strategic committee on Sustainability and Embodied Carbon that will take on the first task of bringing the façade industry up to a knowledge base that will give an understanding of what net zero actually means together with all the other terminology that is associated

(Continued)

(cont'd)

with it. This is the first attempt in the façade industry of a trade organisation attempting to upskill across all disciplines in this sector.

Lendlease are developing carbon calculators with supply chain partners to have consistent and harmonised datasets. Whilst manufacturers Environmental Product Declarations (EPD) are continuously developing to a very high standard there is a very noticeable gap in how to combine the data from separate manufacturers, suppliers and contractors in a cohesive, standardised and harmonised manner. Digital platforms utilising digital twin technology will most likely be the future and there are some excellent examples of progress in this field.

We believe that developing strategic partnerships to collaborate on innovative technology solutions will be one of the most important steps to achieve absolute carbon zero which in Lendlease's case is by 2040.

TB10

Dominic Shillington
Hub Building Envelopes Lead @ Laing O'Rourke,
Sydney, NSW, Australia

The world has been hit by a shock that we were not prepared for. The impact has also been felt by our façade industry where its vulnerabilities have been exposed. These vulnerabilities are evident today in terms of a lack of depth of specialised skills and training in the industry, the fragile supply chain that is reliant on a few key players which are in turn reliant on suppliers that are geographically limited. We must learn from recent global shocks and become more resilient as an industry.

There is hope. This publication describes countless exciting innovations that can change the future. The next innovations should be focused in the short term on how we can rebuild from recent global shocks to create a more resilient and sustainable industry. Then in the medium term we should reinvigorate efforts to prepare our industry for the inevitable next global shock being the effects of climate change.

We commonly look at our industry as including the designers, supply chain and contractors but look past the ultimate Client and users as a key part of this relationship. I see that social innovations will play a major role in how our buildings are delivered to better serve their final users at handover and further into the future. This can be as simple as earlier engagement with the Client and the key designers and contractors tasked to deliver the building. This can ensure the best possible outcome for the end user. Although there are more exciting possibilities for micro-funding of buildings where the initial idea and delivery can be further divorced from vested and commercial interests. For example, community serving buildings where the idea and delivery are driven directly by the end-users interests. Client and end-user involvement can also be further integrated in the design process using rapid prototyping by 3D printing or hyper-real walk throughs using virtual reality technology.

There are currently innovations in our industry on the horizon that can be used to improve resilience of our industry. However, we need to be smart about innovations such as 3D printing and mass customisation. There is the opportunity for this innovation to be used to enable a lower threshold of entry for new manufacturing players in the industry. This could be a useful target of this innovation as opposed to giving further freedom to our Clients and Architects to develop multiple iterations of façade designs to suit the same purpose (and sometimes aesthetic) for the end user. By using new additive manufacturing techniques in combination with a suite of parametrically controlled system types then we can maintain freedom of design, but also enable local manufacturing opportunities. Where system design can be standardised using these techniques, we reduce the barriers of entry as specialist manufacturing techniques are in reach of 3D printing machines.

To avoid the next global shock due to climate change the industry must transform at all stages to achieve net zero or similar carbon targets. Achieving these targets must focus on all aspects of the building cycle. However, particular focus should be given to innovations that reduce the energy use, are able to produce energy, and building envelope designs that consider reduced embodied carbon. When considering whole of life carbon the focus is too often on operational carbon. Whole of life carbon considerations must include embodied carbon. There is nothing stopping reductions in embodied carbon today by using good design principles. For example, we can reduce embodied carbon by; efficient design - can structural spans be reduced; accurately analysing wind loads on the façade; and making use of local materials that have longevity. Material selection is critical as we now realise the added benefit of materials being reused at the end of the building design life when considering the circular economy. The key innovation here is to have the data available to our designers so that informed decisions are made. The development of simple rule-of-thumb tools to compare the embodied carbon of different designs will be pivotal to tackling our climate emergency. There is some exciting progress being made on an academic level regarding these tools and I am keen to see these tools being used on our current projects.

At Laing O'Rourke we are currently actively targeting key innovations that we believe will be beneficial for the future design of building envelopes. We are investigating megapanel solutions that are manufactured off-site. These solutions give us more local manufacture opportunities as the benefits of shorter transport routes are realised. Also, the current challenges we have on

(Continued)

(cont'd)

construction sites particularly at interfaces between façade, structure and service trades are more effectively managed in an offsite environment. Where we also used to be restricted by the size of a shipping container, we can now look at volumetric or panelised design for manufacture and assembly (DfMA) solutions that are only restricted by the size of a semi-trailer truck and the transport route. These solutions will require earlier coordination of designers and contractors that will encourage earlier engagement with the Client. This can only be a good thing, but will require rethinking our standard building contracts to re-evaluate the risk profile including the higher investments in design required earlier in the project lifecycle.

We have also been at the cutting edge of virtual reality tools to facilitate rapid design decisions. This solution minimises material and time wastage compared to more conventional visual mock-up units and physical samples that are often discarded at the end of projects. Technology is not a limitation in making this work, but the perception of the Clients and end users that this tool can replace the physical samples.

I look forward to a closer collaboration with specialists that are working on these exciting innovations. Where they offer real benefits to the end user and improve the resilience of our industry then they should thrive.

TB11

Davina Rooney,
Chief Executive Officer @ Green Building Council of Australia,
Sydney, NSW, Australia

Standing sentinel along shoreline of Sydney Harbour at Barangaroo, the three soaring International Towers designed by Rogers Stirk Harbour + Partners each boasts a beautiful façade of glass fins. Painstaking building information modelling during design ensured each vertical panel was aligned to the sun's path, reducing glare and heat gain. While the façade may not be as breathtaking as the views, the trio of International Towers are an energy-efficient powerhouse with 'world leadership' Green Star ratings.

Just a stone's throw away, the lush green façade of One Central Park turns heads. The 120-metre high vertical garden, featuring 38,000 plants, has been described by the architect, Bertram Beissel of Ateliers Jean Nouvel, as "a flower for each resident and a bouquet to the city". The vertical garden also acts as a carbon sink, reducing the urban heat island effect while enhancing the building's thermal mass and biodiversity. Residents report spotting birdlife that had never ventured into this part of the city before.

Both buildings are real world exemplars that point to a future where façades work harder than ever before.

When Australia's Green Star rating system shot onto the scene in 2003, it sparked a new conversation about the role of building façades in delivering energy efficiency. Since then we have seen a radical rethink on everything from low-glare glass to automatic shading, insulation to solar passive design. But if we are to create buildings that are good for people and the planet, and that meet the <u>World Green Building Council's ambitious 100% net zero carbon targets by 2050</u>, we must think of façades first.

A 'façade first' approach can help us deliver adaptive, resilient, climate-ready buildings that respond to emerging global megatrends reshaping our world. Whether we are accelerating climate action, expanding the circular economy, bringing nature back to our cities or reprioritising human health and wellbeing, buildings will play a central role in addressing these megatrends.

The statistics are well known but worth repeating. Buildings are responsible for 39% of global emissions. Energy demand from our buildings is expected to increase by 50% by 2050. Buildings are also responsible for 50% of global material use – an eye-watering 42.4 billion tonnes – annually. According to the World Health Organization up to 30% of our buildings make us sick, through mould and other sick building symptoms, and that is before we even consider coronavirus.

In the future, a building's skin will open, close and breathe through cellular systems that maximise light and air while reducing energy load. Solar panels won't be bolted on but integrated into to the façade itself. Micro-algae, bred on the building's shell, will be fed into bioreactors to produce biomass and heat. Sophisticated water collection systems will capture and filter every drop of rainwater before it hits the ground. Urban farms will harness the power of hydroponics to grow vast quantities of food while reducing the urban heat island effect.

This future is very different to our present reality. Glittering skyscrapers of glass may offer bird's eye views, but also drive up energy bills and carbon emissions from heat gain and heat loss. The authors of one seminal report, High-Cholesterol Buildings, argue that the insulation value of modern glass-façade skyscrapers is "equivalent to medieval half-timber houses".

Skyscrapers aren't the only sustainability supervillains. Up to 40% of a home's heating energy can be lost and up to 87% of its heat gained through glazing. To achieve certification under the Green Building Council of Australia's new Green Star Homes standard, buildings will need to balance double-glazed windows and doors with good air flow and access to daylight. Around 25% of heat loss in winter is caused by drafts, so the standard will reward airtight designs that maintain a high degree of thermal comfort throughout the year.

How do we set the scene for future-ready façades? Meeting our national Paris Agreement targets requires radical change at scale – and that scale will only be brought about with the right policy frameworks. The best buildings

(Continued)

> **(cont'd)**
>
> aren't delivered by accident or in a vacuum. Our challenge is to set the temperature to ensure they are not exemplars but the norm.
>
> The secret is to set minimum standards and then incentivise industry to do better. Increasing construction codes establishes the baseline and provides a guiding light, while incentives — such as those embedded in green finance mechanisms or planning policies — encourage best practice and beyond.
>
> We must also turn to technology. Building information modelling can streamline the design process and push the performance of the building envelope, as is in evidence at International Towers. Computer-controlled sunshades, like those found at the Architectus and Ingenhoven-designed 1 Bligh Street in Sydney, pack an energy-efficient punch. And 3D printing is rapidly evolving to support versatility in design and the use of sustainable alternatives to concrete.
>
> We can design the future we want, where building façades are not just beautiful. They can also be energy efficient, resilient to a changing climate, responsive during wild weather and adaptive to their local environment. But if we fail to reimagine our buildings, we risk a dystopian future: one where our cities' rich denizens live in a relatively comfortable 'bubble' while the rest of the population suffers in an urban inferno. This would be the ultimate façade — and the façade of the future can't be healthy only for the wealthy.
>
> There is one other facet to this façade story. In Brisbane, a 30-storey residential building designed by Japanese-born and Sydney-based architect Koichi Takada is taking shape. Dubbed Urban Forest, the 382-apartment building will be adorned with 1,000 trees and 20,000-plus native plants. Urban Forest is a naturalist's dream, so it is perhaps unsurprising that Sir David Attenborough has applauded the project on social media. As Urban Forest's organic façade attracts attention far and wide, it becomes a symbol of a greener future. Paradoxically, the façade is far from superficial and, as Takada says, creates "awareness for more green buildings to be built around the world".

2.5 Conclusion

This chapter has presented a critical analysis of the future innovation drivers, trends and constraints in the façade industry, based on the essays received from the 11 selected façade industry experts. The contributions provided an overview on innovation from different perspectives, reflecting differences in geographical context, field of expertise and professional background. This chapter has also critically reflected on these contributions and provided a cohesive snapshot of the future of façade innovation, identifying recurring topics and emerging trends across the sector.

The analysis shows that, in the process of rethinking building skins, sustainability-driven innovation will be at the forefront, receiving a mention from all 11 contributors. Second, the future will see the diffusion of new means of collaboration that are necessary to generate a fruitful, integrated and unified environment in which innovation can thrive.

The chapter 'Façade innovation, an Industry Perspective' contributes to this direction, gathering 11 industry experts from renowned international companies and providing an imaginary roundtable to share their experiences and perspectives on façade innovation. Acknowledgement and thanks are due to all the experts who enthusiastically agreed to participate, enriching the discussion in this chapter with their contributions.

This approach is further expanded in the two sections of the book—Part A: Product Innovation and Part B: Process Innovation— where more than 50 authors from both industry and academia, with different backgrounds and expertise, share their research and vision on the future of building skins.

References

Azari, R., & Abbasabadi, N. (2018). Embodied energy of buildings: A review of data, methods, challenges and research trends. *Energy and Buildings, 168*, 225–235.

Blayse, A. M., & Manley, K. (2004). Key influences on construction innovation. *Construction innovation, 4*.

Brambilla, A., & Sangiorgio, A. (2020). Mould growth in energy efficient buildings: Causes, health implications and strategies to mitigate the risk. *Renewable and Sustainable Energy Reviews, 132*, 110093.

Braungart, M., & McDonough, W. (2009). *Cradle to cradle*. D&M Publishers.

Cardellicchio, L., & Stracchi, P. (2020). Innovation by design: Technological challenges and opportunities behind the Parramatta Powerhouse and Sydney Modern Project. *Architectural Bulletin, 77*(2).

Cole, R. J. (2012). Transitioning from green to regenerative design. *Building Research & Information, 40*(1), 39–53.

Demircioglu, M. A. (2016). *Organizational innovation. Global encyclopedia of public administration, public policy and governance* (pp. 1–5). .

Dixit, M. K., Jose, L., Fernández-Solís, Sarel Lavy, & Charles, H. Culp (2012). Need for an embodied energy measurement protocol for buildings: A review paper. *Renewable and Sustainable Energy Reviews, 16*(2012), 3730–3743.

Dubois, A., & Gadde, L.-E. (2002). The construction industry as a loosely coupled system: Implications for productivity and innovation. *Construction Management & Economics, 20*(7), 621–631.

Dulaimi, M. F. Y., Ling, F. Y., Ofori, G., & Silva, N. D. (2002). Enhancing integration and innovation in construction. *Building Research & Information, 30*(4), 237–247.

Etzkowitz, H. (2003). Innovation in innovation: The triple helix of university-industry-government relations. *Social Science Information, 42*(3), 293–337.

Luo, L., He, Q., Jaselskis, E. J., & Xie, J. (2017). Construction project complexity: Research trends and implications. *Journal of Construction Engineering and Management, 143*(7), 04017019.

MacArthur, E. (2013). Towards the circular economy. *Journal of Industrial Ecology, 2*, 23–44.

Munari, B. (2018). *Design e comunicazione visiva: Contributo a una metodologia didattica* Gius. Laterza & Figli Spa.

Patterson, M., Silverman, B., Kensek, K., & Noble, D. (2016). *The millennium IGU: Regenerative concept for a 1000-year insulated glass unit.*

Rodima-Taylor, D., Olwig, M. F., & Chhetri, N. (2012). Adaptation as innovation, innovation as adaptation: An institutional approach to climate change. *Applied Geography*, *33*(0), 107−111.

Slaughter, E. S. (1998). Models of construction innovation. *Journal of Construction Engineering and Management*, *124*(3), 226−231.

Stahel, W. R., & Reday-Mulvey, G. (1981). *Jobs for tomorrow:* The *potential for substituting manpower for energy.* Vantage Press.

Part A

Product Innovation

Urban overheating mitigation through facades: the role of new and innovative cool coatings

Mattia Manni[1], Ioannis Kousis[1], Gabriele Lobaccaro[2], Francesco Fiorito[3,4], Alessandro Cannavale[5] and Mattheos Santamouris[3]

[1]Department of Engineering, CIRIAF—Interuniversity Research Center on Pollution and Environment 'Mauro Felli', University of Perugia, Perugia, Italy, [2]Department of Civil and Environmental Engineering, Faculty of Engineering, NTNU Norwegian University of Science and Technology, Trondheim, Norway, [3]School of Built Environment, University of New South Wales, Sydney, NSW, Australia, [4]Department of Civil, Environmental, Land, Building Engineering and Chemistry, Polytechnic University of Bari, Bari, Italy, [5]Polytechnic University of Bari, Bari, Italy

Abbreviations

UHI	urban heat island
PCM	phase-change materials
IR	infrared
HR	highly reflective
RR	retroreflective
DGU	double-glazed unit
AS-RR	angular-selective retroreflective
UC	urban canyon

3.1 Introduction

The urban overheating effect is an ongoing environmental concern caused by a synergistic combination of local and global climate change phenomena together with the increasing urbanization (Piselli, Castaldo, Pigliautile, Pisello, & Cotana, 2018; Santamouris, Cartalis, Synnefa, & Kolokotsa, 2015; Zhou, Zhang, Li, Huang, & Zhu, 2016). The most documented associated issue is the urban heat island (UHI) effect (Rahman, Moser, Rötzer, & Pauleit, 2017; Xu, González, Shen, Miao, & Dou, 2018). However, the boundary between UHI and urban overheating concepts is unclear and the two phenomena tend to be generally considered unique. Although both refer to the increase of urban air temperature, the former describes a rising trend that is shown in the spatial distribution of the air temperature, while the latter concern a growing trend in urban air temperature over time. In particular, the UHI observed in several urban districts from the 20th century consists of a layer of stagnant warm air over the heavily built-up areas. Consequently, the air temperature of the city can be far higher than the air temperature experienced in the rural surroundings. Anticyclonic conditions, low wind speed and clear sky conditions strongly contribute to increase the magnitude of the UHI. Conversely, precipitation decreases UHI magnitude due to the increase of thermal admittance in rural areas and the amount of heat exchanged with the surroundings over time (Santamouris, 2015a).

The UHI can be investigated either as surface UHI or as atmospheric UHI, depending on the temperature parameter utilized to characterize the phenomenon: the surface temperature in the former, the air temperature in the latter. Fig. 3.1 shows the spatial pattern of surface UHI intensity throughout the globe (Cui, Xu, Dong, & Qin, 2016). This was found to reach up to 15.0°C, while negative values were observed in South America and Asia.

Extensive monitoring has documented increased ambient temperatures in more than 450 major world cities (Founda & Santamouris, 2017; Santamouris, 2007, 2015b; Santamouris & Feng, 2018). Ambient temperature has been evaluated as the outdoor air temperature, estimated locally. The amplitude of the urban overheating depends on the urban patterns, the landscape and the specifics of the regional climate (Akbari & Kolokotsa, 2016; Brans, Engelen, Souffreau, & De Meester, 2018; Lima, Scalco, & Lamberts, 2019; Lin, Ge, Liu, Liao, & Luo, 2018). In addition, reduced number of trees and permeable surfaces, higher production of

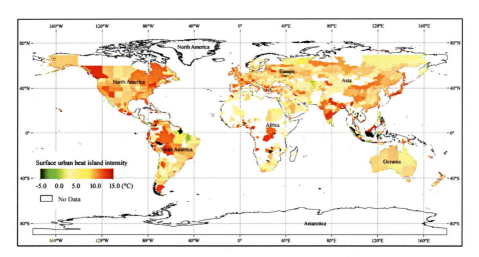

Figure 3.1 Spatial pattern (space unit: province/state) of the surface UHI intensity. Surface UHI intensity is here calculated as the difference between urban and rural mean surface temperature. A negative value indicates that cities are cooler than their surroundings. *UHI*, Urban heat island.
Source: Reproduced with permission from Cui, Y., Xu, X., Dong, J., & Qin, Y. (2016). Influence of urbanization factors on surface urban heat island intensity: A comparison of countries at different developmental phases. *Sustainability*, *8*, 706. https://doi.org/10.3390/su8080706.

anthropogenic heat[1] and properties of the materials used for outdoor pavements and building fabrics contribute to increase the local thermal stress in big cities (Akbari & Kolokotsa, 2016; Oke, 1981; Piselli et al., 2018; Santamouris, 2001; Warren, 2014; Xie, Huang, Wang, & Xie, 2005). A comprehensive analysis on 100 Asian and Australian cities revealed that the local effects of the atmospheric UHI are quite significant. It presents a mean maximum amplitude close to 5°C, but this may vary up to 10°C (Santamouris, 2015c).

It is well demonstrated that UHI exacerbates the intensity of extreme heat events generated by heat waves and global climate change. Indeed, a recent study showed that the peak ambient temperature of inland Western Sydney during the heat wave events in January 2018 was between 6°C and 8°C higher than at the airport located more than 8.0 km away Sydney Central Business District (Santamouris et al., 2020).

Overheating in cities causes a significant energy impact on buildings and electricity generation and on peak electricity needs in cities. In particular, it causes a mean energy penalty (throughout the year) close to 2.4 kW h per square metre, 0.74 kW h per square metre and degree of temperature increase and 237 kW h per person, respectively. Values were calculated in the study of Santamouris (2014) for the municipalities of Athens, Tokyo and Beijing. Furthermore, it also brings an energy penalty close to 70 kW h per person and degree of UHI intensity (Santamouris et al., 2001) while raising peak electricity requirements by around 20 W per person and degree of temperature increase (Santamouris et al., 2015). Besides the penalty on energy, the UHI seriously affects human heat-related mortality and morbidity. It is demonstrated that higher urban temperatures are strongly correlated with increasing health problems and outdoor thermal stress at pedestrian level (Pioppi, Pigliautile, Piselli, & Pisello, 2020).

During the last decades, numerous systems and technologies have been proposed, developed and implemented to mitigate the impact of the UHI effects (Akbari et al., 2016). Technologies involve the use of additional greenery, the utilization of new cool and supercool materials [ie, retroreflective (RR) materials, thermochromics, radiative coolers] for buildings and urban spaces, solar control, water-based systems to reject the excess heat in atmospheric sinks and numerous combined arrangements of the proposed technologies (Akbari & Touchaei, 2014; Chung & Park, 2016; Djedjig, Bozonnet, & Belarbi, 2013; Fabiani, Pisello, Bou-Zeid, Yang, & Cotana, 2019; Lobaccaro et al., 2019). Akbari and Kolokotsa (2016) reviewed the mitigation strategies proposed during the last three decades for tackling the over warming of the urban districts and defined highly reflective (HR) materials and green infrastructures as the most promising technologies. Evidence from different research activities proved that these solutions can reduce the outdoor air temperature by up to 5°C (Radhi, Sharples, & Assem, 2015) in urbanized areas, which consequently reduces the energy requirements for cooling (Xu et al., 2018). Similarly, the evaluation of the current large-scale urban mitigation projects involving several UHI mitigation

---

[1] The term 'anthropogenic heat' refers to the heat released to the atmosphere as a result of human activities. Industrial plants, building heating and cooling, human metabolism and vehicle exhausts contribute to the anthropogenic heat output. Typically, it accounts for around 15–50 W/m² of the local heat balance, but it may be 10 times greater in the centre of large cities in cold climates and industrial areas.

technologies and measures has demonstrated a considerable cooling capacity and a potential to reduce the peak urban temperature by up to 3°C (Pisello, Saliari, Vasilakopoulou, Hadad, & Santamouris, 2018).

Within the framework of the book, which has the building skin as the main research domain, this chapter aims to outline the state of the art of finishing materials applicable to building facades for UHI mitigation. The most common and newest coatings, which have also been considered the most promising to mitigate UHI effects, are described in the following sections along with the research trends driving their developments in the coming years. Furthermore, an overview of their impact on the urban environment (ie, mitigation of air temperature, increase surface reflectance) is reported for each. The main research priorities are also discussed and analyzed.

3.2 Background

Materials utilized in the envelope of buildings or in the open urban fabric largely determine the thermal balance of cities. Finishing materials with high reflectance are traditionally applied to urban surfaces to counterbalance the overheating of cities and settlements. The recent development of HR materials for roofs and pavements has contributed greatly to the increase of the albedo of cities and to the decrease of peak ambient urban temperatures. Such materials have been extensively investigated through numerical and physical modelling and numerous case studies on their applications have been performed (Cotana et al., 2014; Nazarian, Dumas, Kleissl, & Norford, 2019; Santamouris, Synnefa, & Karlessi, 2011). Despite their beneficial thermal performance, reflective materials are known to present significant optical ageing problems. Furthermore, high albedo materials could create problems of glare and contrast at the pedestrian level.

Recent research on new cool and supercool materials for buildings and urban open spaces has yielded noteworthy advances. Intensive recent research has improved the thermal and optical performance of reflective materials, while also providing new technological products such as advanced thermochromic, fluorescent, plasmonic and photonic materials. These materials are characterized by very low surface temperature, decreased release of sensible heat and a high cooling potential. Such innovative materials and structures exhibit subambient surface temperatures and a more appropriate adaption to the transient urban climatic conditions. Achieving subambient temperatures allows drastically reducing of sensible heat release in the atmosphere, which depends on the difference between air temperature and material surface temperature. This aspect can significantly impact the magnitude of the UHI mitigation.

Low surface temperature of materials can be achieved through different strategies such as decreasing heat and solar gains, increasing heat losses and modulating heat gains. The absorptivity of the materials in the visible, near-infrared (IR), or the whole solar spectrum should be the minimum possible to decrease heat and solar gains. Also, materials lose energy through convective, conductive and radiative heat transfer processes. To increase radiative losses, the spectral emissivity should be maximized (particularly in the atmospheric window[2]). Incorporating coatings enhanced with phase-change materials (PCM) in the building skin may control emissivity and allow the prevention of undesirable cooling of the materials and structures, especially during the winter season. In fact, PCM enable energy storage by releasing (or absorbing) energy at phase transition (ie, between solid and liquid) to provide heat (or cooling). Similarly, fluorescent materials present additional radiative losses because of the fluorescent emission in specific wavelengths. When it comes to solution based on the modulation of the heat gains, materials presenting a high sensible heat thermal capacitance or increased storage capacity of latent heat can efficiently modulate heat gains and decrease their surface temperature.

Research aiming to develop innovative materials for UHI mitigation has designed, tested and proposed numerous innovative technological solutions presenting superior optical and thermal properties (up to 97% reflection of the incident solar radiation) as well as a significant mitigation capacity (ambient temperature reduction by up to 12°C) (Santamouris et al., 2011). The most innovative developments are advanced light-coloured, coloured coatings to reflect the IR irradiation, RR layers, PCM, doped surfaces, thermochromic, fluorescent and photonic or plasmonic radiative structures. Fig. 3.2 presents the historical progress of the most innovative technologies to minimize heat gains and maximize losses, while Table 3.1

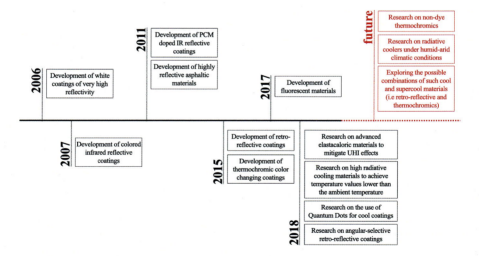

Figure 3.2 Historical development and future progress (in *red*) of the most important materials used for UHI mitigation purposes. *UHI*, Urban heat island.
Source: Reproduced with permission from Santamouris, M., & Yun, G. Y. (2020). Recent development and research priorities on cool and super cool materials to mitigate urban heat island. *Renewable Energy, 161*, 792−807. https://doi.org/10.1016/j.renene.2020.07.109.

---

[2] Not all wavelengths of electromagnetic radiation from the Sun reach the Earth and not all wavelengths emitted by the Earth reach into Space. The atmosphere absorbs some of this energy while allowing other wavelengths to pass through. The places where energy passes through are called 'atmospheric windows'.

Table 3.1 Main optical characteristic of the different material technologies proposed for urban mitigation.

Type of materials	Optical properties of the materials					Mitigation potential	Drawback		
	High solar reflectance in the visible spectrum	High solar reflectance in the IR spectrum	High broadband emissivity	High emissivity in the atmospheric window	High fluorescent emission		Ageing issues	Glaring issues	Winter penalty
Light colour reflective coatings	✓		✓			■□□□	✓	✓	✓
Coloured IR reflective materials		✓	✓			■□□□		✓	✓
Reflective materials with nano-PCM	✓	✓	✓			■■■□	✓	✓	
Dye-based thermochromic materials	✓	✓	✓			■■■□	✓	✓	
Fluorescent materials	✓	✓	✓		✓	■■■■□	✓		✓
Photonic materials and components for daytime radiative cooling	✓	✓		✓		■■■■■			

IR, Infrared; PCM, phase-change materials.
Source: Reproduced with permission from Santamouris, M., & Yun, G. Y. (2020). Recent development and research priorities on cool and super cool materials to mitigate urban heat island. *Renewable Energy*, 161, 792–807. https://doi.org/10.1016/j.renene.2020.07.109.

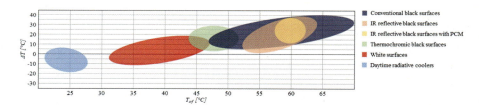

Figure 3.3 Comparative presentation of existing experimental data of various urban mitigation technologies reviewed in the study of Santamouris and Yun (2020). The main axes report quantities regarding the surface temperature (T_{srf}) and the difference between surface and ambient temperature (ΔT).
Source: Reproduced with permission from Santamouris, M., & Yun, G. Y. (2020). Recent development and research priorities on cool and super cool materials to mitigate urban heat island. *Renewable Energy*, *161*, 792–807. https://doi.org/10.1016/j.renene.2020.07.109.

describes their principal optical properties along with their advantages and disadvantages. Reduction of solar absorptance is the main requirement for materials devoted to passive radiating cooling below ambient air temperature: at the same time, such technologies should maximize selective emissivity in the IR region of the spectrum to boost thermal energy dissipation from materials, by activating an effective, passive, radiative sky cooling of excessive heat to the Space. Achieving subambient temperatures under the Sun is only possible if materials possess low absorptivity in the solar range of wavelengths and high emissivity in the atmospheric window (ie, the wavelength interval between 8 and 13 μm). While conventional materials absorb and emit within a broad range of the electromagnetic spectrum, radiative coolers are highly emissive in wavelengths between 8 and 13 μm, where the atmospheric radiation is minimal. In the following sections the optical and thermal properties, as well as the cooling performance of the most frequently considered material technologies for UHI mitigation purposes, are discussed. A comparative presentation of existing experimental data of various UHI mitigation technologies is outlined in Fig. 3.3.

3.3 Highly reflective materials

Materials with high solar reflectance have been investigated as potential surface treatments to maintain sunlit objects at low temperatures (Berdahl et al., 2016). HR coatings can contribute to increasing the solar reflectance of facades (Fig. 3.4), leading to the reduction of solar radiation absorbed by building facades (Doya, Bozonnet, & Allard, 2012; Radhi et al., 2015; Taha, 1997). Exploiting materials with high solar reflectance in the visible range and high IR emissivity allowed a reduction of surface and air temperatures, lowering both energy demand for cooling and electricity energy peaks and improving indoor comfort conditions (Jandaghian & Berardi, 2020). A lower surface temperature of the building also causes a reduction in the transfer of sensible heat toward the interior as well as toward the outdoor

Figure 3.4 Applications of HR materials to building facades, both opaque and transparent, to reduce surface temperature. *HR*, Highly reflective.

environment (Synnefa, Saliari, & Santamouris, 2012). Nonetheless, when applied to building facades, their effectiveness is negatively influenced by factors such as the distance from other buildings and the window-to-wall ratio (Nazarian et al., 2019). HR materials reflect most of the solar irradiation toward nearby buildings by increasing the amount of solar radiation impinging on their facades. This may contribute to exacerbating the surface UHI phenomenon and the amount of mutual multiple solar interbuildings reflections, especially in highly dense built environments.

The influence of HR coatings on urban microclimates has been extensively studied through numerical and physical models by demonstrating the related advantages (ie, materials with low absorption coefficients) and disadvantages (ie, an increased number of solar interbuildings reflections) (Berdahl & Bretz, 1997; Bretz & Akbari, 1997; Bretz, Akbari, & Rosenfeld, 1998; Doya et al., 2012; Prado & Ferreira, 2005; Radhi et al., 2015; Synnefa, Santamouris, & Livada, 2006; Synnefa, Santamouris, & Akbari, 2007a; Synnefa, Santamouris, & Apostolakis, 2007; Taha, 1997).

A study of a typical residential building in Sydney showed that cool roofs guarantee a 3.5% savings of peak cooling demand per 0.10 increase in the roof albedo (Santamouris et al., 2018). Nonetheless, the cool roofs did not contribute significantly to the reduction of peak ambient temperature at ground level; they demonstrated only a maximum calculated decrease of about 0.5°C. In the area of Athens, Greece, the Penn State-NCAR MM5 model was utilized to demonstrate that an increase in building structures' albedo of 0.65 can decrease the air temperature by 2.2°C (Synnefa, Santamouris, & Akbari, 2007b). The Penn State-NCAR MM5 model is a fifth-generation mesoscale model (analyzed domain of 10^3-10^4 km) which allows modelling of atmospheric circulation using fluid dynamic equations (ie, mass, energy and momentum conservation). It is commonly used to conduct weather forecasts and climate projections.

The cool coatings are made with highly near-IR reflective metallic oxide pigments such as titanium oxide, chromium oxide, cobalt oxide and barium oxide as additives. While cool white-coloured coatings show significant cooling potential (0.75–0.90 solar reflectance), it is complicated to develop cool dark-coloured coatings. This is mainly due to the limitations regarding the addition of the light-colored metallic oxides. As reported by Synnefa et al., the solar reflection coefficient of cool black-coloured coatings ranges between 0.12 and 0.27. Among other cool-coloured pigments examined in this study, the highest solar reflection is reported for cool orange-coloured coatings with a solar reflection of 0.63 (Synnefa, Santamouris, & Apostolakis, 2007). In another study by Levinson et al. (2007), the dark-coloured paints with high near-IR reflection property are reported to attain an albedo of up to 0.40. The study performed by Levinson et al. showed that the application of a layer of cool black pigment increased the solar reflection coefficient of the roof shingles from 0.04 to 0.12 and adding a thick white basecoat as a near-IR reflective layer resulted in a further 0.06 increase in the solar reflection coefficient. Transparent cool coatings have also been developed alongside opaque cool coatings. In the review study carried out by Zheng, Xiong and Shah (2019), the state of the art of transparent nanomaterial-based solar cool coatings is provided. These consist of composite materials made up of transparent thin-layered substrates incorporated with nanosized additives for the purpose of reducing solar heat gain and passive cooling in buildings.

HR materials presented some drawbacks due to solar energy gains through the building envelope during winter, meaning that they were not suitable for all seasons. As a matter of fact, the increment of building energy demand for heating can be observed as a direct consequence of the application of HR coatings. As shown in the study of Synnefa et al. (2007a), the application of HR coatings on the building facades may cause a heating penalty in the winter. Thus there is a need for the development of a responsive finishing material that can modify its behaviour according to the outdoor temperature or the available solar radiation. Furthermore, HR coatings may exhibit aging problems that significantly diminish their performance (Mastrapostoli et al., 2016; Tsoka, Theodosiou, Tsikaloudaki, & Flourentzou, 2018).

RR materials were developed during the last decade as a subgroup of the HR ones (Manni, Lobaccaro, Goia, & Nicolini, 2018; Morini et al., 2017, 2018; Rossi, Pisello, Nicolini, Filipponi, & Palombo, 2014; Yuan, Emura, & Farnham, 2016). The RR materials consist of engineered surfaces or surface treatments that are capable of reflecting incident solar radiation back toward the sky and thus reducing mutual radiative effect among surfaces in close proximity (Fig. 3.5). The solar light can be reflected in the opposite direction by a sequence of reflections (in RR prism-array structures) or by a combination of refractions and reflections (in RR bead-embedded layers) because of the geometry pattern of RR layers.

Commercial films, realized with prism-array structures, capsule-lens and bead-embedded layers, were tested on miniature models of urban canyons (UCs) to demonstrate their effectiveness in mitigating UHI effects (Rossi et al., 2014; Sakai, Emura, & Igawa, 2008). The results revealed that a reduction of building energy demand of up to 10% can be achieved by adopting RR facades in place of traditional cool materials (eg, diffuse HR materials) (Han, Taylor, & Pisello, 2015). Also, RR materials were capable of reducing outer surface peak temperature by around 15°C (Zhang, Meng, Liu, Xu, & Long, 2017).

The application of RR layers was also analyzed on transparent building elements (ie, windows, glazed facades), coupled with glazed surfaces [ie, float glass, low-emissivity double-glazed unit (DGU)] (Ichinose, Inoue, & Nagahama, 2017; Inoue,

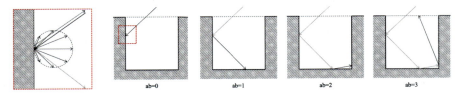

Figure 3.5 On the left, overview of the different ways a surface interacts with incident photons (*thick solid arrow*). The photons can be reflected specularly (*thin solid arrow*), retroreflected (*dotted line*), or diffusively reflected (*all the arrows within the dashed circle*). On the right the visualization of the ray-tracing process depending on the ambient bounces. In this example the sunray exits the UC after three reflections. Conversely, when retroreflection occurs, the sunrays exit the UC after only one reflection (Manni, Bonamente, et al., 2020). *UC*, Urban canyon.

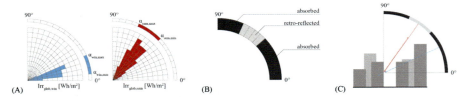

Figure 3.6 Conceptualization process and functioning of AS-RR materials. (A) Angular distribution of the solar irradiation estimated for a generic latitude of the Continental Europe climate in summer and winter season. (B) Absorption/retroreflection pattern implemented for the AS-RR materials. (C) Angular-selective behaviour of the AS-RR materials applied to the façade in a generic urban canyon. *AS-RR*, Agular-selective retroreflective.

Shimo, Ichinose, Takase, & Nagahama, 2017). The float glass window enhanced with heat-ray RR films, which was studied by Yoshida, Yumino, Uchida and Mochida (2015), showed how the peak of the mean radiant temperature is reduced by more than 4°C in comparison with a low-emissivity DGU window.

Despite the proven advantages of the different RR technologies, there are still some aspects that need to be addressed to fully enable the potential of these coating systems. Among them, it is worth mentioning the increased light pollution and the decreased solar loads in winter. Hence, implementing the angular selective behaviour on RR materials represents a fundamental step to enable more effective cool materials (Manni et al., 2018). The concept behind the angular-selective retroreflective (AS-RR) materials is to have surface treatments capable of providing a selective response depending on the angle of incidence of the solar radiation (Fig. 3.6). The AS-RR materials behave as perfect retroreflectors as long as the incident angle of the sunrays is included within the angular range of activation, while they present a Lambertian behaviour when the incident angle of the sunrays is outside the selected angular range, presenting a solar reflectance equal to those of traditional materials. The angular range of activation of AS-RR materials is representative of summer conditions (ie, summer sun elevation angle) and it is generally based on the location where those materials are exploited. It enables solar heat gains during winter, thereby overcoming the main drawback observed for RR materials. Some concepts of AS-RR materials have recently been proposed (Sakai & Iyota, 2017) along with methodologies to define their angular range of activation (Manni et al., 2018; Zinzi, Carnielo, & Rossi, 2015). Numerical analyses estimated a reduction of solar irradiation absorbed by the façade of up to 8% resulting from the application of AS-RR instead of conventional coatings (Manni, Bonamente, et al., 2020; Manni, Cardinali, et al., 2020).

In this framework the main trends in the development of the RR materials highlight that the ongoing research activities aim at (1) implementing RR coatings that perform well in both the IR and visible spectrum; (2) implementing RR surface treatments in historical buildings that do not alter the tone of the facades (Castellani, Morini, Anderini, Filipponi, & Rossi, 2017; Cotana et al., 2015; Morini et al., 2018); (3) implementing the solar retroreflection capability of materials even when the sunrays hit the surface from almost a parallel angle (Yuan, Farnham, & Emura, 2015; Yuan, Emura, & Farnham, 2015); and (4) implementing selective RR materials, optical properties of which vary according to the solar azimuth angle (Manni et al., 2018; Manni, Lobaccaro, Goia, Nicolini, & Rossi, 2019; Manni, Cardinali, et al., 2020; Sakai & Iyota, 2017).

3.4 Thermochromics

In the last decades, materials that can adapt their colour (and their optical properties) to the outdoor temperature have been developed and tested in the built environment (Santamouris & Yun, 2020). These materials, known as thermochromic materials, can alter their colour when reaching a predefined transition-temperature value (Kousis & Pisello, 2020; Santamouris & Yun, 2020). Two main categories of thermochromic materials have been reported to date: (1) dye-based thermochromics and (2) nondye thermochromics (Garshasbi & Santamouris, 2019). The first are mainly based on leuco dye mixtures[3] (Wikipedia Contributors, 2020) and owe their thermochromic effect to the interaction of their components (Fig. 3.7), while molecular rearrangements or optical effects at nanoscale cause such thermochromic effects in nondye thermochromics (Fig. 3.8).

Similarly to HR coatings, the dye-based thermochromics are often prone to ageing issues (Karlessi & Santamouris, 2015). In fact, the prolonged exposure of dye-based thermochromics to ultraviolet radiation negatively influences the reversible thermochromic circle and ultimately destroys it. Such degradation is not observed in the nondye thermochromics that, for this reason, are considered the next generation of advanced cool materials. However, since there is no reported application or

---

[3] A leuco dye is a dye capable of switching from a colourless chemical to a coloured configuration. Such transformations are generally reversible and they can be caused by heat (thermochromism), light (photochromism), or pH (halochromism), depending on the dye typology.

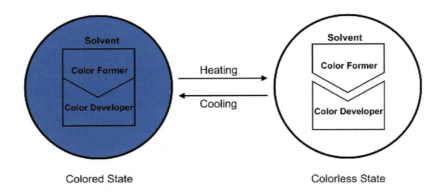

Figure 3.7 Schematic representation of thermochromic mechanism in Leuco dyes.
Source: Reproduced with permission from Garshasbi, S., & Santamouris, M. (2019). Using advanced thermochromic technologies in the built environment: Recent development and potential to decrease the energy consumption and fight urban overheating. *Solar Energy Materials and Solar Cells*, *191*, 21–32. https://doi.org/10.1016/j.solmat.2018.10.023.

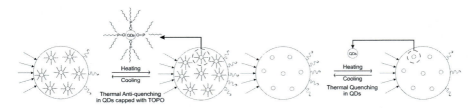

Figure 3.8 Schematic representation of thermal quenching/antiquenching effects in quantum dots.
Source: Reproduced with permission from Garshasbi, S., & Santamouris, M. (2019). Using advanced thermochromic technologies in the built environment: Recent development and potential to decrease the energy consumption and fight urban overheating. *Solar Energy Materials and Solar Cells*, *191*, 21–32. https://doi.org/10.1016/j.solmat.2018.10.023.

experiment in the literature concerning the built environment, this section will henceforth focus exclusively on dye-based thermochromics.

A dye-based thermochromic pigment typically consists of three components: the colour former, the colour developer and the solvent. When thermochromic pigments are added to conventional white coatings, the solar absorption coefficient of the treated surface remains almost unaltered as long as the surface temperature is lower than the transition value (Fig. 3.9) (Karlessi, Santamouris, Apostolakis, Synnefa, & Livada, 2009; Ma, Zhu, & Wu, 2001). Conversely, it lowers when such a threshold is surpassed (Ma, Zhang, Zhu, & Wu, 2002). White Portland cement enhanced with microencapsulated thermochromic materials (Fig. 3.10) showed higher solar absorption below the transition point (24°C) while maintaining the same optical properties of the conventional counterpart above that temperature value (Ma & Zhu, 2009).

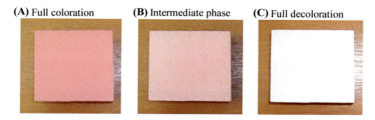

Figure 3.9 The thermochromic sample in three different thermal phases: full coloration, intermediate phase and full decoloration.
Source: Reproduced with permission from Karlessi, T., & Santamouris, M. (2015). Improving the performance of thermochromic coatings with the use of UV and optical filters tested under accelerated ageing conditions. *International Journal of Low-Carbon Technologies, 10*, 45−61. https://doi.org/10.1093/ijlct/ctt027.

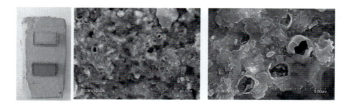

Figure 3.10 From the left: aspect of the reversibly thermochromic mortar at a temperature lower (in the specimen at the bottom, 8°C) and higher (in the specimen at the top and in the base, 50°C) than the transition temperature, a scanning electron microscope image of such a mortar and a zoom of this image.
Source: Reproduced with permission from Perez, G., Allegro, V. R., Corroto, M., Pons, A., & Guerrero, A. (2018). Smart reversible thermochromic mortar for improvement of energy efficiency in buildings. *Construction and Building Materials, 186*, 884−891. https://doi.org/10.1016/j.conbuildmat.2018.07.246.

Such a thermochromic mortar was also characterized by good physical and mechanical properties that can be implemented for energy-efficiency into building infrastructure (Perez et al., 2018). Furthermore, thermochromic components were combined with PCM for building façade application (Soudian, Berardi, & Laschuk, 2020). The resulting plaster was found to have 23% higher solar reflectance when its temperature surpassed the transition value compared to regular cement plaster, while the inclusion of PCM increased the solar absorption for temperatures below the transition value. A thermochromic system based on poly(*N*-isopropyl acrylamide) aqueous solution which is black below the transition temperature (32°C) and opaque above that, was implemented in the study of Ye, Luo, Gao and Zhu (2012). The investigation of different mixtures of thermochromic powder, acrylic emulsion and other assistant components (Zhang & Zhai, 2019) highlighted that the addition of the titanium dioxide compound can further increase the solar reflectance in the colourless phase (above the transition temperature). Furthermore, outdoor measurements demonstrated that these coatings are warmer in winter (and cooler in summer) than conventional counterparts. Fabiani et al. (2019) investigated both experimentally and numerically the application of thermochromic surfaces in UCs and showed that their application may further stabilize the corresponding heat flux and temperature gradient within both warm and cold seasons and thus improve indoor thermal comfort. Thermochromic materials can also result in energy savings. A thermochromic coating implemented with titanium dioxide additives was applied to the building roof in the area of San Francisco and it was capable of decreasing energy consumption by up to more than 40% and of reducing up to 27% energy cost (Hu & Yu, 2019). When coupled with glass and aluminous substrates, thermochromics may decrease the building's overall energy demand both in zones with hot climates (Zheng, Xu, Shen, & Yang, 2015) and in zones with a temperate climate when cooling and heating demands are uniformly distributed on an annual basis (Yuxuan, Yunyun, Jianrong, & Xiaoqiang, 2020).

A smart thermochromic glazed surface that adapts its properties to external temperature stimulus has also been developed by placing Vanadium dioxide thin films on glass. Following this development, Lee, Pang, Hoffmann, Goudey and Thanachareonkit (2013) developed large-area polymer thermochromic laminated windows and tested them in an office building in Berkeley, California (Fig. 3.11). The transition temperature was specifically set according to the ambient temperature values that characterized the local climate. The findings of this study highlighted the importance of the high transmissivity in the visible spectrum (both below and above the threshold) to reduce energy demand for lighting and to improve indoor comfort conditions. Also, Hoffmann, Lee and Clavero (2014) conducted a parametric study on near-IR switching thermochromic glazing systems for commercial buildings and demonstrated that they could represent an alternative energy-efficiency solution when internal loads are dominant in the building energy balance. However, windows typically cover a small fraction of buildings and the resulting energy savings are not significant compared to opaque façade applications.

Figure 3.11 On the left, outdoor view of the south-facing thermochromic window (middle room with the dark windows). On the right, corresponding infrared image showing the surface temperature of the windows (°C) on June 16, 12:02 p.m. ST.
Source: Reproduced with permission from Lee, E. S., Pang, X., Hoffmann, S., & Goudey, H., Thanachareonkit, A. (2013). An empirical study of a full-scale polymer thermochromic window and its implications on material science development objectives. *Solar Energy Materials and Solar Cells, 116,* 14−26. https://doi.org/10.1016/j.solmat.2013.03.043.

Overall, thermochromic coatings may outperform other cooling techniques annually due to their thermoresponsive nature. Ageing issues due to photodegradation, is, however, a critical challenge that needs to be overcome for their wider implementation into the built environment. As such, future research should focus on the exploitation of nondye thermochromic solutions both experimentally and numerically. In fact, recent advances in material science have shown that nanoscale thermochromic structures, such as quantum dots, photonic and plasmonic structures, as well as conjugated polymers and Schiff bases that can modify their optical properties with respect to external stimulus such as heat and light, may not only overcome ageing issues but also achieve higher solar rejection than dye-based thermochromics.

3.5 Radiative coatings

Radiative cooling is considered a technique for the next generation of cool materials and may contribute to the mitigation of urban overheating effects. Radiative cooling is a natural phenomenon that takes place during nighttime (Berdahl, 1984): it refers to the emission of heat from the material that takes place within the range of atmospheric window (8−13 μm). As a result, the materials that strongly emit within such a range during nocturnal hours show a substantial cooling effect and achieve subambient temperatures. However, in the daytime, the cooling effect of natural radiative cooling is counteracted by the high absorption of solar radiation (Feng & Santamouris, 2019). Following this principle, artificial technologies that can perform substantial daytime radiative cooling have recently been proposed. Such technologies incorporate photonic and plasmonic surfaces and polymers and paints (Santamouris & Feng, 2018).

Photonic and plasmonic radiative cooling structures consist of periodic microstructures (or nanostructures) in the form of multilayer films. The first artificial radiative cooling structure was introduced in 2014 (Raman, Anoma, Zhu, Rephaeli, & Fan, 2014). The developed photonic solar reflector, made of seven layers of HfO_2 and SiO_2, could maintain subambient temperatures during the diurnal hours. Its structure was implemented by coupling an integrated high solar reflectivity (nanostructured photonic crystals[4] optimized to minimize the amount of solar radiation absorbed by the surface) with a strong thermal emissivity in the atmospheric transparency window.

This radiative cooler was reported to reflect up to 97% of the incident solar radiation while highly absorbing and emitting at the wave-range of the atmospheric window. Its resulting temperature was up to 4.9°C lower than the ambient temperature. A selective emitter, embodying layers of silicon nitride, amorphous silicon and aluminium was proposed in the study of Chen, Zhu, Raman and Fan (2016). The results highlighted an average daily temperature reduction of 37°C below the

---

[4] Photonic crystals are periodic optical nanostructures that affect the motion of photons in solids. Photonic crystals can be fabricated for one, two, or three dimensions and they allow light flow to be controlled and manipulated.

ambient temperature and a maximum diurnal reduction of 42°C. To minimize heat losses due to convection and air conduction, the emitter was placed within a vacuum chamber and utilized radiation shields and ceramic pegs. Furthermore, the minimization of the solar irradiation was achieved by placing a sunshade mirror and a mirrored cone above the emitter. Thus such results must be interpreted as representative of ideal conditions and they outline the highest performance levels achievable by the material.

A double-layer coating with selective radiative properties (Fig. 3.12) was implemented in the study of Bao et al. (2017). The upper layer was made of TiO_2 nanoparticles to reflect solar radiation (around 90%), while the bottom layer, made of SiC and SiO_2 nanoparticles, emits within the atmospheric window (around 90%). It was theoretically demonstrated that this coating may maintain the surface temperature up to 5°C lower than subambient temperature during the daytime. Furthermore, a relatively simple multilayer structure of inorganic materials (Fig. 3.13) was described in the study of Chae et al. (2020). The stacking sequence and thickness of each layer were optimized through a particle swarm optimization method and achieved an average emission within the atmospheric window of 87% and a solar reflectance of almost 95%. The radiative cooler was found to lower the ambient temperature by 8.2°C under direct sunlight.

Conversely, several research studies focused on the development of painted or polymer coatings. These techniques are characterized by low costs and scalable manufacturing due to their fabrication methods (Zhao, Hu, Ao, Chen, & Pei, 2019). Moreover, they can directly be applied as an additional layer on existing building facades and roofs. An inversion-based method for fabricating a passive daytime radiative cooling coating of hierarchically porous poly(vinylidene fluoride-*co*-hexafluoropropene) [P(VdF-HFP)$_{HP}$] was identified by Mandal et al. (2018). The developed material could maintain an ambient temperature of 6°C under clear sky conditions and low relative humidity. Zhai et al. (2017) developed a translucent and flexible film with daytime radiative cooling properties by randomly dispersing silicon dioxide microspheres within a polymeric matrix. The resulting metamaterial

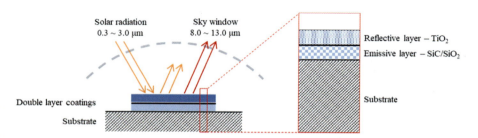

Figure 3.12 Schematic of a double-layer coating design for efficient radiative cooling.
Source: Reproduced with permission from Bao, H., Yan, C., Wang, B., Fang, X., Zhao, C. Y., & Ruan, X. (2017). Double-layer nanoparticle-based coatings for efficient terrestrial radiative cooling. *Solar Energy Materials and Solar Cells, 168*, 78–84. https://doi.org/10.1016/j.solmat.2017.04.020.

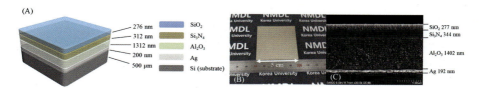

Figure 3.13 (A) Designed inorganic radiative cooler structure by the PSO method. (B) Photograph of a fabricated mirror-like inorganic radiative cooler, logo courtesy of NMDL Korea University. (C) Cross-sectional scanning electron microscopy image of the inorganic radiative cooler. *PSO*, Particle swarm optimization.
Source: Reproduced with permission from Chae, D., Kim, M., Jung, P. H., Son, S., Seo, J., Liu, Y., ... Lee, H. (2020). Spectrally selective inorganic-based multilayer emitter for daytime radiative cooling. *ACS Applied Materials & Interfaces*, *12*, 8073–8081. https://doi.org/10.1021/acsami.9b16742.

reflected shortwave radiation by 96%, while its emittance within the atmospheric window was above 0.93. Similarly, Zhou et al. (2019) implemented a fast solution coating process to develop a cost-effective planar polydimethylsiloxane/metal thermal emitter thin film structure. The temperature reduction was equal to 9.5°C in the laboratory test and to 11.0°C in the on-field test. Due to the inexpensiveness of this film structure, its application over large areas has been considered and recommended by the authors. A polymer-coated fused silica mirror that behaves like a near-ideal reflector of solar radiation (0.9) and near-ideal black body within the range of atmospheric window (0.9–1.0) was implemented in the study of Kou, Jurado, Chen, Fan and Minnich (2017). The radiative cooler showed temperature values 8.2°C and 8.4°C lower than the ambient temperature during daytime and nighttime, respectively. Furthermore, a painting format microsphere-based photonic random media that can be applied to both façade and roof surface for radiative cooling was described in the study of Atiganyanun et al. (2018). The random media comprises low refractive index microspheres of SiO_2 with 1.5-μm diameter and achieved high solar reflectance (0.97) and high absorptivity/emissivity (0.95), within the atmospheric window. The microsphere-based coating was allocated on a black substrate and outdoor experiments were performed. The results showed that the substrate's temperature was 12°C lower than the ambient temperature, during the daytime.

The cooling effectiveness of a radiative cooler depends on meteorological conditions such as humidity levels, wind speed and cloudiness (Carlosena et al., 2020). Suichi, Ishikawa, Hayashi and Tsuruta (2018) investigated both theoretically and experimentally the performance of a radiative cooler made of alternating layers of SiO_2 and polymethyl methacrylate on an aluminium mirror under warm-humid climatic conditions. Experimental results demonstrated that the radiative cooler could achieve a 1.3°C ambient temperature drop when the perceptible water vapour was 1.0 mm. However, in the outdoors experiment, the radiative cooler failed to achieve subambient temperature, mainly due to the high levels of humidity. The cooling performance of a photonic radiative cooler under various climatic conditions was assessed in the study of Bijarniya and Sarkar (2020) and the results showed that hot-humid climatic conditions could substantially weaken it. Similarly, the cooling potential of three different radiative coolers under three different climate conditions (ie, Alice Springs, Sydney, Singapore) was evaluated by Feng, Gao, Santamouris, Shah and Ranzi (2020). It was demonstrated that radiative coolers perform better in hot and arid climates, while the ambient temperature drop is directly proportional to the ambient temperature. Thus an enhancement of the cooling performance of radiative coolers could be achieved by adding an optical grating window on the surface of the radiative cooler, to provide a better control of spectral and directional thermal emission.

Although the aforementioned studies reveal the super cooling potential of radiative cooling structures compared to other UHI mitigation techniques, their implementation into the building skin is still in its infancy. In fact, there are still some issues to be addressed which concern the scalability of the technology as well as its cost-effectiveness and which prevent its implementation in an experimental facility. Further evaluation of radiative cooling surfaces is needed in terms of city-scale modelling under various boundary conditions. Specific focus should be given to improving the performance of radiative coolers under humid-arid climatic conditions. As reported in the study of Santamouris and Feng (2018), the utilization of asymmetric electromagnetic window techniques can allow the transmission of outgoing radiation while reflecting the incoming radiation of the same wavelength. Therefore their application should be regarded as a promising solution for recovering the cooling potential of radiative coolers (Wong, Tso, Chao, Huang, & Wan, 2018). Finally, their scalability should be enhanced through the development of simplified production processes that are also characterized by lower costs.

3.6 Conclusion

Thermal optical properties of surface treatments for building facades and roofs contribute to UHI effects and thus influence building energy demand, the liveability of indoor and outdoor spaces and human health. Therefore the most recent solutions for UHI mitigation that can be applied to facades were examined and reported in this chapter. New and innovative materials such as HR coating, thermochromics and radiative cooling technologies were presented. The benefits of these technologies in mitigating UHI phenomenon were highlighted based on the results achieved by the principal investigators in several experimental and theoretical studies. Their applications in both glazed surfaces and opaque facades were investigated, although windows typically cover a small fraction of the building and the related energy savings are not significant. A summary for each material typology mentioned in this chapter is reported in the following paragraphs.

The HR materials were demonstrated to reduce the ambient temperature when applied to urban surfaces such as roofs, facades and pavements. The reduction of the peak ambient temperature was estimated to be as high as 0.5°C per 0.10 increase in the albedo value, while a decrease of 2.2°C was calculated for the city of Athens if a 0.65 albedo was set (Synnefa et al., 2007a). Nonetheless, the worsening of interbuildings solar reflections phenomenon due to higher reflectivity of urban surfaces, as well as the increase in building energy demand for cooling, led to the development of RR and AS-RR coatings for façade application. These materials allowed a reduction of the outer surface peak temperature by around 15°C (Zhang et al., 2017).

The thermochromics were investigated in various building façade applications ranging from microencapsulated thermochromic materials to mixtures of thermochromics and PCM. Such coatings were found to be capable of increasing the surface solar reflectance by 23% while also lowering the building energy consumption by around 40%. However, the choice of the most adequate transition temperature value is fundamental to achieve the best performance and may affect the material effectiveness in mitigating urban overheating.

The radiative materials enabled a maximum diurnal reduction of 42°C (under ideal conditions that are not observed within the built environment) with photonic and plasmonic radiative cooling structures. Conversely, their exploitation within the urban environment permitted them to lower the ambient temperature by around 5°C. Better performances were observed for painted or polymer radiative coatings, which cooled the ambient temperature by up to 12°C during the daytime.

Despite the proven advantages, these materials still present some drawbacks concerning maintenance, ageing and utilization within an environment characterized by high relative humidity. The effectiveness of HR coatings is worsened by the deposit of dust, while thermochromics' performance level drops after some transition cycles. Furthermore, when radiative coatings have shown lower efficiency in humid environments.

Accordingly, it is worth highlighting that such technologies for façade application need further developments to overcome climate and performance issues and to be widely exploited in the building sector. It is necessary to investigate their scalability as well as to determinate a manufacturing process which may enable greater cost-effectiveness. In addition, ageing and degradation issues should be solved and their adaptability to external stimulus should be enhanced to avoid the reduction of solar heat gains during winter. To achieve these goals, systems coupling two of the described technologies could be considered. In that regard the research group from the CIRIAF research centre, University of Perugia, is planning activities to implement thermal responsiveness in RR materials by adding thermochromics to the substrate layer. System coupling is expected to enable the activation of the retroreflection whenever the transition temperature is surpassed. Similarly, either angular-selective properties (from AS-RR materials) or the thermo-responsiveness (proper of thermochromics) may be implemented on radiative coatings to avoid a heating penalty in the winter.

References

Akbari, H., Cartalis, C., Kolokotsa, D., Muscio, A., Pisello, A. L., Rossi, F., ... Zinzi, M. (2016). Local climate change and urban heat island mitigation techniques—the state of the art. *Journal of Civil Engineering and Management*, 22, 1−16. Available from https://doi.org/10.3846/13923730.2015.1111934.

Akbari, H., & Kolokotsa, D. (2016). Three decades of urban heat islands and mitigation technologies research. *Energy and Buildings*, *133*, 834−842. Available from https://doi.org/10.1016/j.enbuild.2016.09.067.

Akbari, H., & Touchaei, A. G. (2014). Modeling and labeling heterogeneous directional reflective roofing materials. *Solar Energy Materials and Solar Cells*, *124*, 192−210. Available from https://doi.org/10.1016/j.solmat.2014.01.036.

Atiganyanun, S., Plumley, J. B., Han, S. J., Hsu, K., Cytrynbaum, J., Peng, T. L., ... Han, S. E. (2018). Effective radiative cooling by paint-format microsphere-based photonic random media. *ACS Photonics*, *5*, 1181−1187. Available from https://doi.org/10.1021/acsphotonics.7b01492.

Bao, H., Yan, C., Wang, B., Fang, X., Zhao, C. Y., & Ruan, X. (2017). Double-layer nanoparticle-based coatings for efficient terrestrial radiative cooling. *Solar Energy Materials and Solar Cells*, *168*, 78−84. Available from https://doi.org/10.1016/j.solmat.2017.04.020.

Berdahl, P. (1984). Radiative cooling with MgO and/or LiF layers. *Applied Optics*, *23*, 370. Available from https://doi.org/10.1364/ao.23.000370.

Berdahl, P., & Bretz, S. E. (1997). Preliminary survey of the solar reflectance of cool roofing materials. *Energy and Buildings*, *25*, 149−158. Available from https://doi.org/10.1016/S0378-7788(96)01004-3.

Berdahl, P., Chen, S. S., Destaillats, H., Kirchstetter, T. W., Levinson, R. M., & Zalich, M. A. (2016). Fluorescent cooling of objects exposed to sunlight—The ruby example. *Solar Energy Materials and Solar Cells*, *157*, 312−317. Available from https://doi.org/10.1016/j.solmat.2016.05.058.

Bijarniya, J. P., & Sarkar, J. (2020). Climate change effect on the cooling performance and assessment of passive daytime photonic radiative cooler in India. *Renewable and Sustainable Energy Reviews*, *134*, 110303. Available from https://doi.org/10.1016/j.rser.2020.110303.

Brans, K. I., Engelen, J. M. T., Souffreau, C., & De Meester, L. (2018). Urban hot-tubs: Local urbanization has profound effects on average and extreme temperatures in ponds. *Landscape and Urban Planning*, *176*, 22−29. Available from https://doi.org/10.1016/J.LANDURBPLAN.2018.03.013.

Bretz, S., Akbari, H., & Rosenfeld, A. (1998). Practical issues for using solar-reflective materials to mitigate urban heat islands. *Atmospheric Environment (Oxford, England: 1994)*, *32*, 95−101. Available from https://doi.org/10.1016/S1352-2310(97)00182-9.

Bretz, S. E., & Akbari, H. (1997). Long-term performance of high-albedo roof coatings. *Energy and Buildings*, *25*, 159−167. Available from https://doi.org/10.1016/S0378-7788(96)01005-5.

Carlosena, L., Ruiz-Pardo, Á., Feng, J., Irulegi, O., Hernández-Minguillón, R. J., & Santamouris, M. (2020). On the energy potential of daytime radiative cooling for urban heat island mitigation. *Solar Energy*, *208*, 430−444. Available from https://doi.org/10.1016/j.solener.2020.08.015.

Castellani, B., Morini, E., Anderini, E., Filipponi, M., & Rossi, F. (2017). Development and characterization of retro-reflective colored tiles for advanced building skins. *Energy and Buildings*, *154*, 513−522. Available from https://doi.org/10.1016/j.enbuild.2017.08.078.

Chae, D., Kim, M., Jung, P. H., Son, S., Seo, J., Liu, Y., ... Lee, H. (2020). Spectrally selective inorganic-based multilayer emitter for daytime radiative cooling. *ACS Applied*

Materials & Interfaces, *12*, 8073−8081. Available from https://doi.org/10.1021/acsami. 9b16742.

Chen, Z., Zhu, L., Raman, A., & Fan, S. (2016). Radiative cooling to deep sub-freezing temperatures through a 24-h day-night cycle. *Nature Communications*, *7*, 1−5. Available from https://doi.org/10.1038/ncomms13729.

Chung, M. H., & Park, J. C. (2016). Development of PCM cool roof system to control urban heat island considering temperate climatic conditions. *Energy and Buildings*, *116*, 341−348. Available from https://doi.org/10.1016/J.ENBUILD.2015.12.056.

Cotana, F., Rossi, F., Filipponi, M., Coccia, V., Pisello, A. L., Bonamente, E., ... Cavalaglio, G. (2014). Albedo control as an effective strategy to tackle Global Warming: A case study. *Applied Energy*, *130*, 641−647. Available from https://doi.org/10.1016/J.APENERGY.2014.02.065.

Cotana, F. R., Morini, E., Castellani, B., Nicolini, A., Bonamente, E., Anderini, E., & Cotana, F. (2015). Beneficial effects of retroreflective materials in urban canyons: Results from seasonal monitoring campaign. *Journal of Physics: Conference Series*, *655*, 12012.

Cui, Y., Xu, X., Dong, J., & Qin, Y. (2016). Influence of urbanization factors on surface urban heat island intensity: A comparison of countries at different developmental phases. *Sustainability*, *8*, 706. Available from https://doi.org/10.3390/su8080706.

Djedjig, R., Bozonnet, E., & Belarbi, R. (2013). Experimental study of the urban microclimate mitigation potential of green roofs and green walls in street canyons. *International Journal of Low-Carbon Technologies*, *10*, 34−44. Available from https://doi.org/10.1093/ijlct/ctt019.

Doya, M., Bozonnet, E., & Allard, F. (2012). Experimental measurement of cool facades' performance in a dense urban environment. *Energy and Buildings*, *55*, 42−50. Available from https://doi.org/10.1016/J.ENBUILD.2011.11.001.

Fabiani, C., Pisello, A. L., Bou-Zeid, E., Yang, J., & Cotana, F. (2019). Adaptive measures for mitigating urban heat islands: The potential of thermochromic materials to control roofing energy balance. *Applied Energy*, *247*, 155−170. Available from https://doi.org/10.1016/j.apenergy.2019.04.020.

Feng, J., Gao, K., Santamouris, M., Shah, K. W., & Ranzi, G. (2020). Dynamic impact of climate on the performance of daytime radiative cooling materials. *Solar Energy Materials and Solar Cells*, *208*, 110426. Available from https://doi.org/10.1016/j.solmat.2020.110426.

Feng, J., & Santamouris, M. (2019). Numerical techniques for electromagnetic simulation of daytime radiative cooling: A review. *AIMS Materials Science*, *6*. Available from https://doi.org/10.3934/MATERSCI.2019.6.1049.

Founda, D., & Santamouris, M. (2017). Synergies between urban heat island and heat waves in Athens (Greece), during an extremely hot summer (2012). *Scientific Reports*, *7*, 10973. Available from https://doi.org/10.1038/s41598-017-11407-6.

Garshasbi, S., & Santamouris, M. (2019). Using advanced thermochromic technologies in the built environment: Recent development and potential to decrease the energy consumption and fight urban overheating. *Solar Energy Materials and Solar Cells*, *191*, 21−32. Available from https://doi.org/10.1016/j.solmat.2018.10.023.

Han, Y., Taylor, J. E., & Pisello, A. L. (2015). Toward mitigating urban heat island effects: Investigating the thermal-energy impact of bio-inspired retro-reflective building envelopes in dense urban settings. *Energy and Buildings*, *102*, 380−389. Available from https://doi.org/10.1016/j.enbuild.2015.05.040.

Hoffmann, S., Lee, E. S., & Clavero, C. (2014). Examination of the technical potential of near-infrared switching thermochromic windows for commercial building applications. *Solar Energy Materials and Solar Cells*, *123*, 65–80. Available from https://doi.org/10.1016/j.solmat.2013.12.017.

Hu, J., & Yu, X. B. (2019). Adaptive thermochromic roof system: Assessment of performance under different climates. *Energy and Buildings*, *192*, 1–14. Available from https://doi.org/10.1016/j.enbuild.2019.02.040.

Ichinose, M., Inoue, T., & Nagahama, T. (2017). Effect of retro-reflecting transparent window on anthropogenic urban heat balance. *Energy and Buildings*, *157*, 157–165. Available from https://doi.org/10.1016/j.enbuild.2017.01.051.

Inoue, T., Shimo, T., Ichinose, M., Takase, K., & Nagahama, T. (2017). *Improvement of urban thermal environment by wavelength-selective retro-reflective film. Energy procedia* (pp. 967–972). Elsevier. Available from https://doi.org/10.1016/j.egypro.2017.07.447.

Jandaghian, Z., & Berardi, U. (2020). Analysis of the cooling effects of higher albedo surfaces during heat waves coupling the Weather Research and Forecasting model with building energy models. *Energy and Buildings*, *207*, 109627. Available from https://doi.org/10.1016/j.enbuild.2019.109627.

Karlessi, T., & Santamouris, M. (2015). Improving the performance of thermochromic coatings with the use of UV and optical filters tested under accelerated aging conditions. *International Journal of Low-Carbon Technologies*, *10*, 45–61. Available from https://doi.org/10.1093/ijlct/ctt027.

Karlessi, T., Santamouris, M., Apostolakis, K., Synnefa, A., & Livada, I. (2009). Development and testing of thermochromic coatings for buildings and urban structures. *Solar Energy*, *83*, 538–551. Available from https://doi.org/10.1016/j.solener.2008.10.005.

Kou, J. L., Jurado, Z., Chen, Z., Fan, S., & Minnich, A. J. (2017). Daytime radiative cooling using near-black infrared emitters. *ACS Photonics*, *4*, 626–630. Available from https://doi.org/10.1021/acsphotonics.6b00991.

Kousis, I., & Pisello, A. L. (2020). For the mitigation of urban heat island and urban noise island: Two simultaneous sides of urban discomfort. *Environmental Research Letters*, *15*. Available from https://doi.org/10.1088/1748-9326/abaa0d.

Lee, E. S., Pang, X., Hoffmann, S., Goudey, H., & Thanachareonkit, A. (2013). An empirical study of a full-scale polymer thermochromic window and its implications on material science development objectives. *Solar Energy Materials and Solar Cells*, *116*, 14–26. Available from https://doi.org/10.1016/j.solmat.2013.03.043.

Levinson, R., Berdahl, P., Akbari, H., Miller, W., Joedicke, I., Reilly, J., ... Vondran, M. (2007). Methods of creating solar-reflective nonwhite surfaces and their application to residential roofing materials. *Solar Energy Materials and Solar Cells*, *91*, 304–314. Available from https://doi.org/10.1016/j.solmat.2006.06.062.

Lima, I., Scalco, V., & Lamberts, R. (2019). Estimating the impact of urban densification on high-rise office building cooling loads in a hot and humid climate. *Energy and Buildings*, *182*, 30–44. Available from https://doi.org/10.1016/J.ENBUILD.2018.10.019.

Lin, L., Ge, E., Liu, X., Liao, W., & Luo, M. (2018). Urbanization effects on heat waves in Fujian Province, Southeast China. *Atmospheric Research*, *210*, 123–132. Available from https://doi.org/10.1016/J.ATMOSRES.2018.04.011.

Lobaccaro, G., Acero, J. A., Martinez, G. S., Padro, A., Laburu, T., & Fernandez, G. (2019). Effects of orientations, aspect ratios, pavement materials and vegetation elements on thermal stress inside typical urban canyons. *International Journal of Environmental*

Research and Public Health, *16*, 3574. Available from https://doi.org/10.3390/ijerph16193574.
Ma, Y., Zhang, X., Zhu, B., & Wu, K. (2002). Research on reversible effects and mechanism between the energy-absorbing and energy-reflecting states of chameleon-type building coatings. *Solar Energy*, *72*, 511−520. Available from https://doi.org/10.1016/S0038-092X(02)00029-4.
Ma, Y., & Zhu, B. (2009). Research on the preparation of reversibly thermochromic cement based materials at normal temperature. *Cement and Concrete Research*, *39*, 90−94. Available from https://doi.org/10.1016/j.cemconres.2008.10.006.
Ma, Y., Zhu, B., & Wu, K. (2001). Preparation and solar reflectance spectra of chameleon-type building coatings. *Solar Energy*, *70*, 417−422. Available from https://doi.org/10.1016/S0038-092X(00)00160-2.
Mandal, J., Fu, Y., Overvig, A. C., Jia, M., Sun, K., Shi, N. N., ... Yang, Y. (2018). Hierarchically porous polymer coatings for highly efficient passive daytime radiative cooling. *Science*, *362*, 315−319. Available from https://doi.org/10.1126/science.aat9513.
Manni, M., Bonamente, E., Lobaccaro, G., Goia, F., Nicolini, A., Bozonnet, E., & Rossi, F. (2020). Development and validation of a Monte Carlo-based numerical model for solar analyses in urban canyon configurations. *Building and Environment*, *170*, 106638. Available from https://doi.org/10.1016/j.buildenv.2019.106638.
Manni, M., Cardinali, M., Lobaccaro, G., Goia, F., Nicolini, A., & Rossi, F. (2020). Effects of retro-reflective and angular-selective retro-reflective materials on solar energy in urban canyons. *Solar Energy*, *209*, 662−673. Available from https://doi.org/10.1016/j.solener.2020.08.085.
Manni, M., Lobaccaro, G., Goia, F., & Nicolini, A. (2018). An inverse approach to identify selective angular properties of retro-reflective materials for urban heat island mitigation. *Solar Energy*, *176*, 194−210. Available from https://doi.org/10.1016/J.SOLENER.2018.10.003.
Manni, M., Lobaccaro, G., Goia, F., Nicolini, A., & Rossi, F. (2019). Exploiting selective angular properties of retro-reflective coatings to mitigate solar irradiation within the urban canyon. *Solar Energy*, *189*. Available from https://doi.org/10.1016/j.solener.2019.07.045.
Mastrapostoli, E., Santamouris, M., Kolokotsa, D., Vassilis, P., Venieri, D., & Gompakis, K. (2016). On the ageing of cool roofs: Measure of the optical degradation, chemical and biological analysis and assessment of the energy impact. *Energy and Buildings*, *114*, 191−199. Available from https://doi.org/10.1016/j.enbuild.2015.05.030.
Morini, E., Castellani, B., Anderini, E., Presciutti, A., Nicolini, A., & Rossi, F. (2018). Optimized retro-reflective tiles for exterior building element. *Sustainable Cities and Society*, *37*, 146−153. Available from https://doi.org/10.1016/J.SCS.2017.11.007.
Morini, E., Castellani, B., Presciutti, A., Filipponi, M., Nicolini, A., & Rossi, F. (2017). Optic-energy performance improvement of exterior paints for buildings. *Energy and Buildings*, *139*, 690−701. Available from https://doi.org/10.1016/j.enbuild.2017.01.060.
Nazarian, N., Dumas, N., Kleissl, J., & Norford, L. (2019). Effectiveness of cool walls on cooling load and urban temperature in a tropical climate. *Energy and Buildings*, *187*, 144−162. Available from https://doi.org/10.1016/j.enbuild.2019.01.022.
Oke, T. R. (1981). Canyon geometry and the nocturnal urban heat island: Comparison of scale model and field observations. *Journal of Climatology*, *1*, 237−254. Available from https://doi.org/10.1002/joc.3370010304.
Perez, G., Allegro, V. R., Corroto, M., Pons, A., & Guerrero, A. (2018). Smart reversible thermochromic mortar for improvement of energy efficiency in buildings. *Construction*

and Building Materials, *186*, 884−891. Available from https://doi.org/10.1016/j. conbuildmat.2018.07.246.

Pioppi, B., Pigliautile, I., Piselli, C., & Pisello, A. L. (2020). Cultural heritage microclimate change: Human-centric approach to experimentally investigate intra-urban overheating and numerically assess foreseen future scenarios impact. *The Science of the Total Environment*, *703*, 134448. Available from https://doi.org/10.1016/j.scitotenv.2019.134448.

Piselli, C., Castaldo, V. L., Pigliautile, I., Pisello, A. L., & Cotana, F. (2018). Outdoor comfort conditions in urban areas: On citizens' perspective about microclimate mitigation of urban transit areas. *Sustainable Cities and Society*, *39*, 16−36. Available from https://doi.org/10.1016/J.SCS.2018.02.004.

Pisello, A. L., Saliari, M., Vasilakopoulou, K., Hadad, S., & Santamouris, M. (2018). Facing the urban overheating: Recent developments. Mitigation potential and sensitivity of the main technologies. *WIREs Energy and Environment*, *7*, e294. Available from https://doi.org/10.1002/wene.294.

Prado, R. T. A., & Ferreira, F. L. (2005). Measurement of albedo and analysis of its influence the surface temperature of building roof materials. *Energy and Buildings*, *37*, 295−300. Available from https://doi.org/10.1016/j.enbuild.2004.03.009.

Radhi, H., Sharples, S., & Assem, E. (2015). ARTICLE IN PRESS G Model Impact of urban heat islands on the thermal comfort and cooling energy demand of artificial islands—A case study of AMWAJ Islands in Bahrain. *Sustainable Cities and Society*. Available from https://doi.org/10.1016/j.scs.2015.07.017.

Rahman, M. A., Moser, A., Rötzer, T., & Pauleit, S. (2017). Microclimatic differences and their influence on transpirational cooling of Tilia cordata in two contrasting street canyons in Munich, Germany. *Agricultural and Forest Meteorology*, *232*, 443−456. Available from https://doi.org/10.1016/J.AGRFORMET.2016.10.006.

Raman, A. P., Anoma, M. A., Zhu, L., Rephaeli, E., & Fan, S. (2014). Passive radiative cooling below ambient air temperature under direct sunlight. *Nature*, *515*, 540−544. Available from https://doi.org/10.1038/nature13883.

Rossi, F., Pisello, A. L., Nicolini, A., Filipponi, M., & Palombo, M. (2014). Analysis of retro-reflective surfaces for urban heat island mitigation: A new analytical model. *Applied Energy*, *114*, 621−631. Available from https://doi.org/10.1016/j.apenergy.2013. 10.038.

Sakai, H., Emura, K., & Igawa, N. (2008). Reduction of reflected heat by retroreflective materilas. *Journal of Structural and Construction Engineering (Transactions AIJ)*, *73*, 1239−1244. Available from https://doi.org/10.3130/aijs.73.1239.

Sakai, H., & Iyota, H. (2017). Development of two new types of retroreflective materials as countermeasures to urban heat islands. *International Journal of Thermophysics*, *38*. Available from https://doi.org/10.1007/s10765-017-2266-y.

Santamouris, M. (2001). *Energy and climate in the urban built environment*. London: Routledge. Available from https://doi.org/10.4324/9781315073774.

Santamouris, M. (2007). Heat island research in Europe: The state of the art. *Advances in Building Energy Research*, *1*, 123−150. Available from https://doi.org/10.1080/17512549.2007.9687272.

Santamouris, M. (2014). On the energy impact of urban heat island and global warming on buildings. *Energy and Buildings*, *82*, 100−113. Available from https://doi.org/10.1016/j.enbuild.2014.07.022.

Santamouris, M. (2015a). Analyzing the heat island magnitude and characteristics in one hundred Asian and Australian cities and regions. *The Science of the Total Environment*, *512−513*, 582−598. Available from https://doi.org/10.1016/j.scitotenv.2015.01.060.

Santamouris, M. (2015b). Regulating the damaged thermostat of the cities—Status, impacts and mitigation challenges. *Energy and Buildings*, *91*, 43−56. Available from https://doi.org/10.1016/j.enbuild.2015.01.027.

Santamouris, M. (2015c). Analyzing the heat island magnitude and characteristics in one hundred Asian and Australian cities and regions. *The Science of the Total Environment*, *512−513*, 582−598. Available from https://doi.org/10.1016/j.scitotenv.2015.01.060.

Santamouris, M., Cartalis, C., Synnefa, A., & Kolokotsa, D. (2015). On the impact of urban heat island and global warming on the power demand and electricity consumption of buildings—A review. *Energy and Buildings*, *98*, 119−124. Available from https://doi.org/10.1016/j.enbuild.2014.09.052.

Santamouris, M., & Feng, J. (2018). Recent progress in daytime radiative cooling: Is it the air conditioner of the future? *Buildings*, *8*. Available from https://doi.org/10.3390/buildings8120168.

Santamouris, M., Haddad, S., Saliari, M., Vasilakopoulou, K., Synnefa, A., Paolini, R., ... Fiorito, F. (2018). On the energy impact of urban heat island in Sydney: Climate and energy potential of mitigation technologies. *Energy and Buildings*, *166*, 154−164. Available from https://doi.org/10.1016/j.enbuild.2018.02.007.

Santamouris, M., Paolini, R., Haddad, S., Synnefa, A., Garshasbi, S., Hatvani-Kovacs, G., ... Tombrou, M. (2020). Heat mitigation technologies can improve sustainability in cities. An holistic experimental and numerical impact assessment of urban overheating and related heat mitigation strategies on energy consumption, indoor comfort, vulnerability and heat-related mortality and morbidity in cities. *Energy and Buildings*, *217*, 110002. Available from https://doi.org/10.1016/j.enbuild.2020.110002.

Santamouris, M., Papanikolaou, N., Livada, I., Koronakis, I., Georgakis, C., Argiriou, A., & Assimakopoulos, D. (2001). On the impact of urban climate on the energy consumption of buildings. *Solar Energy*, *70*, 201−216. Available from https://doi.org/10.1016/S0038-092X(00)00095-5.

Santamouris, M., Synnefa, A., & Karlessi, T. (2011). Using advanced cool materials in the urban built environment to mitigate heat islands and improve thermal comfort conditions. *Solar Energy*, *85*, 3085−3102. Available from https://doi.org/10.1016/J.SOLENER.2010.12.023.

Santamouris, M., & Yun, G. Y. (2020). Recent development and research priorities on cool and super cool materials to mitigate urban heat island. *Renewable Energy*, *161*, 792−807. Available from https://doi.org/10.1016/j.renene.2020.07.109.

Soudian, S., Berardi, U., & Laschuk, N. (2020). Development and thermal-optical characterization of a cementitious plaster with phase change materials and thermochromic paint. *Solar Energy*, *205*, 282−291. Available from https://doi.org/10.1016/j.solener.2020.05.015.

Suichi, T., Ishikawa, A., Hayashi, Y., & Tsuruta, K. (2018). Performance limit of daytime radiative cooling in warm humid environment. *AIP Advances*, *8*. Available from https://doi.org/10.1063/1.5030156.

Synnefa, A., Saliari, M., & Santamouris, M. (2012). Experimental and numerical assessment of the impact of increased roof reflectance on a school building in Athens. *Energy and Buildings*, *55*, 7−15. Available from https://doi.org/10.1016/j.enbuild.2012.01.044.

Synnefa, A., Santamouris, M., & Akbari, H. (2007a). Estimating the effect of using cool coatings on energy loads and thermal comfort in residential buildings in various climatic conditions. *Energy and Buildings*, *39*, 1167−1174. Available from https://doi.org/10.1016/j.enbuild.2007.01.004.

Synnefa, A., Santamouris, M., & Akbari, H. (2007b). Estimating the effect of using cool coatings on energy loads and thermal comfort in residential buildings in various climatic conditions. *Energy and Buildings*, *39*, 1167−1174. Available from https://doi.org/10.1016/j.enbuild.2007.01.004.

Synnefa, A., Santamouris, M., & Apostolakis, K. (2007). On the development, optical properties and thermal performance of cool colored coatings for the urban environment. *Solar Energy*, *81*, 488−497. Available from https://doi.org/10.1016/j.solener.2006.08.005.

Synnefa, A., Santamouris, M., & Livada, I. (2006). A study of the thermal performance of reflective coatings for the urban environment. *Solar Energy*, *80*, 968−981. Available from https://doi.org/10.1016/j.solener.2005.08.005.

Taha, H. (1997). Urban climates and heat islands: Albedo, evapotranspiration, and anthropogenic heat. *Energy and Buildings*, *25*, 99−103. Available from https://doi.org/10.1016/S0378-7788(96)00999-1.

Tsoka, S., Theodosiou, T., Tsikaloudaki, K., & Flourentzou, F. (2018). Modeling the performance of cool pavements and the effect of their aging on outdoor surface and air temperatures. *Sustainable Cities and Society*, *42*, 276−288. Available from https://doi.org/10.1016/j.scs.2018.07.016.

Warren, C. M. J. (2014). *Heat islands; understanding and mitigating heat in urban Areas20122Lisa Gartland. Heat islands; understanding and mitigating heat in urban areas* (2011). London: Earthscan. 192 pp., ISBN: 978-1-84971-298-9 $64.95. Prop. Manag. 30, 105−106. https://doi.org/10.1108/pm.2012.30.1.105.2.

Wikipedia Contributors (2020). *Leuco Dye*. Wikipedia, Free Encycl.

Wong, R. Y. M., Tso, C. Y., Chao, C. Y. H., Huang, B., & Wan, M. P. (2018). Ultra-broadband asymmetric transmission metallic gratings for subtropical passive daytime radiative cooling. *Solar Energy Materials and Solar Cells*, *186*, 330−339. Available from https://doi.org/10.1016/j.solmat.2018.07.002.

Xie, X., Huang, Z., Wang, J., & Xie, Z. (2005). The impact of solar radiation and street layout on pollutant dispersion in street canyon. *Building and Environment*, *40*, 201−212. Available from https://doi.org/10.1016/j.buildenv.2004.07.013.

Xu, X., González, J. E., Shen, S., Miao, S., & Dou, J. (2018). Impacts of urbanization and air pollution on building energy demands—Beijing case study. *Applied Energy*, *225*, 98−109. Available from https://doi.org/10.1016/J.APENERGY.2018.04.120.

Ye, X., Luo, Y., Gao, X., & Zhu, S. (2012). Design and evaluation of a thermochromic roof system for energy saving based on poly(N-isopropylacrylamide) aqueous solution. *Energy and Buildings*, *48*, 175−179. Available from https://doi.org/10.1016/j.enbuild.2012.01.024.

Yoshida, S., Yumino, S., Uchida, T., & Mochida, A. (2015). Effect of windows with heat ray retroreflective film on outdoor thermal environment and building cooling load. *Journal of Heat Island Institute International*, *48*.

Yuan, J., Emura, K., & Farnham, C. (2015). A method to measure retro-reflectance and durability of retro-reflective materials for building outer walls. *Journal of Building Physics*, *38*, 500−516. Available from https://doi.org/10.1177/1744259113517208.

Yuan, J., Emura, K., & Farnham, C. (2016). Potential for application of retroreflective materials instead of highly reflective materials for urban heat island mitigation. *Urban Studies Research*, *5*, 1−10. Available from https://doi.org/10.1155/2016/3626294.

Yuan, J., Farnham, C., & Emura, K. (2015). Development of a retro-reflective material as building coating and evaluation on albedo of urban canyons and building heat loads. *Energy and Buildings*, *103*, 107−117. Available from https://doi.org/10.1016/j.enbuild.2015.06.055.

Yuxuan, Z., Yunyun, Z., Jianrong, Y., & Xiaoqiang, Z. (2020). Energy saving performance of thermochromic coatings with different colors for buildings. *Energy and Buildings*, *215*, 109920. Available from https://doi.org/10.1016/j.enbuild.2020.109920.

Zhai, Y., Ma, Y., David, S. N., Zhao, D., Lou, R., Tan, G., ... Yin, X. (2017). Scalable-manufactured randomized glass-polymer hybrid metamaterial for daytime radiative cooling. *Science*, *355*, 1062 LP−1061066. Available from https://doi.org/10.1126/science.aai7899.

Zhang, L., Meng, X., Liu, F., Xu, L., & Long, E. (2017). Effect of retro-reflective materials on temperature environment in tents. *Case Studies in Thermal Engineering*, *9*, 122−127. Available from https://doi.org/10.1016/j.csite.2017.02.001.

Zhang, Y., & Zhai, X. (2019). Preparation and testing of thermochromic coatings for buildings. *Solar Energy*, *191*, 540−548. Available from https://doi.org/10.1016/j.solener.2019.09.042.

Zhao, B., Hu, M., Ao, X., Chen, N., & Pei, G. (2019). Radiative cooling: A review of fundamentals, materials, applications, and prospects. *Applied Energy*, *236*, 489−513. Available from https://doi.org/10.1016/j.apenergy.2018.12.018.

Zheng, L., Xiong, T., & Shah, K. W. (2019). Transparent nanomaterial-based solar cool coatings: Synthesis, morphologies and applications. *Solar Energy*, *193*, 837−858. Available from https://doi.org/10.1016/j.solener.2019.10.029.

Zheng, S., Xu, Y., Shen, Q., & Yang, H. (2015). Preparation of thermochromic coatings and their energy saving analysis. *Solar Energy*, *112*, 263−271. Available from https://doi.org/10.1016/j.solener.2014.09.049.

Zhou, D., Zhang, L., Li, D., Huang, D., & Zhu, C. (2016). Climate−vegetation control on the diurnal and seasonal variations of surface urban heat islands in China. *Environmental Research Letters*, *11*. Available from https://doi.org/10.1088/1748-9326/11/7/074009.

Zhou, L., Song, H., Liang, J., Singer, M., Zhou, M., Stegenburgs, E., ... Gan, Q. (2019). A polydimethylsiloxane-coated metal structure for all-day radiative cooling. *Nature Sustainability*, *2*, 718−724. Available from https://doi.org/10.1038/s41893-019-0348-5.

Zinzi, M., Carnielo, E., & Rossi, G. (2015). Directional and angular response of construction materials solar properties: Characterisation and assessment. *Solar Energy*, *115*, 52−67. Available from https://doi.org/10.1016/j.solener.2015.02.015.

The pursuit of transparency

Lisa Rammig[1,2]
[1]Eckersley O'Callaghan, Los Angeles, CA, United States, [2]Delft University of Technology, Delft, The Netherlands

Abbreviations

AM	additive manufacturing
FDM	fused deposition modelling
IGU	insulated glass unit
SG	sentry glass
TSSA	transparent structural silicone adhesive
UV	ultraviolet

4.1 Introduction

Glass is one of the oldest man-made materials known (Schittig et al., 1999) with beads and vessels dating back to 3500 BCE; however, it has only been used in building applications for approximately 1000 years. Even more significantly, its use as a structural component is a recent development, which started with the glass and iron structures developed at the end of the 19th century. In those early glass structures the glass served not only to form an envelope but also to participate in bracing the slender iron framing elements. It took another 100 years before the first all-glass structures were built.

Glass differs from other building materials in the way that it is not defined by its composition but its molecular state. Glass has an amorphous molecular structure as opposed to a crystalline structure metals and other solids are characterized by. This is what makes it isotropic in its properties as well as transparent.

Its transparency is the property that differentiates glass from most other building materials. There are various factors that affect the perception of transparency; however, its basic metric is the quantity of light that passes through a material, which can be described as transmission. In addition to the transmitted light, there is a percentage of light being reflected and absorbed. The absorbed light affects the perception of the colour in the glass, hence playing a significant role in the way we perceive its transparency.

Due to the transparent nature of the material, the way it is connected is always visible. As a result, the connections and connectivity of glass are one of the most important considerations when designing with glass, both technically and architecturally and in particular in structural applications.

4.2 State of the art in glass connections

Although a building material since the beginning of the common era the way glass has been used has undergone significant changes and developments, particularly in its connectivity. While initially not used as a structural component, the development of the structural use of glass can be classified into four main groups:

1. glass to brace a steel frame
   a. informally
   b. formally
2. glass transferring wind loads
3. glass supporting itself
4. glass as a primary structure.

The following sections describe the connectivity of glass associated with typologies associated to groups (2)–(4), following the analysis of the structural use of the material and categorization into functionality groups (Fig. 4.1).

4.3 Glass connections

Due to the transparency of the material and its nature of being brittle, the connectivity becomes the central driver for all design with glass.

Driven by the size limitations of the float glass process, glass elements can only be manufactured, processed, delivered to site and installed in situ in limited sizes. On site, the glass elements are either individually fixed to a loadbearing construction or they are joined together to form a self-supporting structure.

Within the development of glass architecture, the connection typologies might not typically occur in a chronological order, hence they are classified into categories. The main categories shall be:

- discrete connections and
- linear connections.

In addition, a differentiation between mechanical and adhesive connections can be made, with mechanical connections being divided into contact connections and friction connections. That leaves the following categories, which will be discussed with examples of connection typologies:

- bearing connections
- friction connections
- bonded connections.

The pursuit of transparency

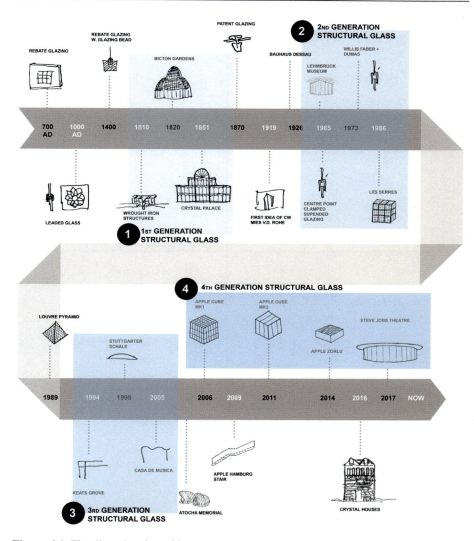

Figure 4.1 Timeline glass in architecture.
Source: From Rammig, L. (2021). Advancing transparency, IOS, Delft.

Traditionally, all the aforementioned connection types form a very visible joint between two glass panels leading to the lightness and openness of a glass enclosure or structure being defined by the connections between traditionally fabricated glass panels or between glass and primary structure (Fig. 4.2).

More recently, the introduction of transparent bonding technology has opened an opportunity for the exploration of fully transparent connections, some of which are explored in the following paragraphs.

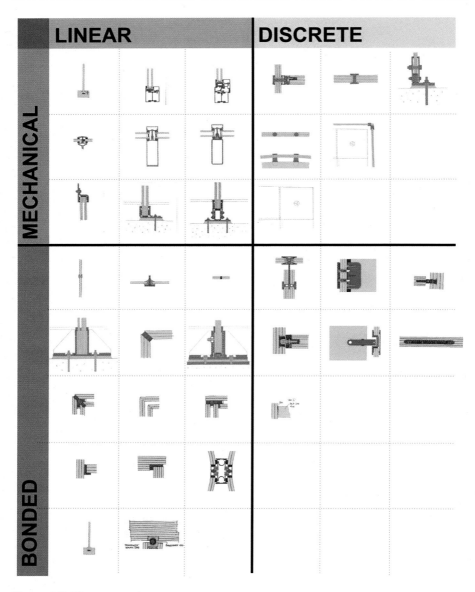

Figure 4.2 Glass connections.
Source: From Rammig, L. (2021). Advancing transparency, IOS, Delft.

As much as the architectural use of glass developed, so did the way it was connected improved and altered over time. Its specific properties require a specific connectivity, which, however, for the use of glass as a structure has only been developed very recently.

4.4 Approach to transparent connections

Fully transparent structural joints have to date primarily been used in experimental structures and small-scale applications. However, few built examples are available and will be discussed in the following sections.

4.4.1 Dry joint

To avoid a metal bolt in the connections between vertical and horizontal structural elements, Dewhurst Macfarlane developed an experimental approach of applying timber detailing and dry joining the beams and columns as a mortise and tenon joint (Fig. 4.3A and B). This idea was executed in a residential building in London, designed by Rick Mather Architects.

Also visible in Fig. 4.3B is the glass spacer used for the insulated glass unit (IGU) instead of a desiccant filled aluminium or stainless steel spacer that is usually sealed with butyl and silicone, leaving the distinct black edge around the glass panel. The glass spacer is bonded to the panel with an ultraviolet (UV) curing glue to achieve the required airtightness. Reportedly, the stiff bond led to a cracking of the glass panels under load and movement though. In a further development of the detail, the dry connection was bonded with a lamination resin, which at the time was still more common than sheet lamination.

4.4.2 Transparent structural silicone adhesive−bonded connections

A novel bonding material that offers great potential to achieve fully transparent glass connections is a silicone material developed by Dow Corning, called transparent structural silicone adhesive (TSSA).

TSSA is an optically clear and high-strength silicone film. With a thickness of 1 mm, this adhesive film is designed to structurally bond glass to metal without any additional dead load support.

(A) (B)

Figure 4.3 (A) Schematic mortise and tenon joint. (B) Close-up of mortise and tenon joint.

As opposed to common structural silicones that are one- or two-component liquid applied and cured at room temperature, TSSA is a one-component material that is provided as sheets and cured under heat and pressure application. Temperatures of 120°C–130°C are applied for approximately 30 minutes simultaneously with applying pressure. This ideally occurs in an autoclave (Dow Corning, 2017).

In addition to its higher strength compared to traditional structural silicone, with a permanent design load approximately 50 times higher (Hayez, 2018), the major difference is the transparency of the material after curing. The film is optically clear with a refractive index very similar to glass (Sitte et al., 2011), which suggests that it offers the potential for connections to become invisible.

When overstressed, the material shows a whitening behaviour, which according to the manufacturer will return to its transparent appearance once the stress is released (Sitte et al., 2011). Given that visually this is an undesirable effect, expected to result in overdimensioning of the connection.

Despite having been tested for a number of years in short- and long-term exposure tests, the project applications of TSSA are still limited. It was intended to be used for point fixings, given its dynamic and static failure strengths that are substantially beyond what is observed for commonly available structural silicone materials.

4.4.3 Glass–metal connections

As per the originally intended use, bonded metal point fittings have been the primary application of the material. The appearance is comparable with the appearance of laminated inserts; however, the use of TSSA allows the connections to be manufactured significantly more economically. TSSA does not require the use of specific metal grades, in many cases titanium, which is necessary for insert lamination, due to its similarity to glass in thermal expansion behaviour. Further to that, the precise machining of the glass drives the cost of laminated inserts (Fig. 4.4A–C).

While many tested applications of the material are still in experimental stage (Santarsiero, 2015), Glas Troesch has provided multiple solutions for project application that have been verified and tested (Kassnell-Henneberg, 2016; Fig. 4.5A–C).

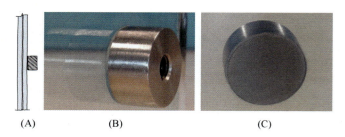

(A) (B) (C)

Figure 4.4 (A) Sketch detail Transparent Structural Silicone Laminate (TSSL) fitting. (B) Close up of Transparent Structural Silicone Laminate bonded point fitting (Sitte). (C) Close up of Transparent Structural Silicone Laminate bonded point fitting through glass (Sitte).

4.4.4 Glass–glass connections

An experimental use of a TSSA (in this case Transparent Structural Silicone LaminateL) was tested in a temporary installation for the Glass Technology live exhibition in Duesseldorf in 2016. In a collaboration with Cricursa and Eckersley O'Callaghan, borosilicate rods were bonded to a curved glass panel with TSSL to form the ladder of a slide (Fig. 4.6A and B). The fully transparent connections were achieved by polishing the rods to fit the shape of the large tube and with the initial application of pressure to get rid of any trapped air between the glass and the silicone sheet material and then autoclaving the assembly to initiate the curing process.

4.4.5 Glass corners

When considering glass corners, transparency is a primary concern, as the corner arrangement of glass allows a wide-angle view out of a building.

Spanning the glass between floors and not relying on fins or mullions to limit deflections allows to provide the view openness from corner to corner; therefore the transparency of the corners becomes more important.

In IGUs, typically the edge seal limits the transparent appearance of the corner, while a monolithic buildup will be limited by the silicone joint. There are various

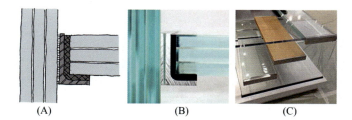

Figure 4.5 (A) Sketch detail TSSL tread connection. (B) Close-up of bonded TSSL metal to glass connection (Hayez, Kassnell-Henneberg). (C) TSSL metal to glass connection (Hayez, Kassnell-Henneberg).

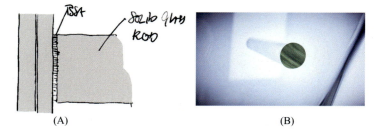

Figure 4.6 (A) Sketch detail TSSA tread connection. (B) Close-up of bonded glass tread connection. *TSSA,* Transparent structural silicone adhesive.

geometrical arrangements that all have a varying effect on the appearance of the corner transparency (Fig. 4.7A−F). A mitered corner can be considered having the lowest visual impact, as no glass edges are visible in this arrangement.

Recent experiments with clear bonding materials explore the possibility to increase the transparency of glass corners. One material that appears to have potential in achieving this is sentry glass (SG). Laminating the corner with SG appears to be a feasible solution, as it provides the desired stiffness as well.

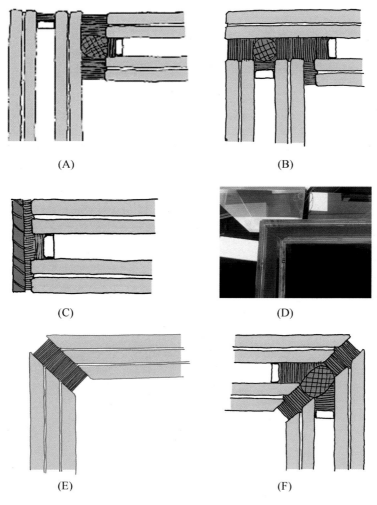

Figure 4.7 (A) Butt joined corner. (B) Stepped glass corner. (C) Stainless steel edge protection. (D) Image structural silicone mitered corner. (E) Sketch detail structural silicone mitered corner. (F) Sketch detail structural silicone mitered corner.

The pursuit of transparency

Experimental studies were undertaken and the approach has been applied in the façade of the ME hotel in London, designed by Foster and Partners (Fig. 4.8A and B).

Subsequent to the development of the window units for the ME hotel by Bellapart and Thiele Glas, this approach has been tested by various other fabricators, including Interpane, Sedak and Agnora. Fig. 4.9A−C shows an SG-laminated corner sample at Sedak's factory in Gersthofen, Germany as per the detail illustrated in Fig. 4.9A.

The spacer runs continuously around the perimeter of the unit and the mitered corner is laminated with SG. Due to its stiffness as a sheet material, the SG cannot be bent around a corner, so an additional sheet of SG is used to connect the mitered surfaces.

Inside out the corner is fully transparent and nearly invisible, while from the outside, the refraction on the glass edge makes both lites in the IGU visible and hence does not lead to a fully clear appearance.

Figure 4.8 (A) ME hotel London Foster + Partners, Image Bellapart. (B) ME hotel London Foster + Partners, Image Bellapart.

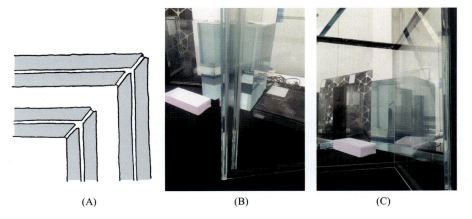

Figure 4.9 (A) SG-laminated corner—sketch detail. (B) SG-laminated corner outside (fabricated by Sedak). (C) SG-laminated corner inside (fabricated by Sedak). *SG*, Sentry glass.

Given the load transfer capabilities of SG, a connection on the edge of the glass should be considered for other structural applications like bonding stair treads to a glass stringer or connecting treads to each other. This concept will be described further next.

4.4.6 Heat-bonded glass connections

The concept of heat bonding to connect glass is not new. It is a traditional craft used to fabricate, repair and customize laboratory ware and other equipment in chemistry laboratories. This craft is known as scientific glass blowing, which as opposed to traditional glass blowing used to produce art pieces, cups and vessels from molten glass uses a prefabricated base material. Typically, this is industrially manufactured borosilicate as tubing and rods which are then fabricated into chemical instruments and laboratory containers. The driver behind the use of glass for these applications is its resistance against chemicals while offering transparency and visibility of chemicals in the containers. Typically, borosilicate is used, due to its better resistance to thermal shock compared with soda lime silicate. This is driven by the linear thermal expansion coefficient that is approximately a third compared to soda lime.

The manual heat bonding process is carried out by first preheating the two components to be connected, then locally bringing them to working temperature at which they are physically connected and cooling them afterwards in a colder flame before they are annealed to release the locked-in residual stress and revealing the final result (Rammig, 2016a,b; Fig. 4.10A−C).

A temperature profile for heating, bonding and cooling phases as well as the following annealing process had to be developed for each geometry a joint typology manufactured.

During the heat bonding process, temperatures were monitored with a thermal camera (FLIR-T 640) throughout the fabrication of connections.

Heat-bonded T connections visually offer a great potential as they allow to omit the appearance of the glass edge that is typically apparent when viewing a glass fin supported façade, both internally and externally.

While distortions around the bond are visible due to the manual manufacturing process, it is assumed that this could be improved using a laser-based welding technique.

Fins could be imagined to be heat bonded to a flat panel to limit deflections as in a typical fin-based façade. Connections could be continuous or local; however, the primary benefit might be in the possibility of achieving a continuous transparent connection.

One of the very common connections in the use of structural glass for building envelopes is as a joint between two panels that works structurally to limit deflections (Fig. 4.11A and B).

Assuming that heat-bonded connections will be limited to fabrication in a factory, particularly if manufactured manually rather than using automated additive or laser technology, panel sizes will still be limited by common handling and transport

The pursuit of transparency

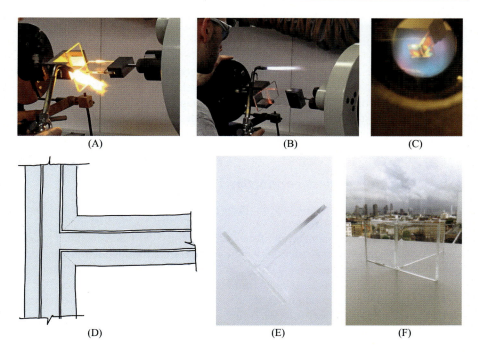

Figure 4.10 (A) large flame to heat up components in lathe. (B) Joint after bonding. (C) Visual stress visualized with polarization filter. (D) Detail heat-bonded glass fin connection. (E) Image HB T-joint. (F) Image HB T-joint.

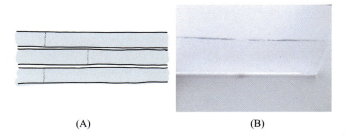

Figure 4.11 (A) Detail heat-bonded butt joint multiply. (B) Heat-bonded butt joint.

restrictions as well as size limitations for lamination. In addition to that, the size of glass components within a building structure will be limited to the size that allows the accommodation of all differential movement while transferring wind and or other live loads.

Based on the principle illustrated for the heat-bonded T connections, in which geometrical stiffness is achieved through a heat-bonded connection, this type of connection is applied to another common application in which a structural connection between glass panels is required. A staircase design with heat-bonded treads was explored as a possibility to achieve fully transparent connections between treads.

As a basis for the design, the heat-bonded corner arrangement illustrated in Fig. 4.12A–G is assumed. Using a multitude of welded corners and arranging them such that they form a ladder while always overlapping with the adjacent corner. The overlapping surfaces could be laminated with an SG interlayer, which would provide the required redundancy of the construction.

4.4.7 3D-printed glass connections

One possible approach to automate the process of heat bonding would be to use an additive manufacturing (AM) process.

The term 'Additive Fabrication' dates back to the 1980s and includes various technologies of layered construction, which as opposed to subtractive methods that involve the removal of material (cutting, milling, drilling, etc.) uses an ' additive' process in which layers of material are used to create an object (Strauss, 2013).

Additive fabrication technologies do not use tools or moulds, but the object is created from a digital model directly, which is why the technology is also referred to as direct manufacturing. This means that the technology is well suited for customization and single-batch fabrication, as the cost for the fabrication of moulds and the additional design complexity coming with a multistep process is not required anymore.

Fused deposition modelling (FDM) works on the principle of melting a polymer rod and depositing it in layers on top of each other to create objects after curing. It could be described as an automated hot glue gun.

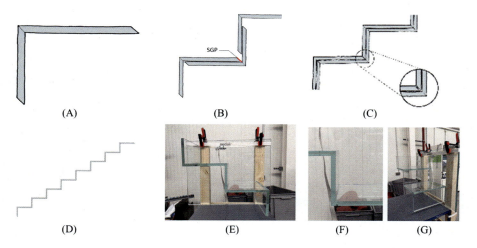

Figure 4.12 (A–G) Heat-bonded corner concept w. SG stair tread lamination (Rammig, 2021; laminated by Sedak). *SG*, Sentry glass.

Given the viscosity of glass at elevated temperatures and its ability to form mono-material bonds, this technology lends itself to the application of an FDM process. Mirroring the welding approach would suggest to use a process in which liquid material is deposited in layers on top of each other creating a bond to each adjacent layer at the same time.

A 3D printer to create glass objects has been developed in 2015 at the Media Lab at the Massachusetts Institute of Technology. A movable platform within a kiln allows to deposit layers of molten glass which flows through a hole in the container at the top of the kiln. Various objects of different sizes were created and initial studies on the performance of the connections of the adjacent layers were carried out. However, the application in glass structures or envelopes is not the main purpose of the research that is primarily centred around the creation of design objects, some of which are shown in Fig. 4.13.

It could be imagined, however, that objects produced in a 3D printing process could form connections when deposited on a flat glass sheet. Recent studies carried out at TU Darmstadt suggest the feasibility of this approach. This approach could be used to form connections or to increase stiffness and hence limit deflection of the sheet material.

None of the results so far however provide transparency in the sense that a distortion-free view through the material is provided. This might be a result of the scale of layers deposited, hence the 'resolution' of the printed object.

At Glasstec 2018 a technology was presented by QSil that would allow for printing at higher resolution (Fig. 4.14).

The specimens shown in Fig. 4.14 suggest that a higher 'resolution' reduces the optical distortion and hence increases the transparency of the objects. To achieve a significantly higher resolution that is comparable to the quality of flat (float) glass, it might be necessary to introduce production technology that is not based on liquefying a strand of material, as this will be limited by flow rates through a nozzle. Based on knowledge of precision welding of other materials, a laser welding technique might be employed to improve the resolution of the AM products.

However, when the comparison with flat glass is not taken into consideration and the traditional quality and transparency concepts that come with it are set aside, the distortions and reflections that are a result of this process add a valuable aesthetic and additional depth to the component.

4.5 Alternative transparency concepts with glass

Traditionally, glass is used as a flat sheet material that is supported linearly or discretely as previously described. That means it is typically designed to resist load in bending, despite the fact that it is significantly stronger in compression than in tension.

Typically, vertical members are used to limit the deflection in flat glass sheets. These can be aluminium or steel mullions as well as glass fins or any other material

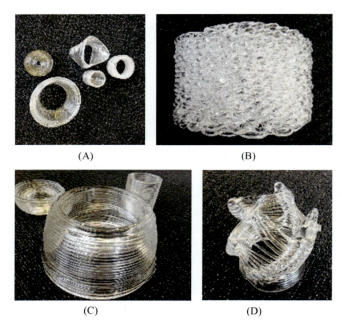

Figure 4.13 (A–D) Examples of objects created with 'G3DP' process at MIT. *MIT*, Massachusetts Institute of Technology.

Figure 4.14 3D-printed objects as presented at Glasstec by QSIL, the Netherlands.

that spans behind the glass. If mullions are not desired and the flat glass is designed in a two-way configuration spanning from floor to floor, the thickness of the glass has to increase and multiply laminates are used.

Due to the colour inherent to the material, the more layers are used, the more tinted the buildup looks (Fig. 4.15).

To reduce the thickness of the glass buildup, an alternative concept is to rely on geometrical stiffness achieved through bending the glass.

The pursuit of transparency

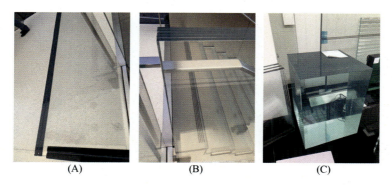

Figure 4.15 (A) 2 × 12-mm low iron = 88%VLT. (B) 5 × 12-mm low iron = 83%VLT. (C) 20 × 12-mm low iron = 57%VLT.

Depending on the nature of the curvature as well as the radius, this allows to significantly reduce the thickness of the glass and hence the clarity of the assembly.

This concept has been applied in a few built examples and case studies.

4.5.1 Vakko Headquarters, Istanbul

The external glazing of the Vakko HQ in Istanbul, designed by Rex is gravity bent (slumped) to increase its stiffness. Over the length and width of the entire panel, an X is formed into the surface that reduces the deflection under wind load and hence allowed to design this façade without vertical mullions. The external deformed glass panel is 19-mm monolithic annealed, while the inner pane of the IGU is flat and fixed with countersunk point fittings, bonded into the monolithic glass. This means the external panel is supported through the IGU edge seal only. The fact that the inner fully tempered panel is point supported with countersunk fittings should be questioned, given that any fracture on the inner panel, the entire unit would be at risk of dislodging (Fig. 4.16).

4.6 Casa de Musica, Porto, OMA/ABT

In the concert hall in Porto designed by OMA, large corrugated glass walls mark the areas of the building that have a connection to the external.

The 25 m × 12.5 m opening is filled with three rows of 4.5-m tall glass panels.

The glass is gravity bent (slumped) into shape in an oven and annealed as part of the process to release locked-in stresses. The annealing process is of significant importance to guarantee that the glass will not fracture due to stress differentials when installed on site and subject to building movement, wind and thermal loads. The structural engineers for the glazed walls (ABT) made use of the geometric stiffness that is gained by corrugating the glass, which omitted the requirement of vertical mullions due to the limited deflection achieved through the geometry.

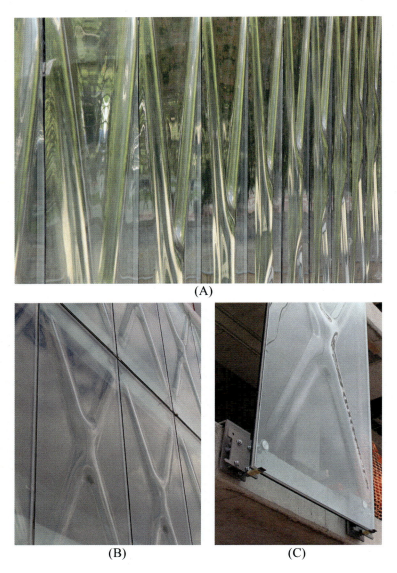

Figure 4.16 (A) Vakko HQ, REX (image A. Betanzos). (B) Vakko HQ, REX (image Betanzos, A.). (C) Vakko HQ, REX (image Betanzos, A.).

The glass panels are base supported in a steel shoe, which is fabricated from bent plates and at the head restrained to a circular steel beam that also provides the dead load support for the panels above.

Given the geometrical stiffness, the wind loads can be resisted with a monolithic 12-mm annealed glass panel, which, however, poses the question of resilience in case of fracture, given that annealed glass could break due to thermal stress, stress concentrations caused by movement of the structure or impact. Laminating the panels appears to be an approach that would have reduced the risk of dislodging; however, it is understood that the approach taken was driven by the assessment that in case of fracture, due to its geometry and capture, the glass would not dislodge.

4.7 K11 Musea, Hong Kong, SO-IL/EOC

As part of a development in Kowloon designed by KPF, the CTF museum façade is designed by SO-IL together with EOC. The museum is set on the eighth floor of the mixed-use development and nontypical for a museum, the façcde along the parameter is fully glazed, to allow transparency out of and into the space. A total of 475 9-m tall glass tubes are set next to each other around the perimeter of the museum to form the envelope. As with the projects previously shown, the geometrical stiffness achieved by curving the glass to a radius of 450 mm, which allows to omit any additional structure despite hurricane loads in Hong Kong (Fig. 4.17).

Most tubes are fabricated of two 12-mm sheets of low iron glass and laminated with SG. Here, SG was used not only due to its load transfer capacity but also its lower viscosity and flow rate in the autoclave allows to better allow for tolerances in the glass surface radius.

4.8 Vidre slide, Duesselorf

Combining the approach outlined earlier with the use of transparent connections, an experimental structure was built to showcase the possibilities of glass processing and the latest connection technologies at Glasstec 2016 in Duesseldorf, Germany.

Based on the glass tubes fabricated for the K11 Musea, the author worked with Cricursa to build the fully transparent glass structure for the Glass Technology live exhibition.

Curving the flat sheets to a very tight radius (450 mm) makes them inherently stiff, meaning that the glass can span 9 m in this triangulated configuration with only two layers of 10-mm glass (Fig. 4.18).

As a comparison, a flat glass would require approximately 10 layers of 10-mm glass, to achieve a similar performance over the span of 9 m.

Adding this amount of thickness to the glass reduces transparency significantly due to the increased absorption of certain wavelengths (ie, red and yellow, will lead to blue and green light being transmitted), so reducing the thickness of the glass by introducing a curvature, in turn increases the light transmission by 20%, compared to the thicker flat buildup. However, visual light transmittance is not the only factor playing a role in the appearance of a transparent object. The colour plays a very important role in the perception of transparency. The smaller the perceivable tint in a glass, the more transparent the glass appears. Even on a glass with a reduced iron content as used in this installation, a tint is existent, which becomes more visible, the more layers of glass are stacked and laminated together (Fig. 4.15).

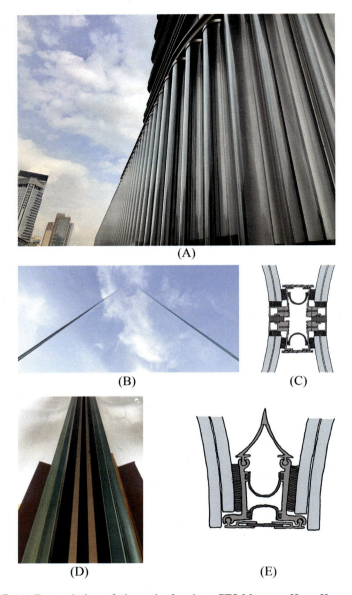

Figure 4.17 (A) External view of glass tube façade at CTS Museum, Hong Kong (SO-IL/EOC). (B) Laminated glass tube prior to unit assembly. (C) Connection detail. (D) Detail of two half tubes after installation. The glass is protected with an adhesive film. (E) Half tube sketch detail.

Figure 4.18 Exhibition of the slide at Glass Technology live.

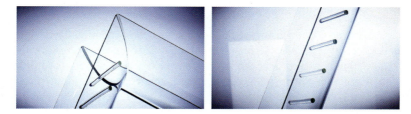

Figure 4.19 Connection details at the top [grey structural silicone (Sikasil SG 500)] and at the treads (TSSA).

The clarity achieved with this approach is significantly higher than using flat glass to span without additional support. This leads to the conclusion that the use of geometrical stiffness might be a useful approach for the design of glass structures, either by introducing stiffness through small amounts of cold bending or for the work with thin and ultrathin glass, which requires a design approach that limits the flexibility of the material. The significantly increased strength of these high-strength ultrathin glasses would allow curvatures to be introduced even through cold bending and might to even larger transparency (Fig. 4.19).

The transparent silicone connections between two pieces of glass achieved for this experimental structure require further testing particularly relating to long-term performance in weathered conditions to evaluate the appropriateness for long-term building applications. Further experimental testing is suggested on the actual connections as currently test data are only available for glass—metal connections.

4.9 Glass as a solid material

Despite the fact that glass in architectural applications is typically used as a thin sheet, examples of the use of glass as a solid material are available. In these applications, glass typically occurs in blocks or bricks, which due to the thickness as well as the small size and increase of joints leads to a reduction of transparency. However, instead of using the glass in bending as is typically the case when designing with sheet material, it is used in compression in which it is much stronger than in bending (Feldmann et al., 2014).

4.10 Atocha Station Memorial Madrid

The memorial for the victims of the terrorist attacks at Atocha station in Madrid in 2004 is designed using approximately 15,000 clear borosilicate bricks that are bonded with a UV adhesive to form a cylinder, which provides light into an underground exhibition space. Within the cylinder, messages left at the station after the attack are engraved into a translucent fabric structure that is suspended from the transparent glass roof.

The cylinder is formed using a glass brick that was specifically designed and fabricated for the project (Schott Duran, 2016). Each brick measures $300 \times 200 \times 70$ mm with a concave and a convex end that allows the bricks to be aligned in a cylinder shape with varying radii and creating a slight interlocking effect. The bricks are bonded using a transparent acrylic adhesive, which is cure with UV light on site. The choice to go with borosilicate glass was made due to its low thermal expansion coefficient compared to soda lime glass.

4.11 Crystal Houses, Amsterdam

The Crystal Houses are designed by MVRDV to resemble a historic façade in Amsterdam's Hooftstraat in a novel interpretation. The brick façade is made entirely of glass, including the window and door frames (Fig. 4.20A–D).

The bricks are made of solid soda lime glass, bonded with a transparent acrylic glue that is cured with UV light on site (Oikonomopoulou et al., 2018). This technique is comparable to the way the Atocha memorial in Madrid was assembled.

The design makes use of the large compressive strength of the glass, which is a very intuitive way to use the material compared to the use of glass as thin sheets, which results in bending stress being induced. To resist wind loads, additional buttresses are used, stiffening the wall (Oikonomopoulou et al., 2018).

4.12 Further development of brick approach at TU Delft

After the construction of the Crystal Houses, Oikonomopoulou and Bristogianni continued the research in glass bricks and developed methodologies of casting bricks out of recycled glass.

The pursuit of transparency

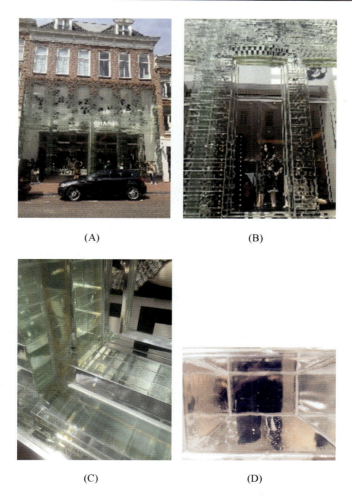

Figure 4.20 (A) Crystal Houses, Amsterdam, MVRDV, ABT, TU Delft. (B) Crystal Houses, Amsterdam, window frames made of glass extrusions. (C) Crystal Houses, Amsterdam, window frame and sill. (D) Crystal Houses, Amsterdam, close-up.

In their ongoing research, they tested melting and casting glass with varying compositions at different temperatures and further developed the casting into interlocking shapes. Examples of the research were exhibited as part of the Glass Technology live exhibition at Gastec in Duesseldorf 2018 (Fig. 4.21).

Various shapes, colours and opacity levels have been explored as well as the cooling rates required to achieve transparent bricks, as well as glass ceramics at slow cooling rates. The research indicates a significant potential for the recycling of glass into building components in the form of bricks and blocks, which can be transparent, translucent or coloured.

Figure 4.21 Interlocking cast glass bricks shown at Glass Technology live, Duesseldorf, 2018.

Figure 4.22 Interlocking cast glass brick arch (Snijder, TU Delft).

Another stream of research at TU Delft investigates the use of interlocking bricks for a bridge design (Fig. 4.22). As opposed to the assembly in the Atocha Memorial and the Crystal Houses, the bridge is designed to be assembled dry, meaning no adhesive will be used to bond the bricks together, but they rely on the geometry of the bricks as well as the bridge to create sufficient friction to keep the glass assembled within an arch.

4.13 Thin glass

The thicker assemblies become, and the more layers of glass are used, the larger the impact of the inherent colour on the appearance of the glass. While a change in the composition can reduce that appearance of tint, low iron glass, in which the iron content is reduced to approximately a quarter compared to traditional soda lime silica, still displays an increased tint when multiple layers are stacked up.

Another way to achieve a clearer appearance in glass assemblies would be to use less material that is, by making it thinner. To be able to use a thinner layer of the material though, its strength needs to be increased to be able to achieve the same structural performance.

The screen industry provides a solution here, an ultrathin aluminosilicate produced in a fusion process, in which the melt flows down vertically. Without exposure to tin or other materials than air, it has a significantly higher bending strength than traditional soda lime silicate when chemically tempered (up to 750 MPa, Corning, 2013). Other thin and ultrathin glass products on aluminosilicate and soda lime silicate basis are available (Schott, AGC, Corning); however, the study described here is based on Corning Gorilla Glass (GG).

At a thickness of 1 mm, GG can be bent to a radius as small as 1 m in a cold bending process (Lenk et al., 2018). Its high strength allows to use very thin layers of material, in turn reducing not only the inherent colour but also the weight, which makes it interesting for application in slender structures. As opposed to glass produced for the construction industry, sizes are limited to typical screen sizes and production is not geared toward custom sizes.

When using a very thin material, considerations of its connectivity have to be adapted to its properties. Although the surface bending strength of the material is very high, it remains fragile to load impact on the edge. This is a common experience with smartphone screens that shatter when impact occurs.

This suggests that load transfer through the edge should be avoided, and connections designed at the face of the glass.

An important consideration further to the limitations on edge connectivity is the fact that in most cases the design of glass envelopes is deflection driven. And despite the increased design strength, an increase in flexibility would mean that the concept of a flat glass panel spanning between two slabs would not be achievable at scale. This should be considered an opportunity though to rethink the way glass is used both structurally and architecturally. Using its flexibility to gain geometrical stiffness without complex heat bending processes appears to be a major advantage over traditional flat glass.

The flexibility of the material offers a range of opportunities, including a reconsideration of windows and other operable units. Typically, these are framed and require specific hardware, that is, window handles and hinges for their operation. The use of thin glass would allow for an operation of a ventilation unit without the use of a frame. A bimetallic strip bonded to the face of the glass, which expands differentially to the glass itself, would lead to a flexing of the material when heated up by the sun.

This would allow automated ventilation based on solar radiation. Various concepts similar to the one described (Fig. 4.23) have been mocked up in a Master course at TU Delft, (Louter, C) and exhibited at Glass Technology live.

Assuming glass to glass connections become more transparent, the transparency of the overall component becomes more important. This suggests that transparent edge seals become more important to increase the overall transparency of a fixed glass façade. When considering kinetic components like windows and doors, which in their functionality play a fundamental role to the performance of the building

Figure 4.23 Lightweight IGU made with ultrathin glass laminates (0.7 mm + 0.7 mm/CAV/ 0.7 mm + 0.7 mm) fabricated by Glas Troesch. *IGU*, Insulated glass unit.

envelope, the development has to move away from traditional framing elements to advance the transparency of these components. Here the thin glass concepts as indicated in Fig. 4.24 and ideas outlined opportunity for a development toward more transparent building envelopes.

Although the use of thin glass has been heavily explored structurally in the last years, this primarily occurred in an experimental context with the focus on the exploration of the structural properties and potential opportunities related to those. However, a knowledge gap can be identified around the connectivity of thin glass and how typologies would have to be adjusted to accommodate its properties. Detailing will have to accommodate the lack of possibilities to support the glass on its edge, as well as through drilled holes, which suggest that the exploration of face bonding techniques with transparent bonds that can accommodate the dead load of the material would be an appropriate field of research.

4.14 Conclusion

Designing buildings and in particular enclosures without the use of glass is generally unimaginable. However, the way the material is used is constantly evolving. Transparency has always been a significant driver for the design with glass and for the evolution of its connectivity. Based on the case studies and research outlined in this chapter, a clear trend away from traditional flat panels that are either point supported or span between floors and the connectivity that comes with this typology can be observed, when transparency is a design concern. Making use of the properties of the material rather than designing against them leads to completely new glass aesthetics.

The pursuit of transparency

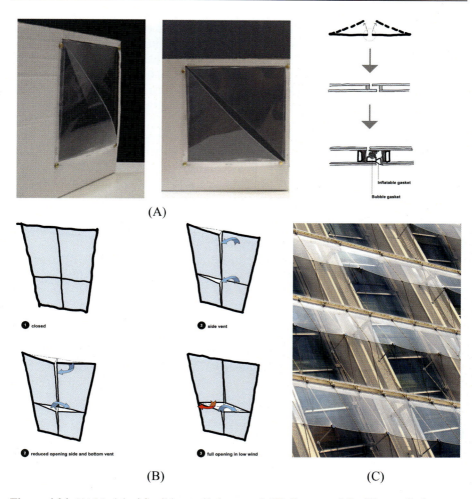

Figure 4.24 (A) Model of flexible ventilation panel. (B) Concept of flexible ventilation panel. (C) Precedent External skin Unilever HQ Hamburg, Behnisch Architekten, 2009 (image Bilow, M).

The transparent bonding technologies (TSSA and heat bonding) outlined in this chapter were tested largely on an experimental basis and their performance assessed through single prototypes (slide and stair treads) and small series (heat-bonded butt glazed and T-beam connections). Even on the experimental scale the potential is evident, that an increase in transparency of the connections and the potential of achieving fully transparent connections increases the visual quality of the glass structures they are applied to. Partly this is attributable to the fact that if a connection is invisible, its functionality is typically questioned and in the case of glass, the perception of safety often relates to the structure connecting the glass or holding it in place. So, if that connection is not visible anymore, the result is not only daunting but also exciting as it sparks curiosity about its functionality.

This was particularly perceivable with the glass slide exhibit that despite its nearly invisible tread connections and single silicone joint keeping the structure up and connected, lead many people to try it both for the test of whether it would hold up and for the personal proof of overcoming the insecurity a fully transparent structure poses.

Despite the fact that the heat-bonded connections can achieve a fully transparent and invisible connection that can be as strong as the parent material when fabricated carefully and homogeneously, throughout the process, nontraditional visual qualities evidently became desirable. Small amounts of distortions caused by a manual process and lack of precision lead to a new quality that adds an additional dimension to the glass, based on the light it reflects. Away from perfect transparency or clarity and lack of connection that one might have expected, the connections add to the depth and multidimensionality of the glass component.

This can equally be observed in the small objects created through 3D printing, which work very well as an object or jewel and which have an aesthetic that might translate beautifully into building enclosures or structures, when the material is used in compression as a shell element similarly to concrete. Playing with structuring the layers where transparency is less desired and smoothing them out where views through the material should be permitted, this approach might provide the opportunity for an enclosure formed from a single material while still addressing various performance requirements a building envelope typically has to achieve.

References

Dow Corning (2017). *TSSA—Transparent structural silicone adhesive, product information, high strength structural glazing silicone film adhesive*.
Feldmann, M., et al. (2014). *Guidance for European structural design of glass components*. Luxembourg: Publications Office of the European Union.
Hayez, V. (2018). *Silicones enabling crystal clear connections*. London: IGS.
Kassnell-Henneberg, B. (2016). Connections in glass, in Challenging glass 5, Ghent. Belis, Bos & Louter (Eds).
Oikonomopoulou., et al. (2018). *The construction of the crystal houses façade: challenges and innovations*. Delft: Springer Glass Structures and Engineering 3.
Rammig, L. (2016a) Residual stress in glass components. In: *'Challenging glass conference 5'*, Ghent.
Rammig, L. (2016b). *Transparency through geometry—A case study*. London: IGS Winter edition.
Rammig, L. (2021). *Advancing transparency*. Delft: IOS.
Santarsiero, M. (2015). *Laminated connections for structural glass applications*. Lausanne: EPFL.
Schittig., et al. (1999). *Glass construction manual*. Basel: Birkhaeuser—Publishers for Architecture.

Schott Duran (2016). *Versatile or multifunctional: DURAN® borosilicate glass 3.3 tubes, rods and capillaries*. Accessed August 2016.

Sitte, S., et al. (2011). Preliminary evaluation of the mechanical properties and durability of transparent structural silicone adhesive (TSSA) for point fixing in glazing. *Journal of ASTM International*.

Strauss, H. (2013). *The potential of additive manufacturing for façade construction*. Delft: IOS.

Advanced fenestration—technologies, performance and building integration

Fabio Favoino[1], Roel C.G.M. Loonen[2], Michalis Michael[3], Giuseppe De Michele[4] and Stefano Avesani[4]

[1]Department of Energy, Polytechnic University of Turin, Turin, Italy, [2]Department of the Built Environment, Eindhoven University of Technology, TU/e, Eindhoven, Eindhoven, The Netherlands, [3]Engineering Department, Glass and Façade Technology Research Group, University of Cambridge, Cambridge, United Kingdom, [4]Energy Efficient Buildings Group, Institute for Renewable Energy, Eurac Research, Bolzano, Italy

Abbreviations

AFS	advanced fenestration systems
AIF	advanced integrated façade
BiPV	building-integrated photovoltaic
BPS	building performance simulation
CCF	closed cavity facades
DGU	double glazing unit
DSF	double-skin façade
DSSC	dye-sensitized solar cells
EC	electrochromic
g-**value**	solar heat gain coefficient
HVAC	heating ventilation and air conditioning
IEQ	indoor environmental quality
IGU	insulated glazing unit
LC	life cycle
LCA	life cycle analysis
PCM	phase-changing material
PV	photovoltaic
PVB	polyvinyl butyral
RES	renewable energy sources
STB	solar thermal blinds
T_{vis}	visible transmittance
U-**value**	thermal transmittance
VIG	vacuum insulation glazing

Rethinking Building Skins. DOI: https://doi.org/10.1016/B978-0-12-822477-9.00038-3
© 2022 Elsevier Inc. All rights reserved.

5.1 Introduction

The building decarbonization agenda is targeting more and more stringent limits aiming at minimizing buildings impact in the context of climate change (Van Berkel, Kruit, Van de Poll, Rooijers, & Vendrik, 2020). Within this context, more recent sustainability trends give emphasis to the large potential of refurbishing existing buildings and their envelope (Luca, 2018). To this purpose, different strategies are put in place at the building level to improve its performance toward this goal, such as (1) reducing building energy use and operational carbon emissions (E. Commission, 2010; The European Parliament and the Council of the European Union, 2012); (2) increasing on-site energy harvesting by means of renewable energy sources (RES) (Sartori, Napolitano, & Voss, 2012); (3) improving energy flexibility by demand side management at building and district level (Jensen et al., 2017); while (4) maximizing indoor environmental quality (IEQ) for the occupant (Xie, Clements-Croome, & Wang, 2017); and (5) reducing construction materials embodied carbon (Chapa, 2019). Fenestration systems, being the transparent and translucent part of the building envelope, have traditionally been identified as the weak point of the building and, at the same time, as a key element to pursue higher building performance in terms of energy use and occupant comfort. Due to the role of the transparent part of the building envelope in regulating heat (including shortwave radiation) and mass transfer, it has a significant impact on building performance (building whole-life carbon and costs, overall energy use and IEQ) (Jin & Overend, 2014) while providing opportunities to respond to different and contrasting building performance requirements and functionalities at the same time (eg, view out, daylight and solar control).

Over time, the main driver for fenestration systems improvements has been the need to reduce heating and cooling energy demand as part of the global sustainability agenda. The fenestration overall thermal transmittance (U-value) trend in Fig. 5.1 shows that significant progress has been made, owning to the introduction of multipane assemblies (eg, double and triple glazing), advanced coating technologies to reduce surface emissivity, innovations in window frames and spacers and the application of inert gases as cavity filling for lower heat conductivity (Kiss & Neij, 2011). In conjunction with lowering the U-values, these developments driven by energy efficiency have also reduced draught and infiltration losses, condensation risks and poor thermal comfort due to radiant asymmetry, as a result of low surface temperatures. On the other hand, the capability of fenestration to control solar radiation in buildings has been developed by focusing on more selective durable and processable thin films, coatings and interlayers to be integrated into either insulated glazing unit (IGU) cavities or laminated within glass layers (Mattox & Mattox, 2007), which has nowadays nearly achieved its theoretical limit in terms of capability to admit light compared to total solar radiation (as shown in Fig. 5.2).

Nowadays the market standard for fenestration systems consists of highly technological elements integrating IGUs (including different types of glass with specialized treatments, coatings, frits, sealings, cavity gases and spacers) with framing elements (different materials, thermal brakes, actuators for opening, etc.). These technological advancements have transformed what was a potential weak point, into an element

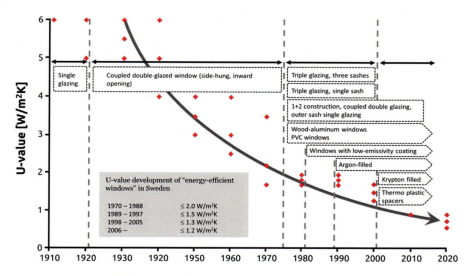

Figure 5.1 Development of *U*-values and window technologies between 1910 and 2020. The overview is based on literature, interviews with window manufacturers and other professionals related to the industry in Sweden.
Source: Adapted from Kiss, B., & Neij, L. (2011). The importance of learning when supporting emergent technologies for energy efficiency—A case study on policy intervention for learning for the development of energy efficient windows in Sweden. *Energy Policy*, 39(10), 6514–6524. https://doi.org/10.1016/j.enpol.2011.07.053.

that could be designed to suit specific needs in various challenging climates, hosting additional complementary technologies and has thereby opened up opportunities for designing largely or fully glazed facades. At this point, however, it is important to realize that energy saving has never been nor will be the reason for designing and installing fenestration systems, which is occupant comfort and multifunctionality. These full set of possible wished functionalities include viewing out/privacy; managing solar radiation; managing glare/daylight; managing and/or controlling heat transfer; increasing thermal storage; managing airflow; and energy generation.

The present chapter is focused on advanced fenestration systems (AFS) that can respond to baseline functionalities (ie, reduce heat transfer, solar control and view out) and can combine these to higher levels aimed at enclosing an occupant with excellent indoor comfort fully under its control while improving building performance toward a positive energy, zero-carbon building objective. In this framework, technologies that are either commercially available or already validated in a representative environment (not in laboratory scale) are reviewed. These are divided into advanced fenestration components and AFS (Fig. 5.3). The former include materials and subcomponents that could be either embedded into a float glass or laminated glass layer (ie, thin films, coatings, interlayer systems used to functionalize and improve the performance of the glazing surface) or within the cavity of an IGU [to exploit this space to add functionalities to the glazing system, such as phase-changing

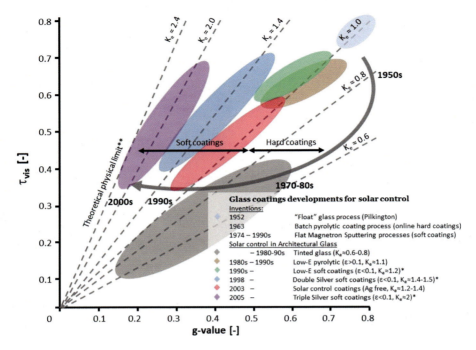

Figure 5.2 Solar control in architectural glazing [*Insulated glazing unit integration for durability reasons; ** ratio between energy within the whole solar spectrum and the one only in the visible region of 380–780 nm, based on AM1.5 solar spectrum (Favoino, Overend, & Jin, 2015)]. The pertaining areas related to different glazing coatings are just indicative of the achievable performance in terms of visible transmission and *g*-value, although these are strictly related to the specific IGU configuration in which the coating is integrated. *IGU*, Insulated glazing unit.

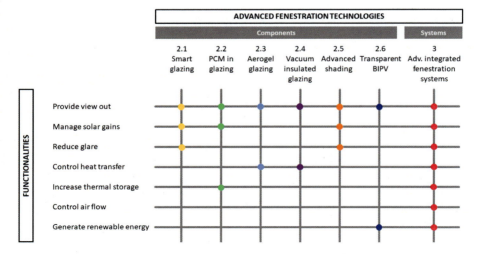

Figure 5.3 Overview of functionalities and advanced fenestration technologies described in this chapter. *AIF*, Advanced integrated façade; *DSF*, double-skin façade; *PCM*, phase-change material; *VIG*, vacuum insulation glazing.

material (PCM), aerogel and shading devices] or within other components of the fenestration system (ie, sealing and spacers, framing system). The second part exploits the synergic use of different fenestration components to provide a façade complete system, which aims at optimizing whole-building energy use (heating, cooling, lighting and ventilation) and providing overall indoor controllable comfort (ie, thermal, visual, aural comfort and indoor air quality). For each of the technologies a brief description will be given, highlighting (1) the main features while providing some example of commercial products and solutions still in the R&D stage; (2) the increase in performance and functionalities; and (3) the challenges for broader market adoption. These are finally summarized in Tables 5.1 and 5.2.

5.2 Advanced fenestration components

5.2.1 Smart glazing

The distinguishing characteristic of smart glazing is its ability to adjust optical properties over time. By modulating the admission of daylight and solar gains to interior spaces, smart glazing can simultaneously influence thermal (eg, overheating in summer) and visual (eg, daylight utilization, glare discomfort and view to outside) comfort conditions. Various functional layers can be used to achieve the mechanism for switching the thermo-optical properties of smart glazing during operation. The most notable ones include chromogenic materials (eg, thin-film metal compounds), liquid crystals and suspended particles. The following considerations are needed when talking about smart glazing for buildings.

1. The control of smart glazing can either be extrinsic or intrinsic (Loonen, Trčka, Cóstola, & Hensen, 2013). In the first case an external signal is used to control the state of smart glazing. This signal can, for example, come from the building energy management system or from user interaction (Luna-Navarro et al., 2019). Electrochromic (EC) and liquid crystal devices are the most common examples in this category. In intrinsically controlled smart windows, on the other hand, the adaptive behaviour is an essential feature of the material, for example, as a function of temperature (ie, thermochromic and thermotropic) or incident light (ie, photoelectrochromic and photovoltachromic).
2. When smart windows modulate their transparency, the net effect is an increase or decrease in either the amount of absorbed or reflected radiation. For certain devices the fenestration system remains specularly transparent, while in other cases, it could exhibit a diffusive behaviour when activated (eg, in thermotropic devices or liquid crystal devices), contributing to a reduced risk of glare discomfort [if translucency is combined with low visible transmittance (T_{vis})] and a more uniform distribution of daylight in the indoor space. Also note that in glazing systems that work with increased absorption to reduce visible light transmittance, additional layers may be needed to reduce the secondary heat flux toward the inside.
3. Some switchable windows can modulate thermo-optical properties across a wide range of wavelengths, while other systems can selectively admit/reflect only in the visible part, nonvisible part or independently in both parts of the solar spectrum (Khandelwal, Schenning, & Debije, 2017). Such spectral selectivity can help to decouple daylighting and thermal issues and, thus lead to better trade-offs among the sometimes competing objectives.

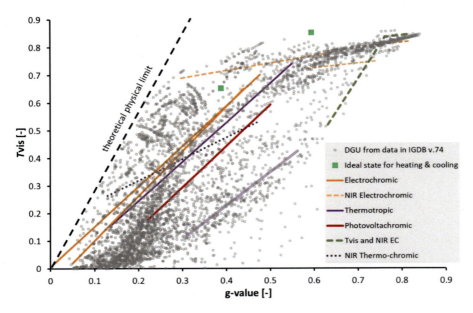

Figure 5.4 Smart glazing properties variation in terms of T_{vis} and g-value (only extreme states) compared to DGU properties (elaborated from glass layers and glass laminate properties in IGDB v.74). *DGU*, Double glazing unit; T_{vis}, visible transmittance.
Source: Adapted from Favoino, F. (2016). Simulation-based evaluation of adaptive materials for improved building performance. In Nano and biotech based materials for energy building efficiency (pp. 125–166). Cham: Springer.

Different types of switchable windows are commercially available on the market (Fig. 5.4, continuous lines) or being investigated in research (Fig. 5.4, dotted lines) (Favoino, 2016). Extensive literature reviews of smart glazing technologies are provided by Baetens, Jelle, and Gustavsen (2010) and Favoino et al. (2015). A comparative analysis by Tällberg, Jelle, Loonen, Gao, and Hamdy (2019) highlights the opportunities and challenges of different types of smart glazing for improving IEQ while reducing the energy demand of buildings.

5.2.2 Glazing integrated phase-change materials

Glazing systems intrinsically contain low thermal mass, resulting in large temperature fluctuations potentially inducing local thermal discomfort and affecting negatively energy use. This could be reduced by use of PCMs within the fenestration system. PCMs undergo a state change (liquid to solid) around room temperatures, storing a large amount of latent heat without significant temperature change. Different kinds of PCMs have been used within building elements and materials, such as organic (paraffins or fatty acids), inorganic (metals and salt hydrates such as $CaCl_2.6H_2$) and eutectics (organic–inorganic mixture), integrated as raw

Figure 5.5 Integration of PCM in glazing systems: (A) as an IGU cavity filler (Goia, 2012) in its solid (a) and liquid (b) phases; (B) macroencapsulated within IGU with transparent insulation and/or light redirection system (Grynning, Goia, Rognvik, & Time, 2013); (C) within glazing shutter or blind (Silva et al., 2016). *IGU*, Insulated glazing unit; *PCM*, phase-changing material.
Source: Schemes in (A) and (C) from Duraković, B. (2020). PCM-based glazing systems and components. In *PCM-based building envelope systems* (pp. 89–119). Cham: Springer.

material, shape stabilized, micro- and macroencapsulated (Silva, Vicente, & Rodrigues, 2016). According to their refractive index in the liquid–solid state (transparent or translucent appearance, respectively) and to the melting temperatures close to thermal comfort range, organic paraffin wax and some salt hydrates are particularly promising for being integrated into a glazing cavity. Ismail, Salinas, and Henriquez (2008), Goia, Perino, and Serra (2013), Goia, Zinzi, Carnielo, and Serra (2015) and Gowreesunker, Stankovic, Tassou, and Kyriacou (2013) conducted different experimental and numerical studies on PCM-filled glazing cavities for double glazing units with direct incorporation of PCM (mainly paraffins and salt hydrates) with melting temperature between 25°C and 35°C (Fig. 5.5A) (Duraković, 2020). The following observation can be summarized from the literature:

1. PCM glazing store solar energy [up to 30% more compared to double glazing units (DGUs)], smooth (up to 50%) and delay (up to 5–17 hours, with optimum thickness of PCM material) peak values of the total heat flux and temperatures, thus improving winter thermal comfort and reducing energy uses in both summer and winter;

2. under stable conditions (in complete liquid or solid state), high T_{vis} values (about 90%) and low T_{vis} values (nearly 40%) can obtained for the liquid (above 30°C−35°C) and solid phases (below 20°C−24°C), respectively, although this is opposite to solar control principles in cooling periods as shown in Fig. 5.5A(a) and (b);
3. during the phase change the transmittance spectra of the PCM are unstable;
4. radiation scattering is dominant in the solid phase, while in the liquid phase mainly absorption takes place;
5. risks of overheating may be a significant factor after the PCM has melted, keeping higher glazing temperatures for a longer time compared to a conventional IGU.

As far as their applicability in various climates, Anisur et al. (2013) stated that PCMs integrated into buildings are more effective in climates (or periods of the year) with a large diurnal temperature swing, due to the possibility provided by the climate to discharge the PCM at night times, as shown by Osterman, Butala, and Stritih (2015) who observed that the maximum potential of the PCMs was in summer, when the difference in air temperature between day and night is large. Conversely, in warm−humid climates (such as Mediterranean coastal areas) the difficulty in discharging the PCM within daily cycles (due to smaller daily temperature variation) could make its implementation less effective (El Loubani, Ghaddar, Ghali, & Itani, 2021). In general, the melting temperature, the quantity and position within the glazing layers depend on local climate characteristics and building design. Despite a promising performance, additional integration issues exist: (1) change in appearance during phase change may limit PCM IGU adoption in glazing areas where view out is required; (2) paraffin presents a large volume change that may induce bowing of the glass panes, ultimately causing leakage; (3) salt hydrates are corrosive and present phase segregation. Different micro- and macroencapsulation methods have demonstrated to avoid segregation and volume change (ie, polymethyl methacrylate and polycarbonate). Manz, Egolf, Suter and Goetzberger (1997) devised a solar translucent wall composed of a transparent insulation material and a macroencapsulated translucent PCM. With a similar strategy, GlassX (2013) commercialized different variants of IGUs with PCM (GlassX Crystal) in which a glass prism element would redirect only low-angle (winter) solar radiation toward to macroencapsulated PCM behind, while the high-angle summer solar radiation would be reflected (to avoid overheating issues in summer) (Fig. 5.5B). More recently, a microencapsulation technique was developed in the framework of the (Harwin, 2020) project to embed a small quantity of PCM material into glass flakes within the transparent polyvinyl butyral interlayer of laminated glass (Scharfe et al., 2019). To separate the thermal storage functionality from the issues related to the quality of daylight and view out, different researchers (Silva, Vicente, Rodrigues, Samagaio, & Cardoso, 2015; Weinläder, Beck, & Fricke, 2005) and a company (ZAE Bayern, 2004) devised a PCM filled internal shading device (aluminium hollow blades, filled with PCM with 25°C−28°C melting temperatures). This PCM shading system (Fig. 5.5C), when tested in experimental tests in free floating environment, showed a significant blind temperature reduction (14°C−16°C), room peak temperature smoothing (up to 2°C−3°C) and shift (from 1:15 to 3 hours), if used in combination with night ventilation for complete discharging of the PCM.

Figure 5.6 Forms of silica aerogels: (A) monolithic, (B) granular and (C) powder.
Source: (A) Retrieved from Merli, F., Anderson, A. M., Carroll, M. K., & Buratti, C. (2018). Acoustic measurements on monolithic aerogel samples and application of the selected solutions to standard window systems. Applied Acoustics, 142, 123−131. https://doi.org/10.1016/j.apacoust.2018.08.008; (B) retrieved from Wei, G., Wang, L., Xu, C., Du, X., & Yang, Y. (2016). Thermal conductivity investigations of granular and powdered silica aerogels at different temperatures and pressures. Energy and Buildings, 118(March), 226−231. https://doi.org/10.1016/j.enbuild.2016.03.008; (C) retrieved and modified from Smirnova, I., & Gurikov, P. (2018). Aerogel production: Current status, research directions and future opportunities. Journal of Supercritical Fluids, 134(December 2017), 228−233. https://doi.org/10.1016/j.supflu.2017.12.037.

5.2.3 Aerogel glazing

Aerogel can be adopted as cavity insulation in multilayer glazing, to provide transparency and view out while presenting good insulation properties. Aerogel is a silica-based, porous material composed of about 4% silica and 96% air (Buratti and Moretti, 2011; Sun, 2017). The microscopic cells of the foam encapsulate air (or gas), creating a bluish haze due to the scattering at the silica−air/gas interfaces. Due to its open-pore microstructure (pores size 20−100 nm), it reduces the heat transfer by means of solid conduction (aerogel is mostly composed of insulating air or gas) and convection (the microstructure prevents net air/gas movement) (Paulos & Berardi, 2020; Sun, 2017). Three types of aerogel are currently used in the construction sector: monolithic, granular and powder (Fig. 5.6) (Merli, Anderson, Carroll, & Buratti, 2018; Wei, Wang, Xu, Du, & Yang, 2016; Smirnova and Gurikov, 2018); only the first two can be integrated in glazing, while the latter is mainly used to reduce thermal conductivity of ceramic and cementitious aggregates (Tsioulou, Erpelding, & Lampropoulos, 2016). According to Baetens, Jelle, and Gustavsen (2011), aerogel is usually used as filling material between the panes of multipane glazing units (Fig. 5.7) (Jelle, Baetens, & Gustavsen, 2015; Berardi, 2015b), as glazing material for skylights, for aerogel-based transparent coatings (Kim & Hyun, 2003) and for improving the thermal performance of window frames (Paulos & Berardi, 2020).

Cuce and Riffat (2015) reported that in double glazing with evacuated conditions, monolithic aerogel can achieve a light transmittance of 0.62 and a U-value of

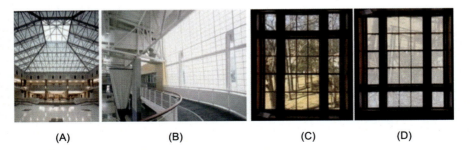

(A) (B) (C) (D)

Figure 5.7 Applications of aerogel insulation in (A) skylights, (B) external walls as a high-performance thermal insulation solution for daylighting, (C) conventional DGU without aerogel and (D) DGU with monolithic aerogel in the cavity. *DGU,* Double glazing unit.
Source: (A) and (B) Retrieved from Jelle, B. P., Baetens, R., & Gustavsen, A. (2015). Aerogel Insulation for building applications. The Sol−Gel Handbook, 3−3, 1385−1412. https://doi.org/10.1002/9783527670819.ch45; (C) and (D) retrieved and modified from Berardi, U. (2015b). The development of a monolithic aerogel glazed window for an energy retrofitting project. Applied Energy, 154, 603−615. https://doi.org/10.1016/j.apenergy.2015.05.059.

0.60 W/m²K, whereas Buratti and Moretti (2012a,b) showed that, compared to double low-e glazing, monolithic aerogel windows guarantee 55% reduction in heat losses with 25% less light transmittance. Granular aerogel has thermal and optical characteristics that are significantly affected by the size of the granules. Smaller particle size leads to a higher reduction in heat losses and light transmittance (Gao, Jelle, Ihara, & Gustavsen, 2014; Gao, Ihara, Grynning, Jelle, & Lien, 2016; Buratti, Merli, & Moretti, 2017; Buratti, Moretti, & Zinzi, 2017). Several experimental works on monolithic and granular aerogels (Buratti and Moretti, 2011, 2012a,b; Buratti, 2003) concluded that between the two types of aerogel, monolithic seems more promising because it has (1) better T_{vis} (above 0.62), (2) very low U-value (about 0.60 W/m²K in double evacuated glazing, with aerogel thickness of 14 mm) and (3) lower thickness and weight compared to conventional DGUs with comparable U-value. Due to its inherent translucent property, aerogel glazing can also reduce glare risk and improve visual comfort (Huang and Niu, 2015a,b; Garnier, Muneer, & McCauley, 2015), although views in/out may be impaired due to haze in monolithic aerogel and to light scattering for granular aerogel (Buratti and Moretti, 2012a,b).

Current research trends focus on the application of granular aerogel in other components of glazing systems (such as spacers and frames) or producing and integrating in IGUs monolithic aerogel of larger sizes. However, two main barriers, namely, manufacturing cost and poor mechanical properties (such as fragility issues when in tension), have so far prevented the widespread use of silica aerogel-based superinsulation components in buildings. A recent EU project investigated the use of monolithic silica aerogel as transparent insulation in windows (Jensen & Schultz, 2007), while ZAE Bayern is developing new glazing element based on granular silica aerogel (ISOTEG project, Reim et al., 2005).

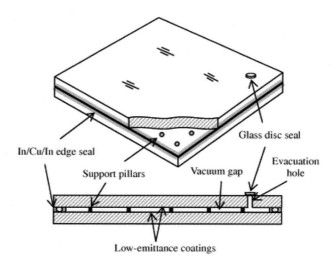

Figure 5.8 Schematic illustration of double evacuated glazing.
Source: Retrieved from Fang, Y., Hyde, T., Eames, P. C., & Hewitt, N. (2009). Theoretical and experimental analysis of the vacuum pressure in a vacuum glazing after extreme thermal cycling. Solar Energy, 83(9), 1723−1730. https://doi.org/10.1016/j.solener.2009.03.017.

5.2.4 Vacuum insulated glazing

Vacuum insulation glazing (VIG) is a highly insulating glazing technology with large potential to affect the energy use (heating, cooling and lighting), as well as the level of thermal comfort in buildings (Kocer, 2017; V-Glass LLC, 2018; Fang et al., 2007; Fang, Eames, & Norton, 2007). Negligible gaseous conduction and convection are achievable by evacuation of the glazing cavity (Cuce and Riffat, 2015; Manz, 2008). Cavity pressure below 0.1 Pa leads to a mean free path for interaction between molecules comparable to the thickness of the cavity (around 0.2 mm) and an equivalent cavity thermal conductance below 0.1 W/m^{2}K (Collins, Fischer-Cripps, & Tang, 1992; Ng, Collins, & So, 2005). The risk of glass deformation due to the pressure difference between the glazing cavity and the surrounding environment ambient is minimized by means of an array of tiny support pillars (typically of 0.25−0.50 mm diameter and 0.1−0.2 mm height) (Fig. 5.8) (Fang, Hyde, Eames, & Hewitt, 2009) of high strength materials (aluminium, stainless steel and Inconel alloy) at equal intervals (20−35 mm) or by using small spheres with radii of roughly 0.25 mm, housed on an indented glass layer, to avoid stress concentration in the glass (Garrison & Collins, 1995; Manz, 2008). Commercial VIGs present *U*-values (at the centre of the panel) between 0.5 and 0.9 W/m^{2}K and a light transmittance above 0.70 (Manz, Brunner, & Wullschleger, 2006 and Manz, 2008). The *U*-value is dependent on the number and size of pillars, array spacing, presence of low emissivity coating and level of vacuum in the cavity (Griffiths et al., 1998 and Kocer, 2019). Further to the integrated low-e coatings, combinations of VIG with other technologies, for example, EC glazing, gas-VIG hybrids and multiple VIG

Figure 5.9 (A) VIG retrofit of Hermitage Museum, Amsterdam; (B) VIG retrofit of Archibald Place, Edinburgh; and (C) Pilkington Spacia VIG windows offer the same thermal performance as conventional double glazing in one-quarter of the thickness and at two-thirds the weight. *VIG*, Vacuum insulation glazing.
Source: Retrieved and adapted from Traditional Building Magazine (2020). Vacuum insulated glass, innovation for historic restoration. Traditional Building Magazine, May 2020, Available at https://www.traditionalbuilding.com/product-report/vacuum-insulated-glass, Accessed date: 10.01.21.

glazing are under investigation (Kocer, 2019). Particularly the coupling of smart glazing features for dynamic solar and daylight control together with high level of insulation achieved by VIG appears to yield promising performance results (Fang and Eames, 2006; Papaefthimiou et al., 2006).

Despite its advantages, VIG presents different drawbacks that need to be resolved to improve a broader market adoption: (1) reduced long-term performance due to outgassing through the edge and from glass surface; (2) reduced performance due to thermal bridges of the pillars (such effect is increasing with the number and diameter of pillars); (3) possible condensation due to the local temperature depression close to the support pillars of the VIG (however, local condensation due to the small temperature nonuniformities would be preceded by significant condensation near the edges) (Collins and Simko, 1998); and (4) high manufacturing costs, due to oven-based manufacturing and solder glass edge seal technique (requiring processing temperatures above 450°C). On one side, metal-based sealing materials have been proposed to overcome the latter issue (Cuce & Riffat, 2015), which provide melting temperatures of around 200°C (Griffiths et al., 1998). On the other, the use of solder glass-free edge seals has been proposed and currently under development (Kocer, 2019). To reduce thermal bridge effects, innovative pillar designs are under research, including pillars made from thermally insulating and high strength materials (Collins, 2019). Finally, as a result of the VIG manufacturing processes, which are oven-based (450°C) hence inherently slow and energy-consuming, its costs are approximately four times higher than DGU (United States DOE, 2019). Therefore research and development interest is focusing in oven-free inline processes or low-temperature solder for sealing the vacuum cavity. Despite its cost, VIG is becoming an attractive option as a high-performance thermally insulating glazing, especially valuable in the retrofit market (due to its high performance with relatively low thicknesses) (Fig. 5.9) (Traditional Building Magazine, 2020).

5.2.5 Complex solar shading systems

In response to the high-performance requirements, such as energy saving, daylight admission/redirection, glare control and mitigation of overheating, light redirecting elements are being more and more included in fenestration systems and/or layer with scattering characteristics. The transmission properties of such systems idiosyncratically depend on the position of the sun and/or wavelength of the incoming radiation. To distinguish these fenestration systems from specular glazing types, they are often referred to as complex fenestration systems, these include prismatic films/structures and venetian blinds.

Prismatic structures are light-redirecting systems used to improve daylighting penetration. Prismatic elements are translucent, reduce the view to outside and generally are fixed systems. Therefore the panels are usually installed in the upper part of the façade opening, above the window area devoted to view out (Mueller, 2019). The miniaturization into microstructures allows to apply such technology as an adhesive film on the glazing surface (Grobe, 2020) which makes it an interesting solution for existing buildings. Several studies showed the efficacy of such systems in reducing the electric lighting consumption (McNeil, Lee, & Jonsson, 2017) and increasing the daylight availability compared to conventional glazing (Mashaly et al., 2017; Tian, Lei, & Jonsson, 2019). Nevertheless, an optimization of the prismatic structure is required to define the best configuration (Agnoli & Zinzi, 2013), mainly according to the site location and façade orientation (Fig. 5.10).

Venetian blinds systems are conventional devices used as solar shading, protection against glare and daylight redirection. They are usually made of a series of horizontal, vertical or oblique lamellas that can be placed inside, outside or between panes of glazing windows. Currently, there is a wide variety of blind systems, which make the use of sophisticated shapes or surfaces coatings (retroreflective coatings), to redirect the light and/or to retroreflect the unwanted light. The effect of retroreflection has been mainly developed focusing on the study of advanced lamella profiles or structures which led to the development of several market

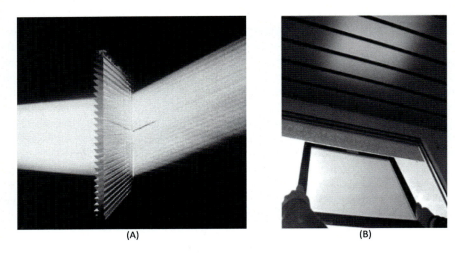

Figure 5.10 Light redirection by refraction in (A) a prismatic panel (LBNL, 2001) and in (B) a prismatic microstructure (Grobe, Wittkopf, & Kazanasmaz, 2017).

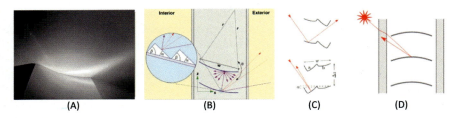

Figure 5.11 Light redirection and retroreflection by (A) and (B) prismatic microstructure on blinds (Schregle et al., 2015), (C) advanced shape blinds RetroLux Therm (Grobe et al., 2017) and (D) retroreflective coating applied on standard blind shape (Papaiz et al., 2020; Grobe, 2018).

products (Koester, 2004; Noback, Grobe, & Wittkopf, 2016; Schregle, Grobe, & Wittkopf, 2015). Nevertheless, the production process of such peculiar shape systems is complex because it is extremely dependent on the machinery setting and material used. Recently, an attempt to apply retroreflecting coatings on standard shape surfaces showed good results in terms of glare protection and daylighting supply compared to reflective metal blinds and addresses the production issues by reducing the process complexity (Papaiz et al., 2020) (Figs 5.11 and 5.12).

Despite the large use of blinds system, the increased technological development made the performances evaluation in building performance simulation (BPS) tools very complex. Standard simulation workflow may not always include sufficient information and provide the flexibility to treat complex optical cases. De Michele et al. (2018) demonstrated that employing simplified models can lead to strong underestimation in terms of daylighting and energy use for the case of high reflective lamellae. On the other hand, the dynamic nature of such systems requires

Standard blinds, Ev = 5903 lux, DGP = 0.56   Retro-reflective blinds, Ev = 1070 lux, DGP = 0.25

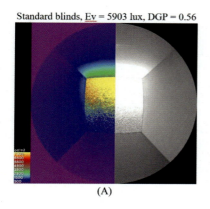
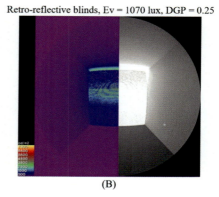

(A)   (B)

Figure 5.12 Comparison of glare control for retroreflective coating applied to standard venetian blinds. (A) Blinds with standard reflective metalized coating and (B) retroreflective coating applied to the same geometry (Grobe, Papaiz, & De Michele, 2016).

adequate control to optimize the functionality of the system itself; the thermal and optical performances of blinds systems are highly variable and depend on the sun position, slat angle and slat surface reflectance. Thus a correct use of this system is essential for obtaining the best performance, and the autonomous control of the system should be preferred for increasing energy saving and daylight availability, as several studies showed (Cuce, 2016; Tzempelikos & Athienitis, 2005; De Michele, Avesani, Belleri, & Gasparella, 2018).

5.2.6 Advanced transparent technologies for solar energy conversion

At component level the main technology on the market coupling transparency and solar energy conversion consists of photovoltaic (PV) technologies. Power generation and semitransparency (ie, light transmission and hence view out and daylighting) are inversely proportional and very customizable depending on the PV technology. As reported by Skandalos and Karamanis (2015), main PV cell technologies for the integration in fenestration systems can be categorized as (1) 'traditional', based on crystalline silicon (c-Si) solar cell and on thin-film PV (amorphous silicon, copper indium gallium selenide and cadmium telluride) and (2) emerging technologies, that is, dye-sensitized solar cells (DSSC) organic solar cells and perovskite. In the first category the PV layer is integrated in the glazing unit, and semitransparency is obtained by removing part of this layer. In the case of thin-film PV the film thickness obtainable can be quite small, achieving a uniform customizable semitransparency, compared to the c-Si PV, where the PV cells' minimum sizes are determined by the silicon wafer. Recent developments focus on a see-thru back-contact solar cell, able to improve aesthetics and design flexibility, while keeping high performances (Bonato, Fedrizzi, D'Antoni, & Meir, 2019). Concerning the emerging technologies, detailed technological reviews can be found

Figure 5.13 Example of semitransparent BIPV technologies. Edited by the authors from the picture in Vasiliev et al. (2019). (A) Onyx Solar a-Si high transparency BIPV panels (http://www.onyxsolar.com/product-services/technical-specifications, accessed on 02.10.2020); (B) Hanergy BIPV panels using a-Si (Hanergy Product Manual 141129 Section 1.3.1 (2018). p. 11. Available online: http://www.slideshare.net/RonaldKranenberg/hanergyproductbrochure141129 (accessed on 02.10.2020); and (C) high-transparency CdTe BIPV panels (Barman, Chowdhury, Mathur, & Mathur, 2018). *BiPV*, Building-integrated photovoltaic.

in Vasiliev, Nur-E-Alam, and Alameh (2019), Tiwari, Tiwari, Carter, Scott, and Yakhmi (2019) and Skandalos and Karamanis (2015). The DSSC PV technology has attracted much interest also for its relatively high efficiency especially under weak illumination, such that a tinted clear power-generating glazing assembly can be obtained. This transparent organic PV is based on organic small molecules or polymers from earth-abundant elements, so that a small ecological footprint could be achieved with the shortest energy payback time possible. Finally, perovskite cells have notable performances and some cells' compositions do not absorb light in the visible spectrum, becoming hence suitable as semitransparent solar generation material (Tiwari et al., 2019).

Integrating the PV cells in the glazing spacer is another solution. A first technological approach is based on the so-called luminescent solar concentrators, which exploit the redirectional properties of suspended particles integrated in the glass, driving the solar radiation to the spacer-integrated conventional strips of PV cells. A second solution exploits spacer-integrated tilted PV cells [eg, example of commercial products reported in Vasiliev et al. (2019) and in Bonato et al. (2019)]. In this configuration the glazing unit remains of the desired colour and transparency. The glazing spacer surface toward the cavity is slightly tilted to host the PV cells and related electronics. Both solutions though present a power generation potential two to three times lowered compared to semitransparent building-integrated photovoltaics (BIPVs). Performance overviews of the set of PV technologies suitable for fenestration systems are given in Vasiliev et al. (2019), Tiwari et al. (2019), Skandalos and Karamanis (2015) and in Cuce (2016). Examples and related performances are reported in Fig. 5.13.

Table 5.1 Characteristics, limitations and research trends for advanced fenestration components.

Technology	Main features	Limitations	Future research avenues
Smart glazing	• Capability to modulate the admission of daylight and solar gains to interior spaces • Particular attention should be paid to the control in synergy with performance objectives and HVAC and lighting systems	• Additional systems for glare control may be necessary • Some technologies have undesired colouration effects	• NIR EC and NIR TC smart glazing • Independently tunable NIR and IR smart glazing • Tuning of reflection rather than absorption to modulate optical transmission • Lower manufacturing • Costs
PCM	• PCM glazing store solar energy • Smooth and delay peak temperatures • Melting temperature and position of the PCM in the buildup should be designed to match local boundary conditions and purpose	• Inconsistency of liquid/solid phase optical properties with requirements • Risk of overheating • May limit view out • Paraffin leakage possible • Salt hydrates (corrosive) might impede phase segregation	• PCM filled internal shading device in combination with natural ventilation • Encapsulation to avoid segregation (micro- and macroencapsulation)
Aerogel	• High thermal and low weight insulator integrated in the cavity • High-quality diffuse light • Reduction of glare effect • Great solar thermal transmittance (benefit for heating-dominated climates)	• Views in/out may be impaired (diffusing behaviour or haze) • High cost • Poor mechanical properties of monolithic • Energy-intensive manufacturing processes	• Integrate additional functionalities (acoustic insulation, glare reduction) • Future applications in other components of glazing systems (spacers, frames) • Reduce energy intensity of production • Minimize the potential hazards of their use and disposal

(Continued)

Table 5.1 (Continued)

Technology	Main features	Limitations	Future research avenues
		• HSE risks (large surface area and porosity, chemicals for hydrophobicity)	
VIG	• Low heat loss, high visible transmittance and slim/low-weight • Indoor glass pane closer to indoor temperature • Particularly promising for existing buildings • Promising combination with switchable glazing	• Reduced long-term performance • High manufacturing costs • Possible outgassing through edge and from the glass surface	• Development of oven-free inline processes or low-temperature solder (decreased cost) • Novel pillars geometries, materials and glass indentation techniques • Flexible and solder-free edge seals
Complex solar shading systems	• Systems that redirect the light improving daylighting penetration • Protection against glare	• Reduce the view to the outside • Production process of peculiar shape systems is complex • Performances evaluation in BPS tools is very complex	• Dynamic nature of such systems with adequate control to optimize the functionality of the system and autonomous control
Semitransparent PVs	• Coupling transparency and solar energy conversion • Trade-off between transparency and solar energy conversion	• Realization of a cost-effective energy-generating clear glazing with a custom-made size • Durability of organic ones • High efficiency to cost ratio	• DSSC could have high potential due to high efficiency under weak illumination • Solar concentrators to overcome reduced power generation • Deposition on flexible transparent substrates for glass lamination

BPS, Building performance simulation; *DSSC*, dye-sensitized solar cells; *EC*, electrochromic; *HSE*, Health safety and environment; *HVAC*, heating ventilation and air conditioning; *NIR*, near infrared; *PCM*, phase-changing material; *PV*, photovoltaic; *TC*, thermochromic; *VIG*, vacuum insulation glazing.

Electricity generator in AFSs is particularly useful to power low energy demanding devices (such as movable solar shadings, sensors and control hubs), in an 'Internet of Electricity' vision (Schellnhuber et al., 2018). Nevertheless, for highly glazed envelopes, the energy generation from fenestration systems could also contribute significantly on the overall building final energy consumption, as reported by studies such as Kapsis and Athienitis (2015) and Li et al. (2009), thanks to both the energy generation itself and the g-value reduction. Current main technical challenge is related to the realization of a cost-effective energy-generating clear glazing with a custom-made size. In particular, colour rendering, transparency and power generation density are the main success parameters. Finally, nontechnical issues include final cost and the stakeholders' acceptance as well as regulation compliance (Vasiliev et al., 2019).

5.3 Advanced integrated fenestration systems

5.3.1 From double-skin façade to advanced integrated fenestration systems

Double-skin facades (DSF) can be considered an evolution of the fenestration for highly glazed high-rise tertiary buildings. In DSF the façade is separated in two glazed layers (the 'skins'), enclosing an air cavity, generally used for hosting and protecting the shading system and for airflow control. Definitions (Streicher, 2007), history, categories (as shown in Fig. 5.14) and main features (Barbosa et al., 2014;

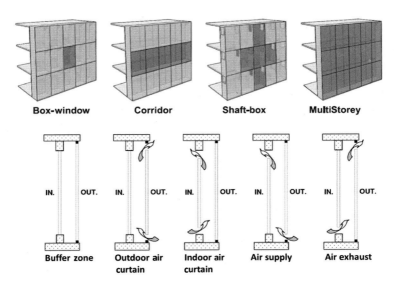

Figure 5.14 DSF categories based on the kind of geometry (top) and based on the airflow (bottom) (Alberto et al., 2017). *DSF*, Double-skin façade.

Shameri et al., 2011; Saelens, 2002) are well documented in literature as well as the discussion about performances and main design parameters relation, as in Alberto, Ramos, and Almeida (2017) and Kim, Cox, Cho, and Yoon (2018).

Given the complexity of the system, tailor-made detailed design must be assisted by dedicated multidisciplinary calculations (Lucchino, Goia, Lobaccaro, & Chaudhary, 2019) to secure the expected performances during operation. A conclusive judgment comparing DSFs to high-performing single-skin facades is far from being found, and it is probably not possible in absolute terms (Lucchino et al., 2019). In fact, main discussions behind the technological choice of the DSF generally flow around the following points: (1) need to protect the shading system from outside weather conditions and reducing its maintenance; (2) exploit the potential of air cavity heating for sunny-cold climates; (3) possibility to increase the façade functionality with integrated components; and (4) budget availability given the hardly justifiable cost—benefit ratio. Nevertheless, during the last decades, DSF has also been developing toward being a design canvas, integrating different technologies (advanced solar shading, solar energy conversion devices, smart glazing, decentralized ventilation system, solar latent thermal energy storage systems and so on), with the aim to achieve dynamic and multifunctional façade for comfortable and zero energy buildings. Advanced integrated façade (AIF) or fenestration systems represent hence the transformation from the DSF to a more complex integrated system interacting systemically with the rest of the building (Goia, Perino, Serra, & Zanghirella, 2010). In this perspective the ability to integrate different technologies and strategies into an AIF elevates the transparent façade system from being a problem to solve, to represent an opportunity to exploit (Favoino, Perino, & Serra, 2019). This can be achieved by means of air flow control, integration of solar energy conversion technologies and water flow integration, as reviewed in the next sections.

Nevertheless, when a maximum transparent appearance has a lower architectural priority, including an opaque area within the AIF system could enable to achieve very high energy and comfort performances, while promoting RES integration and energy independence for the façade operations at the same time. This is achieved by means of plug-and-play façade systems, such as the ACTRESS AIF façade concept (Favoino, Goia, Perino, & Serra, 2016) and the 'solar window block' (Andaloro, Avesani, Belleri, & Machado, 2018). These are prefabricated and energetically autonomous fenestration system, which can provide a high level of IEQ and energy saving, by balancing a transparent part devoted to view out, daylight and solar control with an opaque one embedding high performance insulation, PV panels, electrical and thermal storage for autonomous operations, decentralized ventilation units and control system (Andaloro, Minguez, & Avesani, 2020; Minguez, Gubert, Astigarraga, & Avesani, 2020) (Fig. 5.15).

5.3.2 Airflow control in double-skin façade

The façade air cavity flow is one of the key elements to be considered for securing the expected performances of DSF. Examples in Goia et al. (2010) as prototypes

Figure 5.15 Functional mockup of the solar window block with three PV systems different configurations (PV aside—left, in the sill—centre and overhang—right) (Minguez et al., 2020). *PV*, Photovoltaic.

and in Aelenei et al. (2018) as real facades are described in literature, differing in terms of air flow path, type of ventilation and air velocity (natural and mechanical), depth of air cavity, specification of IGUs enclosing the air cavity and type of shading devices integrated. The main aim of DSFs is to achieve a good degree of management of solar loads while maintaining a good level of indoor thermal comfort, daylight and reducing glare risk. Those targets are achieved managing solar shadings as well as airflow in the DSF cavity by means of dynamically controlling the *g*-value and air-cavity-stored-heat toward the interior or the exterior environment. The high cost and complexity of typical systems, either naturally or mechanical-based ventilation, envisage the used low-tech approach, exploiting autoreactive materials to adapt the DSF performance. In Molter et al. (2017), an example of such concept has been developed based on paraffin-filled thermal cylinders able to expand and open the DSF by pushing outward the outer pane (up to 5 cm) when the temperature rises above a fixed value. When the temperature drops, the cylinders contract, closing the cavity, with a reaction time of only few minutes. Energy savings of 40%–50% in winter and summer can be achieved compared to conventional single-skin façade, but without electronic controls as compare to conventional DSF and no optical difference (Molter et al., 2017). An opposite DSF concept is the one implemented in the so-called closed cavity facades (CCF) unitized system (Zani, Galante, & Rammig, 2020), where the air cavity hosting the shading device (typically between 150 and 250 mm) is sealed between the IGU layers, without the need to doubling up the framing/support system of the two parallel IGU layers. CCF-unitized systems present different functional and operational advantages as compared to DSFs, such as preventing accumulation and settlement of dust and particles in the cavity, increasing the service life of components inside the cavity, reducing maintenance cost, complexity of the airflow control. Although to prevent huge glass climatic loads and condensation risk, due to the thick sealed air cavity, an air-based conditioning system (passive with outside air through a desiccant filled pipe or

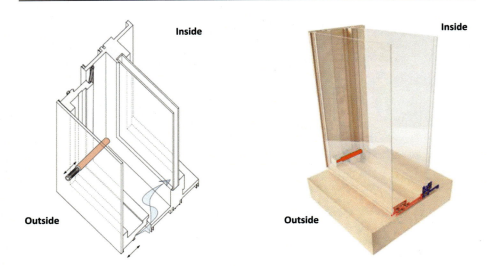

Figure 5.16 Autoreactive façade concept (Molter, Robustheit, & Autoreaktivität, 2019).

active with a mechanically dried air) admits air in the sealed cavity to control cavity pressure (Fig. 5.16).

5.3.3 Solar energy conversion in double-skin façade

Beside the PV fenestration technologies presented in Section 2.6, DSF air cavity itself can be exploited to integrate solar energy conversion technologies within the blind system of the DSF, such as integrating solar concentrator/tracking devices for electricity generation or solar thermal blinds (STB). The Wellsun−Lumiduct concept provides at once energy generation, solar shading, view to the outside, media wall capability functions. The main core component is an optical planar receiver (of around 4×4 cm sizes optic), concentrating beam radiation on a tiny ultrahigh-efficiency solar PV cell while letting diffuse radiation passing through (Morgan, Myrskog, Barnes, Sinclair, & Morris, 2014−2015). A set of receivers composes a module, while a set of those is then mounted on a sun tracking mechanism, resulting in a translucent energy generating layer that can be installed in the DSF cavity. Components and whole system are represented in Fig. 5.14. Simulation results (Saini, Loonen, & Hensen, 2019) show that Lumiduct can lower the cooling load due to unwanted solar gains on sunny days, without reducing the daylight availability on overcast sky conditions. The resulting heating demand is higher compared to the high-performance window without blinds, although electricity generation efficiency of 29% can be reached in a translucent configuration, which is much higher than conventional BIPV solutions. Main weaknesses to be faced are the upfront investment and the compromised view to the outside (Bonato et al., 2019).

STB as described in Denz, Vongsingha, Häringer, and Maurer (2018), Haeringer, Denz, and Vongsingha (2019); and Bonato et al. (2019) use the high

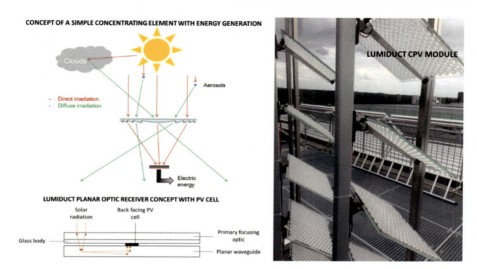

Figure 5.17 At the top left a representation of a simple concentrating system. Direct illumination (red) is concentrated to the PV cell and converted into electrical energy, while diffuse illumination (green) is not concentrated and passes by the system (Bunthof et al., 2016). Bottom left, a simplified scheme of the Lumiduct planar receiver concept [sketched by the authors, based on the description in Saini et al. (2019)]. Right part of the figure shows a set of Lumiduct modules composed by many receiver mounted in the sun tracking (right) resulting in a translucent DSF system. *DSF*, Double-skin façade; *PV*, photovoltaic.
Source: From Roel Loonen.

temperature generated in the DSF cavity in clear sky conditions as solar thermal application. The STB relies on the use of heat pipes, as heat collector, coupled to a tinted metal slat. Slats are tiltable, resulting in a solar capturing and movable shading system. The heat transfer is assured by the connection to a header pipe, located in a dedicated box aside (Fig. 5.15). A demonstrator module has been built (Haeringer et al., 2019) and a first testing campaign shows effective reduction of g-value. Nevertheless, the solar thermal efficiency is highly dependent on the slat tilt angle and the system is cost- and planning-intensive, also because of the number of components of the mechanical system. However, the presence of a design platform leading through the different design options as well as the multibenefit resulting from the adoption of such system in a DSF are keys factors to ease the market penetration of STB (Figs 5.17 and 5.18).

5.3.4 Fluid-flow facades

Adopting a fluid with higher thermal capacity then air, and in which other components could be dispersed, such as water, could disclose additional functionalities as compared to conventional DSF. In fluid-flow glazing the gas enclosed in between

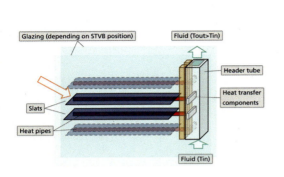

Figure 5.18 Scheme and prototype of the STB main components and solar heat harvesting (Haeringer et al., 2020). *STB*, Solar thermal blinds.

glazed panes is replaced by a circulating fluid that partially absorbs solar heat, limiting the solar transmission and offering the possibility to reuse such heat. Moreover, such layer can be used as a heating or cooling emitter (Heiz, Pan, Su, Le, & Wondraczek, 2018; Heiz, Su, Pan, & Wondraczek, 2018; Li, Lyu, Li, & Qiu, 2020) when used as most internal layer. The visual properties of the glazing system are kept almost unaltered thanks to the highly laminar fluid flow, and the fluid layer becomes as an adaptive coating. Nevertheless, such adaptability requires auxiliary hydraulic equipment, such as a microcirculator and valves, and a heat exchanger, which are normally integrated in an enlarged frame and purposely designer spacers (ie, a transom). The coupling between the direct (absorbed heat) and indirect benefit (usable heat) shows interesting performance figures as calculated by Gutai and Kheybari (2020). A glazing assembly with vacuum- and liquid-glazing proposed by Lyu, Liu, Su, and Wu (2019) could achieve even further energy savings (compared to a nonvacuum glazing). Moreover, the integration of decentralized energy generation devices (eg, a heat pump) is proposed by Su, Fraaß, Kloas, and Wondraczek (2019). Finally, the visual comfort, specifically glare risk, is not controllable by means of this technology, unless coloured metal powder particles are dispersed in the liquid, concentration of which could be controlled by means of a magnetic field, as per the demonstrated concept of the EU Fluid glass project (Baumgärtner, Rodrigo, Stopper, & Von Grabe, 2017). In this perspective the use of switchable, ultrathin suspended particle device has recently been studied, showing the solar shading capability, besides the solar harvesting (Heiz, Pan et al., 2018; Heiz, Su et al., 2018) (Fig. 5.19) (Nikolaeva-Dimitrova, Stoyanova, Ivanov, Tchonkova, & Stoykov, 2018).

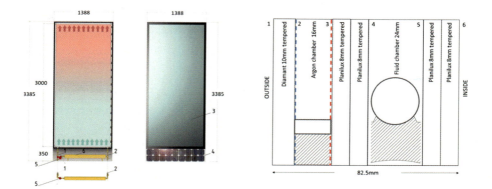

Figure 5.19 Left: scheme of an FFG façade module as in the H2020 Indewag project. (1) Supply pipe; (2) return pipe; (3) glass; (4) PV cells; (5) solar water pump; (6) heat exchanger (Del Ama Gonzalo, Hernandez Ramos, & Moreno, 2017). Right: assembling of an FFG as proposed by the EU H2020 Project 'Indewag'. In red and in blue the coatings position in the two different configurations, respectively, HEATglass and of the COOLglass. *FFG*, Fluid-flow glazing; *PV*, photovoltaic.
Source: Edited by the authors from data in Nikolaeva-Dimitrova, M., Stoyanova, M., Ivanov, P., Tchonkova, K., & Stoykov, R. (2018). Investigation of thermal behaviour of innovative water flow glazing modular unit. Bulgarian Chemical Communications, 50, 21–27.

5.4 Advanced fenestrations performance evaluation to support design and operation decision-making

The recent research and design efforts in AFSs turned the spotlight on the building occupant, trying to overcome the effectiveness and cost limits of the 'energy conservation approach' (Perino & Serra, 2015), which was a necessary initial step toward space heating and cooling reductions, but going beyond this to ensure high building performance across a wide range of physical domains. AFSs in fact offer the possibility to significantly improve the properties and performance of the transparent part of the building envelope, which can be designed and conceived as multi-domain 'selective filter' rather than an 'exclusive barrier' (IEA-ECBCS, 2007). To do so a balance needs to be found between competing performance aspects, by efficiently deploying advanced transparent systems and advanced integrated control strategies, in a synergic and coordinated way with passive design strategies and with building services design and operation. Nevertheless, this is hardly achievable by just ensuring compliance to building regulations and by comparing the different AFS solutions by means of parameters such as U-values and g-value. Moreover, recent studies have highlighted how conventional performance metrics are hardly useful to describe in a comprehensive way the performance of AFS when building integrated (Bianco et al., 2018), and particularly of those involving adaptive elements (ie, smart glazing, PCMs and DSF) (Attia et al., 2018). In fact, these conventional metrics 'just' provide a measure of the fenestration system properties rather than providing a representation of the competing aspects of building performance, while the AFS thermo-physical behaviour is too far from the assumptions under which such metrics can be calculated (static boundary conditions).

For the abovementioned reasons, to reconcile competing performance trade-offs (ie, between energy use, occupant comfort, material efficiency and costs), more comprehensive building performance indicators are particularly needed for AFS, calculated in a representative time frame and compared to appropriate local benchmarks (Ballarini, Corgnati, & Corrado, 2014), so to support multicriteria decision-making. On the other hand, the introduction, the validation and adoption of dynamic AFS performance indicators in representative operational scenarios (Bianco et al., 2018) could support a direct performance-based comparison with different building envelope technologies, where needed. Such informed decision-making requires multidomain and accurate evaluations, describing the multifunctionality of AFS and their dynamic interaction across different physical domains (heat, moisture, airflow, daylight and acoustic); the assessment of AFS control strategies and synergies with other building systems (heating ventilation and air conditioning and lighting) (Loonen, Favoino, Hensen, & Overend, 2017); and the interdependent relationship with building occupants (Luna-Navarro et al., 2020). In this direction, it is of course important also to provide accurate performance prediction of emerging AFS technologies and materials, including adaptive materials and systems (Favoino et al., 2018; Favoino & Overend, 2015), by enlarging modelling possibility for technologies, models of which may not be directly available into specific BPS tools (Loonen et al., 2017). Particularly cosimulation and data-integration could enhance the achievement of these BPS requirements (Taveres-Cachat, Favoino, Loonen, & Goia, 2021). In this direction, different performance-based approaches and examples for AFS are currently available (Favoino et al., 2016; Giovannini et al., 2019; Loonen, 2018), and different tools and research efforts exist to provide user friendliness and seamlessly integration (Wetter & Van Treeck, 2017).

Finally, there is a strong need to broaden the analyses focus to the whole life cycle (LC) and to all the environmentally relevant key performance indicators (not only the embodied carbon), as identified in the LCA (LC analysis) methodology. The vision is toward the decarbonization of the building sector and the façade systems are key technology to support reaching the zero-carbon building targets. This paradigm shift (from energy- to carbon-zero) is necessary, provided the current possibility to achieve very low energy consumptions during operation (for both new and renovated buildings), so that the relevance of all other environmental impacts (along the LC) increases. As an example, embodied carbon was calculated as four times the operational one for the LC of a DSF in hot and arid climate as Iran (Zomorodian and Tahsildoost, 2018). This is confirmed by the findings of Pomponi and D'Amico (2017) for a curtain wall in the United Kingdom. Nevertheless, as reported by Roberts, Allen, and Coley (2020) many different reviews underline that LCA is typically separate from the design process, rather than being an integral part. Practical current challenges to apply an LC-driven façade system design are (1) building information modelling LC-rich information, (2) use of a parametric approach coherent with regionalization, (3) use of LC costing together with LCA and (4) availability of accepted benchmarks.

Table 5.2 Characteristics, limitations and research trends for advanced fenestration systems.

Technology	Main features	Major limitations	Main future research avenues
DSF (airflow)	• Separated in two glazed layers, enclosing an air cavity • Integrates various technologies aiming to achieve a dynamic multifunctional façade • Good degree of management of solar loads while maintaining a good level of indoor thermal comfort, daylight and reducing glare risk	• High cost • Complex systems • Air is difficult to be properly controlled	• Closed cavity façade shows future potential, as they minimize complexity, durability issue and maintenance costs • Management of air flow in a dynamic way, the challenge is to keep complexity low
DSF (solar conversion)	• Integration of solar concentrator within the blind system of the DSF	• High cost • Complex mechanical system • Large cavity might be needed • Compromised outside view • Need for controlling cavity temperature	• STB show significant reduction of g-value which might form the basis of future designs • DSF air cavity is the favourable architectural location to be exploited for the energy conversion, under a multiphysic perspective
Fluid flow façade	• Fluid layer collects solar radiation preventing direct transmission • Auxiliary hydraulic equipment is required	• Glare is not controllable (additional systems required) • High cost • Complex mechanical system	• Coupling with solar conversion systems could be promising (BIPV and BIST) • Increasing reliability and robustness is of high priority

BiPV, Building-integrated photovoltaic; *BIST*, building-integrated solar thermal; *DSF*, double-skin façade; *STB*, solar thermal blinds.

5.5 Conclusion

This chapter presented an overview of and perspective on the development of AFS, from being the weak point in terms of performance of the building envelope, to a high-performing building element integrating different materials and technologies, providing opportunities to achieve multiple desirable features and requirements. At component level, for all the reviewed technologies, research efforts are going in the direction of achieving an improved level of functionality and performance (mainly improving thermal resistance and improving control over amount and direction of solar radiation, either shortwave, long-wave or both) in a material-, energy- and cost-effective way. At a system level the added value and performance obtained by this high level of technological integration need to be balanced with the embedded complexity in the manufacture, design and operations of such AFSs.

The first common consequence of the spreading of AFSs is the need to expand the disciplinary domain, detailing the impacts on the indoor environment in relation to all the other building systems. AFSs finally allow a dynamic and context-oriented control of aspects such as daylight, glare, heat transfer and storage, besides the possibility to increase on-site solar energy conversion. Nevertheless, the high level of complexity, multifunctionality and integration possibilities of AFS need to be reflected by the methods adopted in design, evaluating their performance, by shaping the physical and functional integration with the building, to work in a synergic way with the other building systems and the building strategies adopted toward zero-carbon building targets. This can bring about an increased performance but may represent an increased effort in the design, construction and operation of these elements. Therefore it is important to consider the whole added value of the integration of such components (in terms of energy use, occupant comfort and whole life carbon), as by doing so the transparent façade could enable to disclose a great potential in terms of improved environmental quality (hence value of the built space), and it can enable to find more effective ways to optimize the design, manufacturing and construction processes within a circular economy perspective.

References

Aelenei, L., Aelenei, D., Romano, R., Mazzucchelli, E. S., Brzezicki, M., & Rico-Martinez, J. M. (Eds.), (2018). *Booklet 3.1 'Case studies – Adaptive façade network'*. TU Delft Open, ISBN 978-94-6366-110-2.

Agnoli, S. & Zinzi, M. (2013). *Involucro trasparente ed efficienza energetica*.

Alberto, A., Ramos, N. M. M., & Almeida, R. M. S. F. (2017). Parametric study of double-skin facades performance in mild climate countries. *Journal of Building Engineering*, 12 (May), 87–98. Available from https://doi.org/10.1016/j.jobe.2017.05.013.

Andaloro, A., Avesani, S., Belleri, A. & Machado, M. (2018). Adaptive window block for residential use: Optimization of energy efficiency and user's comfort, In

FAÇADE2018 adaptive! *Adaptive façade network conference*, Luzern, Switzerland, 26−27 November.

Andaloro, A., Minguez, L. & Avesani, S. (2020). Active and energy autonomous window: Renewable energy integration through design for manufacturing and assembly, In *Façade tectonics 2020 world congress*, Los Angeles, United States, 5−27 August.

Anisur, M. R., Mahfuz, M. H., Kibria, M. A., Saidur, R., Metselaar, I. H. S. C., & Mahlia, T. M. I. (2013). Curbing global warming with phase change materials for energy storage. *Renewable and Sustainable Energy Reviews, 18*, 23−30. Available from https://doi.org/10.1016/j.rser.2012.10.014Attia.

Attia, S., Bilir, S., Safy, T., Struck, C., Loonen, R., & Goia, F. (2018). Current trends and future challenges in the performance assessment of adaptive façade systems. *Energy and Buildings, 179*, 165−182.

Baetens, R., Jelle, B. P., & Gustavsen, A. (2010). Properties, requirements and possibilities of smart windows for dynamic daylight and solar energy control in buildings: A state-of-the-art review. *Solar Energy Materials and Solar Cells, 94*(2), 87−105.

Baetens, R., Jelle, B. P., & Gustavsen, A. (2011). Aerogel insulation for building applications: A state-of-the-art review. *Energy and Buildings, 43*(4), 761−769. Available from https://doi.org/10.1016/j.enbuild.2010.12.012.

Ballarini, I., Corgnati, S. P., & Corrado, V. (2014). Use of reference buildings to assess the energy saving potentials of the residential building stock: The experience of TABULA project. *Energy Policy, 68*, 273−284. Available from https://doi.org/10.1016/j.enpol.2014.01.027.

Barbosa, S., & Ip, K. (2014). Perspectives of double skin facades for naturally ventilated buildings: A review. *Renewable and Sustainable Energy Reviews, 40*, 1019−1029.

Barman, S., Chowdhury, A., Mathur, S., & Mathur, J. (2018). Assessment of the efficiency of window integrated CdTe based semi-transparent photovoltaic module. *Sustainable Cities and Society, 37*(September 2017), 250−262. Available from https://doi.org/10.1016/j.scs.2017.09.036.

Baumgärtner, L., Rodrigo, A., Stopper, J., & Von Grabe, J. (2017). Evaluation of a solar thermal glass façade with adjustable transparency in cold and hot climates. *Energy Procedia, 122*, 211−216. Available from https://doi.org/10.1016/j.egypro.2017.07.347.

Bayern Z. A. E. (2004). Tätigkeitsbericht 2004, *Annual report 2004*. ZAE Bayern.

Berardi, U. (2015b). The development of a monolithic aerogel glazed window for an energy retrofitting project. *Applied Energy, 154*, 603−615. Available from https://doi.org/10.1016/j.apenergy.2015.05.059.

Bianco, L., Cascone, Y., Avesani, S., Vullo, P., Bejat, T., Loonen, R., . . . Favoino, F. (2018). Towards new metrics for the characterisation of the dynamic performance of adaptive façade systems. *Journal of Façade Design and Engineering, 6*(3), 175−196. Available from https://doi.org/10.7480/jfde.2018.3.2564.

Bonato, P., Fedrizzi, R., D'Antoni, M., & Meir, M. (2019). State-of-the-art and SWOT analysis of building integrated solar envelope systems. *IEA EBC Annex, 58*. Available from https://doi.org/10.18777/ieashc-task56−2019-0001.

Bunthof, L. A. A., Kreuwel, F. P. M., Kaldenhoven, A., Kin, S., Corbeek, W. H. M., Bauhuis, G. J., & Schermer, J. J. (2016). Impact of shading on a flat CPV system for façade integration. *Solar Energy, 140*, 162−170. Available from https://doi.org/10.1016/j.solener.2016.11.001.

Buratti, C. (2003). Transparent insulating materials: Experimental data and buildings energy saving evaluation. *Sustainable World, 7*, 231−240.

Buratti, C., & Moretti, E. (2011). Transparent insulating materials for buildings energy saving: Experimental results and performance evaluation. In *Proceedings of the third international conference on applied energy*, May (pp. 1421−1432).
Buratti, C., & Moretti, E. (2012a). Experimental performance evaluation of aerogel glazing systems. *Applied Energy*, *97*, 430−437. Available from https://doi.org/10.1016/j.apenergy.2011.12.055.
Buratti, C., & Moretti, E. (2012b). Glazing systems with silica aerogel for energy savings in buildings. *Applied Energy*, *98*, 396−403. Available from https://doi.org/10.1016/j.apenergy.2012.03.062.
Buratti, C., Merli, F., & Moretti, E. (2017). Aerogel-based materials for building applications: Influence of granule size on thermal and acoustic performance. *Energy and Buildings*, *152*, 472−482. Available from https://doi.org/10.1016/j.enbuild.2017.07.071.
Buratti, C., Moretti, E., & Zinzi, M. (2017). High energy-efficient windows with silica aerogel for building refurbishment: Experimental characterization and preliminary simulations in different climate conditions. *Buildings*, *7*(1). Available from https://doi.org/10.3390/buildings7010008.
Chapa, J. (2019). *Bringing embodied carbon upfront, 2019*. <https://www.worldgbc.org/news-media/WorldGBC-embodied-carbon-report-published>.
Collins, R. (2019). *Vacuum insulating glass − Past, present and prognosis*. Glaston: Glass Performance Days.
Collins, R. E., & Simko, T. M. (1998). Current status of the science and technology of vacuum glazing. *Solar Energy*, *62*(3), 189−213. Available from https://doi.org/10.1016/S0038-092X(98)00007-3.
Collins, R. E., Fischer-Cripps, A. C., & Tang, J. Z. (1992). Transparent evacuated insulation. *Solar Energy*, *49*(5), 333−350. Available from https://doi.org/10.1016/0038-092X(92)90106-K.
Cuce, E. (2016). Toward multi-functional PV glazing technologies in low/zero carbon buildings: Heat insulation solar glass − Latest developments and future prospects. *Renewable and Sustainable Energy Reviews*, *60*, 1286−1301. Available from https://doi.org/10.1016/j.rser.2016.03.009.
Cuce, E., & Riffat, S. B. (2015). A state-of-the-art review on innovative glazing technologies. *Renewable and Sustainable Energy Reviews*, *41*, 695−714. Available from https://doi.org/10.1016/j.rser.2014.08.084.
De Michele, G., Avesani, S., Belleri, A., & Gasparella, A. (2018). Advanced shading control strategy for shopping malls: A case study in Spain. *BSO2018 papers, September*, pp. 11−12.
De Michele, G., Loonen, R., et al. (2018). Opportunities and challenges for performance prediction of dynamic complex fenestration systems (CFS). *Journal of Façade Design and Engineering*, *6*(3), 101−115.
Del Ama Gonzalo, F., Hernandez Ramos, J. A., & Moreno, B. (2017). Thermal simulation of a zero energy glazed pavilion in sofia, Bulgaria. New strategies for energy management by means of water flow glazing. *IOP Conference Series: Materials Science and Engineering*, *245*(4). Available from https://doi.org/10.1088/1757-899X/245/4/042011.
Denz, P., Vongsingha, P., Häringer, S. F., & Maurer, C. (2018). Solar thermal energy from opaque and semi-transparent facades-current results from R & D project ArKol Solar thermal energy from opaque and semi-transparent facades − Current results from R & D project ArKol. October.
Duraković, B. (2020). *PCM-based glazing systems and components*. PCM-based building envelope systems (pp. 89−119). *Cham: Springer*.

E. Commission. (2010). Directive 2010/31/EU (2010) of the European parliament and of the Council of 19 May 2010 on the energy performance of buildings (recast). *Official Journal of the European Union*, 13−35.

El Loubani, M., Ghaddar, N., Ghali, K., & Itani, M. (2021). Hybrid cooling system integrating PCM-desiccant dehumidification and personal evaporative cooling for hot and humid climates. *Journal of Building Engineering*, *33*(March 2020), 101580. Available from https://doi.org/10.1016/j.jobe.2020.101580.

Fang, Y., & Eames, P. C. (2006). Thermal performance of an electrochromic vacuum glazing. *Energy Conversion and Management*, *47*(20), 3602−3610. Available from https://doi.org/10.1016/j.enconman.2006.03.016.

Fang, Y., Eames, P. C., & Norton, B. (2007). Effect of glass thickness on the thermal performance of evacuated glazing. *Solar Energy*, *81*(3), 395−404. Available from https://doi.org/10.1016/j.solener.2006.05.004.

Fang, Y., Eames, P. C., Norton, B., Hyde, T. J., Zhao, J., Wang, J., & Huang, Y. (2007). Low emittance coatings and the thermal performance of vacuum glazing. *Solar Energy*, *81*(1), 8−12. Available from https://doi.org/10.1016/j.solener.2006.06.011.

Fang, Y., Hyde, T., Eames, P. C., & Hewitt, N. (2009). Theoretical and experimental analysis of the vacuum pressure in a vacuum glazing after extreme thermal cycling. *Solar Energy*, *83*(9), 1723−1730. Available from https://doi.org/10.1016/j.solener.2009.03.017.

Favoino, F. (2016). *Simulation-based evaluation of adaptive materials for improved building performance. Nano and biotech based materials for energy building efficiency* (pp. 125−166). Cham: Springer.

Favoino, F., & Overend, M. (2015). A simulation framework for the evaluation of next generation responsive building envelope technologies. *Energy Procedia*, *78*, 2602−2607.

Favoino, F., Goia, F., Perino, M., & Serra, V. (2016). Experimental analysis of the energy performance of an ACTive, RESponsive and Solar (ACTRESS) façade module. *Solar Energy*, *133*, 226−248.

Favoino, F., Loonen, R. C., Doya, M., Goia, F., Bedon, C., & Babich, F. (2018). In F. Favoino (Ed.), *Building performance simulation and characterisation of adaptive facades: Adaptive façade network*. TU Delft Open.

Favoino, F., Overend, M., & Jin, Q. (2015). The optimal thermo-optical properties and energy saving potential of adaptive glazing technologies. *Applied Energy*, *156*, 1−15.

Favoino, F., Perino, M. & Serra, V. (2019). Advanced integrated facades: Concept evolution and new challenges. In *Proceedings of the ASHRAE buildings XIV international conference*.

Gao, T., Ihara, T., Grynning, S., Jelle, B. P., & Lien, A. G. (2016). Perspective of aerogel glazings in energy efficient buildings. *Building and Environment*, *95*, 405−413. Available from https://doi.org/10.1016/j.buildenv.2015.10.001.

Gao, T., Jelle, B. P., Ihara, T., & Gustavsen, A. (2014). Insulating glazing units with silica aerogel granules: The impact of particle size. *Applied Energy*, *128*, 27−34. Available from https://doi.org/10.1016/j.apenergy.2014.04.037.

Garnier, C., Muneer, T., & McCauley, L. (2015). Super insulated aerogel windows: Impact on daylighting and thermal performance. *Building and Environment*, *94*(P1), 231−238. Available from https://doi.org/10.1016/j.buildenv.2015.08.009.

Garrison, J. D., & Collins, R. E. (1995). Manufacture and cost of vacuum glazing. *Solar Energy*, *55*(3), 151−161.

Giovannini, L., Favoino, F., Pellegrino, A., Lo Verso, V. R. M., Serra, V., & Zinzi, M. (2019). Thermochromic glazing performance: From component experimental characterisation to whole building performance evaluation. *Applied Energy, 251* (February), 113335. Available from https://doi.org/10.1016/j.apenergy.2019.113335.

GlassX (2013). *GlassX homepage, broschuere online.* <https://docs.google.com/viewerng/viewer?url = https://glassx.jimdo.com/app/download/10112900052/Broschuere_klein_online.pdf?t = 1503651139>. Accessed 07.08.20.

Goia, F. (2012). Thermo-physical behaviour and energy performance assessment of PCM glazing system configurations: A numerical analysis. *Frontiers of Architectural Research; a Journal of Science and its Applications, 1*(4), 341–347. Available from https://doi.org/10.1016/j.foar.2012.10.002.

Goia, F., Perino, M., & Serra, V. (2013). Improving thermal comfort conditions by means of PCM glazing systems. *Energy and Buildings, 60*, 442–452. Available from https://doi.org/10.1016/j.enbuild.2013.01.029.

Goia, F., Perino, M., Serra, V., & Zanghirella, F. (2010). Towards an active, responsive, and solar building envelope. *Journal of Green Building, 5*(4), 121–136.

Goia, F., Zinzi, M., Carnielo, E., & Serra, V. (2015). Spectral and angular solar properties of a PCM-filled double glazing unit. *Energy and Buildings, 87*, 302–312. Available from https://doi.org/10.1016/j.enbuild.2014.11.019.

Gowreesunker, B. L., Stankovic, S. B., Tassou, S. A., & Kyriacou, P. A. (2013). Experimental and numerical investigations of the optical and thermal aspects of a PCM-glazed unit. *Energy and Buildings, 61*, 239–249. Available from https://doi.org/10.1016/j.enbuild.2013.02.032.

Griffiths, P. W., Di Leo, M., Cartwright, P., Eames, P. C., Yianoulis, P., Leftheriotis, G., & Norton, B. (1998). Fabrication of evacuated glazing at low temperature. *Solar Energy, 63*(4), 243–249. Available from https://doi.org/10.1016/S0038-092X(98)00019-X.

Grobe, L. O. (2018). Characterization and data-driven modeling of a retro-reflective coating in RADIANCE. *Energy and Buildings, 162*, 121–133. Available from https://doi.org/10.1016/j.enbuild.2017.12.029.

Grobe, L.O. (2020). Irregular light scattering properties of innovative fenestration for comfortable and energy-efficient buildings. <https://doi.org/10.5281/zenodo.4049475>.

Grobe, L. O., Wittkopf, S., & Kazanasmaz, Z. T. (2017). High-resolution data-driven models of daylight redirection components. *Journal of Façade Design and Engineering, 5*(2), 87–100. Available from https://doi.org/10.7480/jfde.2017.2.1743.

Grobe, L. O., Papaiz, L. & De Michele, G. (2016). Modelling the reflective properties of coated blinds comprising an innovative CFS in Radiance. In *Proceedings of the Radiance workshop 2016.* <http://leader.pubs.asha.org/article.aspx?doi = 10.1044/leader.PPL.21082016.20>.

Grynning, S., Goia, F., Rognvik, E., & Time, B. (2013). Possibilities for characterization of a PCM window system using large scale measurements. *International Journal of Sustainable Built Environment, 2*(1), 56–64.

Gutai, M., & Kheybari, A. G. (2020). Energy consumption of water-filled glass (WFG) hybrid building envelope. *Energy and Buildings, 218*, 110050. Available from https://doi.org/10.1016/j.enbuild.2020.110050.

Haeringer, S. F., Denz, P. R., Vongsingha, P., Bueno, B., Kuhn, T. E., & Maurer, C. (2020). Design and experimental proof-of-concept of a façade-integrated solar thermal venetian blind with heat pipes. *Journal of Façade Design and Engineering, 8*(1), 131–156.

Haeringer, S. F., Denz, P., & Vongsingha, P. (2019). Arkol – Deve*lopment and testing of solar thermal venetian blind*s. January.

Harwin H. (2020) *Project. Harvesting solar energy with multifunctional glass-polymer windows.*

Heiz, B. P. V., Pan, Z., Su, L., Le, S. T., & Wondraczek, L. (2018). A large-area smart window with tunable shading and solar-thermal harvesting ability based on remote switching of a magneto-active liquid. *Advanced Sustainable Systems*, *2*(1), 1700140. Available from https://doi.org/10.1002/adsu.201700140.

Heiz, B. P. V., Su, L., Pan, Z., & Wondraczek, L. (2018). Fluid-integrated glass-glass laminate for sustainable hydronic cooling and indoor air conditioning. *Advanced Sustainable Systems*, *2*(10), 1800047. Available from https://doi.org/10.1002/adsu.201800047.

Huang, Y., & Niu, J. L. (2015a). Energy and visual performance of the silica aerogel glazing system in commercial buildings of Hong Kong. *Construction and Building Materials*, *94*, 57−72. Available from https://doi.org/10.1016/j.conbuildmat.2015.06.053.

Huang, Y., & Niu, J. L. (2015b). Application of super-insulating translucent silica aerogel glazing system on commercial building envelope of humid subtropical climates − Impact on space cooling load. *Energy*, *83*, 316−325. Available from https://doi.org/10.1016/j.energy.2015.02.027.

IEA-ECBCS Annex 44 (2007) *Integrating environmentally responsive elements in buildings, state of the art review*, in: Perino, M. (Ed.).

Ismail, K. A. R., Salinas, C. T., & Henriquez, J. R. (2008). Comparison between PCM filled glass windows and absorbing gas filled windows. *Energy and Buildings*, *40*(5), 710−719. Available from https://doi.org/10.1016/j.enbuild.2007.05.005.

Jelle, B.P., Baetens, R., & Gustavsen, A. (2015). Aerogel Insulation for building applications. The sol−gel handbook, 3−3, 1385−1412. <https://doi.org/10.1002/9783527670819.ch45>.

Jensen, S. Ø., Marszal-Pomianowska, A., Lollini, R., Pasut, W., Knotzer, A., Engelmann, P., ... Reynders, G. (2017). IEA EBC annex 67 energy flexible buildings. *Energy and Buildings*, *155*(2017), 25−34. Available from https://doi.org/10.1016/j.enbuild.2017.08.044.

Jensen, K. I., & Schultz, J. M. (2007). Transparent aerogel Windows: Results from an EU FP5 project. In: 8th International vacuum insulation symposium.

Jin, Q., & Overend, M. (2014). Sensitivity of façade performance on early-stage design variables. *Energy and Buildings*, *77*, 457−466. Available from https://doi.org/10.1016/j.enbuild.2014.03.038.

Kapsis, K., & Athienitis, A. K. (2015). A study of the potential benefits of semi-transparent photovoltaics in commercial buildings. *Solar Energy*, *115*, 120−132. Available from https://doi.org/10.1016/j.solener.2015.02.016.

Khandelwal, H., Schenning, A. P., & Debije, M. G. (2017). Infrared regulating smart window based on organic materials. *Advanced Energy Materials*, *7*(14), 1602209. Available from https://doi.org/10.1002/aenm.201602209.

Kim, D., Cox, S. J., Cho, H., & Yoon, J. (2018). Comparative investigation on building energy performance of double skin façade (DSF) with interior or exterior slat blinds. *Journal of Building Engineering*, *20*(July), 411−423. Available from https://doi.org/10.1016/j.jobe.2018.08.012.

Kim, G. S., & Hyun, S. H. (2003). Synthesis of window glazing coated with silica aerogel films via ambient drying. *Journal of Non-Crystalline Solids*, *320*(1−3), 125−132. Available from https://doi.org/10.1016/S0022-3093(03)00027-9.

Kiss, B., & Neij, L. (2011). The importance of learning when supporting emergent technologies for energy efficiency—A case study on policy intervention for learning for the development of energy efficient windows in Sweden. *Energy Policy*, *39*(10), 6514−6524. Available from https://doi.org/10.1016/j.enpol.2011.07.053.

Kocer, C. (2017). *A novel glass spacer for vacuum insulated glazing*. Glaston: Glass Performance Days.
Kocer, C. (2019). *The past, present, and future of the vacuum insulated glazing technology. Glass Performance Days*. Glaston, Finland: Glass Performance Days.
Koester, H. (2004). *Dynamic daylighting architecture: Basics, systems, projects*.
LBNL (2001). *Daylighting in buildings: A source book on daylighting systems and components*. <http://www.iea-shc.org>.
Li, C., Lyu, Y., Li, C., & Qiu, Z. (2020). Energy performance of water flow window as solar collector and cooling terminal under adaptive control. *Sustainable Cities and Society*, *59*(October 2019), 102152. Available from https://doi.org/10.1016/j.scs.2020.102152.
Li, D. H. W., Lam, T. N. T., Chan, W. W. H., & Mak, A. H. L. (2009). Energy and cost analysis of semi-transparent photovoltaic in office buildings. *Applied Energy*, *86*(5), 722−729.
Littlefair, P., Ortiz, J., & Bhaumik, C. D. (2010). A simulation of solar shading control on UK office energy use. *Building Research & Information*, *38*(6), 638−646.
Loonen, R. C. G. M. (2018). *Approaches for computational performance optimization of innovative adaptive façade concepts* (Issue 2018). <https://research.tue.nl/en/publications/approaches-for-computational-performance-optimization-of-innovati>.
Loonen, R. C. G. M., Favoino, F., Hensen, J. L. M., & Overend, M. (2017). Review of current status, requirements and opportunities for building performance simulation of adaptive facades†. *Journal of Building Performance Simulation*, *10*(2), 205−223. Available from https://doi.org/10.1080/19401493.2016.1152303.
Loonen, R. C., Trčka, M., Cóstola, D., & Hensen, J. L. (2013). Climate adaptive building shells: State-of-the-art and future challenges. *Renewable and Sustainable Energy Reviews*, *25*, 483−493.
Luca, E. (2018). *Sustainability, restorative to regenerative an exploration in progressing a paradigm shift in built environment thinking, from sustainability to restorative sustainability and on to regenerative sustainability COST Action CA16114 RESTORE: REthinking Sustainability TOwards a Regenerative Economy, Working Group One Report: Restorative Sustainability*, in: Brown, M., Haselsteiner, E., Apró, D., Kopeva, D., Luca, E., Pulkkinen, K.-L. & Vula Rizvanolli, B. (Eds).
Lucchino, E. C., Goia, F., Lobaccaro, G., & Chaudhary, G. (2019). Erratum to: Modelling of double skin facades in whole-building energy simulation tools: A review of current practices and possibilities for future developments (Building Simulation, (2019), 12, 1, (3−27), 10.1007/s12273-019-0511-y). *Building Simulation*, *12*, 3−27. Available from https://doi.org/10.1007/s12273-019-0523-7.
Luna-Navarro, A., Blanco Cadena, J. D., Favoino, F., Donato, M., Poli, T., Perino, M., et al. (2019). Occupant-centred control strategies for adaptive facades: A preliminary study of the impact of shortwave solar radiation on thermal comfort. In: Building simulation 2019: 16th Conference of IBPSA (pp. 4910−4917).
Luna-Navarro, A., Loonen, R., Juaristi, M., Monge-Barrio, A., Attia, S., & Overend, M. (2020). Occupant-façade interaction: A review and classification scheme. *Building and Environment*, *177*, 106880. Available from https://doi.org/10.1016/j.buildenv.2020.106880.
Lyu, Y. L., Liu, W. J., Su, H., & Wu, X. (2019). Numerical analysis on the advantages of evacuated gap insulation of vacuum-water flow window in building energy saving under various climates. *Energy*, *175*, 353−364. Available from https://doi.org/10.1016/j.energy.2019.03.101.

Manz, H. (2008). On minimizing heat transport in architectural glazing. *Renewable Energy*, *33*(1), 119−128. Available from https://doi.org/10.1016/j.renene.2007.01.007.

Manz, H., Brunner, S., & Wullschleger, L. (2006). Triple vacuum glazing: Heat transfer and basic mechanical design constraints. *Solar Energy*, *80*(12), 1632−1642. Available from https://doi.org/10.1016/j.solener.2005.11.003.

Manz, H., Egolf, P. W., Suter, P., & Goetzberger, A. (1997). TIM-PCM external wall system for solar space heating and daylighting. *Solar Energy*, *61*(6), 369−379. Available from https://doi.org/10.1016/S0038−092X(97)00086-8.

Mashaly, I. A., Nassar, K., El-Henawy, S. I., Mohamed, M. W. N., Galal, O., Darwish, A., ... Safwat, A. M. E. (2017). A prismatic daylight redirecting fenestration system for southern skies. *Renewable Energy*, *109*, 202−212. Available from https://doi.org/10.1016/j.renene.2017.02.048.

Mattox, D. M., & Mattox, V. H. (2007). (Eds) *50 Years of vacuum coating technology and the growth of the Society of Vacuum Coaters*. Society of Vacuum Coaters.

McNeil, A., Lee, E. S., & Jonsson, J. C. (2017). Daylight performance of a microstructured prismatic window film in deep open plan offices. *Building and Environment*, *113*, 280−297. Available from https://doi.org/10.1016/j.buildenv.2016.07.019.

Merli, F., Anderson, A. M., Carroll, M. K., & Buratti, C. (2018). Acoustic measurements on monolithic aerogel samples and application of the selected solutions to standard window systems. *Applied Acoustics*, *142*, 123−131. Available from https://doi.org/10.1016/j.apacoust.2018.08.008.

Minguez, L., Gubert, M., Astigarraga. A. & Avesani, A. (2020). Validation process of a multifunctional and autonomous solar window block for residential buildings retrofit. *ISES Eurosun*.

Molter, P. L., Bonnet, C., Wagner, T., & Klein, T. (2017). Autoreactive components in double skin facades. In *Advanced Building Skins (ABS)*.

Molter, P. L., Robustheit, T. A., & Autoreaktivität (2019). Temperaturregulierung mittels Dehnstoffelementen. In: *Michael Schumacher, Michael-Marcus Vogt, Luis Arturo Cordón Krumme (Hrsg.): New MOVE Architektur in Bewegung - Neue dynamische Komponenten und Bauteile* (pp. 76−79). Birkhäuser Verlag GmbH 2019.

Morgan, J., Myrskog, S., Barnes, B., Sinclair, M., & Morris, N. (2014−2015). Sunlight concentrating and harvesting device. *United States patent App. 14/196,618; 14/215,913; 14/218,025. 14/217,998; 14/196,291*.

Mueller, H. F. O. (2019). Application of micro-structured sunlighting systems in different climatic zones. *Journal of Daylighting*, *6*(2), 52−59. Available from https://doi.org/10.15627/jd.2019.7.

Ng, N., Collins, R. E., & So, L. (2005). Thermal and optical evolution of gas in vacuum glazing. *Materials Science and Engineering B: Solid-State Materials for Advanced Technology*, *119*(3), 258−264. Available from https://doi.org/10.1016/j.mseb.2004.12.079.

Nikolaeva-Dimitrova, M., Stoyanova, M., Ivanov, P., Tchonkova, K., & Stoykov, R. (2018). Investigation of thermal behaviour of innovative water flow glazing modular unit. *Bulgarian Chemical Communications*, *50*, 21−27.

Noback, A., Grobe, L. O., & Wittkopf, S. (2016). Accordance of light scattering from design and de-facto variants of a daylight redirecting component. *Buildings*, *6*(3), 1−17. Available from https://doi.org/10.3390/buildings6030030.

Osterman, E., Butala, V., & Stritih, U. (2015). PCM thermal storage system for 'free' heating and cooling of buildings. *Energy and Buildings*, *106*, 125−133. Available from https://doi.org/10.1016/j.enbuild.2015.04.012.

Papaefthimiou, S., Leftheriotis, G., Yianoulis, P., Hyde, T. J., Eames, P. C., Fang, Y., . . . Jannasch, P. (2006). Development of electrochromic evacuated advanced glazing. *Energy and Buildings*, *38*(12), 1455−1467. Available from https://doi.org/10.1016/j.enbuild.2006.03.029.

Papaiz, L., Grobe, L., & De Michele, G. (2020). Retroreflective coating for window blinds. In: Proceedings of Façade Tectonics 2020 world congress, August 2020, Los Angeles, CA, USA.

Paulos, J., & Berardi, U. (2020). Optimizing the thermal performance of window frames through aerogel-enhancements. *Applied Energy*, *266*(February), 114776. Available from https://doi.org/10.1016/j.apenergy.2020.114776.

Perino, M., & Serra, V. (2015). Switching from static to adaptable and dynamic building envelopes: A paradigm shift for the energy efficiency in buildings. *Journal of Façade Design and Engineering*, *3*(2), 143−163. Available from https://doi.org/10.3233/fde-150039.

Pomponi, F., & D'Amico, B. (2017). Holistic study of a timber double skin façade: Whole life carbon emissions and structural optimisation. *Building and Environment*, *124*, 42−56.

Reim, M., Körner, W., Manara, J., Korder, S., Arduini-Schuster, M., Ebert, H. P., & Fricke, J. (2005). Silica aerogel granulate material for thermal insulation and daylighting. *Solar Energy*, *79*(2), 131−139. Available from https://doi.org/10.1016/j.solener.2004.08.032.

Roberts, M., Allen, S., & Coley, D. (2020). Life cycle assessment in the building design process—A systematic literature review. *Building and Environment*, 107274.

Saelens (2002). Energy performance assessment of single storey multiple-skin facades, (Dissertation), *KU Leuven*, ISBN: 9789056823702.

Saini, H., Loonen, R. C. G. M., & Hensen, J. L. M. (2019). *Simulation-based performance prediction of an energy-harvesting façade system with selective daylight transmission.* In *Proceedings of the eighth international congress on architectural envelopes simulation-based performance prediction of an energy-harvesting façade system with se. 2018.* <https://pure.tue.nl/ws/portalfiles/portal/101071747>.

Sartori, I., Napolitano, A., & Voss, K. (2012). Net zero energy buildings: A consistent definition framework. *Energy and Buildings*, *48*, 220−232. Available from https://doi.org/10.1016/j.enbuild.2012.01.032.

Scharfe, B., Lehmann, S., Gerdes, T., & Brüggemann, D. (2019). Optical and mechanical properties of highly transparent glass-flake composites. *Journal of Composites Science*, *3*(4), 101.

Schellnhuber, H. J., van der Hoeven, M., Bastioli, C., Ekins, P., Jaczewska, B., Kux, B., et al. (2018). Final report of the high-level panel of the European Decarbonisation Pathways Initiative. https://doi.org/10.2777/636.

Schregle, R., Grobe, L., & Wittkopf, S. (2015). Progressive photon mapping for daylight redirecting components. *Solar Energy*, *114*, 327−336. Available from https://doi.org/10.1016/j.solener.2015.01.041.

Shameri, M. A., Alghoul, M. A., Sopian, K., Zain, M. F. M., & Elayeb, O. (2011). Perspectives of double skin façade systems in buildings and energy saving. *Renewable and sustainable energy reviews*, *15*(3), 1468−1475.

Silva, T., Vicente, R., & Rodrigues, F. (2016). Literature review on the use of phase change materials in glazing and shading solutions. *Renewable and Sustainable Energy Reviews*, *53*, 515−535. Available from https://doi.org/10.1016/j.rser.2015.07.201.

Silva, T., Vicente, R., Rodrigues, F., Samagaio, A., & Cardoso, C. (2015). Development of a window shutter with phase change materials: Full scale outdoor experimental approach.

Energy and Buildings, *88*, 110−121. Available from https://doi.org/10.1016/j.enbuild.
2014.11.053.

Skandalos, N., & Karamanis, D. (2015). PV glazing technologies. *Renewable and Sustainable Energy Reviews*, *49*(2015), 306−322. Available from https://doi.org/10.1016/j.rser.2015.04.145.

Smirnova, I., & Gurikov, P. (2018). Aerogel production: Current status, research directions, and future opportunities. *Journal of Supercritical Fluids*, *134*(December 2017), 228−233. Available from https://doi.org/10.1016/j.supflu.2017.12.037.

Streicher, W., Heimrath, R., Hengsberger, H., Mach, T., Waldner, R., Flamant, G., ... Blomquist, C. (2007). On the typology, costs, energy performance, environmental quality and operational characteristics of double skin facades in european buildings. *Advances in Building Energy Research*, *1*(1), 1−28.

Su, L., Fraaß, M., Kloas, M., & Wondraczek, L. (2019). Performance analysis of multipurpose fluidic windows based on structured glass-glass laminates in a triple glazing. *Frontiers in Materials*, *6*(May), 1−14. Available from https://doi.org/10.3389/fmats.2019.00102.

Sun, Y. (2017). *Saving and daylight comfort glazing system with transparent insulation material for building energy saving and daylight comfort Yanyi Sun, BArch MSc Doctor of Philosophy*. August.

Tällberg, R., Jelle, B. P., Loonen, R., Gao, T., & Hamdy, M. (2019). Comparison of the energy saving potential of adaptive and controllable smart windows: A state-of-the-art review and simulation studies of thermochromic, photochromic and electrochromic technologies. *Solar Energy Materials and Solar Cells*, *200*, 109828.

Taveres-Cachat, E., Favoino, F., Loonen, R., & Goia, F. (2021). *Ten questions concerning co-simulation for performance prediction of advanced building envelopes, submitted to Building and Environment*.

The European Parliament and the Council of the European Union. (2012). Directive 2012/27/EU on energy efficiency, amending directives 2009/125/EC and 2010/30EU and repealing directives 2004/8/EC and 2006/32/EC. *Official Journal of the European Union*.

Tian, Z., Lei, Y., & Jonsson, J. C. (2019). Daylight luminous environment with prismatic film glazing in deep depth manufacture buildings. *Building Simulation*, *12*(1), 129−140. Available from https://doi.org/10.1007/s12273-018-0487-z.

Tiwari, S., Tiwari, T., Carter, S. A., Scott, J. C., & Yakhmi, J. V. (2019). Advances in polymer-based photovoltaic cells: Review of pioneering materials. *Design, and Device Physics*, *44*, 2019.

Traditional Building Magazine (2020). Vacuum insulated glass, innovation for historic restoration. *Traditional Building Magazine*, May 2020, <https://www.traditionalbuilding.com/product-report/vacuum-insulated-glass>, Accessed: 10.01.21

Tsioulou, O., Erpelding, J., & Lampropoulos, A. (2016). *Development of novel low thermal conductivity concrete using aerogel powder. 0−7*.

Tzempelikos, A., & Athienitis, A. K. (2005). The effect of shading design and control on building cooling demand. *Passive and Low Energy Cooling for the Built Environment*, 953−958, *May*.

Van Berkel, P., Kruit, K., Van de Poll, F., Rooijers, F., & Vendrik, J. (2020). *Zero carbon buildings 2050 − Background report*.

Vasiliev, M., Nur-E-Alam, M., & Alameh, K. (2019). Recent developments in solar energy-harvesting technologies for building integration and distributed energy generation. *Energies*, *12*(6). Available from https://doi.org/10.3390/en12061080.

Wei, G., Wang, L., Xu, C., Du, X., & Yang, Y. (2016). Thermal conductivity investigations of granular and powdered silica aerogels at different temperatures and pressures. *Energy*

and Buildings, *118*(March), 226−231. Available from https://doi.org/10.1016/j.enbuild. 2016.03.008.
Weinläder, H., Beck, A., & Fricke, J. (2005). PCM-facade-panel for daylighting and room heating. *Solar Energy*, *78*(2), 177−186. Available from https://doi.org/10.1016/j.solener. 2004.04.013.
Wetter, M. & Van Treeck, C. (2017). *New generation computational tools for building and community energy systems annex 60 final report, 2017*. <https://doi.org/10.4103/0973-1229.86137>.
Xie, H., Clements-Croome, D., & Wang, Q. (2017). Move beyond green building: A focus on healthy, comfortable, sustainable and aesthetical architecture. *Intelligent Buildings International*, *9*(2), 88−96. Available from https://doi.org/10.1080/17508975.2016. 1139536.
Zani, A., Galante, C., Rammig, L. (2020). Thermal performance of closed cavity facades. In *Proceedings of façade tectonics 2020 world congress*, 5 August 2020.
Zomorodian, Z. S., & Tahsildoost, M. (2018). Energy and carbon analysis of double skin facades in the hot and dry climate. *Journal of Cleaner Production*, *197*, 85−96.

Embedding intelligence to control adaptive building envelopes

Fabio Favoino[1], Manuela Baracani[1], Luigi Giovannini[1], Giovanni Gennaro[1] and Francesco Goia[2]
[1]Department of Energy, Polytechnic University of Turin, Torino, Italy, [2]Department of Architecture and Technology, NTNU, Norwegian University of Science and Technology, Trondheim, Norway

Abbreviations

ABE	advanced building envelope
ANN	artificial neural network
DSF	double-skin façade
HVAC	heating, ventilation and air conditioning
IAQ	indoor air quality
ICT	information and communication technology
IoT	Internet of things
MPC	model-predictive control
MBC	model-based control
PMV	predicted mean vote
RBC	rule-based control
RES	renewable energy source

6.1 Introduction

The building envelope plays a major role in actuating strategies aimed at improving the overall building performance to meet the increasingly stringent sustainability targets placed on the built environment (Jin & Overend, 2014). The drive to increase the efficiency of building skins has led, in the last decade, to the development of so-called advanced building envelopes (ABEs) (Perino and Serra, 2015). The name ABEs describes a rather heterogenous group of integrated envelope systems and technologies that ensure high building performance across multiple physical domains by efficiently balancing competing performance aspects [ie, energy use and production by means of renewable energy sources (RES); thermal, visual, acoustic comfort and indoor air quality (IAQ)]. They rely on advanced design methods, advanced material properties and components and advanced integrated control strategies (Taveres-Cachat, Favoino, Loonen, & Goia, 2021).

Adaptive (or responsive) building materials and technologies are among the key elements of those ABEs because they enable an active and dynamic management of

Rethinking Building Skins. DOI: https://doi.org/10.1016/B978-0-12-822477-9.00007-3
© 2022 Elsevier Inc. All rights reserved.

energy and mass transfer between the building and the external environment. These systems can reversibly modulate their thermo-optical properties and operational strategies according to transient boundary conditions and performance requirements to minimize the energy for building climatization and to provide high indoor environmental quality conditions (Favoino et al., 2018). A nonexhaustive list of well-known examples of this category of ABEs includes double-skin facades (DSFs) (Saelens, Carmeliet, & Hens, 2003); switchable glazing (such as electrochromic, liquid crystal and thermochromic glazing) (Baetens, Jelle, & Gustavsen, 2010); operable solar shading (Nielsen, Svendsen, & Jensen, 2011), complex fenestration systems (De Michele et al., 2018); and dynamic insulation (Favoino, Jin, & Overend, 2017).

The shift from a building enclosure that acts as a static shield to an active, responsive and solar-based skin (Goia, Perino, Serra, & Zanghirella, 2010), which continuously tunes its functionalities to achieve optimal energy performance, occupant satisfaction and exploitation of naturally available resource, is clearly a nontrivial process. In such systems the real-time decision-making process that drives their operations in synergy with other building systems adds a new layer of complexity. This is particularly relevant when multiple physical domains (influencing thermal and/or luminous and/or air and mass transfer and/or acoustics aspects at the same time) are simultaneously tackled (Loonen, Favoino, Hensen, & Overend, 2017).

In this chapter, we explore the implications of embedding 'intelligence' to ensure building performance—driven control of adaptive building envelopes. We reflect on the requirements for the design and operation of the control system supporting a decision-making process directly at the building envelope level. By reviewing and exploring the characteristics of these automation systems, we identify which functions should be controlled in adaptive facades and how this real-time management can be efficiently carried out. In addition, we create an overarching framework to describe façade controls addressing energy efficiency, occupant satisfaction and interaction with the occupants. We show that moving toward embedded control systems is a natural evolution to embed intelligence into adaptive building envelopes. Real-world examples of possible control strategies suitable for embedded systems in specific adaptive façade technologies are presented considering the most recent know-how. Current research trends and challenges together with promising future developments are sketched to provide an outlook for embedding intelligence in building envelopes.

6.2 The purpose of automation in adaptive building envelopes

Generally speaking, the purpose of a building envelope characterized by dynamic features is to achieve one or more performance requirements in a more efficient way than conventional 'static' envelopes. When the focus is placed on energy and indoor environmental performance, such general-level requirements can be summarized in the achievement of the following requirements:

- *visual comfort*: by controlling incoming amount and directionality of solar radiation in the visible spectrum to guarantee sufficient daylight and view of the outdoor without causing glare;
- *thermal comfort:* by modulating solar radiation (eg, to avoid overheating in summer or to admit free solar gains and control excessive heat transfer to reduce local radiant asymmetries and draught risk) or by controlling the mass exchange between indoor and outdoor by means of natural ventilation and free cooling.
- *IAQ:* by controlling the amount of fresh outdoor air admitted into the indoor environment.
- *Acoustic comfort*: by modulating sound transmission between outside and inside (especially where natural ventilation is employed).
- Reduction in *energy use for building climatization, ventilation and illumination*: by controlling free solar gains (to decrease electric lighting use and heating energy demand and to reduce cooling energy demand), and by maximizing the *energy harvesting* potential of façade-integrated solar energy systems.

Different dynamic envelope technologies can be adopted to pursue some of these performance requirements. Adaptive building envelopes can be schematized in two main categories according to the control mechanisms, that is, either intrinsic or extrinsic mechanisms (Loonen, Singaravel, Trčka, Cóstola, & Hensen, 2014). Intrinsic indicates that the adaptation mechanism is automatically triggered by a stimulus (eg, surface temperature and solar radiation) activated by a variation in the internal energy of the technology/layer/material (eg, thermochromic, photochromic, phase-change materials and some types of kinetic facades based on shape memory alloys systems). Thus no intervention is required from an external system/user. In there is no requirement for any computational, 'intelligent' decision-making process. In contrast, extrinsic refers to the presence of an external decision-making component that triggers an action within the building envelope system (eg, electrochromic glazing, movable shading devices, kinetic facades) following a sequence that processes input signals and determines an output signal. Such 'intelligent' systems require active management and decision-making relying on sensors, processors (with preset logics) and actuators. Fig. 6.1 provides a nonexhaustive overview of the interconnection between adaptive envelope technologies characterized by extrinsic control mechanisms (numbers), interrelated physical and physiological phenomena, variables that can be controlled (coloured domains), and building performance targets to be pursued (outermost *grey text*). This overview highlights how the highly interlinked system calls for intelligence to manage conflicting requirements across interconnected domains, as we will try to exemplify with concrete cases in the following paragraphs.

For example, active chromogenic technologies (electrochromic layers, suspended particle devices, liquid crystal devices, etc.) and operable shading devices (venetian blinds, roller shades, etc.) can actively modulate the amount of solar radiation entering the indoor environment. Depending on the technology involved, transparency (hence, the incoming solar radiation) may be modulated by reflection, absorption and/or redirection. These different mechanisms may be used to provide privacy or

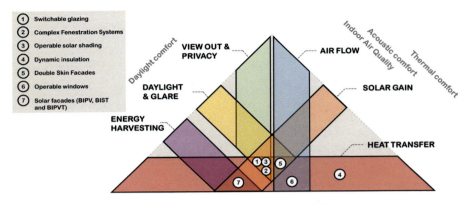

Figure 6.1 Performance objectives for controlling intelligent facades and related technologies.
Source: Original illustration form the authors.

view or controlled to momentarily reduce glare risk while maintaining a sufficient level of daylight and/or reduce the amount of entering solar radiation in summer to avoid overheating while maximizing free solar gains in winter. Nevertheless, depending on local specific conditions (such as climate and latitude, façade exposure and indoor layout), such requirements may be temporarily contradictory, and trade-offs or priorities need to be established in the automated control of such systems. Typical examples of such conflicting tasks are glare risk versus daylight requirements and free solar gains during winter, daylight requirements and overheating risk due to solar gain in summer (Favoino, Fiorito, Cannavale, Ranzi, & Overend, 2016).

Operable windows have an immediate impact on building energy consumption, thermal comfort and IAQ. Window opening is commonly controlled to reduce overheating and cooling energy use (Barbosa, Bartak, Hensen, & Loomans, 2015) by exploiting natural ventilation as a passive cooling technique. In residential buildings, depending on the ventilation control strategy adopted, window opening could reduce the annual cooling energy consumption between 58% and 86% (Sorgato, Melo, & Lamberts, 2016). In addition, operable window control can significantly contribute to thermal comfort improvement and reduction of overheating while improving IAQ (Rabani, Bayera Madessa, & Nord, 2021). Nevertheless, natural ventilation requirements could be, under some circumstances, on conflict with the need of damping outside generated noise (Locher et al., 2018). The conflict between these different requirements may prevent optimal automatic operations if this challenge is not properly addressed through an intelligent control system.

DSFs are complex fenestration systems that actively handle different tasks in a building such as controlling ventilation, solar gains and daylighting to meet different performance targets. This is achieved by controlling the type and path of ventilation in the façade cavity by means of vents and mechanical systems, and the operable solar shading system. DSFs can meet multiple performance requirements within the same enclosure element, combining and enhancing the features of the two separate technologies mentioned earlier, but this comes at the cost of higher automation complexity. Some examples exist in which solar energy harvesting devices are integrated in such a cavity (Bunthof et al., 2016), adding one more layer of complexity in the decision-making process, which then requires considerations about solar energy conversion, thermal energy for space climatization, daylight, glare, etc.

To determine the capability of extrinsic adaptive systems to fulfil building performance targets, the dynamic characteristics related to such automation must be considered, such as

- the dynamicity of climatic and indoor boundary conditions (ie, variation of outdoor temperature, variation of solar geometry and solar radiation intensity, variation of pollutant concentration in outdoor air, variation of outside noise level, variation of occupancy and occupant's requirements, etc.);
- the dynamicity of the physical and physiological phenomena involved, which can be characterized by time constants of milliseconds, to a few seconds, to a few hours (ie, perception of high-pitch sounds and vibration, perception of illuminance and luminance variation, human metabolism variation as a reaction to the thermal environment, mass flow within a room, solar radiation heat storage in building thermal mass, etc.);
- the dynamicity of the adaptive system itself, which is the capability to react and to act on the physical phenomena and their domain of variation (this is a function of a number of possible façade configurations, hence of possible control actions, of the time required for the actuation and for the decision-making related to the automation process).

6.3 Challenges and requirements for the automation of adaptive building envelopes

The control of dynamic envelope systems, especially in comparison with other building systems, requires multiple physical domains with different needs in terms of frequency of control to be addressed. Furthermore, the interaction between the façade and other building components/systems can in some cases be minimal or achieved with easy-to-establish hierarchical functions (eg, daylighting can easily be preferred to artificial lighting), while in other domains the interplay between the façade and the indoor environment needs to be part of the optimal control process (eg, the inertial features of the building in relation to the control of solar gain).

Such complexity is reflected in the need to differentiate between 'open-loop' and 'closed-loop' control logics. If the control action does not affect the sensor input signal for the control itself, this is referred to as a 'nonfeedback' system or 'open-loop' control (ie, control of operable shading device according to solar geometry). In contrast, when the input sensor signal is dependent on the control action itself, this is referred to as a 'feedback' system or 'closed-loop' control (ie, control of operable shading device according to indoor air temperature measured in the room). Closed-loop controls have the advantage of measuring the impact on a certain performance target more directly, as the desired output (in terms of building performance indicator or more simply of sensor signal) is always compared to the actual output, but they have the disadvantage of more complicated implementation.

Combining the multiphysical characteristics of façade control problems with the requirements in terms of frequency and feedback type leads to establishing complex control algorithms that include multiple frequencies and both open-loop and closed-loop controls.

For example, when considering heat transfer control (either solar gain or by another form of heat transfer though the envelope) in relation to the energy balance of a building's zone, the inertial characteristics of the building's zone are to be taken into account to define the optimal time frequency. For this type of control an open-loop feedback is necessary, but the frequency of the control can be relatively lower compared to other domains (ranging from 1 to a few hours), depending on the specific technology, the time of the day (day/night cycle) and building features (Favoino et al., 2016, 2017). Conversely, when airflow is the main target, a higher frequency of control (in the order of a few minutes or less when discomfort due to cold draft is considered) is necessary to implement an effective solution. Here, depending on the objective of the control (energy or comfort), both open- and closed-loop feedback can be adopted.

When assessed in light of the larger scale of the entire building or even neighbourhood, façade control also needs to tackle the complexity of balancing reactive control with predictive control. Some themes in façade control linked to visual and acoustic comfort can clearly be addressed in a suitable way through a reactive action. Conversely, other aspects linked to energy use, optimal solar energy conversion and ventilation, to mention a few, could be tackled with a much greater overall impact on the whole-building efficiency through predictive control, which is tightly integrated into the overall building automated management system.

Daylighting and noise are good examples of façade control areas that can often be based on closed-loop feedback (which are simpler to execute, thus making the control sequence less challenging) but require much shorter frequency (in the order of a few seconds) as they are linked to faster physical phenomena (compared to building heat transfer timescales). Continuous control in the range of seconds can be challenging to accept for the user. A very frequent change of a façade state that can be perceived by the users (eg, the degree of transparency of a glazing or the position of a shading device) to meet the control objectives may, in fact, disrupt their activities (and even become counterproductive in an overall perspective). This latter point pinpoints a whole new topic in envelope automation, which is the need to integrate the user in the control feedback sequence (user-in-the-loop) to ensure user-oriented automated control.

All the challenges mentioned and exemplified earlier are to be tackled considering the interface with the building-level automation architecture. Automated control of a building and its interface with occupants, which goes under the name of building automation (Ahmadi-Karvigh, Becerik-Gerber, & Soibelman, 2019), is conventionally executed through a three-level hierarchical structure typical of any building automation system: a field level, an automation level and a management level (Fig. 6.2). The field level consists of sensors and actuators, some of which are also programmed to interact with building users. Field-level devices are connected to the management level control units by direct connections or fieldbus protocols. The automation layer consists of control units that coordinate field-level measurements, execute control loops and collect short-term historical data. The highest level of the hierarchy, the management layer, is responsible for the overall system management through a centralized, top-down approach.

In this structure, dynamic elements in facades are often seen as the actuators in the three-level control architecture. The reading of the relevant variable for control purposes is usually centralized (with a few control units acquiring little data from a limited number of sensors), while the decision-making process that leads to a control signal to the actuators located in the façade elements (eg, an electric driver to operate a shading device or a chain driver to open a window) usually happens in the upper levels. This approach is economical and, to some extent, functional when the degree of complexity of the actuation and of the control is low; that is, where the control signals are either limited in numbers (few units to be controlled) or the same control signal is sent to a large number of actuators (eg, the control of a shading device is unique for an entire façade orientation and floor). However, when the number of units that requires some optimal operation through control strategies increases or when the type of control needs to be differentiated to meet a set of

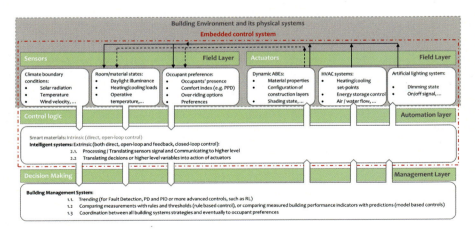

Figure 6.2 Control architecture for building systems, including dynamic ABEs and building systems: the continuous line represents feedback closed-loop control, the dashed line represents smart materials or open-loop controls. *ABE*, Advanced building envelope (Loonen et al., 2017).
Source: Illustration adapted from Loonen, R. C., Favoino, F., Hensen, J. L., Overend, M. (2017). Review of current status, requirements and opportunities for building performance simulation of adaptive facades. *Journal of Building Performance Simulation 10*(2), 205−223.

local performance requirements, especially across multiple domains (ie, thermal and visual comfort), there is a need to upgrade the approach of the control chain and to bring the decision-making process closer to the façade by means of integrating elements of the automation and of the management level within the façade system. This local intelligence enables a series of enhanced characteristics of the control that are of key relevance to ensure effective management of adaptive envelopes, as listed next.

- *performance-oriented* (oriented toward the achievement of a performance goal): the purpose of building envelope automation needs to clearly respond to one or more performance objectives so that the design and operation of such automation will contribute positively toward improving specific building performance aspects (ie, decreasing energy use, increasing share of on-site RES and improving occupant thermal and/or visual comfort) by using input information that corresponds to the specific conditions close to the façade (in contrast to general boundary conditions measured at the building level);
- *integrated*, even if physically separated, *with the automation of other building systems*: coordination with the automation of other building systems [such as heating, ventilation and air conditioning (HVAC) and lighting] will ensure that the performance objective is achieved synergistically with the minimum possible effort (in terms of overall costs related to energy, maintenance, impact on building occupant, etc.);
- *interfaced to building occupant:* building envelope automation, if successfully predicting occupant behaviour and allowing occupant interaction (Luna-Navarro et al., 2020), can effectively combine building energy performance requirements with occupant preferences to achieve overall occupant comfort with minimum expenditure without major occupant disruption or risking occupant override, which might hinder the effectiveness of the control strategy toward the achievement of a performance goal.

6.4 Elements, characteristics and logics for embedding intelligence in adaptive facades

A large building envelope with dynamic features can be intelligently managed if local conditions are taken into considerations. This means local measurement of performance indicators, with sufficient temporal and spatial discretization that suits the specific domain. Furthermore, given the interplay between the envelope and the occupants, as well as the interplay between the envelope and the other systems in the building, the decision-making processes to operate a dynamic envelope should include occupant interaction (Tabadkani et al., 2020) and a synergic approach with the control of the other buildings systems.

Bringing the collection of information and the decision-making process closer to the element that is involved in this control chain (the envelope) seems, therefore, an ideal solution to meeting such requirements. When an adaptive façade system is systematically adopted across the entire building façade, embedded intelligence at the façade level is the key to coordinating local and temporary requirements with long-term ones on the building level, and to truly include the occupant in the control process.

This vision is also supported by the development trend of many other information and communication technology (ICT) and Internet of things (IoT) infrastructures. Looking at other building systems, it is possible to see an increasing number of smart devices, such as thermostats, valves and pumps, which are now available with embedded local intelligence. These devices exhibit the ability to directly communicate with one another and with supervisory systems to enhance the control possibilities and reach a coordinated, optimal performance. Moreover, for such distributed intelligence, to support the control decision-making at the field and automation levels, there is an increasing focus on the use of edge computing infrastructures (performing calculations within local distributed devices translating information from the field layer to the management one and vice versa) as compared to the more traditional centralized computing (performing calculations centrally within the building management system, from the information collected by the field layer, outcomes of which are sent directly to local actuators).

Introducing the concept of distributed intelligence in the building turns the adaptive façade into a self-regulating system. In coordination with the other elements in the building management system, the façade-embedded intelligence acts to regulate the performance of the façade across different domains. Building and embedding systems in the façade means equipping the façade with the following elements:

- sensor(s) that record boundary conditions near the façade;
- microcontroller(s) that convert sensors signals into data, gather/organize data from different sources and use it in a decision-making process;
- communication port(s) that send signals to the façade actuator(s) and send/receive signals with the other intelligent units within the distributed intelligent network for coordinated management.

Recent developments and established trends in IoT make it possible to now imagine, in the coming years, large-scale integration of such systems in building envelopes with a cost/performance compromise that can justify the use of a distributed network of intelligent components. Considering the specific conditions that façade embedded systems should satisfy, it is possible to identify different priorities to develop and design embedded technology for envelope systems (Table 6.1). Compared to other domains, embedded systems for facades need to prioritize easy replacement and maintenance, extended service life, communication with other devices, possibly low-costs for all the components, while other features that can be crucial in different applications (accuracy, redundancy, footprint, computational power and mechanical resistance) can be considered less critical.

The integration of different elements that can compose an embedded system is already a reality in the market of different actuators, for example, electric drivers for the movement of shading devices or so-called chain drivers for mechanical opening of windows. In many cases, these systems include basic sensors, simple microprocessors that control the process variables inside the actuator (eg, the speed of the motor, the start/stop function) and communication ports that allow the actuator to be connected to the automation layer of the conventional building automation

Table 6.1 Priorities for key characteristics of embedded systems for facades.

High priority	Medium priority	Low priority
• Low energy use for system usage (potentially powered through façade-integrated PV systems) • Easy replacement and maintenance • Communication with other devices • Possibilities for user interaction • High system adaptability (software, remote servicing) • High system compatibility with different types of actuators and sensors • Autonomous management of failure (watchdog timer) • Long lifetime (potentially as long as that of the façade/actuators) • Low-cost for both microcontroller and sensors	• Minimal footprint • High computational power • Large memory • High accuracy in measurements of boundary/process variables • High resistance to harsh environmental conditions	• High mechanical resistance (shocks and vibrations) • Redundancy of system • High frequency of processing (in the range below the second) • High system adaptability (physical) once installed

structure. However, what makes these elements far from being local intelligent agents is the lack of an embedded strategy to control the components—in other words they lack the open-loop or closed-loop feedback that automates the action (in synergy with the other components of the building mechanical plants).

To trigger the (intelligent) feedback sequence of a full embedded system, and, thus, to autonomously control an adaptive building envelope, it is necessary to provide the local intelligent agent with adequate information about environmental conditions around the façade (the boundary conditions) and the higher desired performance level to be achieved. Due to the complexity and the cost of measuring actual building performance (ie, occupant comfort, building energy use), the measurement of physical quantities in the indoor environment can be used as a proxy for their prediction (ie, air temperature for thermal comfort or for energy use or illuminance on desk level for glare). Nevertheless, this can be affected by large inaccuracies that can arise from

- differences between local conditions of the measurement and global behaviour of the environment;
- nonlinear dependency between controlled variables (eg, incoming solar radiation, air mass flow, sound pressure, outdoor air CO_2 concentration, just to mention a few) and output parameters (air temperature, thermal comfort, visual comfort, IAQ, building energy use),

and their dependency from disturbances of highly dynamic boundary conditions (eg, climatic conditions, occupancy and plug loads) and
• lag between control action and related feedback on the environmental variable output.

For these reasons, models are more and more often employed in building automation to reduce prediction inaccuracies, both at the control logic and at the decision-making level. The recent ISO 52016:3 'Energy performance of buildings — Energy needs for heating and cooling, internal temperatures and sensible and latent heat loads — Part 3: Calculation procedures regarding adaptive building envelope elements' (under development during the time of writing this chapter) further defines how the decision-making is performed within the management layer for adaptive building envelopes. Particularly, for actively controllable façade systems, it distinguishes between manual, manual motorized operations and automated operations. Within the latter category, different levels of decision-making complexities are envisioned (Table 6.2), replicating the two higher classes of building automation of the ISO 16484−1:2010 (classes A and B): (1) 'rule-based control' (RBC, either simple and integrated); (2) 'model-based controls' (MBC).

RBC strategies are the most common control decision-making adopted in practice (Oldewurtel et al., 2012), both in overall building automation and in the first examples of embedded systems for adaptive envelopes. These strategies are simple and relatively low cost (in terms of software and hardware) and consist of a set of if-then rules based on the comparison of sensors measurements with specific thresholds. RBC is not able to predict the effect of the control decision on building performance. This is because it is only based on the information about the current state of the façade, the boundary conditions and related indoor environmental parameters. The design of an RBC is usually built upon heuristics, that is, expert knowledge of the dynamics of the building envelope system and how these could affect specific building performance parameters (Yoon, Park, & Augenbroe, 2011).

MBCs, conversely, exploit the prediction of the impact of the control action on the indoor environment to carry out decision-making with the aim of maximizing a specific building-level performance and improving the performance of closed-loop control. However, this comes at the cost of increased complexity. In fact, compared to the general requirements already presented, MBCs rely on additional characteristics to support the control decision-making system:

• accuracy: building models need to provide accurate information about the building performance parameters for which maximization is being pursued, which requires a certain level of calibration by means of locally available information; furthermore, a certain level of spatial discretization is necessary to implement reliable controls. This emphasizes, on the one hand, the need to distribute sensing and data processing, and on the other hand, the capability of this approach to accurately measure and predict disturbances (eg, complex sky measurements and weather forecasting, occupancy and endogenous loads measurement and forecasting);
• speed: such information needs to be available in a time frame which is compatible with the dynamics of the actuation of the ABE and of the domains the ABE aims to influence (ie, few seconds as far as glare/daylight control purposes to the order of hours as far as heat transfer phenomena are concerned), which emphasize the need for fast calculations

Table 6.2 ISO 52016:3 control types based on management layer complexity.

Field level/actuation	Management layer/control type		Characteristics
Environmentally activated building components (ie, smart glazings such as thermochromics)	1. Passive		According to the boundary conditions
Actively controllable building components (ie, electrochromic glazing, liquid crystal glazing, suspended particle devices (SPD) glazing, solar shading and double-skin facades)	2. Manual		Controlled by occupant with manual action on the actuator element
	3. Manual motorized		Controlled by occupant with remote electrical control
	4. Automated	4.1 Simple rule-based	Based on feedback from one physical quantity at the time (ie, external solar radiation, temperature or internal temperature)
		4.2 Integrated rule-based	Based on decision tree of rules based on feedback from the combination of different sensors, with different priorities (ie, occupancy, glare, daylight and energy)
		4.3 Model-based control	Performance-oriented control aimed at optimizing a certain performance objective, based on a model (physical, reduced order or black-box model)

and forecasting since a certain time lag can impact the effectiveness of the control (ie, lag between field-level measurement and actuation, and/or vice versa, lag between actuation and impact on physical/physiological variables).

These additional requirements result in significant cost and effort in the design and implementation of MBC strategies and may hinder its cost effectiveness (Wang & Ma, 2008).

Local intelligence in adaptive facades can benefit from both RBC and MBC and combinations of them. Thus it is not possible to identify in absolute terms whether one or another is the best choice to execute the feedback loop in the decision-making process at the façade level. From the perspective of an integrated, synergic automation system where the adaptive façade is a distributed node of intelligence, high-level decision-making functions (those usually associated with the management layer), such as calculating optimal entering solar loads and HVAC configurations by thermal zones, should probably be carried out by the supervisory, building-level intelligence using MBCs. Decentralized, low-level controllers at façade-level can determine, with more accuracy and less computational time, using either RBCs or MBCs, the optimal configuration of the local actuators (Mork, Xhonneux, & Müller, 2020), by, for example, calculating the specific local façade configuration that can provide the optimal level of entering solar radiation without hindering visual and/or thermal comfort of occupants next to the façade.

Using the three cases already mentioned in this chapter (chromogenic glazing, operable windows and DSFs), we will demonstrate in the next section, through a combination of literature review and our personal experience with some research projects, how control of such systems is currently done using either RBCs and MBCs, and how these controls are evolving in the current research, development, and innovation panorama, moving toward embedding intelligence directly at the façade level.

6.5 Current examples from real-life implementation and research activities on controls for adaptive envelopes

6.5.1 Chromogenic technologies and solar shading

Control of shading devices and smart glazing in combination with artificial lighting is an example of façade automation that has reached a certain technology readiness level that allows its integration with current building automation systems with relatively established workflows and schemes (Fig. 6.3). Real-life performance of operable shading technologies and smart glazing is highly determined by the control strategy adopted (Favoino et al., 2016). The influence of the control strategy on the performance of these technologies has been addressed by several studies (Correia da Silva, Leal, & Andersen, 2012; Gugliermetti & Bisegna, 2003; Lee & Tavil, 2007; Tzempelikos & Shen, 2013; Zinzi, 2006) that have highlighted the sensitivity

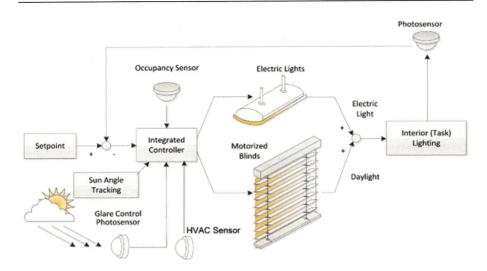

Figure 6.3 Illustrative scheme of the elements of actuation for blind operation (Shen, Hu, & Patel, 2014).
Source: Original illustration in Shen, E., Hu, J., Patel, M. (2014). Energy and visual comfort analysis of lighting and daylight control strategies. *Building and Environment 78*, 155–170.

of both indoor comfort levels and of building energy use not just to the variation of the properties of chromogenic technologies and shading devices but also to their control settings. Assimakopoulos, Tsangrassoulis, Santamouris, and Guarracino (2007) estimated a deviation of energy consumption of 24% for winter heating, 39% for summer cooling, 20% for winter electricity and 63% for summer electricity due to different control strategies of operable shading (daylight-based time scheduled or fuzzy logic controls). As far as occupants' preferences regarding active control of solar radiation are concerned, most studies show that occupants tend to operate them to avoid glare (Gunay, O'Brien, Beausoleil-Morrison, & Gilani, 2017; Van Den Wymelenberg, 2012; Zhang & Birru, 2012), suggesting that in the design of a control logic for dynamic envelope components glare protection should be prioritized over other aspects, such as thermal comfort, energy performance and view to the outside. In the perimeter zone, it is demonstrated how the incoming solar radiation could significantly impact occupant thermal comfort (Luna-Navarro et al., 2019). Considering the contribution of the solar radiation in the human body heat balance it is straightforward that this quantity impacts on how smart glazing should be controlled.

RBCs are currently the most widespread for the management of incoming solar radiation in buildings (Kunwar, Cetin, & Passe, 2018). The most common parameters used in RBC strategies are (1) solar geometry (Lee, DiBartolomeo, & Selkowitz, 1998; Zhang & Birru, 2012); (2) outdoor horizontal or vertical illuminance (Kim, Park, Yeo, & Kim, 2009; Yun, Yoon, & Kim, 2014); (3) outdoor horizontal or vertical irradiance (Gugliermetti & Bisegna, 2003; Wienold, Frontini,

Embedding intelligence to control adaptive building envelopes

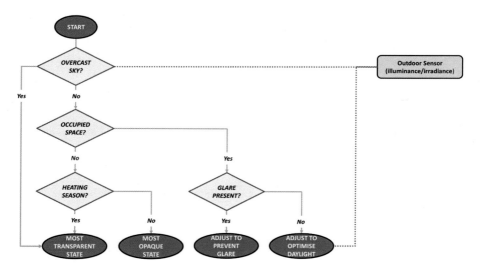

Figure 6.4 Illustrative scheme of a typical heuristic rule-based control decision tree for smart glazing.
Source: Original illustration by the authors.

Herkel, & Mende, 2011); (4) work plane illuminance (Lee & Tavil, 2007; Olbina & Hu, 2012); (5) presence and number of occupants (Kim et al., 2009; Lee, Claybaugh, & LaFrance, 2012); (6) indoor air temperature (Loonen et al., 2014; Nicoletti, Carpino, Cucumo, & Arcuri, 2020); and (7) the presence of heating or cooling loads (DeForest, Shehabia, Selkowitz, & Milliron, 2017; Jonsson & Roos, 2010). The industry standard to control smart glazing and operable shading combines these parameters into tree-structured RBC strategies, defining priorities between glare, daylight and energy use aspects by making the use of different measurements as proxies for such building performance aspects (ie, solar geometry and intensity, cloud coverage and indoor and outdoor temperatures), as shown in Fig. 6.4.

Simplified numerical models can be adopted to inform an RBC to evaluate complex parameters, for which a direct measurement is challenging outside a laboratory setting (eg, glare metrics, operative temperatures and air velocity). This approach is based on models derived from measurements of some environmental quantities that are statistically correlated to the quantities that cannot directly be measured in real implementation. This strategy takes the name of model-enhanced RBCs (Tzempelikos & Shen, 2013; Xiong & Tzempelikos, 2016) and represents a more advanced control logics than a simple RBC. To account for the delayed effect of solar radiation on internal building temperatures, fuzzy logics and proportional−integral−derivative, controls based on environmental conditions and occupants' preferences (Assimakopoulos et al., 2007; Kristl, Košir, Trobec Lah, & Krainer, 2008), have been also evaluated. Finally, to correctly account for the combined effect of modulating solar radiation, and of the climate and occupant disturbances, and to optimize building energy use and/or indoor temperature profiles, different researchers have explored the potential of adopting model-predictive controls (MPCs) (Dussault & Gosselin, 2017; Favoino et al., 2016; Firląg et al., 2015; Le, Bourdais, & Guéguen, 2014; Huchuk, Burak Gunay, O'Brien, & Cruickshank, 2016). According to Favoino et al. (2016), such MPC strategies could yield a significant reduction in the building energy use compared to RBC strategies for conditions in which modulation of the solar radiation can balance heat losses/gains through the building envelope. In fact, in temperate climates and in shoulder seasons, the improvements enabled by the use of MPC strategies compared to state-of-the art heuristic RBCs are in the range of 20%–50%, as confirmed by the study of Gehbauer, Blum, Wang, and Lee (2020), while in more extreme climate conditions, in which energy uses are higher, this difference can drop significantly. As far as the modulation of solar radiation is concerned, MPC-enhanced RBC strategies can provide a cost-effective solution, mimicking the MPC optimal control with a low decrease in operational performance (May-Ostendorp, Henze, Corbin, Rajagopalan, & Felsmann, 2011), as shown by Piscitelli et al. (2019), to implement the receding-horizon control of electrochromic glazing in a simpler way by means of rule extraction from a well-trained MPC simulation dataset.

6.5.2 Operable windows

Window opening is delegated to direct occupant action in the majority of buildings. Nevertheless, in the last decade many studies have focused on the topic of window opening automation and its impact on the competing requirements of winter and summer thermal comfort, energy use reduction, IAQ and acoustic comfort. Most of the studies focus on the development and/or evaluation of RBC strategies to automate natural ventilation for the purpose of reducing building cooling energy use (without increasing heating energy), as per Sorgato et al. (2016), or summer overheating risk (Barbosa et al., 2015), which usually making the use of simple rules based on indoor and outdoor temperature difference and adopting internal air temperature as a proxy for global thermal comfort. Psomas, Fiorentini, Kokogiannakis, and Heiselberg (2017), Sykes (2017) and Fu and Wu (2015) adopted instead a global comfort model (based on either predicted mean vote (PMV) or operative temperature, hence adaptive comfort) to elaborate and implement RBC strategies. Sykes (2017) demonstrated that the advantage of adopting a comfort model in the rule-based decision-making can lead to up to a 56% reduction in the percentage of people dissatisfied with the thermal environment. In the study of Fu & Wu (2015), a PMV-driven RBC control is compared to an adaptive comfort-driven RBC to automate the window opening in a hybrid ventilation system, resulting in an increase of 15%–20% of the time in which a comfortable thermal condition is reached. Nevertheless, such complex indicators or comfort (PMV and/or operative temperatures) are very sensitive to local conditions beyond air temperature (radiant field and air velocity as well for PMV), which are not easy to continuously measure in real implementations. In these cases the use of models to inform the control decision-making is of great importance.

When it comes to IAQ, natural ventilation enabled through openable windows plays a fundamental role in reducing CO_2 concentration, fine particulates and polluting substances. Ren et al. (2021) demonstrated that the concentration of CO_2 is very well correlated to the opening of windows by the occupants. Therefore automation of window opening based on specific CO_2 thresholds is of particular importance to maintain IAQ (Jung, Hong, Oh, Kang, & Lee, 2019). Mass flow across building openings is highly affected by climate uncertainties (ie, wind variability, solar radiation impact on stack effect). This flow impacts the heat balance and the temperatures inside the building, but this effect is particularly difficult to predict when inertial phenomena are present (for buildings with significant thermal mass). Furthermore, building occupancy (which can, in some cases, be highly variable) also plays a role as it is a driver to set the requirements for ventilation to achieve IAQ control.

Forecasting the capability of both external and internal effects can play a significant role in the automation of window opening that aims at balancing energy use, thermal comfort and IAQ requirements (Sykes, 2017). Particularly, in the study of Sykes (2017), in comparison to a strategy based on RBC, when MPC was adopted to avoid overheating in a single zone, the amount of hours above 25°C was reduced by 39%, 28% and 15% for low, medium and high thermal mass constructions, respectively. Walker, Lombardi, Lesecq, and Roshany-Yamchi (2017) adopted an MPC approach to demonstrate the potential to automate window opening of a mix-mode ventilated building to simultaneously optimize thermal comfort and indoor-air quality. The potential of a distributed MPC approach (in which local controllers optimize window openings of three different zones) was evaluated against a more classical architecture with a centralized MPC. Both MPC approaches were able to reduce the thermal discomfort and percentage of CO_2 to the minimum extent, while compared to RBC, the centralized MPC controller reduced heating energy use by 30%, but the distributed one reduced it by nearly 24% with one-fourth of the computational effort. This shows that such an approach is compatible with an embedded controller and decentralized intelligence, as demonstrated by Fiorentini, Serale, Kokogiannakis, Capozzoli, and Cooper (2019) who implemented such an MPC approach through an embedded single board controller that performed the tasks of the field and the automation layer. Given how much window opening remains a behaviour closely linked to the sensitivity of the occupants (Yun & Steemers, 2008), more and more applications have emerged that use machine learning models to predict the window opening behaviour of occupants (Dai, Liu, & Zhang, 2020) and use such models in synergy with automation (Fig. 6.5).

6.5.3 Double-skin facades

Regardless of the control of the DSF cavity solar shading, the common practice for the operation of conventional DSFs is to control cavity openings according to the indoor and outdoor boundary conditions, switching mainly between a thermal buffer mode (which increases the level of separation between indoor and outdoor environment) and an alternative air path which is aimed at a specific purpose, that is,

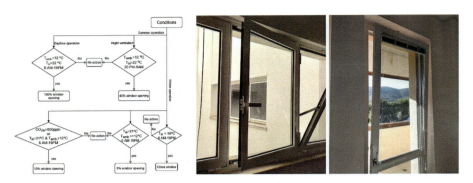

Figure 6.5 Rule-based control scheme for operable windows (Rabani et al., 2021) and different opening types for operable windows (Day et al., 2020).
Source: From Rabani, M., Bayera Madessa, H., Nord, N. (2021). Achieving zero-energy building performance with thermal and visual comfort enhancement through optimization of fenestration, envelope, shading device and energy supply system. *Sustainable Energy Technologies and Assessments 44*, 101020 https://doi.org/10.1016/j.seta.2021.101020 is an Open Access CC-BY publication.

removing unwanted solar gains (outdoor air curtain mode, for cooling dominated climates and buildings) or preheating the ventilation air (supply air, for heating dominated climates and buildings). Commonly used decision-making algorithms for DSF ventilation are based on RBCs, making the use of specific thresholds related to weather conditions (ie, outdoor air temperature, incident solar radiation), façade-related physical variables (ie, air cavity temperature) and/or indoor conditions (ie, indoor air temperature, occupancy). As for solar shading concerned, the most adopted solar shading control option in DSFs are by far the ones preventing direct solar radiation within the indoor environment when occupied, such as the cut-off control for venetian blinds (Jain & Garg, 2018), which can, however, significantly reduce the daylight availability. Choi, Joe, Kwak, and Huh (2012), for example, proposed an RBC for a multistory DSF, based on occupancy, cavity air temperature and indoor air temperature, which reached marginal improvements in terms of heating energy use compared to a benchmark DSF with no automation (thermal buffer mode). In Wigginton and Harris (2002), among all the different built DSF case studies, only very simple RBCs are implemented. For example, the automation of the ventilation air path (between outdoor air curtain or thermal buffer), of the multistory DSF façade of the Helicon building (in London, the United Kingdom), is only based on the difference between air cavity and indoor temperature, while incident solar irradiation and interior illuminance threshold levels at some reference points are adopted to control the cavity venetian blind tilt angle.

Increased flexibility can be achieved if multiple ventilation air paths are included in the design of a single DSF at the same time (ie, outdoor air curtain, supply air, thermal buffer and exhaust air) by allowing multiple ventilation openings toward the indoor and outdoor environment. This, coupled with the operation of the cavity shading device, could make the DSF an even more comprehensive environmental regulator able to balance competing building performance aspects if properly and effectively automated. In this case, MBC would be the preferred strategy to enable a suitable real-time management of such a complex system. Park, Augenbroe, Sadegh, Thitisawat, and Messadi (2004) adopted a lumped model to find the optimal DSF configuration for real-time control (actuating on blind slat angle, ventilation mode and cavity air-path). One of the features of such a model is its capability to dynamically react to the environmental data through real-time optimization algorithms in terms of energy, visual comfort and thermal comfort. This model was consequentially adopted by Park and Augenbroe (2013) to investigate the differences between local (room-level) and centralized (building-level) optimization of MBC of DSFs. The results show no significant difference between the two optimization approaches in terms of energy savings, suggesting that local decentralized controllers could be as effective as coordinated decision-making at the building level for the specific case study. Other examples showed the possibility of adopting black-box data-driven models based on artificial neural networks (ANNs) to inform the control of a DSF in combination with an air handling unit (Seo, Yoon, Mun, & Cho, 2019). In particular, the ANN-informed MBC for the control of a multistorey DSF in a cooling-dominated climate showed a 5% cooling energy use improvement compared to a simple RBC based on cavity and indoor air temperatures. Recently, Gennaro, Goia, De Michele, Perino, and Favoino (2021) investigated the potential improvements of MBC as applied to a single story DSF compared to state-of-the-art heuristic RBC strategies (these latest ones are shown in Fig. 6.6). Within this case study, MBC is adopted to optimize building loads (no receding horizon control is adopted) and visual comfort by implementing all levels of automation within a single board controller (field, automation and management layer performing the optimization). In the temperate climate under analysis (Torino, Italy), the MBC was able to reduce overall energy use between 12% and 18% compared to benchmark RBCs. Additionally, compared to cut-off venetian blind control, the MBC was able to decrease the amount of time with desk level illuminance below 100 lx from

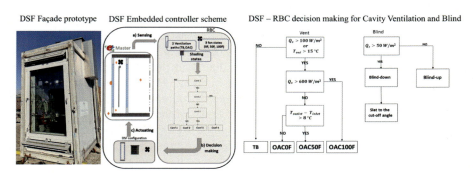

Figure 6.6 Double-skin façade RBC control scheme for embedded controller. *RBC*, Rule-based control.
Source: Original illustration by the authors.

22% to 9% and nearly double the time above 300 lx (autonomous daylight) from 45% to 84%.

6.6 Conclusion

Embedding intelligence in building envelopes is the natural development when the building skin evolves from a passive shield to an adaptive, dynamic system that, in coordination with other components of the building mechanical plant, contributes to delivering a comfortable and healthy indoor environment with minimal energy use. The performance of adaptive facades with extrinsic control is heavily dependent on the logics and processes that are carried out to manage the dynamic features of the envelope. Conventional building automation schemes, currently adopted to manage dynamic envelope components such as shading devices or openable windows, are hardly suitable to carry out efficient management of large-scale adaptive facades targeting multiple and competing performance requirements.

The ongoing digital revolution and ICT innovation, therefore, become the driving elements that enable more advanced control functions, and consequently more intelligent adaptive facades (Böke, Knaack, & Hemmerling, 2020). The technical possibilities in façade automation technologies are booming today by the reductions in terms of both size and costs of electronic components. Diffuse sensors and microprocessors enable the possibility to acquire large sets of distributed data and embed distributed software and calculation capabilities (Arnesano et al., 2019). In combination with the possibility of comprehensive networking of smart components in the building's mechanical plants, this gives unprecedented (theoretical) control over the physical environment through dynamic building envelope technologies.

Bringing reliable (measured and modelled) information on the boundary conditions around the façade, both concerning the present time and the future within a certain horizon, makes it possible to exploit to a greater degree the possibilities offered by dynamic features of adaptive facades (Gehbauer et al., 2020; Piscitelli et al., 2019). This new paradigm in control requires a change from 'simple' rule-based sequences to more advanced MBCs. This opens up a completely new interaction between the different nodes of a network of intelligent façade units and building elements. The interaction between the local (façade-integrated) controller and the other autonomous controllers in the building (Mork et al., 2020), and with the occupant, will be one of the major research trajectories to ensure distributed optimal control across multiple domains and levels of the building.

New possibilities offered by ICT will play a crucial role in enabling more sophisticated control sequences. Complex models and computationally expensive processes are unlikely to be affordable in decentralized microcontrollers, and this might become the bottleneck in the adoption of model-based feedback loops. Balancing edge computing and cloud computing in the right way can represent a viable solution to enable advanced control strategies right at the façade level without the need to deploy embedded systems that can be prohibitively expensive (in both energy use and investment costs). How to exploit edge computing versus cloud computing is a completely innovative field in the domain of adaptive facades and will likely be a very rich area of research and development in the coming years.

In addition to sensing technology, communication technologies and computational technologies, application and advances in machine learning and data mining will play a crucial role in moving toward an intelligent, autonomous façade. This is of particular interest when moving from rule-based to MBC, and especially when the user is included in the control sequence. The use of advanced statistical modelling for predictive control of buildings is already an ongoing research trend at the whole-building level (eg, Chen, Tong, Zheng, Samuelson, & Norford, 2020), and this includes application areas that are of interest for facades (eg, daylighting: (Ayoub, 2020)). Using these techniques to enable a better user-façade interaction while assuring personalized controls that include user's preference (Xiong, Tzempelikos, Bilionis, & Karava, 2019) is a necessary research direction to be pursued to ensure that effective control of adaptive facades is truly intelligent, also from the occupants' perspective.

References

Ahmadi-Karvigh, S., Becerik-Gerber, B., & Soibelman, L. (2019). Intelligent adaptive automation: A framework for an activity-driven and user-centered building automation. *Energy and Buildings, 188*, 184–199.

Arnesano, M., Bueno, B., Pracucci, A., Magnagni, S., Casadei, O., & Revel, G.M. (2019). Sensors and control solutions for Smart-IoT façade modules In: Proceedings of the 2019 IEEE International Symposium on Measurements and Networking (M&N), Catania 8–10 July 2019, art. no. 8805024.

Assimakopoulos, M. N., Tsangrassoulis, A., Santamouris, M., & Guarracino, G. (2007). Comparing the energy performance of an electrochromic window under various control strategies. *Building and Environment, 42*(8), 2829–2834.

Ayoub, M. (2020). A review on machine learning algorithms to predict daylighting inside buildings. *Solar Energy, 202*, 249–275.

Baetens, R., Jelle, B. P., & Gustavsen, A. (2010). Phase change materials for building applications: A state-of-the-art review. *Energy and Buildings, 42*, 1361–1368.

Barbosa, R. M., Bartak, M., Hensen, J. L. M., & Loomans, M. G. L. C. (2015). Ventilative cooling control strategies applied to passive house in order to avoid indoor overheating. In: *Building simulation 2015: 14th international conference of IBPSA, December 7–9, Hyderabad, India* (pp. 1142–1148). [2637] International Building Performance Simulation Association (IBPSA).

Böke, J., Knaack, U., & Hemmerling, M. (2020). Automated adaptive façade functions in practice-case studies on office buildings. *Automation in Construction, 113*, 103113.

Bunthof, L. A. A., Kreuwel, F. P. M., Kaldenhoven, A., Kin, S., Corbeek, W. H. M., Bauhuis, G. J., & Schermer, J. J. (2016). Impact of shading on a flat CPV system for façade integration. *Solar Energy, 140*, 162–170.

Chen, Y., Tong, Z., Zheng, Y., Samuelson, H., & Norford, L. (2020). Transfer learning with deep neural networks for model predictive control of HVAC and natural ventilation in smart buildings. *Journal of Cleaner Production, 254*, 119866.

Choi, W., Joe, J., Kwak, Y., & Huh, J. H. (2012). Operation and control strategies for multi-storey double skin facades during the heating season. *Energy and Buildings, 49*, 454–465.

Correia da Silva, P., Leal, V., & Andersen, M. (2012). Influence of shading control patterns on the energy assessment of office spaces. *Energy and Buildings, 50*, 35–48.

Dai, X., Liu, J., & Zhang, X. (2020). A review of studies applying machine learning models to predict occupancy and window-opening behaviours in smart buildings. *Energy and Buildings, 223*, 110159. Available from https://doi.org/10.1016/j.enbuild.2020.110159.

Day, J. K., McIlvennie, C., Brackley, C., Tarantini, M., Piselli, C., Hahn, J., & Pisello, A. L. (2020). A review of select human-building interfaces and their relationship to human behavior, energy use and occupant comfort. *Building and Environment, 178*, 106920.

De Michele, G., Loonen, R., Saini, H., Favoino, F., Avesani, S., Papaiz, L., & Gasparella, A. (2018). Opportunities and challenges for performance prediction of dynamic complex fenestration systems (CFS). *Journal of Façade Design and Engineering, 6*(3), 101–115.

DeForest, N., Shehabia, A., Selkowitz, S., & Milliron, D. J. (2017). A comparative energy analysis of three electrochromic glazing technologies in commercial and residential buildings. *Applied Energy, 192*, 95–109.

Dussault, J. M., & Gosselin, L. (2017). Office buildings with electrochromic windows: A sensitivity analysis of design parameters on energy performance, and thermal and visual comfort. *Energy and Buildings, 153*, 50–62.

Favoino, F., Fiorito, F., Cannavale, A., Ranzi, G., & Overend, M. (2016). Optimal control and performance of photovoltachromic switchable glazing for building integration in temperate climates. *Applied Energy, 178*, 943–961.

Favoino, F., Jin, Q., & Overend, M. (2017). Design and control optimisation of adaptive insulation systems for office buildings. Part 1: Adaptive technologies and simulation framework. *Energy, 127*, 301–309.

Favoino, F., Loonen, R. C., Doya, M., Goia, F., Bedon, C., & Babich, F. (2018). In F. Favoino (Ed.), *Building performance simulation and characterisation of adaptive facades: Adaptive façade network*. TU Delft Open.

Fiorentini, M., Serale, G., Kokogiannakis, G., Capozzoli, A., & Cooper, P. (2019). Development and evaluation of a comfort-oriented control strategy for thermal management of mixed-mode ventilated buildings. *Energy and Buildings, 202*, 109347.

Firląg, S., Yazdanian, M., Curcija, C., Kohler, C., Vidanovic, S., Hart, R., & Czarnecki, S. (2015). Control algorithms for dynamic windows for residential buildings. *Energy and Buildings, 109*, 157–173.

Fu, X., & Wu, D. (2015). Comparison of the efficiency of building hybrid ventilation systems with different thermal comfort models. *Energy Procedia, 78*, 2820–2825. Available from https://doi.org/10.1016/j.egypro.2015.11.640.

Gehbauer, C., Blum, D. H., Wang, T., & Lee, E. S. (2020). An assessment of the load modifying potential of model predictive controlled dynamic facades within the California context. *Energy and Buildings, 210*, 109762.

Gennaro G., Goia F., De Michele G., Perino M., & Favoino F. (2021). Embedded single-board controller for Double Skin Façade: A co-simulation virtual test bed. In: *Proceedings of international conference of building performance simulation association BS2021*, 1–3 Sep 2021, Bruges, BE.

Goia, F., Perino, M., Serra, V., & Zanghirella, F. (2010). Towards an active, responsive, and solar building envelope. *Journal of Green Building, 5*(4), 121–136.

Gugliermetti, F., & Bisegna, F. (2003). Visual and energy management of electrochromic windows in Mediterranean climate. *Building and Environment, 38*, 479–492.

Gunay, H. B., O'Brien, W., Beausoleil-Morrison, I., & Gilani, S. (2017). Development and implementation of an adaptive lighting and blinds control algorithm. *Building and Environment, 113*, 185−199.

Huchuk, B., Burak Gunay, H., O'Brien, W., & Cruickshank, C. A. (2016). Model-based predictive control of office window shades. *Building Research & Information, 44*(4), 445−455.

Jain, S., & Garg, V. (2018). A review of open loop control strategies for shades, blinds and integrated lighting by use of real-time daylight prediction methods. *Building and Environment, 135*, 352−364.

Jin, Q., & Overend, M. (2014). Sensitivity of façade performance on early-stage design variables. *Energy and Buildings, 77*, 457−466.

Jonsson, A., & Roos, A. (2010). Evaluation of control strategies for different smart window combinations using computer simulations. *Solar Energy, 84*(1), 1−9.

Jung, W., Hong, T., Oh, J., Kang, H., & Lee, M. (2019). Development of a prototype for multi-function smart window by integrating photovoltaic blinds and ventilation system. *Building and Environment, 149*, 366−378. Available from https://doi.org/10.1016/j.buildenv.2018.12.026.

Kim, J. H., Park, J. K., Yeo, M. S., & Kim, W. (2009). An experimental study on the environmental performance of the automated blind in summer. *Building and Environment, 44*(7), 1517−1527.

Kristl, Ž., Košir, M., Trobec Lah, M., & Krainer, A. (2008). Fuzzy control system for thermal and visual comfort in building. *Renewable Energy, 33*(4), 694−702.

Kunwar, N., Cetin, K. S., & Passe, U. (2018). Dynamic shading in buildings: A review of testing methods and recent research findings. *Current Sustainable/Renewable Energy Reports, 5*(1), 93−100.

Le, K., Bourdais, R., & Guéguen, H. (2014). From hybrid model predictive control to logical control for shading system: A support vector machine approach. *Energy and Buildings, 84*, 352−359.

Lee, E. S., Claybaugh, E. S., & LaFrance, M. (2012). End user impacts of automated electrochromic windows in a pilot retrofit application. *Energy and Buildings, 47*, 267−284.

Lee, E. S., DiBartolomeo, D. L., & Selkowitz, S. E. (1998). Thermal and daylighting performance of an automated venetian blind and lighting system in a full-scale private office. *Energy and Buildings, 29*(1), 47−63.

Lee, E. S., & Tavil, A. (2007). Energy and visual comfort performance of electrochromic windows with overhangs. *Building and Environment, 42*(6), 2439−2449.

Locher, B., Piquerez, A., Habermacher, M., Ragettli, M., Röösli, M., Brink, M., ... Wunderli, J. M. (2018). Differences between outdoor and indoor sound levels for open, tilted, and closed windows. *International Journal of Environmental Research and Public Health, 15*(1), 149.

Loonen, R. C., Favoino, F., Hensen, J. L., & Overend, M. (2017). Review of current status, requirements and opportunities for building performance simulation of adaptive facades. *Journal of Building Performance Simulation, 10*(2), 205−223.

Loonen, R. C. G. M., Singaravel, S., Trčka, M., Cóstola, D., & Hensen, J. L. M. (2014). Simulation-based support for product development of innovative building envelope components. *Automation in Construction, 45*, 86−95.

Luna-Navarro, A., Blanco Cadena, J. D., Favoino, F., Donato, M., Poli, T., Perino, M., & Overend, M. (2019). Occupant-centred control strategies for adaptive facades: A preliminary study of the impact of shortwave solar radiation on thermal comfort. In: *Building simulation 2019: 16th conference of IBPSA* (pp. 4910−4917).

Luna-Navarro, A., Loonen, R., Juaristi, M., Monge-Barrio, A., Attia, S., & Overend, M. (2020). Occupant-façade interaction: A review and classification scheme. *Building and Environment, 177*, 1−13.

May-Ostendorp, P. T., Henze, G. P., Corbin, C. D., Rajagopalan, B., & Felsmann, C. (2011). Model-predictive control of mixed-mode buildings with rule extraction. *Building and Environment, 46*(2), 428−437.

Mork, M., Xhonneux, A., & Müller, D. (2020). Hierarchical Model Predictive Control for complex building energy systems. *Bauphysik, 42*, 306−314. Available from https://doi.org/10.1002/bapi.202000031.

Nicoletti, F., Carpino, C., Cucumo, M. A., & Arcuri, N. (2020). The control of venetian blinds: A solution for reduction of energy consumption preserving visual comfort. *Energies, 13*(7), 1731. Available from https://doi.org/10.3390/en13071731.

Nielsen, M. V., Svendsen, S., & Jensen, L. B. (2011). Quantifying the potential of automated dynamic solar shading in office buildings through integrated simulations of energy and daylight. *Solar Energy, 85*, 757−768.

Olbina, S., & Hu, J. (2012). Daylighting and thermal performance of automated split-controlled blinds. *Building and Environment, 56*, 127−138.

Oldewurtel, F., Parisio, A., Jones, C. N., Gyalistras, D., Gwerder, M., Stauch, V., ... Morari, M. (2012). Use of model predictive control and weather forecasts for energy efficient building climate control. *Energy and Buildings, 45*, 15−27.

Park, C. S., & Augenbroe, G. (2013). Local vs. integrated control strategies for double-skin systems. *Automation in Construction, 30*, 50−56.

Park, C. S., Augenbroe, G., Sadegh, N., Thitisawat, M., & Messadi, T. (2004). Real-time optimization of a double-skin façade based on lumped modeling and occupant preference. *Building and Environment, 39*(8), 939−948.

Perino, M., & Serra, V. (2015). Switching from static to adaptable and dynamic building envelopes: A paradigm shift for the energy efficiency in buildings. *Journal of Façade Design and Engineering, 3*, 143−163. Available from https://doi.org/10.3233/FDE-150039.

Piscitelli, M. S., Brandi, S., Gennaro, G., Capozzoli, A., Favoino, F., & Serra, V. (2019). Advanced control strategies for the modulation of solar radiation in buildings: MPC-enhanced rule-based control. In: *Proceedings of the 16th IBPSA Conference, Rome, Italy* (pp. 869−876), Sept. 2−4, 2019, https://doi.org/10.26868/25222708.2019.210609.

Psomas, T., Fiorentini, M., Kokogiannakis, G., & Heiselberg, P. (2017). Ventilative cooling through automated window opening control systems to address thermal discomfort risk during the summer period: Framework, simulation and parametric analysis. *Energy and Buildings, 153*, 18−30. Available from https://doi.org/10.1016/j.enbuild.2017.07.088.

Rabani, M., Bayera Madessa, H., & Nord, N. (2021). Achieving zero-energy building performance with thermal and visual comfort enhancement through optimization of fenestration, envelope, shading device, and energy supply system. *Sustainable Energy Technologies and Assessments, 44*, 101020. Available from https://doi.org/10.1016/j.seta.2021.101020.

Ren, J., Zhou, X., An, J., Yan, D., Shi, X., Jin, X., & Zheng, S. (2021). Comparative analysis of window operating behavior in three different open-plan offices in Nanjing. *Energy and Built Environment, 2*, 175−187. Available from https://doi.org/10.1016/j.enbenv.2020.07.007.

Saelens, D., Carmeliet, J., & Hens, H. (2003). Energy performance assessment of multiple-skin facades. *HVAC and R Research, 9*, 167−185.

Seo, B., Yoon, Y. B., Mun, J. H., & Cho, S. (2019). Application of artificial neural network for the optimum control of HVAC systems in double-skinned office buildings. *Energies, 12*(24), 4754.

Shen, E., Hu, J., & Patel, M. (2014). Energy and visual comfort analysis of lighting and daylight control strategies. *Building and Environment, 78*, 155−170.

Sorgato, M. J., Melo, A. P., & Lamberts, R. (2016). The effect of window opening ventilation control on residential building energy consumption. *Energy and Buildings, 133*, 1−13. Available from https://doi.org/10.1016/j.enbuild.2016.09.059.

Sykes, J. S. (2017). Control of naturally ventilated buildings: A model predictive control approach (Doctoral dissertation). University of Sheffield.

Tabadkani, A., Roetzel, A., Li, H. X., & Tsangrassoulis, A. (2020). A review of occupant-centric control strategies for adaptive facades. *Automation in Construction, 122*, 103464.

Taveres-Cachat, E., Favoino, F., Loonen, R., & Goia, F. (2021). Ten questions concerning co-simulation for performance prediction of advanced building envelopes. *Building and Environment, 191*, 107570.

Tzempelikos, A., & Shen, H. (2013). Comparative control strategies for roller shades with respect to daylighting and energy performance. *Building and Environment, 67*, 179−192.

Van Den Wymelenberg, K. (2012). Patterns of occupant interaction with window blinds: A literature review. *Energy and Buildings, 51*, 165−176.

Walker, S. S., Lombardi, W., Lesecq, S., & Roshany-Yamchi, S. (2017). Application of distributed model predictive approaches to temperature and CO_2 concentration control in buildings. *IFAC-PapersOnLine, 50*(1), 2589−2594.

Wang, S., & Ma, Z. (2008). Supervisory and Optimal Control of Building HVAC Systems: A Review. *HVAC&R Research, 14*(1), 3−32.

Wienold, J., Frontini, F., Herkel, S., & Mende, S. (2011). Climate based simulation of different shading device systems for comfort and energy demand. In: *Proceedings of building simulation 2011* (pp. 2680−2687).

Wigginton, M., & Harris, J. (2002). *Intelligent skins*. Gray Publishing.

Xiong, J., & Tzempelikos, A. (2016). Model-based shading and lighting controls considering visual comfort and energy use. *Solar Energy, 134*, 416−428.

Xiong, J., Tzempelikos, A., Bilionis, I., & Karava, P. (2019). A personalized daylighting control approach to dynamically optimize visual satisfaction and lighting energy use. *Energy and Buildings, 193*, 111−126.

Yoon, S. H., Park, C. S., & Augenbroe, G. (2011). On-line parameter estimation and optimal control strategy of a double-skin system. *Building and Environment, 46*(5), 1141−1150.

Yun, G., & Steemers, K. (2008). Time-dependent occupant behaviour models of window control in summer. *Building and Environment, 43*, 1471−1482. Available from https://doi.org/10.1016/j.buildenv.2007.08.001.

Yun, G., Yoon, K. C., & Kim, K. S. (2014). The influence of shading control strategies on the visual comfort and energy demand of office buildings panel. *Energy and Buildings, 84*, 70−85.

Zhang, S., & Birru, D. (2012). An open-loop venetian blind control to avoid direct sunlight and enhance daylight utilization. *Solar Energy, 86*(3), 860−866.

Zinzi, M. (2006). Office worker preferences of electrochromic windows: A pilot study. *Building and Environment, 41*(9), 1262−1273.

Biomimetic adaptive building skins: design and performance

Aysu Kuru[1,2], Philip Oldfield[1], Stephen Bonser[3] and Francesco Fiorito[1,4]
[1]School of Built Environment, University of New South Wales, Sydney, NSW, Australia, [2]School of Architecture, Design and Planning, University of Sydney, Sydney, NSW, Australia, [3]School of Biological, Earth and Environmental Sciences, University of New South Wales, Sydney, NSW, Australia, [4]Department of Civil, Environmental, Land, Building Engineering and Chemistry, Polytechnic University of Bari, Bari, Italy

Abbreviations

Bio-ABS biomimetic adaptive building skins
IR infrared
SHGC solar heat gain coefficient

7.1 Introduction

Living systems optimize their ability to survive through evolution. The transfer of biological adaptations can offer novel approaches for designing buildings to improve performance, especially in terms of facades, as facades act as thermal, acoustic and visual barriers (López et al., 2017). New technologies such as biomimetic adaptive building skins (Bio-ABS) are emerging that adapt to changing environmental conditions, host multiple functions, offer local controls for occupants' needs, provide efficient strategies for functional requirements, improve comfort and reduce building operational carbon emissions (Al-Obaidi, Ismail, Hussein, & Rahman, 2017). The adoption of biological mechanisms is known as biomimetics, a design process inspired by nature (Pohl and Nachtigall, 2015). Using biomimetics to discover the overlapping concepts in nature and technology benefits developing efficient strategies, since most organisms have advanced adaptations to survive with minimal reliability on external mechanisms (Knippers and Speck, 2012).

Environmental regulation in buildings (eg, thermal regulation, light regulation) often requires hosting multiple functions, such as providing daylight and limiting solar gains (Aelenei et al., 2016). However, most studies are limited in informing the transfer of multifunctional concepts from nature into building design (Svendsen and Lenau, 2019a). Although there are many examples of Bio-ABS, there are far fewer examples with measured environmental performance benefits that quantify the opportunities for energy savings, improved indoor environmental quality and comfort (Hosseini, Mohammadi, Rosemann, Schröder, & Lichtenberg, 2019). A significant challenge and reason why there is a lack of quantitative evidence is the required expertise in a range of disciplines to design and measure the performance of Bio-ABS, such as architecture, engineering, physics and biology (John et al., 2005; Vincent et al., 2006a).

Rethinking Building Skins. DOI: https://doi.org/10.1016/B978-0-12-822477-9.00018-8
© 2022 Elsevier Inc. All rights reserved.

This chapter focuses on the design processes of Bio-ABS and performance benefits of recent applications. In doing so, it draws attention to challenges, limitations and future opportunities in the field. The significant challenge of the required expertise in a range disciplines is addressed here through the development of a categorization for biomimetic design levels. This categorization aims at serving an initial discussion of how diverse disciplines can provide a conceptual basis for developing biomimetic designs. An opportunity is to learn from natural design principles regarding how living systems are formed through the evolutionary process, which show great potential to develop optimum strategies for Bio-ABS. More research and development focusing on customized digital tools, material, product and system innovation to realize complex Bio-ABS technologies are needed.

7.2 Biomimetic adaptive building skins

7.2.1 Definition and significance of biomimetic adaptive building skins

Buildings currently consume roughly 75% of the world's primary energy and release around 60% of the world's total greenhouse gases (Santamouris et al., 2018). Reducing building carbon emissions is an important goal to overcome these global challenges. Building performance is heavily influenced by the skin, being the barrier between the interior and exterior (Halliday, 2013; Knaack et al., 2014). Adaptive building skins can improve the building performance and, therefore, help reducing building carbon emissions (Pesenti et al., 2018; Giovannini et al., 2019). There are diverse types of adaptive building skins, including automatically controlled shutters, shape changing façade components, interactive facades with robotic design and acclimated kinetic envelopes (Del Grosso & Basso, 2010; Loonen, Trcka, Cóstola, & Hensen, 2013; Ramzy & Fayed, 2011; Wigginton & Harris, 2002; Hasselaar, 2006; Pan and Jeng, 2010).

Biomimetics is used extensively in developing adaptive building skins. This has resulted in the emergence of a new concept: Bio-ABS (Kuru, Oldfield, Bonser, & Fiorito, 2019). Examples of Bio-ABS include Flectofin, a shading device prototype inspired by the flapping mechanism of the bird-of-paradise flower (Lienhard et al., 2011; Schleicher, Lienhard, Poppinga, Speck, & Knippers, 2015). Another Bio-ABS example is the temperature-triggered Air Flow(er) that replaces the façade and provides shading and ventilation functions through smart materials embedded in its design (O'Payne and Johnson, 2009). Some other examples of Bio-ABS include the Adaptive Solar Façade, Biomimetic Façade based on Animal Fur and Gymnasium Façade that are described in detail in the following section of this chapter (Fig. 7.1) (Payne and Johnson, 2013; Park, 2014; Schleicher, Lienhard, Poppinga, Speck, & Knippers, 2015; Webb, Aye and Green, 2018; Sheikh and Asghar, 2019).

7.2.2 Bio-driven design concepts in building design

Biology has long been an inspiration for innovative engineered designs and presents a novel basis for technological thinking. Advanced through the evolutionary processes, biological solutions can be complex and highly responsive to changing functional requirements. Learning from biological functions, bio-driven buildings have

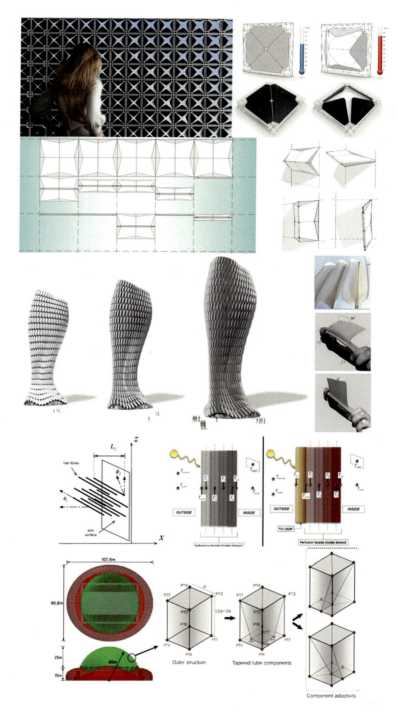

Figure 7.1 *Examples of Bio-ABS*. From top to bottom: The Air Flow(er), Adaptive Solar Façade, Flectofin, Biomimetic Façade based on Animal Fur and Gymnasium Façade. *Bio-ABS*, Biomimetic adaptive building skins.
Source: Copyright agreements provided by Andrew Payne at LIFT Architects, Wajiha Tariq Sheikh, Matthew Webb, Jongjin Park and Elsevier.

the potential to replace conventional and static systems to improve performance in a new and dynamic manner (Lepora, Verschure, & Prescott, 2013). Through using nature's adaptations, it is possible to develop building designs that offer structurally, mechanically, environmentally and material-efficient alternatives (Fayemi, Wanieck, Zollfrank, Maranzana, & Aoussat, 2017).

Biology has been integrated with building design through various forms of bio-driven design concepts. Definitions include bionics, biomorphism, biophilia, biomimicry, bioinspiration and biomimetics (Speck, Speck, Horn, Gantner, & Sedlbauer, 2017) (Fig. 7.2). Bio-driven design has initiated with the concept called bionics (Reap et al., 2005). Then, the transfer of natural forms, known as biomorphism, has become popular (Mazzoleni, 2013). Along with the use of natural forms, living systems have been adopted as parts of a design, mostly by involving green spaces, known as biophilia. The presence of plants, for instance, provides numerous psychological benefits to occupants and improves the indoor air quality (Makram, 2019). Other bio-driven design concepts termed biomimicry, bioinspiration and biomimetics are used interchangeably; however, they have slight distinctions in the meaning:

- *biomimicry* is copying nature in design;
- *bioinspiration* is a design inspired by nature; and
- *biomimetics* is the transfer of biological functions into designs (Fayemi et al., 2017).

Recently, new bio-driven design concepts have emerged, including

- *biornametics* is using natural patterns in architecture; and
- *electrosynbionics* is creating engineered devices inspired by nature to perform an electrical function (generation, use and storage of electricity) (Dunn, 2020; Gebeshuber, Gruber, & Imhof, 2015).

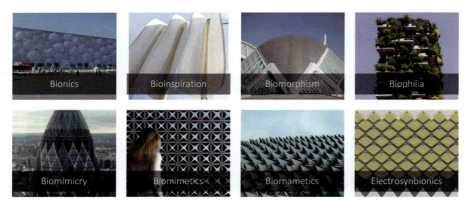

Figure 7.2 *Bio-driven design concepts.* Water Cube in Beijing (bionics), Flectofin (bioinspiration), City of Arts and Sciences in Valencia (biomorphism), Bosco Verticale in Milan (biophilia), the Gherkin in London (biomimicry), the Air Flow(er) (biomimetics), Esplanade Theatres by the Bay in Singapore (biornametics), shading/energy generating skin (electrosynbionics).
Source: Copyright agreements provided by Elsevier, Andrew Payne (LIFT Architects), Lidia Badarnah. Free copyright license is provided by Unsplash for the photographs taken by Kim Yong-Kwan, Victor Garcia, Alvaro Ibanez and Viktor Forgacs.

7.3 Designing biomimetic adaptive building skins

7.3.1 Biomimetic design processes and levels

Several biomimetic design processes have been developed to biomimetic systems. These processes follow either a top-down or a bottom-up approach, depending on how they initiate. For example, top-down processes start with a technical problem such as managing excessive solar gains. Bottom-up processes start with a biological solution; for example, examining the water conservation properties of cacti. Various works in the literature have reviewed existing biomimetic design processes (Al-Obaidi, Ismail, Hussein, & Rahman, 2017; Badarnah & Kadri, 2015; Kuru, Oldfield, Bonser, & Fiorito, 2020; Schleicher, Lienhard, Poppinga, Speck, & Knippers, 2015). The main conclusions drawn out are the following:

- limitations in systematically addressing technical problems;
- limitations in mapping the multifunctional characteristics of biological solutions; and
- challenges in developing biomimetic strategies that potentially require expertise in multi-disciplinary fields.

An important element in biomimetic design processes is the biomimetic design level. Several definitions of biomimetic design levels are presented by researchers (Fig. 7.3) (Gebeshuber & Drack, 2008; Kuru, Oldfield, Bonser, & Fiorito, 2020; López, 2017; Pohl & Nachtigall, 2015; Zari, 2012; Biomimicry Institute, 2018b). What informs the design level is the level at which the biological and technical domains belong. For instance, if the biological input is an animal's body form, the output will be the abstraction of this form into a design. This example takes place at a morphological level, as an animal's body shape would have evolved mostly as a morphological adaptation.

Biomimetic levels can be 'scale-based' and 'process-based', as defined in this work. Scale-based biomimetic design levels provide information regarding the scale; a system can be developed from biological strategies. For example, the thorny outer layer of a lizard species can be transferred into a ridged surface material at a smaller scale. This means that the spatial scale it has evolved may inform its translation into a design. Process-based biomimetic design levels refer to the functional processes of an organism. For instance, evapotranspiration for homeostasis is process-based. This means that its translation into a design will be the process of evaporative cooling, independent from the scale.

7.3.2 Future opportunities in developing Bio-ABS: natural design principles

Organisms are formed by natural design principles and some of these principles in nature are provided next:

- *Adaptability* is the ability to adapt to chemical and physical changes. For example, the Namibian beetle collects water by condensation with a special behavioural adaptation (Mazzoleni, 2013).

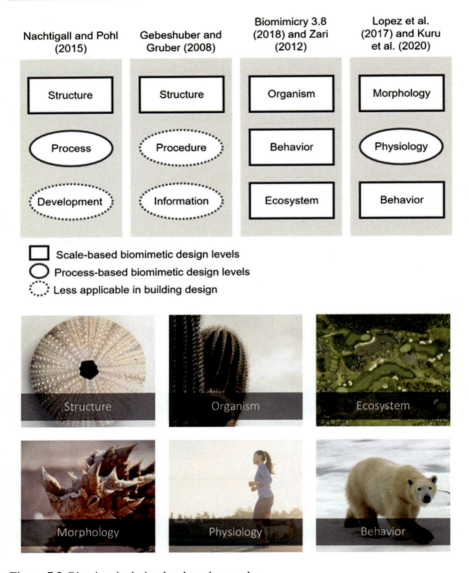

Figure 7.3 Biomimetic design levels and examples.
Source: Free copyright license is provided by Unsplash for the photographs taken by Milada Vgerova, Richard Brutyo, Stephanie Greene, Andrew Tanglao and Dan Bolton.

- *Multifunctionality* is the ability to host multiple functions. For example, the Menelaus blue butterfly has multifunctional wings; the wings as the flight bodies and nanostructured ridges giving colour without pigments (Giraldo and Stavenga, 2016).
- *Hierarchy* is having multilayered structures. For example, corals have integrated structural and material properties through hierarchy (Fratzl and Weinkamer, 2007).

• *Heterogeneity* is having differentiated forms. For example, the Saharan silver ant has uniquely shaped triangular hairs for cooling through heterogeneity (Shi et al., 2015).

Natural design principles inform the scale, process and context, in which the biological strategies are developed, and, thus, can help realizing biomimetic designs applicable to Bio-ABS. An interesting aspect of natural design principles is that they are the result of one another. For example, through hierarchy, organisms can achieve multifunctionality. An example of this in nature is the outer covering of animals' bodies comprising the skin and fur. The skin is an organ that is at a larger scale and the fur is a special tissue at a smaller scale over the skin. They host different functions, and through the hierarchical order of scales, they achieve multifunctionality (Fratzl et al., 2013).

Natural design principles are rarely explored in Bio-ABS; however, their implementation is promising to develop innovative facades. For example, the principle of hierarchical biological structures is transferred into innovative constructions, combining the properties of structure and material together. Through digitally and empirically tested timber shell modules, researchers have developed pavilions with inseparable structure and material (Goel et al., 2009; Reichert et al., 2015). As such, the structure of the pavilions is embedded with material properties as opposed to modern constructions with structural elements separated. Heterogeneity is also explored in designing multifunctional Bio-ABS. As such, a Bio-ABS that hosts multiple functions with differentiated forms has been developed with smart materials (Kuru, Oldfield, Bonser, & Fiorito, 2020). Much prospect still exists for future opportunities in adopting the natural design principles in Bio-ABS to develop innovative systems addressing global environmental challenges in a new and creative fashion.

7.4 Application of biomimetic adaptive building skins

7.4.1 Application of biomimetic design processes

Most biomimetic design processes are based on several aspects of the field of biomimetics to perform the transfer of concepts from life sciences into engineered designs. For instance, interdisciplinary collaboration is used among researchers and practitioners to develop structurally and material-wise biomimetic pavilions in the University of Stuttgart and University of Freiburg through the application of biomimetic design as annual research pavilions showcased to public (Knippers et al., 2018). The application of biomimetic design processes addresses several key aspects of biomimetics. These aspects include the following:

- interdisciplinary collaboration or knowledge exchange;
- biological inspiration and observation;
- the use of analogies between biology and technology;
- adoption and comparison of similarity and scale; and
- abstraction and translation of design principles (Kuru, 2020).

Several biomimetic design processes include a database that collects biological inspirations to be used as part of the method accompanying it. Most of these databases are open source and can be accessed by practitioners, designers, researchers and students. Some of these databases are AskNature, as part of the Biomimicry Institute (Biomimicry Institute, 2018a); BioTRIZ, as part of the TRIZ method (Vincent et al., 2006a; Bogatyrev and Bogatyreva, 2009); Pinnacles, as part of BioGen (Badarnah and Kadri, 2015), and the Multi-Biomechanisms Database, as part of the Multi-Biomechanism Framework (Kuru, Oldfield, Bonser, & Fiorito, 2020). The methods without the use of a database have other means of investigating biological strategies. These ways include developing a linguistic approach to identify commonalities between biology and technology, translating concepts in biology to technology, using a systematic analogical transfer mechanism to find biological inspiration (Chakrabarti, Siddharth, & Dinakar, 2017; Gamage & Hyde, 2012; Garcia-holguera, Clark, Sprecher, & Gaskin, 2016; Hirtz, Stone, McAdams, Szykman, & Wood, 2002; Nagel et al., 2008; Schleicher, Lienhard, Poppinga, Speck, & Knippers, 2015; Fayemi et al., 2014; Shu, 2010).

Biomimetic design processes are defined as bottom-up and top-down approaches depending on the initial step taken in the method. If the process starts with observing a biological model, it is a bottom-up approach; if the process starts with defining a technical challenge, it is a top-down approach (Badarnah and Kadri 2015). Both top-down and bottom-up approaches end with a common final step named developing a biomimetic design strategy. Regardless of how the process has started, the biological inspiration is usually linked with a technical challenge to improve the current state in engineered designs, or to provide an alternative to existing designs. The final step of biomimetic design processes tends to have distinctions. These differences are related to the implementation and validation of the biomimetic strategy. Depending on the biomimetic design process, the biomimetic design outcome may result in a conceptual model, a list of design principles driven from biological inspirations, as an analogy to be applied as a design product, or also to a read to implement technology. For example, some processes involve a more abstract phase for concept generation and principle extraction, while others may involve testing and prototyping with a product-like strategy with higher real-world world uptake (Kuru, 2020).

For example, a very commonly used biomimetic design process is the Biomimicry 3.8. Design Spiral by the Biomimicry Institute applied together with the open-source AskNature database (Biomimicry Institute, 2018a). AskNature provides an online public platform for interdisciplinary collaboration and inspiration from the various principles found in nature to tackle numerous technical functions and challenges in the built environment, engineering, robotics, medical sciences as well as many other disciplines. AskNature contains over 1600 biological strategies as they call it, categorized by functions. AskNature provides a Biomimicry Taxonomy to organize how organisms meet different challenges, with step-by-step examples on how to use and browse the data. This relies on verbs and their synonyms to be used in their search engine. The steps applied in the design process and taught by Biomimicry 3.8 are called the 'Design Spiral'. The biomimetic strategy developed as the result of the process involves validation against Life's Principles to perform a sustainability assessment. Biomimicry Design Spiral offers both top-down and bottom-up approaches and its steps for a top-down approach are provided next:

- *Define technical function and context*: This step involves the identification of a technical problem to be solved as a function within a specific context. Examples include a shading device as a technical function and excessive solar gains as a technical problem in buildings.
- *Biologize the challenge*: This step involves the establishment of an analogical correspondence between the technical function and a potential biological function. Examples include special morphologies that prevent excessive sunlight penetration.
- *Discover biological models:* This step involves discovering and finding biological models and living systems that perform the biologized function. Examples include lizards with thorny skin to prevent overheating by refracting light.
- *Abstract biological strategies:* This step involves abstracting and transferring the biological strategies into conceptual design principles. Examples include abstracting the thorny skin structure of a lizard into light refracting microscaled structures.
- *Emulate biomimetic strategies:* This step involves emulating or copying the biological functions into biomimetic strategies. Examples include developing a special surface covered with the light refracting microscaled structures.
- *Evaluate fitness to context:* This step involves evaluating and assessing the biomimetic strategy in terms of its suitability to the predefined technical function within the context. Examples include an outdoor shading structure with a special surface comprising the light refracting microscaled structures (Biomimicry Institute, 2018b).

Newly developed biomimetic design processes aim at addressing varying needs and diverse technical functions. For example, multifunctionality in technical systems is an important property that helps addressing multiple needs at a time (Svendsen and Lenau, 2019a). A newly developed process, the multibiomechanism approach, concentrates on the mechanisms of biological systems as interdependent properties. The process is a top-down approach and focuses on technical problems to be solved through learning from nature. The process follows similar phases to existing top-down approaches with an additional step on achieving multifunctionality. The framework comprises four steps: (1) identifying a technical problem; (2) selecting a biological solution; (3) achieving multifunctionality; and (4) developing a biomimetic strategy, as listed in Fig. 7.4. The first, second and the last steps are similar to those steps of the Biomimicry 3.8. Design Spiral. The distinctive step, achieving multifunctionality, aims at addressing conflicting situations in technical functional requirements through the integration of multiple functions.

The way the multibiomechanism approach addresses multifunctionality is through the implementation of the natural design principles of hierarchy and heterogeneity, as explained in the previous section of this chapter (Kuru, Oldfield, Bonser, & Fiorito, 2020). This is done by investigating the multiscaled and geometrically differentiated properties of the selected biological mechanisms regarding hierarchy and heterogeneity. Multiple biological mechanisms, whether they belong to the same organism or not, can be selected in such a way as to develop a

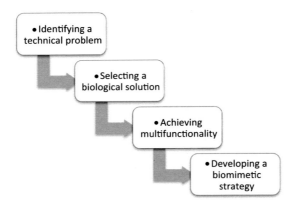

Figure 7.4 The multibiomechanism approach's top-down process steps.

multifunctional biomimetic design. The selected biological mechanisms respond to conflicting requirements of heat, light, air, water and energy regulation. They do so, by having the individual biological mechanisms triggered, sensed and actuated by diverse and multiple stimuli to be able to respond to conflicting requirements in the environment simultaneously. The way in which this is translated in this methodology is through the adoption of diverse and multiple stimuli. For example, a hinge that is triggered to sense and actuate a movement based on the changes in the thermal environment (eg, temperature) is embedded in the design of a biomimetic output together with a surface structure that can reflect and refract light by changing its nanoscaled ridges positions based on the changes in the luminous environment (eg, light).

7.4.2 Demonstration of a biomimetic adaptive building skin case study

A Bio-ABS case study named Adaptive Solar Façade is inspired by the leaves of *Oxalis oregana*, known as Redwood sorrel, hosts a folding movement as a behavioural and adaptive characteristic (Sheikh and Asghar, 2019). The leaves of the plant were explored in terms of their biomimetic design levels at organism and behaviour levels (Fig. 7.5). At the organism level the leaves' physical appearance involving the hierarchy of leaf veins and divergent angles of these veins was examined. At the behaviour level the adjustment of leaf angles according to the light intensity and detection and tracking of the sun path were examined. Through the abstraction of the form, folding movement and triggering mechanism of the leaves, a modular façade that tracks the sun's path that provides shading with a particular folding movement was developed (Sheikh and Asghar, 2019).

The modular Adaptive Solar Façade has a square-like form with two folding creases that intersect with each other with a right angle. The vertical fold provides a shading in the form of an overhang, while the horizontal fold provides a shading

Biomimetic adaptive building skins: design and performance 191

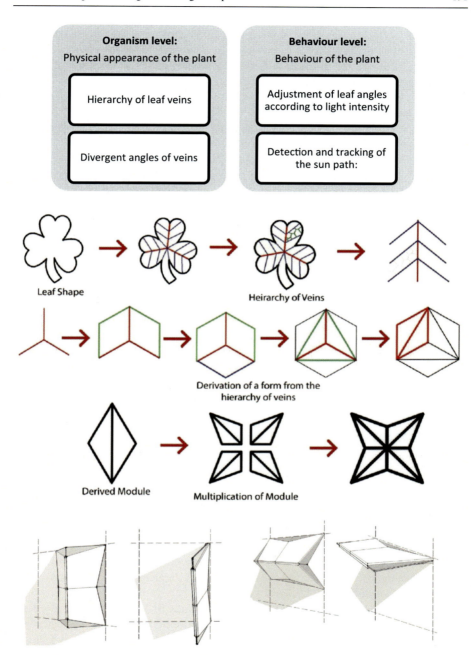

Figure 7.5 Adaptive Solar Façade and its design inspiration *Oxalis oregana*.
Source: Copyright agreements provided by Wajiha Tariq Sheikh (Sheikh and Asghar, 2019).

in the form of a louver. These intersecting folding patterns are derived from the leaves that have a similar form as observed in the hierarchy of their veins. The primary vein in the centre of the leaf creates a vertical folding pattern, and the intersecting secondary veins create a horizontal folding pattern. The difference between the plant's leaves and the façade component is the shape of the basic design module. The leaf has an organic shape, while the façade is designed as a rigid square-based module. The triggering element to activate the folding behaviour of the plant and the moving mechanism of the façade module is the same, that is, the changes in the light exposure depending on the sun path and angle.

7.5 Performance benefits of biomimetic adaptive building skins

7.5.1 Performance evaluation and benefits of Bio-ABS

The performance targets of Bio-ABS include reducing the energy use, improving visual and thermal comfort, improving the indoor air quality and acoustic performance, achieving structural and mechanical efficiency. Bio-ABS can respond and adapt to internal and external changes, including solar radiation, temperature, wind, humidity, precipitation and noise (Aelenei et al., 2016). For example, the lily mechanism responds to solar radiation through providing shading and it is inspired by the petal movements of the lily flower (Schleicher, Lienhard, Poppinga, Speck, & Knippers, 2015).

Various physical domains, including thermal, air flow, optical and energy, combined with different sets of simulation tools or experimental setups, help measuring the potential environmental benefits. The methods to measure the performance of Bio-ABS depend on the performance targets. For example, properties related to air flow of a ventilating Bio-ABS can be measured through computational fluid dynamics simulations or laboratory experiments. Bio-ABS is an emerging field, and the realization of such façade systems may be challenging from various aspects, including complex construction processes, the use of expensive materials and products, as well as the complexity associated with their application in buildings. Therefore digital modelling and simulation of Bio-ABS are significantly important to predict the performance and benefits. Regarding the digital modelling and simulation of Bio-ABS, it is important to simultaneously consider and calculate the changing behaviour in time, in various scales and the responses to multiple physical domains (AFN, 2018).

Simulation tools used to evaluate the performance of Bio-ABS are categorized as application-oriented and general purpose, depending on the modelling features of the tool (Loonen et al., 2016). Application-oriented tools have integrated features, while general purpose tools require customized input data to model Bio-ABS. Some of the most common digital tools are EnergyPlus, TRYSNS, IES VE and ESP-r. The use of EnergyPlus is often assisted with user-friendly graphical interfaces, including Open Studio with SkecthUp, and parametric modelling tools such as Grasshopper or Dynamo. Through these graphical interfaces, modelling processes of complex Bio-ABS are becoming a familiarized approach. However, the integration of multiple indoor and outdoor climatic variables to control the adaptive behaviour of Bio-ABS still remain mostly unexplored (Favoino, Goia, Perino, & Serra, 2014; Kuru, Oldfield, Bonser, & Fiorito, 2019).

An important aspect that influences the performance of Bio-ABS is the operation, defined as passive or active (Kuru, Oldfield, Bonser, & Fiorito, 2019). Mostly, active operation is controlled through indoor or outdoor sensors and passive operation is controlled through the embedded capacity of materials. With active controls, sensors that detect changes in environmental variables will trigger the movement of Bio-ABS. With passive controls, movement will be stimulated by an intelligent system. As such, one of the most common passive controls is through employing smart materials that respond to various types of energy, for example, thermal, optical and mechanical (Ritter, 2007). The types of operation can be integrated in simulation tools through time schedules, scripts and codes to predict the operation of Bio-ABS with various control strategies. For example, an example of a photovoltachromic glazing is triggered by solar irradiance levels and changes its solar heat gain coefficient (SHGC) value for an improved control of the optical environment (Favoino, Fiorito, Cannavale, Ranzi, & Overend, 2016; Cannavale et al., 2018).

The assessment of Bio-ABS to improve the performance of a building is promising, yet the implementation of Bio-ABS in buildings is underdeveloped. For example, recent studies have indicated that less than half (46%) of a selection of 52 Bio-ABS applications have measured environmental performance benefits (Kuru, Oldfield, Bonser, & Fiorito, 2019). However, there are some examples with performance measurements that indicate what Bio-ABS can achieve regarding environmental and energy efficiency when compared against conventional facades. For instance, the Gymnasium Façade is a parametric envelope design of Bio-ABS, adaptable to changing solar radiation levels inspired by the optics of reflecting eyes (Table 7.1). The performance of the Gymnasium Façade is implemented on a stadium and its performance is measured through daylight simulations to reflect sunlight rays as suitable for the building typology. The façade achieves satisfactory illuminance levels of 6369.81 lx at 12:00 p.m. and 6341.9 lx at 3:00 p.m. that are higher than the target for sports matches which is 1000 lx (Parr, 2013). Another example, the Desert Snail Envelope, is inspired by the adaptations of the desert snail, including an insulating air cushion, a reflective shell coating and reduced conduction through shading (Al Amin and Taleb, 2016). Installed on an office building, it achieves a decrease in energy consumption by 3%−19% as compared with a conventional approach (Al Amin and Taleb, 2016).

Another example that measures the performance benefits of Bio-ABS is the Biomimetic Façade based on Animal Fur. It is a façade system designed by transferring the principles of animal fur and blood perfusion, developed through a mathematical model and simulated to measure heat transfer. It achieves 17.7% reduction in annual energy consumption, a cooling of greater than 50 W/m^2 and reduced mean surface temperatures in the occupied zone by 2.8°C (Webb, Aye and Green, 2018). The Adaptive Solar Façade is designed by transferring the adaptations of a sun tracking leaf into a position changing shading device for office retrofits. It achieves

Table 7.1 Performance of biomimetic adaptive building skins (Bio-ABS) examples.

Bio-ABS example	Performance target	Performance output	Performance benefit
Gymnasium Façade (Parr, 2013)	Visual comfort	Illuminance levels and solar radiation	$\approx$ 600 illuminance, 95%–97% solar radiation
Desert Snail Envelope (Al Amin and Taleb, 2016)	Thermal comfort	Visible and infrared (IR) solar radiation, cooling load, average shading factor	90% visible and 95% IR reflectance, 19% cooling load, 48%–77% shading
The Biomimetic Façade based on Animal Fur (Webb, Aye and Green, 2018)	Energy efficiency and thermal comfort	Annual energy consumption, cooling demand and mean surface temperature	17.7% reduction in annual energy consumption, 50 W/m^2 cooling demand and reduce temperatures by 2.8°C
Adaptive Solar Façade (Sheikh and Asghar, 2019)	Energy efficiency and visual comfort	Energy load, daylighting levels	32% reduction in energy loads, 50% of interior space achieves the recommended daylighting between 500–750 lx
A multifunctional Bio-ABS (Kuru, Oldfield, Bonser, & Fiorito, 2020)	Thermal comfort	Discomfort hours	23.18% in the 90% and 5.06% decrease in the 80% acceptability limits

The Gymnasium Façade, the Desert Snail Envelope, the Biomimetic Façade based on Animal Fur, Adaptive Solar Façade and a Multifunctional Bio-ABS; with their achieved environmental and energy performance benefits (Kuru, Oldfield, Bonser, & Fiorito, 2020; Parr, 2013; Sheikh & Asghar, 2019; Webb, Aye, & Green, 2018; Al Amin and Taleb, 2016).

a reduction in energy loads by 32% and the recommended daylighting levels (between 500 and 750 lx) for the 50% of the interior (Sheikh and Asghar, 2019). Lastly, a multifunctional Bio-ABS inspired by the biological functions of the golden barrel cactus to provide shading and ventilation triggered by smart materials embedded in its design. Its performance is analysed for a naturally ventilated school building for reducing the discomfort hours for thermal comfort and achieved 23.18% in the 90% and 5.06% decrease in the 80% acceptability limits in adaptive thermal comfort (Kuru, Oldfield, Bonser, & Fiorito, 2020).

7.5.2 Challenges and opportunities in the performance evaluation of Bio-ABS

Despite the increasing interest in Bio-ABS, there is a further need for evaluating the empirical performance of Bio-ABS and realizing them as built applications. Bio-ABS have been developed since the modern era with Frei Otto one of the pioneers in the field in the early 20th century (Otto et al., 1967). At the time, most works have focused on structural advances, analysis and efficiency. However, the current uptake of Bio-ABS mostly centres around environmental performance. A study by the authors here assessed projects developed in academia and industry in the last decade and found that half of Bio-ABS are theoretical and only a third of projects analyse thermal or visual comfort, while less than 10% measure energy. Empirical performance analysis of Bio-ABS is also limited, with only 15% of the studies conducting laboratory or on-site experiments for prototypes or full-scale applications (Kuru, Oldfield, Bonser, & Fiorito, 2019). There are various challenges associated with the lack of information available on the performance benefits of Bio-ABS. The challenges are as follows:

- the required expertise in diverse disciplines, including building physics;
- knowledge in computer simulations;
- limited case studies demonstrating performance assessment of Bio-ABS as exemplary studies; and
- availability and access to laboratory facilities and equipment and funding for prototyping and developing full-scale applications.

The most important reason for these challenges is that the field of Bio-ABS is still emerging; therefore further research and development in the field is needed. A challenge related to measuring the environmental benefits of Bio-ABS is similar to those of adaptive building skins; that is the limited methodologies and case studies in modelling and simulating complex façade systems in the existing software and tools. This difficulty applies also to empirical analysis, which presents challenges to build such complex systems in controlled conditions as prototypes or develop them as real-world applications (Loonen et al., 2016). Another attribute is the vast amount of shape morphing shading devices realized as Bio-ABS (Fiorito et al., 2016; Kuru et al., 2018). Studies have found that most Bio-ABS are monofunctional and a metaanalysis of 52 most promising works in the field stated that only 13.4% of those are multifunctional, meaning they respond to more than one environmental stimulus (Kuru, Oldfield, Bonser, & Fiorito, 2019). One major reason for this conclusion is the limitations in the currently developed and used biomimetic design frameworks addressing how to achieve multifunctionality learning from multifunctional principles in nature (Badarnah, 2017; Kuru, Oldfield, Bonser, & Fiorito, 2020). The integration of multiple functions beyond single-functioned shading systems is needed to explore and measure the performance benefits of a variety of Bio-ABS. Further work is needed in the field for both digital and physical processes to design, develop, model and evaluate Bio-ABS.

7.6 Conclusion

This chapter suggested that the existing design paradigm of facades benefits shifting toward the use of Bio-ABS technologies in buildings for improved performance. It provided a definition of Bio-ABS as well as discussing the existing design processes to develop Bio-ABS and environmental and energy performance benefits of Bio-ABS. It did so, to present the challenges and limitations in the field, and discussing the future uptake. The future needs in the area of Bio-ABS is suggested to focus on investigation and translation of natural design principles for improved performance and efficiency, as well as measuring the environmental performance of Bio-ABS to draw out the environmental or energy benefits. Major conclusions are associated with transferring natural design principles for increased efficiency, measuring the digital and physical performance benefits of Bio-ABS, strengths in diverse-disciplinary collaborations to address the different required expertise and integrating multifunctionality in developing Bio-ABS.

Moreover, most Bio-ABS are found to be developed in academia which may result in a gap between practice and research in the field. Industrial projects are 26.9% of the developed Bio-ABS (Kuru, Oldfield, Bonser, & Fiorito, 2019). Reasons for the limited availability of industrial projects include the complex fabrication processes of such façade systems, the challenges and uncertainty in material selection and development, relatively higher cost of realizing such facades and the lack of demand and potentially familiarity with the concept. To overcome these challenges, a keen interest in designing and evaluating the performance of Bio-ABS can have a positive impact on the industry and the potential development of the field. Understanding the performance benefits and innovative aspects of Bio-ABS may help the industry to shift toward using the available technology to realize Bio-ABS and promote developing new systems for better suitability.

References

Aelenei, D., Aelenei, L., & Vieira, C. P. (2016). 'Adaptive façade: Concept, applications, research questions', *energy procedia SHC 2015. International Conference on Solar Heating and Cooling for Buildings and Industry*, *91*, 269–275. Available from https://doi.org/10.1016/j.egypro.2016.06.218.

AFN (2018) Building performance simulation and characterisation of adaptive facades – Adaptive façade network. In F. Favoino et al. (Ed.) *TU Delft open for the COST action 1403 adaptive façade network*.

Al Amin, F. and Taleb, H. (2016) Biomimicry approach To achieving thermal comfort in a hot climate. In *Proceedings of SBE16 Dubai-UAE 17–19 January* (pp. 1–8). Available from: http://content.buid.ac.ae/events/Proceedings/SBE16D140.pdf.

Al-Obaidi, K. M., Ismail, M. A., Hussein, H., & Rahman, A. M. A. (2017). Biomimetic building skins: An adaptive approach. *Renewable and Sustainable Energy Reviews*, *79*, 1472–1491. Available from https://doi.org/10.1016/j.rser.2017.05.028.

Badarnah, L. (2017). Form follows environment: Biomimetic approaches to building envelope design. *Buildings*, *7*(40). Available from https://doi.org/10.3390/buildings7020040.

Badarnah, L., & Kadri, U. (2015). A methodology for the generation of biomimetic design concepts. *Architectural Science Review*, *58*(2), 120−133. Available from https://doi.org/10.1080/00038628.2014.922458.

Biomimicry Institute (2018a) *AskNature: A project of biomimicry 3.8*. Available from: http://www.asknature.org.

Biomimicry Institute (2018b) *Biomimicry.net: Biomimicry 3.8*. Available from: http://biomimicry.net/.

Bogatyrev, N. & Bogatyreva, O. (2009) TRIZ evolutionary trends in biology and technology: Two opposites, pp. 1−4. doi: 10.1002/mds.20450.

Cannavale, A., Ayr, U. & Martellotta, F. (2018) 'Innovative electrochromic devices: Energy savings and visual comfort effects'. In *Energy procedia 73rd conference of the Italian Thermal Machines Engineering Association (ATI 2018), 12−14 September 2018, Pisa, Italy*, 148, 900−907.

Chakrabarti, A., Siddharth, L., & Dinakar, M. (2017). Idea Inspire 3.0—A tool for analogical design. *Research into design for communities*, 475−485. Available from https://doi.org/10.1007/978-981-10-3521-0.

Del Grosso, A. E., & Basso, P. (2010). Adaptive building skin structures. *Smart Materials and Structures*, *19*(124011). Available from https://doi.org/10.1088/0964-1726/19/12/124011.

Dunn, K. E. (2020). The emerging science of electrosynbionics. *Bioinspiration and Biomimetics*, *15*(3). Available from https://doi.org/10.1088/1748-3190/ab654f.

Fayemi, P. et al. (2014) Bio-inspired design characterisation and its links with problem solving tools. In International *design conference − desi*gn 2014, Dubrovnik − Croatia, May 19−22, 2014, 132 (pp. 1−10).

Favoino, F., Fiorito, F., Cannavale, A., Ranzi, G., & Overend, M. (2016). Optimal control and performance of photovoltchromic switchable glazing for building integration in temperate climates. *Applied Energy*, *178*, 943−961.

Favoino, F., Goia, F., Perino, M., & Serra, V. (2014). Experimental assessment of the energy performance of an advanced responsive multifunctional façade module. *Energy and Buildings*, *68*, 647−659. Available from https://doi.org/10.1016/j.enbuild.2013.08.066.

Fayemi, P. E., Wanieck, K., Zollfrank, C., Maranzana, N., & Aoussat, A. (2017). Biomimetics: Process, tools and practice. *Bioinspiration and Biomimetics*, *12*(011002). Available from https://doi.org/10.1088/1748-3190/12/1/011002.

Fiorito, F., Sauchelli, M., Arroyo, D., Pesenti, M., Imperadori, M., Masera, G., & Ranzi, G. (2016). Shape morphing solar shadings: A review. *Renewable and Sustainable Energy Reviews*, *55*, 863−884. Available from https://doi.org/10.1016/j.rser.2015.10.086.

Fratzl, P., & Weinkamer, R. (2007). Nature's hierarchical materials. *Progress in Materials Science*, *52*, 1263−1334. Available from https://doi.org/10.1016/j.pmatsci.2007.06.001.

Fratzl, P., Dunlop, J., & Weinkamer, R. (2013). *Materi*als design inspired by natu*re: F*unction through inner architectu*re*. Cambridge: RSC Publishing − Royal Society of Chemistry.

Gamage, A., & Hyde, R. (2012). A model based on biomimicry to enhance ecologically sustainable design. *Architectural Science Review*, *55*(3), 224−235. Available from https://doi.org/10.1080/00038628.2012.709406.

Gebeshuber, I. C., & Drack, M. (2008). An attempt to reveal synergies between biology and mechanical engineering. *IMECHE, Part C: Journal of Mechanical Engineering*, *222*(7), 1281−1287.

Garcia-holguera, M., Clark, O. G., Sprecher, A., & Gaskin, S. (2016). *Ecosystem biomimetics for resource use optimization in buildings*. Taylor & Francis. Available from http://doi.org/10.1080/09613218.2015.1052315.

Gebeshuber, I. C., Gruber, P., & Imhof, B. (2015). *Biornametics: Architecture defined by natural patterns. Encyclopedia of nanotechnology*. Springer. Available from http://doi.org/10.1007/978-94-007-6178-0.

Giovannini, L., Favoino, F., Pellegrino, A., Lo Verso, V. R. M., Serra, V., & Zinzi, M. (2019). Thermochromic glazing performance: From component experimental characterisation to whole building performance evaluation. *Applied Energy*, *251*, 113335.

Giraldo, M. A., & Stavenga, D. G. (2016). Brilliant iridescence of Morpho butterfly wing scales is due to both a thin film lower lamina and a multilayered upper lamina. *Journal of Comparative Physiology A: Neuroethology, Sensory, Neural, and Behavioral Physiology*, *202*, 381−388. Available from https://doi.org/10.1007/s00359-016-1084-1.

Goel, A. K., Rugaber, S., & Vattam, S. (2009). Structure, behavior, and function of complex systems: The structure, behavior, and function modeling language. *Artificial Intelligence for Engineering Design, Analysis and Manufacturing.*, *23*(1), 23−35. Available from https://doi.org/10.1017/S0890060409000080.

Halliday, S. (2013). *Sustainable construction*. New York: Routledge.

Hasselaar, B. L. (2006) Climate adaptive skins: Towards the new energy-efficient façade. In *Proceedings of the 1st international conference on the management of natural resources, sustainable development and ecological hazards* (pp. 351−360).

Hirtz, J., Stone, R. B., McAdams, D. A., Szykman, S., & Wood, K. L. (2002). A functional basis for engineering design: Reconciling and evolving previous efforts. *Research in Engineering Design − Theory, Applications, and Concurrent Engineering*, *13*(2), 65−82. Available from https://doi.org/10.1007/s00163-001-0008-3.

Hosseini, S. M., Mohammadi, M., Rosemann, A., Schröder, T., & Lichtenberg, J. (2019). A morphological approach for kinetic façade design process to improve visual and thermal comfort: Review. *Building and Environment*, *153*, 186−204. Available from https://doi.org/10.1016/j.buildenv.2019.02.040.

John, G., Clements-Croome, D., & Jeronimidis, G. (2005). Sustainable building solutions: A review of lessons from the natural world. *Building and Environment*, *40*(3), 319−328. Available from https://doi.org/10.1016/j.buildenv.2004.05.011.

Knaack, U., et al. (2014). *Facades: Principles of Construction* (2nd ed.), Berlin: Birkhauser.

Knippers, J., & Speck, T. (2012). Design and construction principles in nature and architecture. *Bioinspiration and Biomimetics*, *7*, 015002. Available from https://doi.org/10.1088/1748-3182/7/1/015002.

Knippers, J. et al. (2018) The ITECH approach: Building(s) to learn. In C Mueller. & S. Adriaenssens (Eds.) *Proceedings of the IASS symposium 2018, creativity in structural design, July 16−20, 2018, MIT, Boston, USA*.

Kuru, A. (2020). *Biomimetic adaptive building skins: An approach towards multifunctionality*. The University of New South Wales.

Kuru, A. et al. (2018) Multi-functional biomimetic adaptive facades: Developing a framework. In *FAÇADE 2018 final conference of COST TU1403 'Adaptive Facades Network' Lucerne, Switzerland, November 26/27* (pp. 231−240).

Kuru, A., Oldfield, P., Bonser, S., & Fiorito, F. (2019). Biomimetic adaptive building skins: Energy and environmental regulation in buildings. *Energy and Buildings*, *205*(109544). Available from https://doi.org/10.1016/j.enbuild.2019.109544.

Kuru, A., Oldfield, P., Bonser, S., & Fiorito, F. (2020). A framework to achieve multifunctionality in biomimetic adaptive building skins. *Buildings*, *10*(114). Available from https://doi.org/10.3390/buildings10070114.

Lepora, N. F., Verschure, P. F. M. J., & Prescott, T. J. (2013). The state of the art in biomimetics. *Bioinspiration and Biomimetics*, *8*(013001). Available from https://doi.org/10.1088/1748-3182/8/1/013001.

Lienhard, J., Schleicher, S., Poppinga, S., Masselter, T., Milwich, M., Speck, T., & Knippers, J. (2011). Flectofin: A hingeless flapping mechanism inspired by nature. *Bioinspiration and Biomimetics*, *6*(4). Available from https://doi.org/10.1088/1748-3182/6/4/045001.

Loonen, R. C. G. M., et al. (2016). Review of current status, requirements and opportunities for building performance simulation of adaptive facades. *Journal of Building Performance Simulation*, *1493*, 1–19. Available from https://doi.org/10.1080/19401493.2016.1152303.

Loonen, R. C. G. M., Trcka, M., Cóstola, D., & Hensen, J. L. M. (2013). Climate adaptive building shells: State-of-the-art and future challenges. *Renewable and Sustainable Energy Reviews*, *25*, 483–493. Available from https://doi.org/10.1016/j.rser.2013.04.016.

López, M., et al. (2017). How plants inspire facades. From plants to architecture: Biomimetic principles for the development of adaptive architectural envelopes. *Renewable and Sustainable Energy Reviews*, *67*, 692–703. Available from https://doi.org/10.1016/j.rser.2016.09.018.

Makram, A. (2019). Nature-based framework for sustainable architectural design – biomimetic design and biophilic design. *Architecture Research*, *9*(3), 74–81. Available from https://doi.org/10.5923/j.arch.20190903.03.

Mazzoleni, I. (2013). *Architecture follows nature*. Davis, CA: Taylor & Francis.

Nagel, R. L., Midha, P. A., Tinsley, A., Stone, R. B., McAdams, D. A., & Shu, L. H. (2008). Exploring the use of functional models in biomimetic. *Journal of Mechanical Design*, *130*, 1–13. Available from https://doi.org/10.1115/1.2992062.

O'Payne, A., & Johnson, J. K. (2009). FireFly interactive PrototyPes for architectural design. *Architectural Design*, *83*, 144–147.

Otto, F., Trostel, R., & Schleyer, F. K. (1967). *Tensile structures: Design, structure, and calculation of buildings of cables, nets, and membranes*. MIT Press.

Pan, C. A. & Jeng, T. (2010) 'A robotic and kinetic design for interactive architecture'. In *Proceedings of the SICE annual conference 2010* (pp. 1792–1796).

Park, J. J. (2014). Bio-inspired parametric design for adaptive stadium facades. *Australasian Journal of Construction Economics and Building Conference Series*, 27–35.

Parr, D. R. G. (2013). Biomimetic lessons for natural ventilation of buildings a collection of biomimicry templates including their simulation and application. In N. F. Lepora, et al. (Eds.), *Living machines* (pp. 421–423). Berlin Heidelberg: Springer-Verlag. Available from http://doi.org/10.1007/978-3-642-39802-5-54.

Payne, A. O., & Johnson, J. K. (2013). Firefly: Interactive prototypes for architectural design. *Architectural Design*, *83*, 144–147.

Pesenti, M., Masera, G., & Fiorito, F. (2018). Exploration of adaptive origami shading concepts through integrated dynamic simulations. *Journal of Architectural Engineering*, *24*(4), 04018022. Available from https://doi.org/10.1061/(ASCE)AE.1943-5568.0000323.

Pohl, G., & Nachtigall, W. (2015). Biomimetics for architecture & design: Nature – Analogies – Technology. *eBook*. Springer.

Ramzy, N., & Fayed, H. (2011). Kinetic systems in architecture: New approach for environmental control systems and context-sensitive buildings. *Sustainable Cities and Society*, *1*, 170–177. Available from https://doi.org/10.1016/j.scs.2011.07.004.

Reap, J., Baumeister, D., & Bras, B. (2005). Holism, biomimicry and sustainable engineering. *Energy Conversion and Resources*, *2005*, 423–431. Available from https://doi.org/10.1115/IMECE2005-81343.

Reichert, S., Menges, A., & Correa, D. (2015). Meteorosensitive architecture: Biomimetic building skins based on materially embedded and hygroscopically enabled responsiveness. *Computer-Aided Design*, *60*, 50–69. Available from https://doi.org/10.1016/j.cad.2014.02.010.

Ritter, A. (2007). *Smart materials in architecture, interior architecture and design*. Berlin: Birkhauser.

Santamouris, M., Haddad, S., Saliari, M., Vasilakopoulou, K., Synnefa, A., Paolini, R., ... Fiorito, F., et al. (2018). On the energy impact of urban heat island in Sydney: Climate and energy potential of mitigation technologies. *Energy and Buildings*, *166*, 154–164. Available from https://doi.org/10.1016/j.enbuild.2018.02.007.

Schleicher, S., Lienhard, J., Poppinga, S., Speck, T., & Knippers, J. (2015). A methodology for transferring principles of plant movements to elastic systems in architecture. *Computer-Aided Design*, *60*, 105–117. Available from https://doi.org/10.1016/j.cad.2014.01.005.

Sheikh, W. T., & Asghar, Q. (2019). Adaptive biomimetic facades: Enhancing energy efficiency of highly glazed buildings. *Frontiers of Architectural Research*, *8*(3), 319–331. Available from https://doi.org/10.1016/j.foar.2019.06.001.

Shi, N. N., Tsai, C. C., Camino, F., Bernard, G. D., Yu, N., & Wehner, R. (2015). Keeping cool: Enhanced optical reflection and radiative heat dissipation in Saharan silver ants. *Science*, *349*(6245), 298–301. Available from https://doi.org/10.1126/science.aab3564.

Shu, L.H. (2010). A natural-language approach to biomimetic design, pp. 507–519. doi: 10.1017/S0890060410000363.

Svendsen, N. & Lenau, T. A. (2019a) How does biologically inspired design cope with multi-functionality? In *Proceedings of the international conference on engineering design, ICED, 2019-Augus(August)* (pp. 349–358). doi: 10.1017/dsi.2019.38.

Speck, O., Speck, D., Horn, R., Gantner, J., Sedlbauer, K. P., et al. (2017). The signs of life in architecture biomimetic bio-inspired biomorph sustainable? An attempt to classify and clarify biology-derived technical developments. *Bioinspiration and Biomimetics*, *12* (011004).

Vincent, J. F. V., et al. (2006a). Biomimetics: Its practice and theory. *Journal of the Royal Society Interface*, *3*, 471–482. Available from https://doi.org/10.1098/rsif.2006.0127.

Webb, M., Aye, L., & Green, R. (2018). Simulation of a biomimetic façade using TRNSYS. *Applied Energy*, *213*, 670–694. Available from https://doi.org/10.1016/j.apenergy.2017.08.115.

Wigginton, M., & Harris, J. (2002). *Intelligent skins*. Oxford: Butterworth-Heinemann.

Zari, M. P. (2012). Ecosystem services analysis for the design of regenerative built environments. *Building Research and Information*, *40*(1), 54–64. Available from https://doi.org/10.1080/09613218.2011.628547.

Building integrated photovoltaic facades: challenges, opportunities and innovations

Francesco Frontini[1], Pierluigi Bonomo[1], David Moser[2] and Laura Maturi[2]

[1]University of Applied Sciences and Arts of Southern Switzerland (SUPSI), Manno, Switzerland, [2]Energy Efficient Buildings Group, Institute for Renewable Energy, Eurac Research, Bolzano, Italy

Abbreviations

AC	alternating current
aSi	amorphous silicon photovoltaic solar cells
BIM	building information modelling
BIPV	building-integrated photovoltaics
BIST	building-integrated solar thermal
CAM	computer aided manufacturing
CSD	computer aided design
cSi	crystalline solar cells
DC	direct current
DSSC	dye-sensitized solar cells
F2F	file to factory
IEA	International Energy Agency
LSC	luminescence solar concentrator
NPV	net present value
O&M	operation and maintenance
OLED	organic light emitting diode
PVPS	Photovoltaic Power Systems Programme
PVSS	photovoltaic shading systems
PV	photovoltaic
PVT	photovoltaic-thermal
RES	renewable energy sources
SHC	solar heating and cooling
ST	solar thermal

8.1 Introduction

The sector of solar building envelopes embraces a rather broad range of technologies—building-integrated photovoltaics (BIPV), building-integrated solar thermal

(BIST) collectors and photovoltaic (PV)-thermal collectors—that actively harvest solar radiation to generate electricity or usable heat (Frontini, Scognamiglio, Graditi, & Pellegrino, 2013; Meir, 2019; Wall et al., 2012). The focus of this chapter is a critical review of the current knowledge and challenges of solar facades that actively generate electricity thanks to the use and integration of BIPV technologies. The current status of heat-generating solar facades, such as BIST systems, can be found in the study of Meir (2019) as part of International Task 56 of the International Energy Agency (IEA) Solar Heating and Cooling (SHC) programme.

The transformation of buildings to solar buildings is a tangible 'cause' of innovation in both contemporary architecture and solar technologies, as the use of active facades is much more than a technical possibility: it is a true new opportunity in building skin aesthetics, ethics and technology. The influx of new materials and technologies in the building field has profoundly marked the history of construction with innovative archetypes: in our opinion, as happens with new materials, this 'solar' transformation will cause a profound change of perspective in design, building skin engineering and industrial approaches, as well as in the whole construction process.

BIPV will play an essential role in a new era of distributed power generation. BIPV systems (as both roof and façade applications) represent a powerful and versatile technology, able to produce renewable energy where the sun is available, to meet the ever increasing demand for zero- (or even positive-) energy or zero-carbon buildings in the coming years. While some critical policy challenges exist, the value of generating power directly where it is used today matches the twin values of aesthetic needs and high flexibility of design and manufacturing, which may help to mitigate the barriers inherent in BIPV applications in construction field (Corti, Bonomo, Zanetti, Polo Lopez, & Frontini, 2018).

A BIPV system serves as building envelope component and power generator simultaneously. BIPVs have a great advantage over nonintegrated PV plants in that since there is need neither for the allocation of land nor stand-alone PV systems (Jelle & Breivik, 2012). The on-site electricity produced by BIPV systems can reduce total building material costs by achieving cost competitiveness (Corti, Capannolo, Bonomo, De Berardinis, & Frontini, 2020), for example, by optimizing end user costs (yielding compelling savings in terms of costs of mounting, especially since BIPV systems do not require additional specific assembly components such as brackets and rails) and creating new business models (Macé et al., 2020). A BIPV system simply makes zero-emission electricity out of sunlight. All these advantages have caused a worldwide growing interest in BIPV products and a dynamic market trend in recent years (Frontini et al., 2017).

A data collection was conducted by reviewing databases such as ScienceDirect, Scopus and ResearchGate and also thanks to the authors' experiences in European and international collaborative projects that address the topic of reducing the cost and increasing the competitiveness of solar facades in buildings. Moreover, research and innovation together with various industries served as a collection of different insight details to define the challenges and the need to rethink solar façade envelopes.

In this chapter the authors focus on the transformation of facades into an active skin systems thanks to the use of PV technologies, which can be considered one of the main drivers, and also an enabler, of decarbonization of the EU heating sector through electrification (Thomaßen, Kavvadias, & Jiménez Navarro, 2021).

8.2 Challenges and needs in rethinking solar photovoltaic facades

Solar facades in general are multifunctional building elements that, along with satisfying construction requirements as primary function, use incoming solar radiation to

- generate on-site renewable energy (thermal or electric) and
- improve daylight control and manage solar gains while also reducing the heating and cooling demands of buildings (eg, in the case of transparent systems).

New materials and advances in manufacturing/assembling processes occurring both in PV or construction field can result in the development of new solar façade concepts, the progressive improvement of existing technologies, or even new applications of consolidated concepts. Many solar façade solutions can adapt their behaviour or characteristics to the local climate conditions to better meet a specific performance goal. This can be done by utilizing the possibility of integrating mechanical and automated actuators, material sciences and IT systems (Tabadkani, Roetzel, Li, & Tsangrassoulis, 2021). By doing so, they mediate between changing environments, offering opportunities for energy savings and improvements of indoor environmental quality. The façade element is designed to be more than a barrier to the external weather, becoming a multifunctional system offering advanced daylight control and solar protection or renewable energy source (RES) generation. To appeal to architects and building designers, many solar façade producers have invested in marketing products in a range of colours (or even textures, transparency levels and materials), installation options and sizes that allow for both seamless integrations and standout installations. However, this increase in options often comes at the expense of lower efficiency of solar energy conversion (Eder et al., 2019; Frontini, Bonomo, Saretta, Weber, & Berghold, 2016).

Although available solar thermal (ST) and PV technologies have a proven history of successful integration (Bonomo, Frontini, & Chatzipanagi, 2015; Frontini et al., 2015, 2017; Jelle & Breivik, 2012; Scognamiglio, 2017) in creating solar façade envelopes, their applications are still confined to exemplary buildings or lighthouse projects (Devetaković et al., 2020; Frontini, Zanetti, Bonomo, Andreas, & Studer, 2020). Architects, construction companies and building owners struggle to integrate these technologies into the design of new buildings (Heinstein, Ballif, & Perret-Aebi, 2013) and even more so into the renovation of existing buildings (Aguacil, Lufkin, & Rey, 2019). Several research projects (Machado et al., 2017, 2019; Ulbikas, Galdikas, & Stonkus, 2016) and international projects (BFIRST, 2016; ConstructPV, 2017; SMART-Flex, 2016) and studies by the IEA (under the Photovoltaic Power Systems Programme [PVPS] and SHC programme) conducted in the last decades reveal the need to overcome different challenges (Prieto, Knaack, Auer, & Klein, 2017), such as to rethink the design, the engineering, the integration and the operation of solar facades to be more cost-effective and more easily integrated.

These challenges and needs can be grouped into two main categories/families:

- technological challenges and product design/engineering,
- formal and architectural challenges.

While the first is purely related to the development and optimization of new materials and technologies to improve the performance, the energy generation, the aesthetics (in terms of the integrability of materials), the lifetime and product manufacturing, the second is strongly related to the formal aspect of the façade and to the capacity to integrate the new materials, components and façade systems in the building design. Moreover, the integration of active facades within the construction process is still complex, mainly because it involves different actors and stakeholders.

These challenges include (1) the ability of the product or system to meet certain performances (electrical, thermal, etc.), (2) aesthetics (the capability to meet certain architectural targets), (3) technical integration (the capability to avoid complex assembly and integration into the façade construction system), (4) durability/safety and lifetime, (5) energy integration (how solar facades are integrated into and support the building's energy system) and (6) availability and awareness.

A summary of these aspects is presented in Table 8.1.

In the following sections the challenges and needs presented in Table 8.1 are briefly discussed as a basis for the rethinking of PV facades.

8.2.1 Solar façade performance levels

One of the main challenges and market needs is to improve the performance and the capacity of the façade to produce renewable electricity or ST energy. As demonstrated by numerous studies and projects, PV and ST are the easiest technologies to be integrated in facades, suitable for meeting the net zero-energy buildings target (Buonomano, De Luca, Montanaro, & Palombo, 2016; Jelle & Breivik, 2012; Scognamiglio, 2017). When incorporated in building elements, PV solar cells may reduce their energy performance levels not only due to nonoptimal working conditions (such as higher temperatures) or nonoptimal orientations (due mainly to the building design and available surfaces) but also due to aesthetic needs. Indeed, as demonstrated by Frontini, Bonomo et al. (2016), Frontini, Saretta, and Bonomo, (2016) and by Saretta, Bonomo, and Frontini (2018), front glass treatments to hide PV solar cells, while providing colours to the BIPV module, can result in module efficiency losses of from 10% up to 60%.

Table 8.1 Main challenges and needs to rethink solar facades.

		Challenges and needs				
	1. Performance	2. Aesthetics	3. Technical complexity (integration)	4. Durability and lifetime	5. Energy integration	6. Availability
Specification and objectives	Improve energy production	Improve overall aesthetics	Avoid complex assembly (integration) and operation modes	Improve durability of components and systems	Ease of integration in BEMS	Complete value chain coverage
	Allow energy management	Allow for customization of appearance	Modularity and plug-and-play components to ease integration (prefabrication)	Devise long-term maintenance strategies and end-of-life solutions	Energy control	Awareness
	Allow energy storage	Various designs (form and material)	Standardization of components	Allow for replacement and retrofit	Different uses of energy	Sufficient production capacity
	Increase cost-effectiveness of components	Low defects and visual issues			Autonomous system	

BEMS, Building energy management system.

8.2.2 Façade aesthetics and technical complexity

In architecture field the formal and aesthetic acceptance of technology is as important as the functional one (Sánchez-Pantoja, Vidal, & Pastor, 2018).

The use of solar technologies in buildings has great potential for application if it is integrated since the early design phase of the building process, where most of the design choices are made with wide freedom of possibilities. Conversely, considering or installing these systems at the end of the process involves a higher economic cost, a risk of nonefficient solutions, and it can result in an aesthetically less attractive building. However, how can the aesthetics of a solar façade be defined? Are there quantitative measures to assess? As described by Sánchez-Pantoja et al. (2018) and Probst and Roecker (2012), the goal of high aesthetic value is met when solar technology becomes part of the building product, creating a unique element. Even if the definition of integration remains highly subjective, different studies have shown how the architectural 'integrability' of solar modules can be considered from all three architectural points of view: *functional* (preserving/ensuring the core envelope functions), *constructive* (relating to the building envelope as a whole and guaranteeing construction safety and continuity) and *formal* (providing the desired aesthetic). As a positive side effect, this will also bring a reduction in technical complexity that helps the building planner or the façade designer to properly design the façade, thus optimizing time and resources.

8.2.3 Durability and maintenance

The main issue regarding the service life of BIPV façade systems is the need for maintenance, which would have an impact on the operating cost of the building and on the façade durability. The possible answers to this challenge are focused on two aspects as barriers to the integration of current technologies: the lifetime of the components needs to be improved, together with the provision of detailed information on the performance of aging components to convince stakeholders and minimize economic risks. Moreover, even with enhanced durability, these components will need maintenance, so the possibility to easily maintain, repair, or even replace several parts should be considered a key-value in the product engineering, along with end-of-life scenarios for the different components.

8.2.4 Energy integration

Energy flexibility in buildings is a topic that is gaining momentum with the introduction of new European directives, such as the EU Directive 2018:2001 (2018), which introduces the concept of collective self-consumption (Article 21) and the creation of renewable energy communities (Article 22), enabling buildings to manage their energy demand and production according to their needs.

Thus buildings are more than just stand-alone energy units. Buildings are becoming increasingly active elements of the energy network by consuming, producing, storing and supplying energy, thus transforming the EU energy market and shifting from centralized, fossil fuel—based national systems toward decentralized, renewable, interconnected and variable energy systems.

This shifting paradigm is affecting the way the building envelope is conceived and designed, exploiting roof and façade surfaces to maximize the match between solar harvesting and building consumption.

8.2.5 Availability of products

Availability of products and systems refers both to the number of product types available on the market and to the production capacity offered by manufacturers. For many years the limited number of products available has been identified as one of the key barriers to the diffusion of BIPV systems, as it was not able to address all architecture and construction needs. However, as demonstrated by BIPV in Eurac (2020) and by Frontini et al. (2020), nowadays the number of available products has increased drastically, even if the number of companies able to produce a large number of products in Europe is still limited.

8.3 Innovation in solar facades—archetypes of innovation

Starting from the foregoing analysis, this section presents possible innovations in solar PV façade products and systems to answer and contribute to overcoming the main challenges identified earlier. The chapter is organized into two parts, identifying the main factors and evolutionary traits in the transparent part of the façade [ie, organic materials, dye-sensitized solar cells (DSSC) technologies, luminescence solar concentrator (LSC) windows and active PV shading devices] and in the opaque part of the façade envelope (ie, lightweight solutions, flexible products and coloured products). PV technology can be used in different parts of the facades, as depicted in Fig. 8.1.

The chapter will also include an investigation into new possibilities for solar façade modules as unitized and prefabricated systems, covering both opaque and transparent solutions.

A comprehensive review of the design of building integrated PV systems and technologies has been addressed by Shukla, Sudhakar, Baredar, and Mamat (2017).

8.3.1 Innovation in transparent PV facades

Transparent facades not only absorb and reflect incident solar radiation but also transfer solar heat gained into the building while providing outside visibility for the occupants. If such facades transform part of the incident sunlight into electricity, they are called transparent and translucent active solar facades (Quesada, Rousse, Dutil, Badache, & Hallé, 2012).

Figure 8.1 Solar façade archetypes. Photovoltaic technologies can be integrated in both the transparent part of the facades (as window elements) and in the opaque part of the envelope as rain screen elements or as part of a prefabricated system.
Source: Reproduced with permission from SUPSI.

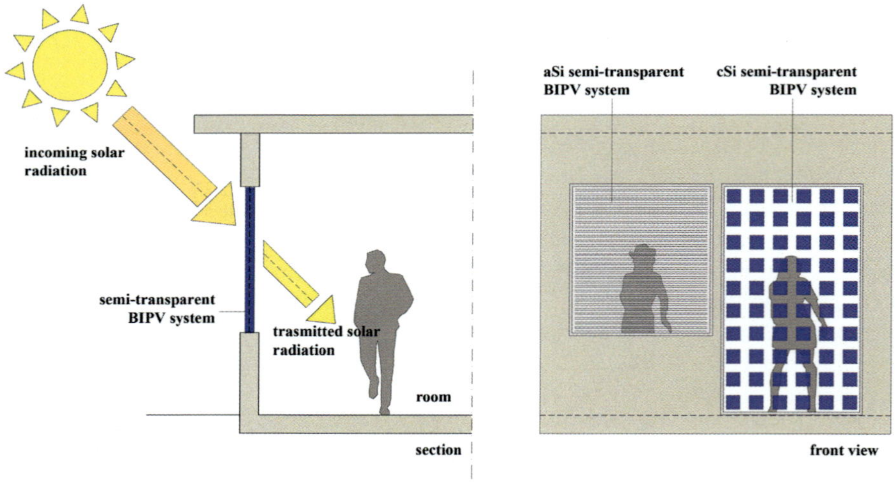

Figure 8.2 Schematic diagram of two types of semitransparent building-integrated photovoltaic system using aSi or cSi. *aSi*, Amorphous silicon photovoltaic solar cells; *cSi*, crystalline solar cells.
Source: Reproduced with permission from SUPSI.

A semitransparent BIPV system is integrated into the building envelope, generating electricity and allowing daylight to enter the interior spaces (Fig. 8.2). Different solar PV cell technologies are available today with different power efficiencies and optical properties.

Crystalline solar cells (cSi) are widely used in glazed facades. They offer different design opportunities and different levels of transparency. cSi solar cells can be placed at different distances within a laminated glass pane. Studies by Olivieri et al. (2015) have shown how the cell cover ratio of the glass has a direct impact on the total heat gain coefficient of the windows. Different design approaches are possible to make this solution more attractive to architects and planners, such as the one developed within the EU SMART-flex project, which uses a flexible and semiautomated machine to optimize the position of solar cells on the glass according to the desired design and customization while also speeding up production time to reduce manufacturing costs. Flexibility and automation of the manufacturing line is crucial to reducing BIPV production costs and ensuring optimal product quality. Further improvements are under development, for example, in the H2020 BIPVBOOST project (https://bipvboost.eu/project/). Semiautomation in module production and cell placement can guarantee flexible design as required by building performance needs. For example, semitransparent BIPV facades can be designed by arranging solar cells only in those parts of the glazing where solar protection is needed, so improving the view of the outside by leaving the central part of the glass façade fully transparent and the upper and lower areas more opaque. For these solutions the choice of glass laminate and lamination foils is also very important to be able to meet the required mechanical performance levels of the façade. A particular application of this technology is represented by bifacial solar cells. Bifacial facades take the advantage of the recent development of bifacial PV cells, which are able to harvest the sunlight from both sides (front and rear). This application is very interesting in a double-skin configuration, where the solar cells can benefit from the gap between the module and the inner wall, which allows backside albedo, and natural or forced ventilation to give additional power to the module (Fig. 8.3). In this way a second skin glass element with bifacial PV can be added in the building envelope to also exploit the light reflected by the inner façade (Fig. 8.4).

Amorphous silicon PV solar cells (aSi) can also be integrated into laminated glass. Thanks to the laser scribing process, aSi modules can be made semitransparent with different visible light transmittance levels of up to 30%–40%. As demonstrated by Liao and Xu (2015), see-through aSi PV glazing is more beneficial than traditional glazing when it is applied in shallow rooms with large windows or high ceilings, since the lower light and solar transmittance of aSi PV glazing contribute to considerably reducing the cooling load while not impacting the energy needs for artificial lighting. This advantage is diminished once the room depth exceeds 10 m, due to the increase in energy consumption of artificial lighting with aSi PV glazing.

Complementary to cSi and aSi technologies, a specific solution that seeks to solve the problem of uniformity, darkness of colours and visibility is that of LSCs (Norton et al., 2011). These are transparent dyed slabs of vitreous or polymeric glassy materials, doped with tiny amounts of fluorescent dyes. The fluorescent dyes absorb only part of the solar radiation, reemitting it at a larger wavelength. The reemitted light, exploiting the total internal reflection of the glassy material (Fig. 8.5), is partially trapped in the slab and wave-guided toward the edges, where it is concentrated on small solar cells (Aste et al., 2019).

LSC panels can be integrated in double- or triple-glazed units to provide maximum thermal comfort while guaranteeing sufficient daylight. A solution also exists

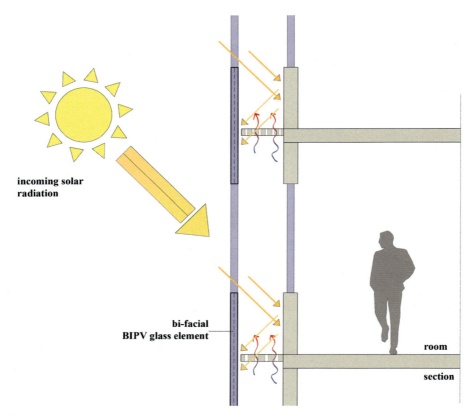

Figure 8.3 Advantages of using bifacial modules on a vertical façade in a double skin configuration: backside albedo and natural or forced ventilation.
Source: Reproduced with permission from SUPSI.

Figure 8.4 The southern façade of the CSEM research organization building in Neuchâtel is constructed with photovoltaic bifacial solar cells.
Source: Reproduced with permission from SUPSI.

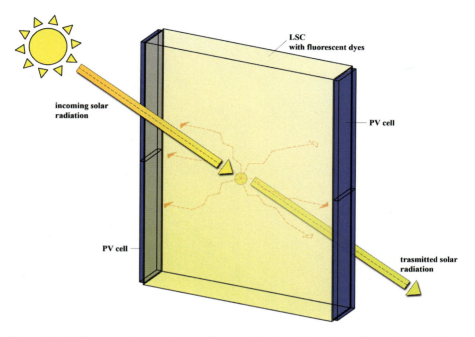

Figure 8.5 LSC concept for window application. The incoming solar radiation is trapped in the LSC dyes and partially redirected toward the perimeter of the window where the photovoltaic solar cells (PV-cell) are placed. Part of the solar radiation is retransmitted to the interior of the room. *LSC*, Luminescence solar concentrator; *PV*, photovoltaic.
Source: Reproduced with permission from SUPSI.

to keep the LSC element colourless (Albano et al., 2020) so as to be easily integrated in highly glazed buildings. Different sensors can also be integrated and powered to continuously measure light intensity, temperature, pressure and air quality, to enhance occupant comfort. This solution is already being experimented by Delft University of Technology groups and the start-up company Physee with the PowerWindow solution, by ENI and Politecnico di Milano with the LSC Smart Windows project, and by Glass2Power and the University of Milano-Bicocca.

This new breed of smart windows is able to provide a multifunctional contribution in terms of energy saving and renewable energy production, since it is characterized by the following features:

- solar electricity production
- indoor daylight improvement
- solar control
- thermal insulation

A possible concept proposed by Aste et al. (2019) is presented in Fig. 8.6. The Smart Windows solution is designed to be mounted mainly in new or retrofitted commercial buildings, where consumption for artificial lighting is generally

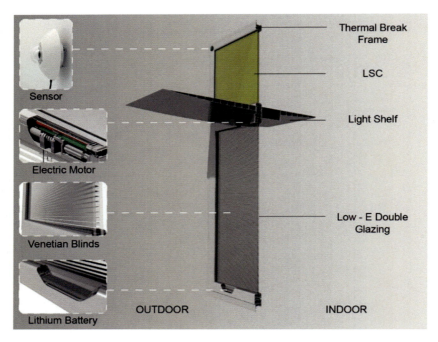

Figure 8.6 In the Smart Window concept, solar control is obtained through adjustable Venetian blinds, powered by integrated LSCs (which also contribute to daylight improvement) and driven by a dedicated control logic. *LSC*, Luminescence solar concentrator.
Source: From Aste, N., Buzzetti, M., Del Pero, C., Fusco, R., Leonforte, F., & Testa, D. (2019). Triggering a large scale luminescent solar concentrators market: The smart window project. *Journal of Cleaner Production, 219*, 35–45.

significant, despite the extensive use of glass surfaces, and where glare and overheating often occur.

Among semitransparent solar modules, those based on DSSC are worth mentioning (Fig. 8.7). These have a relatively low production cost compared to silicon-based solar cells and provide a range of colours for building design. As such, DSSCs are easy to fabricate since they are insensitive to environmental contaminants and processable at ambient temperature. These unique features are favoured in roll-to-roll processing, which is a continuous, low-cost manufacturing method of printing dye-sensitized solar cells on flexible substrates (Gong, Sumathy, Qiao, & Zhou, 2017). Furthermore, DSSCs work better even during darker conditions, such as at dawn and dusk or in cloudy weather. This capability to effectively utilize diffused light makes DSSCs an excellent choice for indoor applications like windows and sunroofs. However, its commercialization has been limited by its low power generation efficiency and durability (Lee & Yoon, 2018).

Semitransparent or translucent solar façade concepts using these PV technologies are able to reduce the solar heat gain coefficient both in curtain-wall and in double-

Figure 8.7 View of full-scale DSSC BIPV mock-up. *BIPV*, Building-integrated photovoltaics; *DSSC*, dye-sensitized solar cells.
Source: From Lee, H. M., & Yoon, J. H. (2018). Power performance analysis of a transparent DSSC BIPV window based on 2 year measurement data in a full-scale mock-up. *Applied Energy*, *225*, 1013–1021. https://doi.org/10.1016/j.apenergy.2018.04.086.

skin façade solutions, but they are not able to adapt to dynamic solar condition. Indeed, aSi or DSSC glazing is not able to change and adapt their transparency during the day and according to the season. To achieve this objective, PV modules can also be used as PV shading systems (PVSSs) either as fixed elements or as movable devices. PVSSs combine the benefits of shading systems with renewable solar energy harvesting strategies, since the light that is stopped from entering the interior is converted to electricity (Taveres-Cachat, Lobaccaro, Goia, & Chaudhary, 2019). Accurate simulation studies need to be done to properly design the system and avoid constant over-shading by the PV blinds and enhance solar harvesting while controlling solar energy and glare to guarantee the desired thermal and visual comfort.

Large shading systems are the ones used most often to date (Fig. 8.8), as demonstrated by Bahr (2014) or Mandalaki, Zervas, Tsoutsos, and Vazakas (2012) where 13 types of south-facing shading solutions mostly used in office buildings are analyzed, and by different installations already available worldwide and collected by Zhang, Lau, Lau, and Zhao (2018). Reducing the penetration of solar radiation and sunlight into the interior space during the cooling periods and admitting the needed solar energy during the heating periods are desirable. Therefore movable shading systems are also used. Previous studies of Zhang et al. (2018) identified 24 common types of architectural shading devices that can be equipped with PV technologies. This study demonstrates that thanks to the reduction in cost of available PV technologies and the increase in demand for BIPVs, these solutions represent an interesting opportunity for façade designers. However, available solutions are still limited compared to products using PV-façade cladding or semitransparent BIPV windows and PV-roof systems (Frontini et al., 2017).

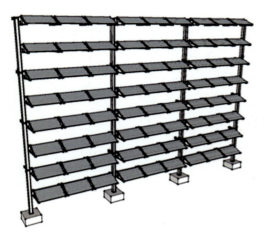 

Figure 8.8 Fixed large photovoltaic shading systems are widely used in buildings. They can be movable, like the one shown on the left, or fixed, and they can use both cSi and thin-film photovoltaic technologies.
Source: From Bahr, W. (2014). A comprehensive assessment methodology of the building integrated photovoltaic blind system. *Energy and Buildings*, *82*, 703−708. https://doi.org/10.1016/j.enbuild.2014.07.065 and Taveres-Cachat, E., Lobaccaro, G., Goia, F., & Chaudhary, G. (2019). A methodology to improve the performance of PV integrated shading devices using multi-objective optimization. *Applied Energy*, *247*, 731−744. https://doi.org/10.1016/j.apenergy.2019.04.033.

To this extent, great opportunity is offered by lightweight PV material such as $Cu(In, Ga)Se_2$ (CIGS). CIGS offers the advantage of flexible, curved shapes and a lightweight structure when compared to traditional cSi modules, but it has a lower energy conversion efficiency per square metre. However, this is not a problem for buildings with large glazed areas or multistory buildings that require shading systems and where the transparent area is sufficiently large in comparison with a roof or an opaque cladding. Therefore integrating PV modules into a dynamic shading system offers the possibility to fine-tune different functions, generate electricity and balance energetic performance with architectural expression (Nagy et al., 2016) while also guaranteeing perfect visual and daylight comfort (Fig. 8.9).

Today, two concepts represent the frontier of innovation of dynamic active shading devices. The first is external movable shading, developed by the Polytechnic of Zurich ETHZ, and the second is the integrated windows solution developed by SUPSI (Fig. 8.10). These two solutions exploit the potential of thin-film technology to control sunlight and solar gains while contributing on the architectural appearance of the building façade. The two concepts are different: the first is able to adapt the blind position according to user needs and the comfort desired thanks to a soft-pneumatic actuator (Nagy et al., 2016) developed by the ETHZ team, while the second, developed by SUPSI, is designed to guarantee a long service life and improved visual comfort thanks to the integration of the PV blind in an insulating glazing

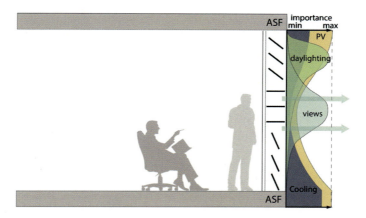

Figure 8.9 The mediation of solar radiation has the potential to reduce heating and cooling demands while simultaneously distributing daylight according to the occupants' desires. Dynamic photovoltaic facades can transmit light differently according to user needs.
Source: From Jayathissa, P., Zarb, J., Luzzatto, M., Hofer, J., & Schlueter, A. (2017). Sensitivity of building properties and use types for the application of adaptive photovoltaic shading systems. *Energy Procedia*, *122*, 139–144. https://doi.org/10.1016/j.egypro.2017.07.319.

Figure 8.10 Two different movable photovoltaic blind solutions. The picture on the left shows the Advance Fenestration system as described by Jayathissa, Caranovic, Hofer, Nagy, and Schlueter (2018). The picture on the right shows the iWin prototype developed by SUPSI where the blind is integrated in a double-glazed unit.
Source: Courtesy Jayathissa and SUPSI (https://www.iwin.ch).

unit that protects the lamella from the outdoor weather conditions (iWin—innovative Windows, 2019).

To be properly used, dynamic systems have to implement intelligent algorithms that are able to orientate the blinds to the most energy-efficient position, thus finding the optimum balance between PV generation and daylight control to minimize heating, cooling and lighting demands.

8.3.2 Innovation in opaque PV façade

Following the development of BIPV, an architectural language based on standard solar panels accompanied the first age of integration in opaque surfaces, mainly roofs. However, these conventional PV elements displayed all the limitations of a functional approach, so the drastic change of today's second age involves a change in technology that has to comply with key market-driven demands, such as aesthetics, flexibility, building skin performance, durability and cost-effectiveness. This trend can be referred to as 'camouflaged' and 'customized' PV, where the key material is glass, but not in its commonly held embodiment of a transparent and dematerialized skin (Saretta, Bonomo, & Frontini, 2017). Under the banner of 'invisible PV', there is currently a promising joint effort between PV and the glass industry with the aim of combining high production of solar energy with attractive visual design aesthetics. No technical limits seem to be applicable to the revolutionary flexibility of glass design. The main customization techniques, typically considering the layering of a glass-based module (Fig. 8.11), can be applied to glass, intermediate foils and PV cells (Eder et al., 2019).

Patterns and sketches can be obtained by treating the outer glass surface (eg, by sandblasting) which, in turn, can be combined with a glass colour to dissimulate the solar cells behind it. A design or colour on the front glass can be obtained with a silk screen printing process that deposits a special ink on the glass surface, such as digital ceramic-based printing (Ertex-solar, 2020) or, alternatively, by stabilizing the colour at high temperature (Sunage, 2021) with mono- or multichromatic scales

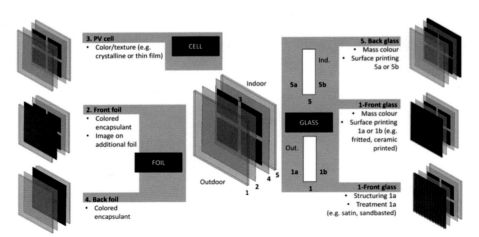

Figure 8.11 Customization alternatives to obtain colour and graphic effects on opaque BIPV claddings based on glass. *BIPV*, Building-integrated photovoltaics.
Source: Reproduced with permission from SUPSI.

used to obtain high-resolution images or prints (Kaleo Solar, 2020). By combining the satin finishing on the outer glass surface with silkscreen printing on the inner side, a resulting coloured matte surface can make the glass opaque and active. Scattering and reflection filters have also been developed, and they can be added as internal foils to reflect and diffuse the visible spectrum, thus providing a coloured appearance (Solaxess, 2020, SwissINSO, 2020). All of these techniques are progressively being developed to find the best compromise between visual effect and efficiency of PV production, and they represent a dynamic branch of the 'active glass' industry which is the current frontier of BIPV for the coming years. This innovation trend aims to facilitate the transition to active buildings while providing endless possibilities for aesthetic variation (Solar Visual, 2020; SolarLab, 2020). Cost-efficient mass customization is expected to further support the market pull in the EU and worldwide, including aspects of quality, reliability and safety (Saretta et al., 2018). However, further aspects also contribute toward the wide range of possibilities such as the construction innovation of solar skins. For example, the flexibility and lightness of thin film solar strips introduced the possibility to radically overcome some issues of PV integration by simply embedding PV in tents and ultrathin and multiperformance membranes. Some promising trends are emerging for ultrathin solar modules which can be applied on low-load-bearing roofs, ventilated facades and streetlights (Flisom, 2020). Inkjet 'roll to roll' printing, in combination with organic light-emitting diodes, will in the future also supply flexible and eco-friendly supports, allowing the building to act as a self-sufficient organic system, offering a spectacular application of PV for media functions and digital art (Zanetti et al., 2017).

Together with the abovementioned technologies, to increase the efficiency of the PV cell and to reduce the production cost, many research activities are focused on Perovskite solar cells. Perovskite solar cells are at the stage of research projects and experiments demonstrated efficiencies higher than 25% when deposit on cSi cells as reported by Green et al. (2020) but record efficiency has been proved to be up to 29.15% in silicon-based tandem cells (NREL, 2021). However, no modules have been offered commercially to date since long-term stability suffers from the high sensitivity to humidity and moisture.

Other trends of innovation focus on coupling active PV claddings with sandwiched façade systems, including thermal insulation layers, which are aimed at obtaining a totally dry installation, ensuring simplicity and efficiency of mounting for a unitized kit with thermal protection, fire prevention and sound insulation (PIZ, 2020). Lastly, a significantly improved plug-and-play mounting process stimulating better solutions to replace and repair elements with a reduction in transport and construction costs, ease of mounting and unmounting during the life cycle and operation and maintenance (O&M) (Tulipps, 2020).

The solar window block system (Fig. 8.12) is an example of a prefabricated, insulated system that can integrate PV modules, dynamic and automated shading systems and decentralized ventilation units with the aim of maximizing indoor comfort without adding electrical loads to the building. These systems provide valuable opportunities for enhancing the energy performance of existing buildings,

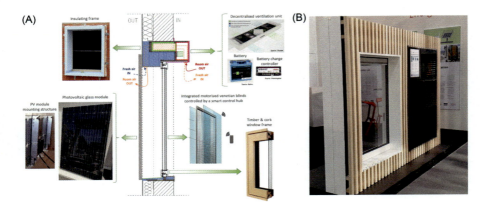

Figure 8.12 Solar window block: conceptual schema (A) and prototype picture (B).
Source: Reproduced with permission from H2020 EnergyMatching project. Eurac Research.

minimizing the impact on building occupants as well as allowing the effective integration of active components to maximize the exploitation of RESs. As part of the EU H2020 EnergyMatching project, a set of multifunctional window block prototypes was developed for residential building retrofitting (Adami et al., 2020; Andaloro, Avesani, Belleri, & Machado, 2018; Andaloro, Minguez, & Avesani, 2020; Lovati, Adami, & Moser, 2018).

The innovation in these trends lies both in the component technology and in the fact that the building image becomes a clear manifesto of a new 'solar age' in architecture. The morphogenesis of a building, the overlap between energy, performance and construction spatial conception, the rules governing the structure and the materiality of the building skin, all contribute to show the increasingly revolutionary key: innovation above all. Solar is becoming architecture.

8.4 Integrating the solar façade in the building energy system

PV is one of the main technologies that can support the transition toward a low-carbon energy system, promoting on-site energy production and enhancing self-consumption, if integrated into the overall building/district energy system and coupled with electric or thermal storage.

Solar facades with PV integration, thus, become part of a broader system that can be conceived as shown in Fig. 8.13 to optimize overall energy use within a building district. The buildings can be interconnected to optimize and maximize the use of the energy that has been harvested in the district through an electricity system that controls interaction with the external parts (electricity grid, PV supply, building loads and electrical vehicle loads) as well as the energy systems in the other buildings.

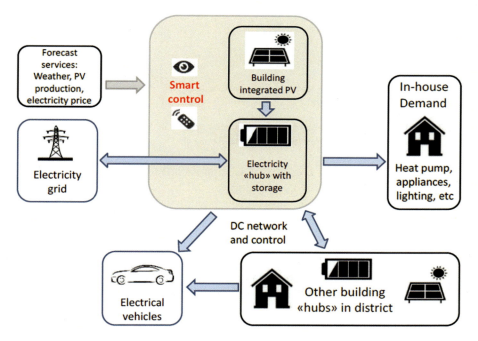

Figure 8.13 Conceptual schema of "EnergyMatching" district energy system (**Huang, Lovati, Zhang, & Bales, 2020**).
Source: From Huang, P., Lovati, M., Zhang, X., & Bales, C. (2020). A coordinated control to improve performance for a building cluster with energy storage, electric vehicles and energy sharing considered. *Applied Energy*, *268*, 114983. https://doi.org/10.1016/j.apenergy.2020.114983.

It is, thus, essential when designing a solar envelope to take into account its integration within the overall energy system (including heat pumps, electrical and thermal storage and all electrical appliances) to find the best match between on-site energy production and consumption. New methodological approaches are needed to support solar façade design toward energy-matching optimization.

8.4.1 Methodological approaches for solar envelope design toward energy-matching optimization

Current approaches for dimensioning PV systems for building envelope integration are often based on minimum thresholds fixed by national standards that define the minimum nominal power to be installed based on building dimensions or other parameters. For example, for single residential units, it is common practice to install around 3−6-kW p systems according to the house dimensions as indicated by the local regulations. A slightly more advanced approach is to size the PV system to produce the same amount of energy that is consumed annually by the building

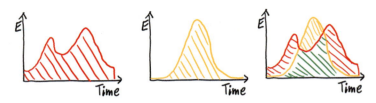

Figure 8.14 Conceptual schema of energy-matching approach.
Source: Reproduced with permission from Eurac Research.

(Fig. 8.14). In this case the commonly used approach is to install the PV system on the most irradiated surfaces of the building.

In any case, to better exploit the available building envelope area toward energy-matching optimization, it is essential to develop new methodological approaches and hourly based simulation tools that can support solar envelope design (Lovati et al., 2019).

An example of hourly based simulation tool toward energy-matching optimization was developed as part of the H2020 EnergyMatching project (Maturi et al., 2019). The EnergyMatching Tool is conceived as an early design tool to support architects and designers in the design of BIPV systems to optimize the capacity and placement of the BIPV system on the building envelope, taking into consideration techno-economic and energy-matching parameters.

Optimization based on energy-matching parameters opens interesting opportunities toward PV integration in facades since, even if facades usually have a lower annual yield (kW h/kW p) compared to the roof, they can be profitable when load-matching (ie, the contemporaneity between consumption and production) is considered (Adami, Lovati, & Moser, 2019; Lovati et al., 2019).

In fact, PV system dimensioning and positioning is optimized to fit with the hourly building electricity demand, considering, for example, heat pumps, appliances and lighting consumption. The hourly electric consumption of the building is crucial to determining the optimal PV capacity: if a PV system is too small, it has no impact on the demand profile (hence on the net present value), while if it is too big most of the electricity produced is not self-consumed. When there is a good balance between production and consumption, the PV orientation is essential to optimize production−consumption matching.

8.4.2 Solar envelope integration in the renewable electricity distribution system

Fig. 8.15 shows the overall concept of a renewable electricity distribution system for a district, based on a direct current (DC) grid. The system is based on a bidirectional converter that converts energy between alternating current (AC) and DC. The DC grid utilizes smart DC/DC converters to manage energy flow from PV modules, energy storage with batteries and electrical vehicle chargers, allowing for flexible installations and efficient power management. It can store overproduced energy in

Figure 8.15 Overall concept of renewable electricity distribution system for a district.
Source: From Huang, P., Lovati, M., Zhang, X., Bales, C., Hallbeck, S., Becker, A., ... Maturi, L. (2019). Transforming a residential building cluster into electricity prosumers in Sweden: Optimal design of a coupled PV-heat pump-thermal storage-electric vehicle system. *Applied Energy*, *255*, 113864. https://doi.org/10.1016/j.apenergy.2019.113864.

batteries if export to the AC grid is unfavourable and it can charge the batteries from the AC grid to control power peaks for the lowest possible grid fee.

This kind of system enables energy sharing toward effective energy communities, allowing sharing of PV, energy storage and electrical loads among different users within a DC grid:

- By sharing a PV installation a property owner can utilize the most suitable space for PV but transfer the energy to the building with the highest demand, thus increasing self-consumption.
- By sharing an energy storage the capacity of the energy storage can be reduced compared to the one each building would need to install for its own energy storage.
- By sharing electrical loads the AC grid subscription does not have to be increased and/or the system will be able to support more DC loads, for example, electrical vehicle charging stations.

Sharing of resources also leads to the possibility to maintain uninterruptable power during a power outage.

Further research and development is needed in two main areas: first the optimization of system performance through advanced control strategies that are implemented in each component and coordinated by a multilevel control hierarchy steering the overall energy system (Huang et al., 2019; Lovati, Zhang, Huang, Olsmats, & Maturi, 2020), and second the development of economy services and effective business models based on the overall system, which is made possible by the monitoring of all energy flows (Casalicchio, Manzolini, Prina, & Moser, 2020; Delponte et al., 2015; Radl, Fleischhacker, Revheim, Lettner, & Auer, 2020; Schram, Louwen, Lampropoulos, & van Sark, 2019).

8.5 Further opportunity: digitalization of solar facades as a driver for innovation

The building industry and the energy sector have begun a fundamental transformation toward a fully digitized transition which, in the case of solar building skins, raises the possibility of strong synergies. Digitalization will be an enabler for distributed PV generation in the built environment through the adoption of a more open and collaborative workflow based on data-sharing among process stages, disciplines and stakeholders. From design to O&M, there are many possibilities for supporting process quality and improving collaboration, coordination and communication while also reducing extra costs, making the European BIPV value chain more competitive. Process innovation will not be separated from product innovation in BIPV, since parametric planning methods and mass-customized process chains, differently from other industrial models mainly based on serial production, can take the advantage of a digital approach. The quality and cost of a BIPV skin in fact is not only related to product manufacturing but also to fragmented process inefficiency and poor coordination between the construction and electrical disciplines. The building process can today completely be digitized, evolving from computer-aided design to building information modelling (BIM) and computer-aided manufacturing and 'file to factory', encompassing all communication and data management technologies and services. Solar optimized 3D facades can be created and optimized by taking into account design freedom, solar potential and efficient fabrication and installation needs thanks to parametric tools that ensure interfaces between product requirements, energy needs and construction requirements in a digital process chain. The design and production of complex and multifunctional building skins, especially in the case of customizable and multifunctional BIPV systems, will be susceptible of higher quality control thanks to collaborative workflow and information sharing among owners, real estate managers, designers, engineers, manufacturers, installers and facility managers. Digitalization through process automation will support better integration between disciplines (eg, construction and electrical) and a high level of preassembly of components to reduce costs of both manufacturing and labour for BIPV installation. Moreover, the possibility of applying further approaches, such as machine learning techniques involving data collected from different stages, introduces the possibility of finding relations between events that occur at every stage of the whole lifecycle and which can affect PV production or maintenance costs. The adoption of a BIM-based approach can, for example, ensure the reduction of unforeseen clashes between electrical and building parts (Alamy et al., 2019; Bonomo, Saretta, & Frontini, 2018). A 'Digital Twin' of the building, digitally constructed and shared along the whole process, could, thus, be used for planning, engineering, construction and operation of the built asset by matching energy and construction needs. This complex domain encompasses different scales, from components engineering to specialist simulations of solar and energy production, for assessing the urban solar potential for existing urban areas (Lobaccaro, Lisowska, Saretta, Bonomo, & Frontini, 2019) in synergy with urban policies to support decentralized renewable energy production and refurbished building stock (Saretta, Bonomo, & Frontini, 2020). The development of methods, models and tools oriented to optimally support a digitalized environment will be a crucial aspect of overcoming the current barriers that still exist in this sector (cost reduction, performance verification across lifetime, standards compliance, etc.) toward a digitally augmented building that can adapt in real time to its users' needs while also contributing to the achievement of demanding energy efficiency targets. Collaborative platforms designed to reduce project risks and allow project teams to collaborate more effectively along the whole process and value chain will represent innovation to support a wider market implementation of solar technologies and higher long-term quality along the building life cycle.

8.6 Conclusion

An analysis of the most important technologies (emerging or nearer to market) able to 'activate' the façade envelopes by means of PV technology is presented in this chapter.

Different approaches are possible for both the transparent part of the envelope and the opaque part. While crystalline solar cell technologies today represent the most used technologies, next-generation thin-film technologies are gaining some momentum thanks to their flexibility and light weight.

Different challenges and solutions capable of overcoming these barriers in the diffusion of solar PV facades are presented. Dynamic shading elements represent one of the most promising technologies due to their multifunctional solar protection and solar harvesting, both as window elements and as external shading devices able to control the light penetration while producing considerable electricity.

Different possibilities are today available in terms of rain screen facades, thanks to the implementation of different techniques for changing the texture, colours and patterns of the front glass and to 'camouflage' the solar cells.

Customization of the facades via the use of window block solutions have been shown to be a strategic choice to provide valuable opportunities for enhancing the energy performance of existing buildings while minimizing the impact on building occupants, as well as allowing the effective integration of active components to maximize exploitation of RESs.

Some further steps still need to be taken by the construction industry first of all through more flagship demonstration projects that increase visibility and long-term experience with BIPV. Moreover, the different European projects presented are currently working on new industrial facilities for customized and highly automated production thanks in part to the increased role of digitalization. Especially, the digitalization of the entire value chain would be very helpful for the construction sector in general and solar PV technology in particular, as it requires smooth and quality-assured communication between the different stakeholders and throughout the entire construction process. New approaches and tools are needed to identify the most valuable facades, especially in dense urban environments. As such, the exploitation of their potential is a priority strategy toward the implementation of energy-efficient buildings and neighbourhoods.

References

Adami, J., Lovati, M., & Moser, D. (2019). Evaluation of the impact of multiple PV technologies integrated on roofs and facades as an improvement to a BIPV optimization tool, in: *Advanced building skins 2019*. Bern, Switzerland.

Adami, J., Maturi, L., Minguez, L., Gubert, M., Astigarraga, A., & Lovati, M. (2020). Preliminary design of a BIPV multifunctional window block for residential buildings, in: *15th Conference on advanced building skin*. Bern, Switzerland.

Aguacil, S., Lufkin, S., & Rey, E. (2019). Active surfaces selection method for building-integrated photovoltaics (BIPV) in renovation projects based on self-consumption and self-sufficiency. *Energy and Buildings, 193*, 15−28. Available from https://doi.org/10.1016/j.enbuild.2019.03.035.

Alamy, P., Nguyen, V., Saretta, E., Bonomo, P., Román, E.; Tecnalia, S., ... Alonso, P. (2019). BIM—A booster for energy transition and BIPV adoption, in: *Conference proceedings of the 36th European photovoltaic solar energy conference and exhibition*.

Albano, G., Colli, T., Nucci, L., Charaf, R., Biver, T., Pucci, A., & Aronica, L. A. (2020). Synthesis of new bis[1-(thiophenyl)propynones] as potential organic dyes for colorless luminescent solar concentrators (LSCs). *Dyes and Pigments, 174*, 108100. Available from https://doi.org/10.1016/j.dyepig.2019.108100.

Andaloro, A., Avesani, S., Belleri, A., & Machado, M. (2018). Adaptive window block for residential use: Optimization of energy efficiency and user's comfort, in: *Adaptive façade network conference*, Luzern. Luzern, Switzerland.

Andaloro, A., Minguez, L., & Avesani, S. (2020). Active and energy autonomous window: Renewable energy integration through design for manufacturing and assembly, in: *Açade tectonics 2020 world congress*. Los Angeles, United States.

Aste, N., Buzzetti, M., Del Pero, C., Fusco, R., Leonforte, F., & Testa, D. (2019). Triggering a large scale luminescent solar concentrators market: The smart window project. *Journal of Cleaner Production, 219*, 35−45. Available from https://doi.org/10.1016/j.jclepro.2019.02.089.

Bahr, W. (2014). A comprehensive assessment methodology of the building integrated photovoltaic blind system. *Energy and Buildings, 82*, 703−708. Available from https://doi.org/10.1016/j.enbuild.2014.07.065.

BFIRST [WWW document] (2016). Build. fibre-reinforced Sol. Technol. https://cordis.europa.eu/project/id/296016. Accessed 25.10.20.
BIPV in Eurac [WWW document] (2020). https://bipv.eurac.edu/en/products.html. Accessed 25.10.20.
Bonomo, P., Frontini, F., & Chatzipanagi, A. (2015). Overview and analysis of current BIPV products: New criteria for supporting the technological transfer in the building sector. *VITRUVIO—International Journal of Architectural Technology and Sustainability*, *1*, 67−85. Available from https://doi.org/10.4995/vitruvio-ijats.20150.4476.
Bonomo, P., Saretta, E., Frontini, F. (2018). Towards the implementation of a BIM-based approach in BIPV sector, in: *13th Conference on advanced building skins* (pp. 789−798). Bern, Switzerland: Economic Forum.
Buonomano, A., De Luca, G., Montanaro, U., & Palombo, A. (2016). Innovative technologies for NZEBs: An energy and economic analysis tool and a case study of a non-residential building for the Mediterranean climate. *Energy and Buildings*, *121*. Available from https://doi.org/10.1016/j.enbuild.2015.08.037.
Casalicchio, V., Manzolini, G., Prina, M. G., & Moser, D. (2020). A multi-objective optimization to assess economic benefit distribution and impact of energy communities, in: *37th European photovoltaic solar energy conference and exhibition* (pp. 2032−2037). https://doi.org/10.4229/EUPVSEC20202020-7DO0.9.4.
ConstructPV [WWW document], 2017. Constr. Build. with Cust. size PV Modul. Integr. opaque part Build. Ski. https://cordis.europa.eu/project/id/295981/it. Accessed 25.10.20.
Corti, P., Bonomo, P., Zanetti, I., Polo Lopez, C., & Frontini, F. (2018). Overcoming barriers for the BIPV diffusion at urban and building scale, in: *Tatus-Seminar «Forschen Für Den Bau Im Kontext von Energie Und Umwelt»*. BRENET.
Corti, P., Capannolo, L., Bonomo, P., De Berardinis, P., & Frontini, F. (2020). Comparative analysis of BIPV solutions to define energy and cost-effectiveness in a case study. *Energies*, *13*, 3827. Available from https://doi.org/10.3390/en13153827.
Delponte, E., Marchi, F., Frontini, F., López, P., Fath, K., & Batey, M. (2015). BIPV in EU28, from niche to mass market: An assessment of current projects and the potential for growth through product innovation, in: *EUPVSEC 2015*. https://doi.org/10.4229/EUPVSEC20152015-7DO.15.4.
Devetaković, M., Djordjević, D., Radojević, M., Krstić-Furundžić, A., Burduhos, B.-G., Martinopoulos, G., ... Lobaccaro, G. (2020). Photovoltaics on landmark buildings with distinctive geometries. *Applied Sciences*, *10*, 6696. Available from https://doi.org/10.3390/app10196696.
Eder, G., Peharz, G., Trattnig, R., Bonomo, P., Saretta, E., Frontini, F., ... Zanelli, A. (2019). *COLOURED BIPV market, research and development IEA PVPS Task 15, subtask E report T15−07 : 2019*.
Ertex-solar [WWW document] (2020). URL https://www.ertex-solar.at/en/. Accessed 17.02.21.
EU Directive 2018:2001 (2018).
Flisom [WWW document] (2020). URL https://www.flisom.com/. Accessed 17.02.21.
Frontini, F., Bonomo, P., Saretta, E., Weber, T., & Berghold, J. (2016). Indoor and outdoor characterization of innovative colored BIPV modules for façade application, in: *32nd European photovoltaic solar energy conference and exhibition* (pp. 2498−2502). https://doi.org/10.4229/EUPVSEC20162016-6DO.8.2.
Frontini, F., Chatzipanagi, A., Bonomo, P., Verbene, G., Donker van den, M., Folkerts, W. (2015). *BIPV product overview for solar facades and roofs*. SUPSI.

Frontini, F., Saretta, E., & Bonomo, P. (2016). Colored BIPV glass modules: The 'price' of aesthetics, in: BRENET (Ed.), 19. *Status-Seminar—Forschen Für Den Bau Im Kontext von Energie Und Umwelt.* Zurich.

Frontini, F., Scognamiglio, A., Graditi, G., & Pellegrino, M. (2013). From BIPV to building component, in: *EUPVSEC 2013*. https://doi.org/10.4229/28thEUPVSEC2013-5CO.8.1.

Frontini, F., Zanetti, I., Bonomo, P., Andreas, H., & Studer, D. (2020). Solarchitecture.ch [WWW document]. *Solar architecture—Sun as a building materials.* URL https://www.solarchitecture.ch. Accessed 25.10.20.

Frontini, F., Zanetti, I., Bonomo, P., Saretta, E., Donker van den, M., Vossen, F., & Folkerts, W. (2017). *Building integrated photovoltaics: Product overview for solar building skins* (2017th ed.). SUPSI.

Gong, J., Sumathy, K., Qiao, Q., & Zhou, Z. (2017). Review on dye-sensitized solar cells (DSSCs): Advanced techniques and research trends. *Renewable and Sustainable Energy Reviews, 68*, 234−246. Available from https://doi.org/10.1016/j.rser.2016.09.097.

Green, M. A., Dunlop, E. D., Hohl-Ebinger, J., Yoshita, M., Kopidakis, N., & Hao, X. (2020). Solar cell efficiency tables (version 56). *Progress in Photovoltaics: Research and Applications, 28*, 629−638. Available from https://doi.org/10.1002/pip.3303.

Heinstein, P., Ballif, C., & Perret-Aebi, L.-E. (2013). Building integrated photovoltaics (BIPV): Review, potentials, barriers and myths. *Green, 3*. Available from https://doi.org/10.1515/green-2013-0020.

Huang, P., Lovati, M., Zhang, X., & Bales, C. (2020). A coordinated control to improve performance for a building cluster with energy storage, electric vehicles, and energy sharing considered. *Applied Energy, 268*, 114983. Available from https://doi.org/10.1016/j.apenergy.2020.114983.

Huang, P., Lovati, M., Zhang, X., Bales, C., Hallbeck, S., Becker, A., ... Maturi, L. (2019). Transforming a residential building cluster into electricity prosumers in Sweden: Optimal design of a coupled PV-heat pump-thermal storage-electric vehicle system. *Applied Energy, 255*, 113864. Available from https://doi.org/10.1016/j.apenergy.2019.113864.

iWin—innovative Windows [WWW document] (2019). http://www.iwin.ch. Accessed 17.02.21.

Jayathissa, P., Caranovic, S., Hofer, J., Nagy, Z., & Schlueter, A. (2018). Performative design environment for kinetic photovoltaic architecture. *Automation in Construction, 93*, 339−347. Available from https://doi.org/10.1016/j.autcon.2018.05.013.

Jayathissa, P., Zarb, J., Luzzatto, M., Hofer, J., & Schlueter, A. (2017). Sensitivity of building properties and use types for the application of adaptive photovoltaic shading systems. *Energy Procedia, 122*, 139−144. Available from https://doi.org/10.1016/j.egypro.2017.07.319.

Jelle, B. P., & Breivik, C. (2012). State-of-the-art building integrated photovoltaics. *Energy Procedia, 20*, 68−77. Available from https://doi.org/10.1016/j.egypro.2012.03.009.

Kaleo Solar [WWW document] (2020). http://www.kaleo-solar.ch/. Accessed 17.02.21.

Lee, H. M., & Yoon, J. H. (2018). Power performance analysis of a transparent DSSC BIPV window based on 2 year measurement data in a full-scale mock-up. *Applied Energy, 225*, 1013−1021. Available from https://doi.org/10.1016/j.apenergy.2018.04.086.

Liao, W., & Xu, S. (2015). Energy performance comparison among see-through amorphous-silicon PV (photovoltaic) glazings and traditional glazings under different architectural conditions in China. *Energy, 83*, 267−275. Available from https://doi.org/10.1016/j.energy.2015.02.023.

Lobaccaro, G., Lisowska, M. M., Saretta, E., Bonomo, P., & Frontini, F. (2019). A methodological analysis approach to assess solar energy potential at the neighborhood scale. *Energies*, *12*. Available from https://doi.org/10.3390/en12183554.

Lovati, M., Adami, J., & Moser, D. (2018). Open source tool for a better design of BIPV + battery system: An applied example, in: *35th European photovoltaic solar energy conference and exhibition* (pp. 1641−1646). https://doi.org/10.4229/35thEUPVSEC20182018-6CO.4.1.

Lovati, M., Salvalai, G., Fratus, G., Maturi, L., Albatici, R., & Moser, D. (2019). New method for the early design of BIPV with electric storage: A case study in northern Italy. *Sustainable Cities and Society*, *48*, 101400. Available from https://doi.org/10.1016/j.scs.2018.12.028.

Lovati, M., Zhang, X., Huang, P., Olsmats, C., & Maturi, L. (2020). Optimal simulation of three peer to peer (P2P) business models for individual PV prosumers in a local electricity market using agent-based modelling. *Buildings*, *10*, 138. Available from https://doi.org/10.3390/buildings10080138.

Macé, P., Stridh, B., Frieden, D., Woess-Gallasch, S., Frederiksen, K., Kenergy, D., ... Cea, F. (2020). Development of BIPV business cases—Guide for stakeholders. *IEA PVPS Task 15*.

Machado, M., Alonso, R., Frontini, F., Bonomo, P., Weiss, I., Alonso, P., ... Haller, A. (2019). BIPVBOOST project—Bringing down costs of building-integrated photovoltaic (BIPV) solutions and processes along the value chain, enabling widespread implementation in near zero energy buildings (nZEBs) implementation, in: *36th European photovoltaic solar energy conference and exhibition* (pp. 2037−2038). https://doi.org/10.4229/EUPVSEC20192019-7DV.2.36.

Machado, M., Challet, S., Weiss, I., Medina, E.R., Espeche, J.M., Noris, F., ... V.K. Nguyen (2017). Supporting market uptake of building-integrated photovoltaic technologies with the PVSITES project, in: *33rd European photovoltaic solar energy conference and exhibition* (pp. 2882−2887). https://doi.org/10.4229/EUPVSEC20172017-7DV.1.29.

Mandalaki, M., Zervas, K., Tsoutsos, T., & Vazakas, A. (2012). Assessment of fixed shading devices with integrated PV for efficient energy use. *Solar Energy*, *86*, 2561−2575. Available from https://doi.org/10.1016/j.solener.2012.05.026.

Maturi, L., Giona, S., Moser, D., Lollini, R., Lovati, M., Alonso, P., ... Metayer, S. (2019). Energymatching project—Adaptable and adaptive RES envelope solutions to maximize energy harvesting and optimize EU building and district load matching, in: *36th European photovoltaic solar energy conference and exhibition* (pp. 1827−1831). https://doi.org/10.4229/EUPVSEC20192019-6BV.4.13.

Meir, M. (2019). *State-of-the-art and SWOT analysis of building integrated solar envelope systems*. https://doi.org/10.18777/ieashc-task56-2019-0001.

Nagy, Z., Svetozarevic, B., Jayathissa, P., Begle, M., Hofer, J., Lydon, G., ... Schlueter, A. (2016). The adaptive solar façade: From concept to prototypes. *Frontiers of Architectural Research*, *5*, 143−156. Available from https://doi.org/10.1016/j.foar.2016.03.002.

Norton, B., Eames, P. C., Mallick, T. K., Huang, M. J., McCormack, S. J., Mondol, J. D., & Yohanis, Y. G. (2011). Enhancing the performance of building integrated photovoltaics. *Solar Energy*, *85*, 1629−1664. Available from https://doi.org/10.1016/j.solener.2009.10.004.

NREL (2021). Best research-cell efficiency chart [WWW document]. https://www.nrel.gov/pv/cell-efficiency.html. Accessed 04.04.21.

Olivieri, L., Frontini, F., Polo López, C. S., Pahud, D., Caamaño-Martín, E., Polo-López, C., ... Caamaño Martín, E. (2015). G-value indoor characterization of semi-transparent photovoltaic elements for building integration: New equipment and methodology. *Energy and Buildings*, *101*, 84–94. Available from https://doi.org/10.1016/j.enbuild. 2015.04.056.

PIZ [WWW document] (2020). https://www.piz.it/en/. Accessed 17.02.21.

Prieto, A., Knaack, U., Auer, T., & Klein, T. (2017). Solar facades—Main barriers for widespread façade integration of solar technologies. *Journal of Façade Design and Engineering*, *5*, 51–62. Available from https://doi.org/10.7480/jfde.2017.1.1398.

Probst, M. M., & Roecker, C. (2012). Criteria for architectural integration of active solar systems IEA Task 41, subtask A. *Energy Procedia*, *30*, 1195–1204. Available from https://doi.org/10.1016/j.egypro.2012.11.132.

Quesada, G., Rousse, D., Dutil, Y., Badache, M., & Hallé, S. (2012). A comprehensive review of solar facades. Transparent and translucent solar facades. *Renewable and Sustainable Energy Reviews*, *16*, 2643–2651. Available from https://doi.org/10.1016/j.rser.2012.02.059.

Radl, J., Fleischhacker, A., Revheim, F. H., Lettner, G., & Auer, H. (2020). Comparison of profitability of PV electricity sharing in renewable energy communities in selected European countries. *Energies*, *13*, 5007. Available from https://doi.org/10.3390/en13195007.

Sánchez-Pantoja, N., Vidal, R., & Pastor, M. C. (2018). Aesthetic perception of photovoltaic integration within new proposals for ecological architecture. *Sustainable Cities and Society*, *39*, 203–214. Available from https://doi.org/10.1016/j.scs.2018.02.027.

Saretta, E., Bonomo, P., & Frontini, F. (2017). Active BIPV glass facades: Current trends of innovation, in: *GPD glass performance days 2017*.

Saretta, E., Bonomo, P., & Frontini, F. (2018). BIPV meets customizable glass: A dialogue between energy efficiency and aesthetics, in: WIP (Ed.), *35th European photovoltaic solar energy conference and exhibition*. Brusselles. https://doi.org/10.4229/35thEUPVSEC20182018-6AO.8.2.

Saretta, E., Bonomo, P., & Frontini, F. (2020). A calculation method for the BIPV potential of Swiss facades at LOD2.5 in urban areas: A case from Ticino region. *Solar Energy*, *195*, 150–165. Available from https://doi.org/10.1016/j.solener.2019.11.062.

Schram, W., Louwen, A., Lampropoulos, I., & van Sark, W. (2019). Comparison of the greenhouse gas emission reduction potential of energy communities. *Energies*, *12*, 4440. Available from https://doi.org/10.3390/en12234440.

Scognamiglio, A. (2017). Chapter 6—Building-integrated photovoltaics (BIPV) for cost-effective energy-efficient retrofitting. In F. Pacheco-Torgal, C.-G. Granqvist, B. P. Jelle, G. P. Vanoli, N. Bianco, & J. Kurnitski (Eds.), *Cost-effective energy efficient building retrofitting* (pp. 169–197). Woodhead Publishing. Available from https://doi.org/10.1016/B978-0-08-101128-7.00006-X.

Shukla, A. K., Sudhakar, K., Baredar, P., & Mamat, R. (2017). BIPV in Southeast Asian countries—Opportunities and challenges. *Renewable Energy Focus*, *21*, 25–32. Available from https://doi.org/10.1016/j.ref.2017.07.001.

SMART-Flex [WWW document] (2016). *Demonstr. Ind. scale Flex. Manuf. SMART Multifunct. Photovolt. Build. Elem.* URL https://cordis.europa.eu/project/id/322434. Accessed 25.10.20.

Solar Visual [WWW document] (2020). https://www.solarvisuals.nl/. Accessed 17.02.21.

SolarLab [WWW document] (2020). https://solarlab.ch/. Accessed 17.02.21.

Solaxess [WWW document] (2020). http://www.solaxess.ch. Accessed 17.02.21.

Sunage [WWW document] (2021). https://www.sunage.ch. Accessed 17.02.21.
SwissINSO [WWW document] (2020). https://www.swissinso.com/. Accessed 17.02.21.
Tabadkani, A., Roetzel, A., Li, H. X., & Tsangrassoulis, A. (2021). Design approaches and typologies of adaptive facades: A review. *Automation in Construction, 121*, 103450. Available from https://doi.org/10.1016/j.autcon.2020.103450.
Taveres-Cachat, E., Lobaccaro, G., Goia, F., & Chaudhary, G. (2019). A methodology to improve the performance of PV integrated shading devices using multi-objective optimization. *Applied Energy, 247*, 731−744. Available from https://doi.org/10.1016/j.apenergy.2019.04.033.
Thomaßen, G., Kavvadias, K., & Jiménez Navarro, J. P. (2021). The decarbonisation of the EU heating sector through electrification: A parametric analysis. *Energy Policy, 148*, 111929. Available from https://doi.org/10.1016/j.enpol.2020.111929.
Tulipps [WWW document] (2020). https://tulipps.com. Accessed 17.02.21.
Ulbikas, J., Galdikas, A., & Stonkus, A. (2016). Smart-FLeX solution way forward for cost competitive BIPV production?, in: *32nd European photovoltaic solar energy conference and exhibition* (pp. 2790−2791). https://doi.org/10.4229/EUPVSEC20162016-6AV.5.28.
Wall, M., Probst, M. C. M., Roecker, C., Dubois, M.-C., Horvat, M., Jørgensen, O. B., & Kappel, K. (2012). Achieving solar energy in architecture-IEA SHC Task 41. *Energy Procedia, 30*, 1250−1260. Available from https://doi.org/10.1016/j.egypro.2012.11.138.
Zanetti, I., Bonomo, P., Frontini, F., Saretta, E., van den Donken, M., Vossen, F., & Folkers, W. (2017). *BIPV status report 2017* (2017th ed.). SUPSI.
Zhang, X., Lau, S.-K., Lau, S. S. Y., & Zhao, Y. (2018). Photovoltaic integrated shading devices (PVSDs): A review. *Solar Energy, 170*, 947−968. Available from https://doi.org/10.1016/j.solener.2018.05.067.

Unitized Timber Envelopes: the future generation of sustainable, high-performance, industrialized facades for construction decarbonization

9

Eugenia Gasparri
School of Architecture, Design and Planning, The University of Sydney, Sydney, NSW, Australia

Abbreviations

BIM	building information modelling
BIPV	building integrated photovoltaic
BIST	building integrated solar thermal
CLT	cross-laminated timber
DfD	design for disassembly
DfMA	design for manufacturing and assembly
GHG	greenhouse gas
LB	load bearing
nLB	non-load bearing
NCC	national construction code
PVC	polyvinyl chloride
TF	timber-framed
UCW	unitized curtain wall
UTE	unitized timber envelope

9.1 Introduction

The current climate crisis has forced our society and communities to reconsider the way they operate and behave to guarantee our ecosystem's sustainable development for future generations to survive and thrive. Greenhouse gas (GHG) emissions are among the main causes for global warming, with the building construction industry being one of the major global contributors. In 2019 the International Energy Agency estimated that buildings and construction sector combined are responsible for 36% of global energy consumption and about 40% of global CO_2 emissions (IEA, 2019). Besides the population is rapidly growing and is expected to reach 9.7 billion in 2050, that is, about 2 billion more people over the next 30 years. At the same time, the urbanization process is rapidly increasing and soon enough most of the humankind will be living in urban areas (United Nations, 2019). But is the construction industry ready to take on such a rapid growth while considerably reducing its environmental impact to stay on track with sustainable development goals?

The effects of urbanization in the architecture of our most populated cities worldwide are already evident. In fact, urban densification together with high land prices is contributing to significantly increase the number of multistorey buildings in metropolitan areas. Another emerging trend includes the integration of different activities in buildings, to account for the needs of an ever-changing, diverse and dynamic society. Mixed-use tall buildings will soon be the new normal in construction. In this context a shift toward a more industrialized and modular approach would create immense opportunities to boost productivity for the construction industry to keep pace with rapid floor area expansion and, at the same time, increase buildings' adaptability to change (Barbosa et al., 2017; Ribeirinho et al., 2020). Sustainable development strategies must be set in place to concurrently optimize the process while reducing the sector's environmental impact by favouring the use of sustainable resources and materials, minimizing buildings' energy demand and consumption, reducing waste and drastically shifting toward the adoption of circular economy behavioural models.

This chapter builds up on the previous considerations and discusses new trends and opportunities in the field of industrialized façade technologies for multistorey buildings. At first, it reflects on the current practice by highlighting the inadequacy of most common façade technologies in responding to the climate emergency, due to their poor energy and environmental performance. Subsequently, it highlights the potential of mass timber construction in contributing to the sector decarbonization. To conclude, it presents the results of a 2-year (2017–19) industry-embedded research project carried out at the University of Sydney, investigating the design of modular high-performance timber envelopes for industrialized multistorey buildings in Australia. The newly developed, innovative unitized timber envelope (UTE) makes the use of sustainable resources, guarantees excellent energy performance and construction efficiency, and favours circularity through modular design, assembly and disassembly (Gasparri & Aitchison, 2019).

9.2 Tall buildings and facades

The building envelope plays a fundamental role when it comes to buildings' energy demand and consumption; thus performance improvements are critical to achieve global sustainability targets (IEA, 2020). Also, in tall buildings the façade constitutes a significant proportion of the overall building cost, particularly in residential developments (Barton & Watts, 2013).

Nowadays, most high rises around the world make the use of modular façade systems, also known as unitized curtain walls (UCWs), which have become key features in tall building for the advantages they provide in terms of floor plate design optimization and installation efficiency (Herzog et al., 2004; Knaack et al., 2007). Indeed, these systems mainly consists of aluminium frames with large infill glass panels that create uninterrupted view and maximize the ingress of daylight within buildings, thus allowing the realization of larger and more cost-effective floor plates. UCW units generally include a spandrel area (typically insulated) with the purpose of concealing building's structural or service components. Moreover, they are equipped with perimeter gasket seals that allow consecutive façade units to mate together working in series to guarantee joints weather tightness at interfaces. Therefore they can be installed at building edges from the construction floor with no need for scaffolding or any other access to the façade from the outside. This contributes to considerably boost construction efficiency, minimize costs and risks and increase site safety by limiting exposure to work at height. Commonly, unitized façade single elements and consist of insulated glass panels supported by aluminium-extruded frames equipped with gasket seals. These systems generally span between floor plates (being free from load bearing function) and quite limited in dimensions (about 1- to 1.5-m wide) due to both logistics and structural-related reasons. They get to construction site on pallets, ready to be off-loaded from the delivery truck, and lifted to the reference construction floor by crane. From there, each unit is installed from the inside of the building, usually via monorail perimeter beam or spider crane placed a couple of floors above (see Fig. 9.1).

Despite some of the benefits UCWs might offer, the extensive use of glass in facades means to compromise with building's sustainability standards and energy efficiency (Kim, 2011; Yi et al., 2017). These systems are characterized by

Figure 9.1 Daramu House in Sydney, Australia. (Left) Installation process and (Right) connection-to-floor detail of a glass façade unit.
Source: Author's original.

high embodied energy, being glass and aluminium both materials with high environmental impact, and carbon intensive production processes. They guarantee poor thermal performance that favours indoor heat gain in summer and high thermal loss in winter, contributing to the increase in GHG emissions. Moreover, the use of reflective coatings as a strategy to control radiation through glazed surfaces may often contribute to urban overheating, disturbances of the building surroundings and/or bird migration endangerment (Link, 2020).

Recently, field experts and political leaders from around the world have been urging to consider the ban of high energy demanding 'glass skyscrapers' as a response to the climate crisis (Gaviola, 2019; Tapper, 2019). Urban and building policy makers are also moving in this direction by gradually introducing stricter limitations on transparent surfaces allowed by national building codes, toward a better optimized window-to-wall ratio. So far, this has mainly resulted in a mere reduction of the transparent surface of a curtain wall in favour of an increase spandrel area, which unfortunately would still not guarantee adequate energy performance for these systems.

This change of trajectory, instead, constitutes a fertile ground to rethink facades in multistorey buildings, opening new opportunities to the development of high-performance, sustainable alternatives to UCW technologies. When it comes to high energy performance, multilayered envelope technologies are 'the way to go'. They are made of distinct layers, each of them expressly fitted to fulfil one (or more) function, and all together working as an integrated system to guarantee optimal performance for the specific design context. However, the greater flexibility and adaptability of multilayered envelopes come together with a higher number of variables and, therefore, a higher complexity. Each layer may vary from project to project for a number of different reasons (eg, market availability, cost, ecological choices, needed performance, climatic context, construction culture and building use). This complexity has made it difficult so far to come up with fully optimized, industrialized systems that could still maintain a good level of adaptability and be used across a variety of projects and contexts.

This chapter emphasizes the potential for industrialized multilayered envelopes to enhance sustainability in construction by reducing building's energy consumption and addressing emerging needs for adaptability and climate resilience. The author's research introduces an innovative envelope design proposal for the multi-storey building market, which inherits the benefits of UCW systems and makes the use of timber technologies to maximize envelope's energy and environmental performance.

9.3 The 'Rise' of mass timber

Mass timber technologies, such as cross-laminated timber (CLT) and glue-laminated timber (GLT or Glulam), are today increasingly gaining market share in large-scale or multistorey building developments, over more traditionally used structural materials like concrete and steel (Dangel, 2016; Kuzmanovska et al., 2018; Ramage et al., 2017). The reasons behind this shift are manifold. Timber structures not only guarantee excellent structural performance, but their use in construction is often associated with considerable site productivity enhancement as well as reduced environmental impacts (Li et al., 2019). As a matter of fact, timber is a natural renewable resource that acts as carbon storage and its fabrication process is low energy intensive when compared to most of other construction materials, making it by far the most sustainable solution toward construction decarbonization goals. Timber is also lightweight, easy to be cut into shape, modified, assembled into parts or components, moved or transported and eventually disassembled. All this encourages the adoption of industrialized approaches to construction and favours circularity toward zero-emission and zero-waste timber buildings in future.

Australia's first timber tower, the Forte Living 10-storey apartment building, was completed in Melbourne in 2013 by Lendlease. Since then, the interest toward timber buildings has been growing considerably and many other best practice examples have been raising across all major Australian centres, both for residential and commercial use. The acquisition of new knowledge and experience in the field of mass timber construction, along with the recent implementations to the Australian National Construction Code (NCC), have been pushing the boundaries of reachable height and open the path to future generations of tall timber towers. The rapid increase of the average building height inevitably challenges other aspects of construction, such as the system of building enclosures for quick, easy and safe installation (Gasparri & Mazzucchelli, 2016; Gasparri et al., 2015). This has consistently been reflected in an increased use of UCWs over the last decade, moving away from multilayered timber envelope systems with punched windows (Kuzmanovska et al., 2018). In fact, despite prefabricated timber envelopes are not uncommon, they can be prefabricated only up to a certain extent leaving the completion of the joints between envelope components to be finished on site, either via scaffolds or mobile platforms. Of course, as previously discussed, the advantages offered by currently available UCW systems in terms of construction efficiency and safety have to compromise with other factors, particularly the envelope energy performance and environmental impact.

9.3.1 Envelope prefabrication in timber buildings

Prefabrication of timber envelopes is widely diffused across Europe as a strategy to minimize the exposure of timber structures to unfavourable weather conditions during site activities, thus mitigating the risk of water-related damages. Timber envelope panelized components are usually fabricated in large dimensions to further enhance site productivity by reducing the number of lifting and installation operations. The level of prefabrication of these systems can highly vary, from low level, including only the wall structure (eg, stud wall frame assembly or precut CLT panel), to high level, comprising most of the functional layers instead, such as the structure, insulation, weather membranes and barriers and external cladding. However, evidence in

Figure 9.2 (Left) LCT ONE in Dornbirn, Austria. Installation of prefab envelope panels via crane and scaffolds. (Right) Nexity Ywood « L'Ensoleillée II » building in Aix-en-Provence, France. Installation of prefab envelope panels via crane and mobile platforms.
Source: Reproduced with permission from (Left) http://www.creebuildings.com ©DarkoTodorovic|Photography|adrok.net. (Right) ©Nexity, Yann Bouvier.

the field shows that, no matter the level of envelope prefabrication, timber envelopes weatherproofing at panels' interface has always been completed from the outside of the building (Andaloro et al., 2019; Gasparri & Aitchison, 2019; Kaufmann et al., 2018), unlike UCW systems. Fig. 9.2 show two best practice examples in the field timber envelope prefabrication, the LifeCycle Tower One using a timber-framed wall system that comes to site preassembled to the building vertical structure (mass timber columns), while the Ywood « L'Ensoleillée II » building using a load bearing CLT-based envelope system.

Two timber projects only have so far attempted to push the boundaries of prefabrication for multilayered opaque envelopes: the Brock Commons in Vancouver and the Fenner Hall in Canberra (see Fig. 9.3). These two buildings have much in common. They both are students' accommodations located in university campuses, showing a powerful commitment in embracing academia's value toward education and innovation. They adopted mass timber as structural material and stand out as best practice examples in pushing the sustainability agenda in construction. They explored new concepts and ideas, experimented with the use of new systems and technologies and contributed to the industry growth by bringing new products to the market and knowledge to the field. Both projects showcased the use of pioneering unitized multilayered envelopes, assembled off-site in large-size panelized elements and able to be installed in their final position with no need for scaffolding or any action from the outside of the building. Both systems use steel-stud frame technology missing the opportunity for a more sustainable design approach. The next two sections shortly present the two projects and discuss their envelope systems'

Figure 9.3 (Left) Brock Commons student accommodation in Vancouver, Canada. Scaffold-free installation of façade elements (Centura Building Systems Ltd). (Right) Fenner Hall student accommodation in Canberra, Australia. Scaffold-free installation of façade elements (CSR Inclose).
Source: Reproduced with permission from (Left) Steven Errico, courtesy naturallywood.com. (Right) CSR Building Products Ltd'.

technical features and characteristics.[1] They introduce the reader to the topic of unitized multilayered envelopes and provide a background to the authors' research presented in Section 9.4, proposing a highly innovative system for sustainable and efficient multistorey construction.

Brock Commons | *The University of British Columbia (UBC)—Vancouver, Canada*

Brock Commons is a student accommodation, completed in 2017 at the UBC campus in Vancouver. At the time of construction, the building was the tallest residential timber hybrid tower worldwide, which comprises 17 storeys of mass timber construction above a concrete podium. The project was part of the *Tall Wood Building Demonstration Initiative* to demonstrate the viability in the use of mass timber for multistorey construction and, at the same time, contribute to timber technologies market expansion through innovation in design and construction practices. Indeed, the Brock Commons building served as a case study to experiment on a variety of new technical solutions, responding to DfMA (design for manufacturing and assembly) principles, to maximize performance and optimize construction efficiency. The project has also pioneered the use of a digital twin for construction

---

[1] General information and qualitative data about each building were collected from various sources, such as interviews, conference papers or presentations, project reports, construction photographs and/or videos available on the websites of the architects, engineers, builders, developers and/or clients involved in either project. The author also had the opportunity to participate in site visits for research purposes during each project's construction stage.

planning and management, as a collaborative tool for knowledge integration. The structural system is highly industrialized and consists of an efficient two-way slab diaphragm (5-ply CLT) directly supported by columns without the need for downstand beam. This reduces the number of elements to be installed on site and optimizes the building interstorey height. The adoption of newly designed steel connectors allows for vertical loads to be transferred between columns and bypass floor panels, otherwise subject to possible damage due to compression perpendicular to grains. Connectors come to site preassembled to columns and are designed to ease and guide floor panels installation to speed up the process.

The "quasi-unitized" façade system consists of multilayered insulated panels with punched windows. Prefabricated wall panels comprise a steel-stud frame structure equipped with sheathing board, weatherproofing membrane, external rigid insulation and a rain screen system with a wood−fibre laminate cladding. Differently from purely unitized facades, the Brock Commons envelope system requires the adoption of flashing and seals at joints' interfaces to ensure weather tightness (see Fig. 9.4). Caulking was also applied from the inside to prevent leakages and reduce the risk of joint failure, as shown in Fig. 9.4. To allow for joints completion, a few functional layers were added on site from the building interior, namely, batt insulation, vapour barrier and drywall.

Large-size panels were assembled horizontally at the factory on special jigs for tolerance control then transported vertically to site on specific racks (see Fig. 9.5). Site installation happened via crane and did not require the use of scaffolding. Floors were equipped with L-beam connectors strengthening floor edges and providing support for façade panels to be hanged onto. Indeed, the façade system is non-load bearing but subject to horizontal loads only, other than its own weight. Adjustments and tolerances' management are guaranteed by the ad hoc L-beam connector; however, the special corner detail and allowance for movement added a layer of complexity to the overall installation process (see Fig. 9.6). More information and details on the Brock Commons envelope system can be found in the author's previous publications on the topic (Gasparri & Aitchison, 2019).

Figure 9.4 Brock Commons façade system. (Left) Top view: flashing and seals at horizontal joints. (Right) Rear view: caulking application at joint interfaces.
Source: Reproduced with permission from Graham Finch at RDH Building Science.

Figure 9.5 Brock Commons façade system. (Left) Production line: horizontal assembly includes window and cladding installation. (Right) Transportation to site: vertical on racks.
Source: Reproduced with permission from (Left) Centura Building Systems Ltd'. (Right) KK Law, courtesy naturallywood.com.

Figure 9.6 Brock Commons façade system. (Left) Inside view: planar standard elements and L-beam connector at floor edges. (Right) Outside view: corner special elements and L-beam connector at floor edges.
Source: Reproduced with permission from (Left) Steven Errico, courtesy naturallywood.com. (Right) KK Law, courtesy naturallywood.com.

Fenner Hall | *Australian National University (ANU)—Canberra, Australia*
The Fenner Hall student accommodation was completed in 2018 as part of the ANU Kambri precinct strategic redevelopment in Canberra. The building consists of a seven-storey high mass timber structure atop a double-height concrete podium and features several innovations in the design and construction of mass timber

Figure 9.7 Fenner Hall façade system. (Left) Large scale mock-up. (Right) Transportation to site: vertical on racks.
Source: Author's original.

buildings in the Australian context. The extensive use of CLT structures and prefabrication strategies allowed to minimize the building carbon footprint considerably[2].

The Fennel Hall building excels for the pioneering use of unitized multilayered facades with punched windows and brick cladding. As per the Brock Commons' example, this façade system comprises a steel-stud frame structure equipped with sheathing board, weatherproofing membrane, external insulation, and a rain screen system. This time, a brick-on-rail cladding system was adopted to provide the building with a look typical of Australian architecture. Here too the interior lining was completed on site after facades installation from the inside of the building. However, this time no flashing, seals or caulking was necessary as façade panels were equipped with thermally broken extruded frames and perimeter rubber gaskets, which allowed for adjacent panels to simply mate together to ensure joints' weathertightness.

The proprietary façade system was fabricated off-site in exceptionally large dimensions (up to 13 m in lengths) and transported vertically to site on specific racks (see Fig. 9.7). The manufacturing process was labour intensive, with little or no involvement of automated processes. The walls' assembly mainly took place on horizontal jigs, but windows installation and brick mortar application happened once façade panels were tilted in a vertical position, which understandably required an extra coordination effort (see Fig. 9.8). This decision was particularly strategic with respect to the cladding system as it was necessary for bricks to properly settle by gravity on their rail support system before filling the gaps in between them. Site installation did not require the use of scaffolding, and thanks to the gasket-based weatherproof system any action at joint's interfaces from either the exterior or interior of the building. L-brackets were installed at floor edges for façade panels to be hooked onto and for site tolerances adjustments. The turnaround time for each panel ranged between 20 and 30 minutes, allowing for the envelope to come up in record time.

---

[2] Carbon footprint reduction equivalent to 2.5 T CO_2-e/m^2 (-48%) with respect to the reference design proposal, as certified by The Footprint Company.

Figure 9.8 Fenner Hall façade system. Production line at CSR Inclose factory. (Left) Horizontal assembly stage. (Right) Vertical assembly stage for windows and cladding completion.
Source: Author's original.

9.4 The case for unitized timber envelopes

This section presents the author's research in the field of high-performance modular envelope design, carried out at the University of Sydney as a part of an industry-embedded research project investigating the future of industrialized construction and multistorey timber buildings in Australia. This study builds upon the author's doctoral research, exploring new design pathways for envelope prefabrication in mass timber construction (Gasparri, 2016). The applied and collaborative nature of this research project, conducted in partnership with the industry partners Lendlease and Design Make, allowed for the definition of the first-of-its-kind unitized timber envelope (UTE) system, which inherits from and overtakes UCW advantages in terms of construction efficiency while guaranteeing high sustainability and energy performance standards typical of multilayered envelopes (see Fig. 9.9).

Research boundaries and success criteria were discussed and agreed upon with industry partners during the initial stage of the project. The general aim was to develop an innovative system for the multistorey building market, adaptable to both residential and commercial uses and able to respond to the macro area objectives listed later through an integrated design approach.

a. *Design*: The system is configurable in terms of performance and functions as well as compatible with a wide range of cladding types and configurations, so to be integrated repeatedly across a variety of building types, contexts and locations.
b. *Factory*: The system is modular and factory-fitted. DfMA principles in terms of part count reduction and ease/number of activities must be addressed. Design for disassembly (DfD) is another fundamental principle to be considered to enable circularity.
c. *Site*: The system maximizes construction site efficiency and safety allowing for quick, easy and scaffold-free installation. The use of large panel size was strategically considered to further speed-up the construction process and optimize production, transportation and handling processes.

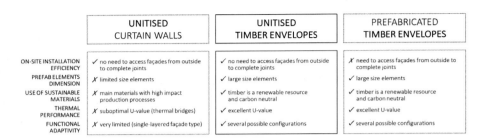

Figure 9.9 Summary of advantages related to the use of unitized timber envelopes.
Source: Author's original.

d. *Use*: The system is durable and guarantees high energy performance standards across the whole building life. Programmed maintenance was carefully addressed during the design phase, so for the system to be resilient and adaptable in time.

The adoption of a system-based design approach would positively impact all the project phases by cutting down inefficiencies in terms of waste of resources, time and cost. It also provides significantly greater control over the whole process thanks to information systematization and knowledge management.

9.4.1 Design methodology

This work belongs to the applied design research field and followed an incremental-iterative approach (Research *through* Design), alternating and reiterating processes of design, modelling, simulation, prototyping, validation and finally integration of findings and feedback. Indeed, the design process adopted a mixed methodology consisting of both exploratory and descriptive research phases.

The final system design with all its possible configurations was then embedded into a building information modelling (BIM) model, to verify design adaptability and efficiency via parametrization and validate technical and architectural design outcomes, both at component and building scale.

9.4.2 Design process

The UTE system is constituted of three main parts, as exemplified in Fig. 9.10.

1. *Wall (configurable)*: It includes the structural system and all other necessary functional layers, such as built-in insulation, sheathing boards, fire barriers, weatherproofing membrane. It can be fabricated in large-size elements and is designed to allow for a defined set of configurations to be used in different contexts and/or locations (eg, different load transfer, energy performance requirements).
2. *Frame (standardized)*: It is an engineered system running around the *Wall* perimeter and comprises a custom-made aluminium-extruded profile integrating rubber gaskets. Frames of adjacent façade panels are specifically designed to mate together ensuring joints'

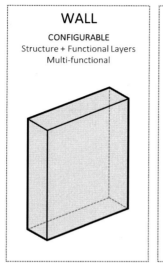 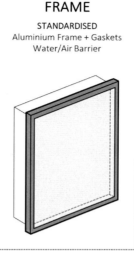 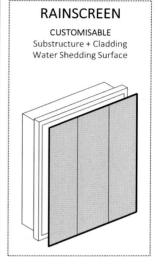

Figure 9.10 UTE system design concept—Wall, Frame, Rainscreen.
Source: Author's original.

weathertightness at interfaces, as in the case of UCWs. In fact, the proposed design allows for the two systems (UTE and UCW) to be compatible (see Fig. 9.14).
3. *Rainscreen (customizable)*: It protects the *Wall* behind from getting wet, acting as first barrier to wind-driven rain. It also has an important architectural function, defining the aspect and aesthetic of the building outer skin. The UTE system allows for the integration of commercially available rainscreen systems or newly designed one. Different cladding materials and configurations can be adopted. The only constraint is for the cladding substructure to be fixed to the structural elements in the *Wall* (eg, following timber studs spacing).

Dimensional limitations are determined through design and based on structural considerations, transportation constraints, construction efficiency objectives and architectural design needs. The design choices and process are explained later, following a bottom-up approach (from the single parts to the whole system).

Wall design process

The design of a multilayered envelope is a rather complex process that requires high coordination among the various design experts involved. In fact, not only each layer must satisfy its own specific function or performance requirement, but it must also be compatible and work organically with other layers as part of a system. The three main steps to design a multilayered envelope are the following: (1) functional analysis and requirement specification, (2) selection of the appropriate materials/ products, (3) layer positioning within the wall cross-section.

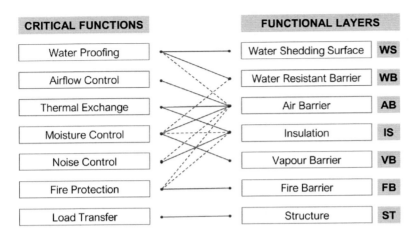

Figure 9.11 Functional analysis scheme of an opaque envelope component.
Source: Author's original.

The design of the *Wall* and its set of possible configurations were guided by several concurrent factors, such as industry partners' inputs, factory's production line characteristics, literature and best practice examples, author's own experience in the field and Australian construction practice and building regulation. For the sake of clarity, design choices in terms of material selection and layer positioning are briefly described based on the functional layers' classification in Fig. 9.11. It is fundamentally highlighting the importance of following an integrated design approach, where all the different functional requirements are considered holistically and not independently from one another.

- *Water shedding surfaces:* A rainscreen approach was considered the most appropriate for dealing with both timber and prefabrication. In fact, it prevents most of the water from getting in contact with the timber-based wall assembly and, at the same time, allows an element of 'play' with the cladding arrangement (materiality, colour, pattern, spacing, orientation, etc.) so to either hide or emphasize the horizontal and vertical joints at modular panels interfaces.
- *Weather barrier*: Air and waterproofing are guaranteed by a weathertight membrane with high vapour permeability, to keep the wall structure dry and minimize the risk of interstitial condensation.
- *Insulation*: Rock mineral wool was the selected material as it combines a well-balanced set of positive characteristics, as per good thermal resistance, vapour permeability, acoustic performance, fire protection and relatively moderate cost. However, the use of natural insulation products can be a better choice to embrace sustainability targets, whenever this does not compromise fire safety requirements.
- *Vapour barrier*: A moisture open design strategy was adopted. Therefore no vapour barrier is used in the wall layering. The design choice was further supported by the in-depth investigation on timber-based wall moisture safety across Australian climates (Brambilla & Gasparri, 2020; Gasparri et al., 2018).

Unitized Timber Envelopes

Figure 9.12 Exploded views of (Left) CLT-based wall and (Right) TF-based wall. CLT, Cross-laminated timber; TF, timber-framed.
Source: Author's original.

- *Fire barrier*: Encapsulation strategy has been adopted to comply with the Australian NCC under specific conditions,[3] with one or more layers of fire-rated boards on both sides of the structure. The number of boards depends on the required fire resistance level that may vary mainly according to different factors, for instance the building classes, type and height. In the case of CLT-based walls it would be worthwhile to explore different fire strategies (eg, combined use of sacrificial charring layer and sprinkler system) to allow for mass timber exposure to the interior space, maximizing advantages in terms of aesthetics and indoor comfort. Rainscreen cavities are provided with fire barriers at each floor to control fire and smokes spread through chimney effect.
- *Structure*: Both cross-laminated timber (CLT) and timber-framed (TF) wall structural technologies are considered possible structural options for the envelope. Also, the two timber technologies can either be load bearing (LB) and non-load bearing (nLB), for a total of four structural configurations for the envelope system (CLT_LB, CLT_nLB, TF_LB, TF_nLB). Fig. 9.12 shows the exploded view for both structural systems.

Frame design process

In modular envelopes, joints' technical design is challenging as it needs to account for differential movements of façade elements, construction tolerances, as well as guarantee the envelope weathertightness despite the discontinuity of weather control layers at element interfaces. The design of UTE's joint takes inspiration from UCW's pressure equalized joint system that uses, as aforementioned, an aluminium-extruded perimetral frame equipped with gaskets to prevent air and water exchanges between the indoor and outdoor environments. Differently from UCW's frames that are meant to take horizontal loads acting on facades, the UTE's one is free from structural function being loads taken by the

---

[3] Fire safety is of outmost importance when it comes to façade design in multistorey building, particularly when combustible materials are involved. However, fire safety requirements may considerably vary across different contexts and countries. A broader discussion and deep investigation on fire-related design strategies must be dealt with on a case-by-case scenario or be addressed through further studies with a specialist focus on the topic.

wall structure itself. This results in a much smaller, cheaper and optimized aluminium profile, main task of which is to accommodate gaskets extrusions to ensure the continuity of the weather barrier and keep timber elements far from water. They also allow for a plug-and-play installation of modular façade elements through an ad hoc aluminium profile (stack joint) specifically designed to be independent from the UTE main standard frame, so to be taken apart from the inside of the building and allow from programmed maintenance (or replacement) of gaskets at the end-of-life (see Figs 9.13 and 9.14). Indeed, this is a topic of major relevance in façade design where the life span of the various components can vary greatly and strongly affect the performance of the whole system (eg, aluminium frames and gaskets). A circular design approach that closely looks at systems' adaptability and resilience should be a fundamental, inescapable part of good design practice moving forward.

The first step of the frame design process was to look at market-ready profiles to be modified and adapted to the UTE system. In fact, the use of bespoke frames was soon abandoned for the following two reasons:

1. Time and cost effectiveness. The use of both aluminium and rubber gasket bespoke extrusions would have incurred in high cost with respect to engineering design and new extrusion dies fabrication.
2. The opportunity for the UTE system to be compatible with a UCW commercial product. The use of fully opaque modular elements with fully glazed ones can bring several advantages particularly with respect to fabrication cost and efficiency. In fact, there would be no more need for punched windows to be included within multilayered envelopes, which understandably (refer to the case studies presented in Section 9.3.1) requires a lot of effort in coordination and specialized labour, as that is a work that cannot be fully automated in a factory setting.

The design of the frames' extrusion involved an attentive analysis of commercially available systems and their characteristics. Two main UCW systems were considered for the research purpose:

A. Male−female joint type, relying on the full aluminium frame section equipped with minimal gaskets for joints' weathertightness. Mainly diffused on the Australasian and North American markets.
B. Female−female joint type, relying on a sophisticated front gasket system for joints' weathertightness. Mainly diffused on the European market.

System B soon emerged as a more appropriate and advantageous solution for the following main reasons: more efficient and standardized fabrication process, material and cost optimization, weathertightness reliability at the four-ways joint. The two systems are illustrated in Fig. 9.15 and their main differences summarized in Table 9.1.

The newly designed UTE frame system is composed by three main parts:

1. *Aluminium extrusions*: They are fixed through screws into a guiding slot specifically created on TF wall perimetral studs. Screw fixings were preferred to glued ones as they allow for the system's various parts to be easily taken apart without damaging them. Transoms and mullions profiles are alike, so timber slots processing and aluminium frames installation process are error-proof. As mentioned earlier, the frame is provided with an

Figure 9.13 Rendered section of the UTE CLT-based wall in a load bearing configuration (with integrated punched window). (Left) During installation and (Right) after interior works. *CLT*, Cross-laminated timber; *UTE*, unitized timber envelope.
Source: Author's original.

Figure 9.14 Rendered section of the UTE CLT-based wall in a non-load bearing configuration (compatible with glazed unitized curtain walls). (Left) During installation and (Right) after interior works. *CLT*, Cross-laminated timber; *UTE*, unitized timber envelope. *Source:* Author's original.

Unitized Timber Envelopes

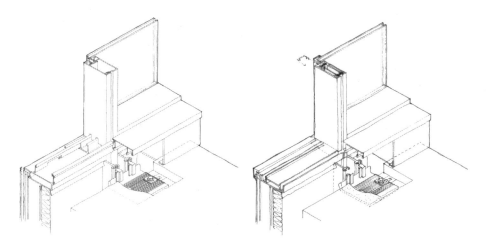

Figure 9.15 Systems A and System B: UCW typical details.
Source: Reproduced with permission from Troy Donovan, Prism Facades (Sydney, Australia).

Table 9.1 Systems A and System B: UCW main differences.

System A	System B
Male–female joint types imply that both mullion and transom profiles are necessarily different from one another. Thus four different extrusions die for fabrication.	*Female–female joint* types imply that both mullion and transom profiles may be identical. Thus one dies only for fabrication.
Joints' water drainage is managed through the full frame section, leaving not much chances for optimization.	*Joints' water drainage* is managed through the front gasket system, while the rest of the frame section serves mainly structural purposes. This allowed to design compact and cost-effective aluminium extrusions for the UTE system, being them free from structural function.
The four-ways joint weathertightness relies on the combined effect of the alignment clip between adjacent units and significant use of silicon to solve discontinuities.	*The four-ways joint* weathertightness relies on a continuous barrier, through smart and effective gasket system overlaps.

UTE, Unitized timber envelope.

independent ad hoc aluminium profile at the stack joint, function of which is threefold; it acts as a barrier to water ingress by supporting the horizontal top-transom gasket, guides façade elements installation, allows for programmed maintenance of the gasket system from the inside of the building, without the need for replacing other parts or engaging in extensive renovation work.

2. *PVC extrusions*: They are fixed to aluminium frames and host the first-barrier gasket system. They also span over the rainscreen cavity and provide a frame to the cladding, preventing vulnerable edges from getting damaged during handling and transportation phases. The top extrusion can sometime be metallic and/or coupled with a plasterboard layer to prevent the spread of smokes and fire between floors via the rainscreen cavity.
3. *Rubber gasket system*: It is made of one horizontal gasket that runs across the entire building perimeter and two vertical ones, lengths of which are dependent on wall height. In some cases the bottom transom may also be provided with an additional extra gasket. The horizontal top-transom gasket has a complex shape that creates two barriers against water penetration and is supported by the aforementioned aluminium profile (stack joint). Gaskets are the last items to be installed on walls and are easily integrated onto their frames through special clip-on slots.

Rainscreen design process

In the field of prefabricated envelopes, designers always need to compromise with specific design boundaries, such as façade elements' dimensional flexibility and other limitations related to their structural integrity or weight. The use of lightweight materials might seem preferable particularly when dealing with large-size panels. However, robust cladding systems would be less vulnerable during transportation and on-site handling operations. The integration of construction joints within the overall building façade design is another challenge (or perhaps opportunity) that architects must deal with when selecting or designing the cladding system.

The approach behind the design of the UTE cladding system aims at maximizing customization opportunities within a standardized system. In fact, all UTE wall configurations are identically provided with a metallic vertically oriented substructure (thermally broken), fixed to the framed wall structure. This allows for a standard, repetitive, easy and quick assembly process. Designers can use that support to select or design their cladding system. Four different cladding options were designed in detail to show the system potential in terms of flexible design (see Fig. 9.16). The whole system integration at the building scale is illustrated throughout the next section.

Figure 9.16 Examples of different cladding materials and arrangements on the same substructure. In order left to right: high pressure laminate, reinforced porcelain, timber lamellas, bricks-on-rails.
Source: Author's original.

9.4.3 Design integration

The BIM design integration aimed at testing system complexity in terms of information management and validating technical and architectural design outcomes, both at the component and building scale. In fact, aside from the standard planar wall unit, the system should account for a series of different geometries or configurations that are typical of every building design, as per the building edges or the façade topping at roof level. It is important to mention that the design of the system does not yet provide a solution for balconies or loggias integration within the envelope, which may constitute a significant limitation for residential developments.

Four different panel types (standard planar, standard planar w/punched window and two corners options) for each structural configuration (CLT_LB, CLT_nLB, TF_LB, TF_nLB) were fully modelled in three dimensions via Revit BIM software (see Fig. 9.17). The combination of the 4 different panel types led to the development of 34 different joint details.

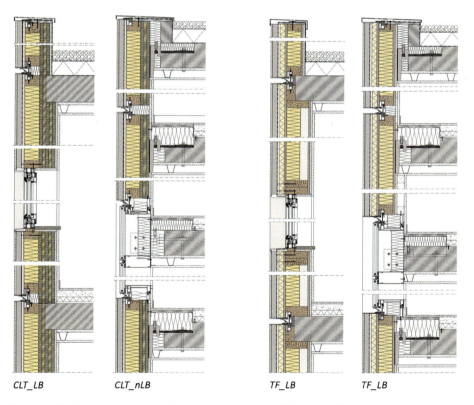

Figure 9.17 BIM catalogue of vertical joint details for the UTE system. In order left to right: CLT_LB, CLT_nLB, TF_LB, TF_nLB. *BIM*, Building information modelling; *CLT*, cross-laminated timber; *LB*, load bearing; *nLB*, non-load bearing; *TF*, timber-framed; *UTE*, unitized timber envelope. *Source:* Author's original.

Figure 9.18 Possible design outcomes for an aged-care building that uses the UTE system.
UTE, Unitized timber envelope.
Source: Reproduced with permission from Building design by arch. Iorillo.

Each façade panel type (parent) was designed following a parametric approach and the different parts (children) within one panel were linked via a 'parent-child relationship'. This allowed for automated design adaptation of the whole façade system according to preset dimensional rules, mainly based on performance requirements and fabrication constraints.

Once all the possible configurations were fully modelled, the UTE system was "tested" for design efficiency at the building scale, as shows Fig. 9.18. All the different wall panel types and structural arrangements were integrated in a four-storey aged-care facility used as a case study building. This phase allowed to investigate both architectural flexibility and joint integration with the overall building façade design, via integration of various cladding technologies and design approaches. For example, vertical lamellas were strategically spaced with the aim of hiding vertical joint interfaces, while horizontally oriented ones were used to camouflage horizontal joints. Other arrangements were instead emphasizing joints and using them as a modular pattern to play with the overall envelope building design.

An important lesson that came out from the design integration into a BIM digital environment refers to the inadequacy of such tools to cope with systems' design complexity, including performance assessment or design for fabrication. They are mainly capable of organizing and storing information 'statically'. However, they lack functionalities for knowledge management, which is fundamental when dealing with industrialized design and automation (Day et al., 2019; Montali et al., 2018; Montali et al., 2019).

9.5 Conclusion

This chapter argued for the need for more energy-efficient and sustainable envelope technologies in tall buildings, as a fundamental step to reduce the environmental impact of the construction sector and face the climate emergency. The widespread adoption of unitized curtain walls (UCWs) in high-rises largely contributes to the increase in greenhouse gas emissions and buildings' carbon footprint. Recently, new opportunities seemed to be developing in the field of industrialized multilayered envelopes to reduce building energy consumption and address emerging needs for adaptability and climate resilience. Drawing on international best practice examples, the author presented the result of a 2-year industry-embedded research on the design of innovative unitized timber envelopes (UTEs) for multistorey buildings.

The system consists of large-size multilayered timber panels, preassembled off-site to allow for a quick, safe and scaffold-free site installation. The use of a timber structure makes it highly sustainable and offer opportunities to pursue regenerative design strategies and lead the sector decarbonization. The multilayered nature of the system guarantees excellent energy performance and high potential for adaptability to different climatic contexts and design needs. Indeed, the newly designed timber envelope can be integrated across a variety of building types (eg, concrete or steel buildings) and structural typologies (eg, LB or nLB). Large-size panels with openable punched windows might be particularly valuable for the whole residential market, including aged-care housing, student accommodations and hotels. The use of UTE fully opaque façade elements integrated with UCW fully glazed ones might instead be more suitable for commercial use, such as offices where the envelope transparency is usually higher. Further studies and design investigations might include the integration of balconies, as well as renewable energy generation technologies (eg, BIPV, BIST) and air-treatment appliances.

The UTE system is also designed for durability, programmed maintenance and easy disassembly, in line with circular economy principles. Moreover, opaque multilayered envelopes offer opportunities to be 'upgraded' in time, for instance to respond to more stringent future performance requirement regulations and standards. Thus, the adoption of such a system has huge potential to benefit the whole value chain and positively impact the different stages of a building life cycle, including design, fabrication, construction, use and maintenance, end-of-life. The system demonstrates international relevance and high potential for commercialization across different construction markets other than the Australian; the North American and European being highly promising given the interest in mass timber technologies and rapid uptake of industrialized construction trends. Potential future steps toward commercialization might include both technological and business-related investigations, such as full-scale system testing and validation, as well as business case and specific market analyses.

Acknowledgement

This research was developed as part of CRC-P50578 in conjunction with Lendlease and DesignMake. Thanks to our industry partners and to everyone who have provided feedback, collaborated and informed this research.

References

Andaloro, A., Gasparri, E., Avesani, S. & Aitchison, M. (2019) *Market survey of timber prefabricated envelopes for new and existing buildings, Powerskin* conference. Munich, Delft: TU Delft Open.

Barbosa, F., Woetzel, J., Mischke, J., Ribeirinho, M. J., Sridhar, M., Parsons, M., ... Brown, S. (2017). *Reinventing construction: A route to higher productivity.* McKinsey Global Institute.

Barton, J., & Watts, S. (2013). Office vs. residential: The economics of building tall. *CTBUH Journal(II)*.

Brambilla, A., & Gasparri, E. (2020). Hygrothermal behaviour of emerging timber-based envelope technologies in Australia: A preliminary investigation on condensation and mould growth risk. *Journal of Cleaner Production*, 276.

Dangel, U. (2016). *Turning point in timber construction: A new economy.* Basel: Birkhaüser Verlag AG.

Day, G., Gasparri, E., & Aitchison, M. (2019). Knowledge-based design in industrialised house building: A case-study for prefabricated timber walls. *Lecture Notes in Civil Engineering*, 989–1016.

Gasparri, E., & Aitchison, M. (2019). Unitized timber envelopes. A novel approach to the design of prefabricated mass timber envelopes for multi-storey buildings. *Journal of Building Engineering*, 26.

Gasparri, E. & Mazzucchelli, E.S. (2016) *Façade prefabrication in tall CLT buildings: Time, cost and operation quality analysis through Building Information Modelling, Advanced Building Skins.* Bern, Switzerland.

Gasparri, E. (2016) Prefabricated external wall system for tall cross-laminated timber buildings. Design of unitized wood based façade assemblies for fast-track construction and quality assurance (PhD Politecnico di Milano).

Gasparri, E., Brambilla, A. & Aitchison, M. (2018) Hygrothermal analysis of timber-based external walls across different Australian climate zones. In: *WCTE 2018—World conference on timber engineering.*

Gasparri, E., Lucchini, A., Mantegazza, G., & Mazzucchelli, E. S. (2015). Construction management for tall CLT buildings: From partial to total prefabrication of façade elements. *Wood Material Science and Engineering*, 10(3), 256–275.

Gaviola, A. (2019) Glass skyscrapers have turned entire cities into energy vampires, VICE.2019 [Online]. Available online: https://www.vice.com/en/article/ywy9ek/condos-and-office-towers-are-the-biggest-source-of-greenhouse-gas-emissions-in-canada. Accessed 22.02.21.

Herzog, T., Krippner, R., & Lang, W. (2004). *Facade construction manual.* Basel: Birkhauser-Publishers for Architecture.

IEA. (2019). *Global status report for buildings and construction.* Paris: IEA.

IEA. (2020). *Building envelopes.* Paris: IEA.

Kaufmann, H., Krötsch, S., & Winter, S. (2018). *Manual of multi-storey timber construction.* Munich: Detail Business Information.

Kim, K.-H. (2011). A comparative life cycle assessment of a transparent composite façade system and a glass curtain wall system. *Energy and Buildings*, 43(12), 3436–3445.

Knaack, U., Klein, T., Bilow, M., & Auer, T. (2007). *Facades: Principles of construction.* Birkhäuser Verlag AG: Basel.

Kuzmanovska, I., Gasparri, E., Monne, D.T. & Aitchison, M. (2018) Tall timber buildings: Emerging trends and typologies. In: *WCTE 2018—World conference on timber engineering*.

Li, J., Rismanchi, B., & Ngo, T. (2019). Feasibility study to estimate the environmental benefits of utilising timber to construct high-rise buildings in Australia. *Building and Environment*, *147*, 108−120.

Link, J. (2020) Lethal glass landscapes, Landscape Architecture Magazine 2020 [Online]. Available online: https://landscapearchitecturemagazine.org/2020/02/13/lethal-glass-landscapes/. Accessed 22.02.21.

Montali, J., Overend, M., Pelken, P. M., & Sauchelli, M. (2018). Knowledge-based engineering in the design for manufacture of prefabricated façades: Current gaps and future trends. *Architectural Engineering and Design Management*, *14*(1−2), 78−94.

Montali, J., Sauchelli, M., Jin, Q., & Overend, M. (2019). Knowledge-rich optimisation of prefabricated façades to support conceptual design. *Automation in Construction*, *97*, 192−204.

Ramage, M., Foster, R., Smith, S., Flanagan, K., & Bakker, R. (2017). Super tall timber: Design research for the next generation of natural structure. *The Journal of Architecture*, *22*(1), 104−122.

Ribeirinho, M. J., Mischke, J., Strube, G., Sjödin, E., Blanco, J. L., Palter, R., . . . Andersson, T. (2020). *The next normal in construction: How disruption is reshaping the world's largest ecosystem*. McKinsey & Company.

Tapper, J. (2019) Experts call for ban on glass skyscrapers to save energy in climate crisis, *The Guardian.2019* [Online]. Available online: https://www.theguardian.com/environment/2019/jul/28/ban-all-glass-skscrapers-to-save-energy-in-climate-crisis. Accessed 22.02.21.

United Nations. (2019). *World urbanization prospects: The 2018 revision (ST/ESA/SER.A/420)*. New York: United Nations.

Yi, J.-S., Kim, Y.-W., Lim, J. Y., & Lee, J. (2017). Activity-based life cycle analysis of a curtain wall supply for reducing its environmental impact. *Energy and Buildings*, *138*, 69−79.

Industrialized renovation of the building envelope: realizing the potential to decarbonize the European building stock

Thaleia Konstantinou and Charlotte Heesbeen
Department of Architectural Engineering + Technology, Chair Building Product Innovation, Faculty of Architecture and the Built Environment, Delft University of Technology, Delft, The Netherlands

10.1 Introduction

It is estimated that 85%−95% of the buildings that exist today will still be standing in 2050 (European Commission, 2020), accounting for almost 40% of energy consumption in the European Union (EU) (Tsemekidi Tzeiranaki et al., 2020). The role of the existing building stock is instrumental in the energy transition and the goals for carbon neutrality of the built environment (Filippidou & Jimenez Navarro, 2019). Renovation is an integral part of the building's life, as the different components reach the end of their service life (Brand, 1994). Every year, 11% of the EU existing building stock undergoes some level of renovation. However, renovation works that address the building's energy performance are at a rate as low as 1%, with a deep renovation that achieves energy reduction over 60% being at 0.2% (European Commission, 2020).

The upgrade of the existing buildings stock can result in significant energy savings, improved health and comfort of the occupants, elimination of fuel poverty and job creation (BPIE, 2011). Nevertheless, to tackle this potential, both the number of renovated buildings and resulting energy savings need to increase (Artola, Rademaekers, Williams, & Yearwood, 2016). The annual rate of renovated buildings to the total building stock varies from 0.4% to 1.2% in the different European Union Member States (European Commission, 2020). This rate will need to double to reach the EU's energy efficiency and climate objectives for no greenhouse gases net emissions in 2050 (European Commission, 2019).

An effective renovation plan must significantly improve the current energy performance toward a zero-energy level. Interventions that reduce the building's energy demand and generated power are essential to attain this goal. Nevertheless, most improvements in residential buildings currently consist of basic maintenance and shallow renovation. Broader or deeper energy renovation measures that result in higher energy savings are required (Filippidou, Nieboer, & Visscher, 2016).

Several studies, such as BPIE (2011), Meijer, Straub, and Mlecnik (2018), have been looking at barriers to renovation implementation and upscaling. Next to financial, institutional and regulatory barriers, there are also informational barriers related to the lack of common direction among the main stakeholders and the lack of overview of which building types and renovation activities to prioritize (Jensen, Maslesa, Berg, & Thuesen, 2018).

Particularly when tackling deep retrofitting actions, higher complexity and costs are incurred compared to lower impact energy retrofitting solutions. Large-scale building renovation is still considered a difficult task (Filippidou & Jimenez Navarro, 2019). This perception can be attributed to the large number of retrofitted components, their interconnection and the integration of renewable energy sources (Avesani et al., 2020). Moreover, renovation projects are complex to carry out because of the many actors involved (D'Oca et al., 2018) and high risks for contractors due to the lack of lean methods and risk-sharing models (Bystedt et al., 2016).

Industrialization can trigger a virtuous circle between higher demand for energy-efficiency renovation and falling costs for smarter and more sustainable products (European Commission, 2020). From this perspective, industrializing the renovation process helps to overcome some of the financial and stakeholders' issues. It can bring the costs down, utilizing the large numbers of the buildings to be renovated to lead to economies of scale. Additionally, it offers opportunities for end users to benefit from high-quality, affordable products and institutional real estate owners to benefit from affordable customization. Industrialization is an effective strategy for improving the construction industry's productivity (Hong, Shen, Mao, Li, & Li, 2016), which is an important consideration given the enormous renovation task at an European level (BPIE, 2013). Furthermore, prefabrication of the retrofitting components can achieve high-performance results while minimizing on-site construction time (IEA Annex 50, 2012).

Industrialization is not a new concept in the building industry. The idea of industrialized building systems has been developed since the interwar years and the Modern movement (Moe & Smith, 2012). In recent years, prefabrication in the whole building market is currently undergoing significant growth (Tumminia et al., 2018). However, in renovation, the holistic application of prefabricated modules is often only used in subsidized demonstration cases (BPIE, 2016) and not experience a high market uptake (Bystedt et al., 2016).

Considering the need to upscale renovation of the building stock and the potential of industrialized renovation to achieve that this chapter discusses current practices of industrializing the building envelope's renovation. First, we determine the role of the building envelope as an integral part of deep renovation strategies. Subsequently, the definitions and application of industrialized techniques in the design and construction of renovation are investigated, particularly regarding the renovation process and design concepts. Finally, the chapter gives an outlook on critical aspects for the future of industrialized building envelope renovation.

10.2 The importance of the building envelope for deep renovation

Before discussing industrialized renovation in more detail, we need to establish a common vocabulary about the type of interventions this building activity comprises. Different terms, such as major renovation, deep renovation, refurbishment and sustainable renovation, can be encountered. Those terms have in common that they deal with an existing building, and they need to consider the existing condition, function, users, architectural characteristics and performance (Jensen et al., 2018).

The depth of renovation is defined by the level of savings on energy, specifying as deep such renovations that achieve energy savings of 60%—90% (BPIE, 2011; European Commission, 2019/786). Deep renovation, in particular, has the potential to be the preferred solution from an ecological and economic point of view. As opposed to deep renovation, superficial renovations significantly increase the risk to miss the climate targets of savings in energy and CO_2 emissions to remain untapped (Hermelink & Müller, 2011). Despite having a higher initial investment, deep renovation also generates higher energy savings (BPIE, 2011). As a result, deep renovation, toward zero-energy standards in existing buildings, is set as a priority in long-term renovation strategy (DIRECTIVE, 2018/844/EU).

Jensen et al. (2018) are using the term sustainable building renovation 'as the renovation of existing buildings that results in upgraded buildings, which are more sustainable in terms of environmental, social and economic aspects after the renovation than before— or at least in relation to two of these aspects'. Based on this definition, reducing energy consumption and the related CO_2 emissions of a building is considered part of a sustainable renovation. It improves its environmental impact while lowering the cost of energy use for the occupants.

In current practice, renovation is a term widely used to express a range of construction activities related to interventions onto existing buildings. They range from simple repairs and maintenance, restricted to replacement or repair of defective components, to adaptive conversion and reuse, which affect the load-bearing structure and interior layout. Giebeler et al. (2009) places renovation works close to maintenance and cosmetic repairs that do not add new components. On the other hand, refurbishment refers to defective or outdated parts, components or surfaces being repaired or replaced, with no major changes in the load-bearing structure (Giebeler et al., 2009). The upgrade of fire protection, acoustics and thermal performance can be achieved through the building's refurbishment. Additionally, during the refurbishment, buildings can be retrofitted with technologies for energy generation from renewable sources. Retrofits are defined as the strengthening, upgrading or fitting of extra equipment to a building once the building is completed (Gorse, Johnston, & Pritchard, 2012). In this sense, refurbishment and retrofits are very similar as activities, as they both address replacing and upgrading building components. For this chapter's discussion the term 'renovation' will be further used, as it is a term widely used in the building industry and policy documents. It is considered to encompass measures that refurbish or retrofit building components.

Regarding deep renovation in particular, which aims to save over 60% in energy demand, a combination of measures is needed. The renovation strategy should address different building components. Regarding technology to be used for sustainable and energy-efficiency building renovations, the amount and sophistication of building materials, technical installations or services has escalated over the past decade (Jensen et al., 2018). The primary interventions, as defined by Pacheco-Torgal et al. (2017), include the following:

- heating and cooling demand reduction, such as thermal insulation, multiple-pane windows and increase of airtightness;
- energy-efficient equipment and low energy technologies, such as thermal storage and heat recovery; and
- renewable energy supply using technologies, such as heat pumps, photovoltaic (PV) panels and solar collectors.

It is then clear that the building envelope is essential as it is the medium to apply those measures, combining both passive and active measures (Kilaire & Stacey, 2017; Konstantinou, 2014; Pacheco-Torgal et al., 2017). Not only can the building fabric prevent heat losses but also it can incorporate technologies such as PV panels and other building services. Moreover, the building envelope is a component with a shorter life span than the building structure (Brand, 1994). Retrofitting the building envelope to tackle physical signs of deterioration and upgrading its performance extends the building's life span.

Based on the previous definitions and type of interventions, the present chapter uses the term 'renovation' as a building activity that applies a combination of measures on an existing building, including the building envelope, with the aim, but not limited to, of improving energy efficiency.

10.3 Degrees of industrialized renovation

Industrialization in the context of building products refers to items produced in a repetitive process. Richard (2005) defined industrialization as 'catering a large market, by investing in technologies that simplify production and also reduce cost'. This principle commonly applies on a small scale, such as screws and standardized building products, such as bricks or doors.

The term *industrialized construction* is often used interchangeably with prefabricated construction; however, it is not strictly the same. Industrialization aims at simplifying the production process of complex goods using strategies and techniques that require an investment divided over a high production volume and, therefore, marginal per end product. Bricks, for example, are an industrial interpretation of adobe architecture. The fired modules are easier to transport and handle and more robust than earth. This optimization of the production process resulting in higher quality requires an investment, for example, for the acquisition of moulds and a kiln. Therefore it is essential to balance the upscaled production with the actual demand to justify the initial investment.

Under the umbrella of industrialization, five different degrees can be found (Richard, 2005). *Prefabrication* can be seen as a subset of industrialized construction. The other degrees are *mechanization, automation, robotics* and *reproduction*. The definition of those degrees does not imply a hierarchical classification but different type of activities. The first four degrees described in Table 10.1 are applied in practice separately or combined to build the industrial character of a production process. Reproduction is as an overarching degree that aims at maximizing replicability. The separate degrees can be seen as complementary, independently or building upon each other.

Particularly in building envelope renovation, the different industrialization degrees can be found in literature examples and case studies. The next table is an introduction to the different degrees of industrialization and some example activities.

In current renovation practices, mechanization and automation are increasingly applied. The degrees of industrialization are often combined to deliver the final result. Renovation as a part of state-of-the-art construction activity is already largely industrialized. However, the possibilities to expand the degree of industrialization in the renovation are important, particularly given the need to increase the depth and rate of renovation of the building stock. Thus industrialized renovation should aim at going beyond standard practices. Within this chapter's scope, we refer to industrialized renovation as the renovation that increases the energy efficiency of the existing building stock while aiming to maximize reproduction, through an effective combination of all degrees of industrialization, particularly with the application of prefabricated components.

10.4 Industrialized renovation process and state of the art

Industrialized renovation follows a similar process to other building renovation projects. This process includes the project requirements' definition, concept design, final strategy design, execution and, finally, the renovated building (Konstantinou, 2014). However, in the industrialized renovation, several steps in the decision-making and the design and engineering are specific to the degrees of industrialization, particularly regarding prefabrication, and reproduction. This section presents the process that relates to the industrialized renovation. Moreover, it elaborates on industrialized renovation design concepts, as illustrated in state-of-the-art examples.

10.4.1 Industrialized renovation process

Manufacturing and installation benefit from the industrialization of the renovation in terms of cost reduction, replicability and productivity. However, a large share of the effort to achieve a high level of reproduction occurs during the design and engineering of the components. The design is optimized and standardized, modularity

Table 10.1 Degrees of industrialization in renovation (Richard, 2005).

Degree of industrialization	Description	Example of related activities in renovation
Prefabrication	Building components or complete modules off-site (in the factory) before being transported to the site and become an integral part of the building	• Construction of façade units in the factory, with integrated components, such as windows and ventilation pipes • Retrofitting with sandwich panels, prefabricated in a factory • Retrofitting of building services with preassembled configurators
Mechanization	Machinery is employed to ease the work done with human intervention. This is already a widely adopted degree of industrialization	• Using CAD software to draw and communicate design • The use of hand tools, such as a hammer or a drill • Using a crane for heavy and large object lifting
Automation	Tooling completely takes over a repetitive production task without the need for a tool or workpiece adjustment by human intervention	• Optimization with the use of parametric design generation • Sheet metal forming in a press brake and automatic tool changing
Robotics	Tooling has the flexibility to perform diversified tasks without human intervention	• Scanning a renovation object with a drone • CAM tools, including a multiaxis milling machine or robotized bricklaying • Bricklaying with a robot arm
Reproduction	Simplified multiplication, with the use of the degrees of industrialization	• Inventory method for potential renovation objects on an urban scale using publicly available photographic data • Retrofit system that provides adaptability through a variable choice of highly efficient energy technologies and intelligent controls • Standardized interfaces applicable to varying modules for a prefabricated façade solution

Source: Adapted form Richard, R.-B. (2005). Industrialised building systems: Reproduction before automation and robotics. *Automation in Construction, 14*(4), 442−451. https://doi.org/10.1016/j.autcon.2004.09.009.

can be introduced to customize the end product, and functional and material synergies are created. To this end the chapter looks into the renovation process and identifies the key aspects of industrialized renovation that influence the decisions and when they are considered.

In an effort to systematize and facilitate decision-making during the construction projects, different phases have been identified (Cooper et al., 2005; Klein, 2013; RIBA, 2020). The exact number of phases and subphases might vary in the different publications, but there is consensus on the main broad stages. The stages are the preproject, which defines the need for the project; the preconstruction, when an appropriate design solution is developed; construction, which implements the solution; and postconstruction, which aims at monitoring and maintenance of the project.

The renovation process, which researchers have also specified (Ferreira, Pinheiro, & Brito, 2013; Konstantinou, 2014; Ma, Cooper, Daly, & Ledo, 2012), is still a construction project and the phases mentioned earlier apply as well. However, since renovation is dealing with an existing building, the preproject phase includes the analysis and diagnostic of the building to define the intervention's scope. Moreover, the existing occupants that might continue to occupy during construction have an important role in the execution phase, for example, with regards to time planning. Industrialized renovation follows the same phases, but some subphases are specific or more essential compared to on-site renovation construction, particularly with regards to the existing building analysis, the renovation design and the components' production (Aldanondo et al., 2014). Fig. 10.1 shows an overview

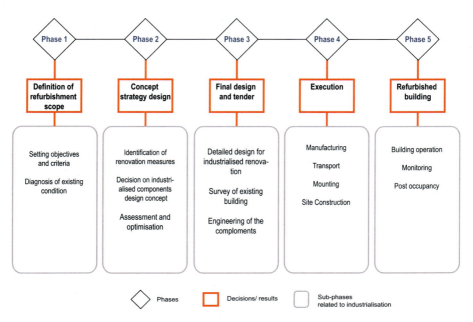

Figure 10.1 The renovation process, adapted from Konstantinou (2014), includes the subphases that are important for renovation industrialization.

of the renovation process phases, including the important activities for industrialized renovation.

The decision to develop an industrialized renovation concept determines the concept design and the measures' specifications. There are several designs and construction concepts, as will be discussed in Section 10.4.2. The design of an industrialized renovation concept does not start from scratch, as is often the case in the on-site renovation. Once the system has been developed, reproduction for the different projects is possible to a large extend. The design focuses on adapting the industrialized concept to the project specifics, geometry and energy performance objective. The design's replicability, which is defined in the concept design phase, facilitates the standardization of construction details and the supply chain, which occurs in the final design and tender phase. A significant step of the industrialized construction is the detailed survey of the existing building.

Based on the experience of the research project EASEE ('Envelope Approach to improve Sustainability and Energy efficiency in Existing buildings'), where insulating prefabricated panels, for a total of 28 different typologies and with different textures were applied, Brumana et al. (2016) have organized the renovation process into the following steps:

- laser scanning technologies and thermographic survey campaign of the existing building's envelope, to identify nonhomogeneous parts of the building structure and support the design and the installation of the prefabricated panels,
- use of building information modelling (BIM) tools in the design of the panels and the anchoring systems, toward the automation in the installation of the elements and the energy performance evaluation over time,
- architectural and executive design of the renovation and the manufacturing of the prefabricated panels,
- installation to the existing building facades utilizing steel profiles and realization of the finishing works and
- monitoring campaign before and after the renovation, to evaluate thermal performance.

Next to the building dimensions, the components' size is determined by the product design, manufacturing and transportation capacity and mounting method. There can be certain limitations in terms of size; typically, the factory can provide a maximum width, height and panel thickness according to the equipment and facilities dimensions (Fig. 10.2) and the vehicles that will transfer the panels to the building site.

Finally, panels can be hung directly onto the façade using brackets if the façade has the load-bearing capabilities. Otherwise, a concrete block is added at the bottom of the lowest panel onto the façade. The panels are stacked on top of each other and the load is transferred through them. The panels still need support from the façade for wind and rain load, however. Different anchoring systems are possible, such as the examples in Fig. 10.3.

Figure 10.2 The manufacturing process of the sandwich panel in the Rc panels factory, Overijssel, the Netherlands. The panel's max length is determined to be 13 m (Rc Panels, 2021). *Source:* Nunez, R.

Figure 10.3 (A) The panels are connected to wooden posts, attached to the existing structure with steel U profiles. (B) Anchors on the back of the façade panel, to be connected on rails attaching to the existing building. (C) The installation of the prefabricated panel on the concrete block that bears the extra weight. Hem, France, Net-zero retrofit, under construction, Vilogia.
Source: Energiesprong International by Samyn O. (2018) https://www.flickr.com/photos/150184035@N07/albums/72157695874102660/with/43094497810/. Licensed under CC BY 2.0.

10.4.2 Design and construction principles of industrialized renovation

A number of prefabricated renovation concepts focus on the envelope optimization, limiting construction time while reducing energy demand through the use of component prefabrication (Astudillo et al., 2018; Bruno & Grecea, 2017; Bumanis & Pugovics, 2019; Bystedt et al., 2016; Callegari, Spinelli, Bianco, Serra, & Fantucci, 2015; EnergieSprong, 2014; Malacarne et al., 2016; Pihelo, Kalamees, & Kuusk, 2017; Stroomversnelling, 2014; TESEnergyFaçade, 2014). Those concepts show both similarities and differences in their approach to industrialized renovation.

The first step toward the industrialization of the construction is the decomposition of the building in different elements, which are then produced and prefabricated off-site (MORE-CONNECT, 2019). The design and construction principles of those elements can be used to classify industrialized renovation in different categories. They determine the type and size of the component, the functions to be integrated and industrialization degrees. The following sections explain the various design principles in more detail and present examples of industrialized renovation state of the art. Those examples are selected to illustrate the principles and they are by no means exhaustive. Overlaps in the characteristics and scope of the following categories can be found. The classification criterion is the building envelope construction principle, as a starting point in the design of the industrialized renovation.

Prefabricated sandwich panels

Sandwich insulation panels, also referred to as structural insulating panels, are common construction components, both in new construction and renovation. They consist of two layers of rigid panels bonded to either side of a lightweight core. The panels are typically made from oriented strand board (OSB), particleboard, plywood panels or cement-bonded particleboard or metal sheets. Those panels are preassembled and, depending on their size and functionality, they can also include prefitted windows and doors, services and finishes (Mayer, 2021). In this respect the panels' construction employs manufacturing techniques like automation and prefabrication. In renovation, they are used for improving the thermal performance of walls and roof (Fig. 10.4).

The 2ndSkin renovation concept, which is based on prefabricated sandwich façade panels to reach zero-energy dwellings (Konstantinou, Guerra-Santin, Azcarate-Aguerre, Klein, & Silvester, 2017), proposes floor-high sandwich panels, featuring new windows and integrated service installations. This sandwich panels are delivered as separate parts by the insulation product manufacturer, as in the

Figure 10.4 (A) Prefabricated roof panels used in the construction of the 12 zero-energy dwellings, in Vlaardingen, the Netherlands. (B) CompletRC RC Panels (Rc Panels, 2021) with weather finishing of brick tiles.
Source: Nunez, R.

Figure 10.5 (A) sample of the prefabricated sandwich panel. (B) Left and right panel attached to the substructure, through timber sticks. (C) Testing different cladding material.

sample in Fig. 10.5A, which are then assembled in their final configuration together with the windows in the factory. They are transported to the building site in one piece, to minimize connections between the pipes. Different cladding options are possible, applied after the panel installation (Fig. 10.5C).

Timber-frame panels

Timber-frame panels are similar to the sandwich panels since they also consist of two boards enclosing an insulated core. However, they are a distinct category as their strength comes from a framework of timber beams (Fig. 10.6) with insulation in between. Windows are also incorporated in the framework to make sure that their weight is transferred down properly. The sheathing boards, typically wooden boards, plywood boards or OSB, ensure the element's stability. Moisture and water proofing foils are also included in the panes, before the external finish.

Timber-frame panels have been increasingly used in renovation projects (Coupillie, Steeman, Van den Bossche, & Maroy, 2017; Loebus, Ott, & Winter, 2014; Ochs, Siegele, Dermentzis, & Feist, 2015; Stroomversnelling, 2019). Since the panel does not rely on the existing backing wall for stability, it is possible to replace the wall. This approach was applied by the MORE-CONNECT (2019) project, in one of the pilots, where the existing walls and balconies were removed before the new panel's connection.

Energiesprong is a whole-house renovation concept that applies the timber-frame construction principle. Over 4000 net-zero-energy houses, both new built and retrofitted, have been constructed in the Netherlands using this concept, and the first 10 performance guaranteed that net-/near-zero energy retrofits have been completed in both the United Kingdom and France (Energiesprong, 2019). The façade technical concept consists of large-scale, timber-frame insulated panels that arrive on the building site prefabricated. The openings are already integrated into the panels, based on third scanning of the existing building. The cladding material is also pre-applied to the panel, as shown in Fig. 10.7. Different options for materials are possible to match the renovation design intent.

Figure 10.6 Timber frames, stacked in the factory, before the insulation and sheathing boards are applied.

Figure 10.7 (A) The installation of the prefabricated panel on the existing façade. Hem, France, Net zero retrofit, under construction, Vilogia (B) Longueau—Rénovation thermique E = 0.
Source: (A) Energiesprong International by Samyn, O. (2018) https://www.flickr.com/photos/150184035@N07/albums/72157695874102660/with/43094497810/. (Licensed under CC BY 2.0.). (B) Energiesprong International by Singevin, F. (2018) (https://www.flickr.com/photos/150184035@N07/albums/72157701746144094j). Licensed under CC BY 2.0.

Modular façade

Modular façade (MF) refers to the exterior finish of a building made by separate, often prefabricated units (modules). The units' system, also referred to as unitized facades, is assembled in a repetitive manner on or off-site (Knaack, Klein, Bilow, & Auer, 2007). The modules should have standardized interfaces for future maintenance and upgrade (Du, Huang, & Jones, 2019). Considering the definition of a modular product as a function-oriented design that can be integrated into different systems for the same functional purpose without or with minor modifications' (Chang & Ward, 1995 in Gershenson et al., 2003), the system also provides the possibility to integrate modules with different functions. The multifunctionality is particularly important in the case of building envelope renovation when the existing building needs to be retrofitted both with passive and active measures.

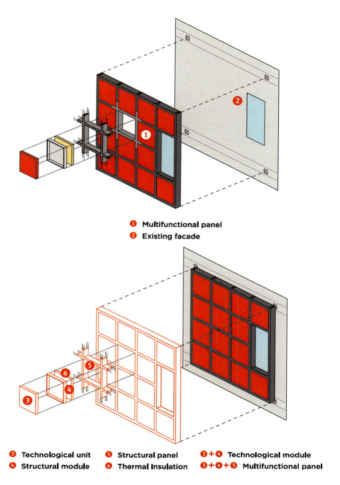

Figure 10.8 Principle of the MEEFS system, including the structural panel and the technological modules. *MEEFS*, multifunctional energy-efficient façade system (Ochoa & Capeluto, 2015).
Source: Ochoa, C. E., & Capeluto, I. G. (2015). Decision methodology for the development of an expert system applied in an adaptable energy retrofit façade system for residential buildings. *Renewable Energy, 78,* 498−508. https://doi.org/10.1016/j.renene.2015.01.036, p. 499, Fig. 10.1.

The approach was demonstrated by the MEEFS, which stands for 'multifunctional energy-efficient façade system'. The system relies on industrialized production of the standardized panels that integrate different technological modules, allowing personalized configurations for each façade typology, orientation and local climate conditions. The façade system comprises a structural panel, made of fibre-reinforced polymer, which acts as the frame for the modules of different technologies, such as building-integrated photovoltaic systems (BIPV) and solar thermal collectors, green facades and shading (Paiho, Seppa, & Jimenez, 2015). The principle is illustrated in Fig. 10.8.

Prefabricated rainscreen facades

Ventilated façade, also referred to as rainscreen systems (greenspec, 2013) or dry-cladding systems (Thorpe, 2010), comprises the outer skin, the air cavity, the substructure and the insulation layer. The outer skin or panel is called the 'rainscreen', as it forms the primary rain barrier. It does not prevent the passage of air through open joints between the panelling components (Konstantinou, 2014). This type of construction is often using in renovation as it gives the possibility to upgrade the thermal resistance of the envelope, chance the external finishing and prevent moisture accumulation in the existing structure (Borodulin & Nizovtsev, 2021).

Industrialized construction of ventilated facades offers possibilities to go beyond typical construction and integrate different type of products and components off-site. One example of a renovation system based on the ventilated façade concept is the off-site prefabricated rainscreen façade, developed by the project BuildHEAT (Avesani et al., 2020). It employs a substructure that acts as the frame to host both active and passive components, connects to the existing structure and retains the external cladding. Such components include thermal insulation, PV panels, pipes and ducts for energy distribution and ventilation (Avesani, Ilardi, Terletti, Rodriguez, & Fedrizzi, 2019). The frame is preassembled off-site in floor-to-floor-height panels (Fig. 10.9).

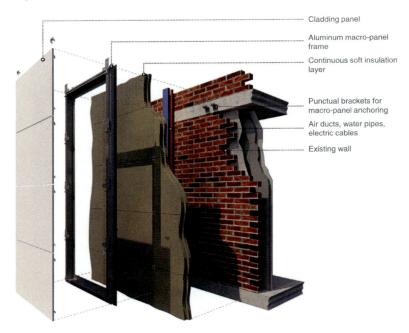

Figure 10.9 Visualization of the BuiltHEAT façade final concept and its layers (Avesani et al., 2020).
Source: Avesani, S., Andaloro, A., Ilardi, S., Orlandi, M., Terletti, S., & Fedrizzi, R. (2020). Development of an off-site prefabricated rainscreen façade system for building energy retrofitting. *Journal of Façade Design and Engineering, 8*. https://doi.org/10.7480/jfde.2020.2.4830, p. 47, Fig. 10.2, licenced under CC by 4.0.

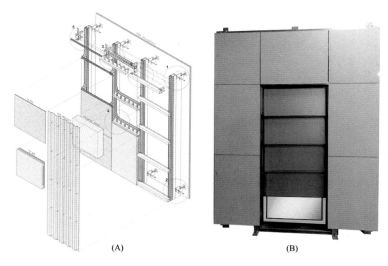

(A)   (B)

Figure 10.10 (A) Different systems to be integrated into the substructure of the BRESAER renovation system, (B) mock-up of the system (Aguirre et al., 2018).
Source: Aguirre, I., Azpiazu, A., Lacave, I., Álvarez, I., & Garay, R. (2018). BRESAER. In *Breakthrough solutions for adaptable envelopes in building refurbishment VIII international congress on architectural envelopes, San Sebastian-Donostia, Spain*; Figs 10.2 and 10.3, © EURECAT and LKS KREAN.

Capeluto (2019) describes how the BRESAER system, which stands for breakthrough solutions for adaptable envelopes in building refurbishment, can adjust the components to allow for adaptability to different climates and requirements. It consists of a lightweight structural mesh that integrates active and passive prefabricated solutions. Following the idea of rainscreen facades, an aluminium frame supports components such as lightweight industrialized ventilated façade module, insulation panels made of ultrahigh-performance fibre-reinforced concrete and dynamic windows with sun protection (Fig. 10.10A and B). All the systems are exchangeable and removable, facilitating the maintenance and the adaptation of the façade system (Aguirre, Azpiazu, Lacave, Álvarez, & Garay, 2018).

Preassembled configurations

As discussed in Section 10.2, renovation and particularly deep renovation needs to comprise a variety of technologies, from thermal insulation to renewable energy and upgrade of the building services components. Industrialization supports the effective integration of components in the building envelope, limiting the installation time, space requirements and occupants' disturbance. The different components to be retrofitted, such as internal and external heat pumps units and the water buffer tank, can be preassembled and placed in prefabricated constructions, which are then transferred to the construction site and installed on top or on the side of the existing building envelope.

Figure 10.11 Preassembled building services unit placement on the roof. Factory Zero, the Netherlands, energy modules.
Source: Energiesprong International, 2019, https://www.flickr.com/photos/150184035@N07/albums/72157690034665123. Licensed under CC BY 2.0.

An interesting concept that follows this principle was developed by the company Factory Zero. The concept's main innovation is the integration of the building services in the Climate Energy Module (Factory Zero, 2020), which can be installed during the roof renovation, as seen in Fig. 10.11. It combines an insulated roof panel with a plastic hood that hosts the heat pump outdoor unit and the foils and binders to connect to the rest of the roof construction.

Overview of design concepts

The previous principles refer to design and construction concepts that apply degrees of industrialization to upgrade the building envelope as a whole or parts of it. They can be combined with each other or with on-site renovation technologies. Table 10.2 presents an overview of the design concepts, providing a brief decription of the principle, indicative reference projects and degree of industrialisation.

Two overarching categories can be identified; components consisting of an insulation core and sheathing layers and components that consist of different modules. Modularity is a common characteristic of all concepts, to different levels and sizes. The special mention of MF as a distinct category refers to the different function of each module. The prefabricated rainscreen façade can also be considered a subset of the MF. However, it differentiated from the type of substructure that is based on the construction principle of the ventilated façade. Finally, it is worth highlighting that the preassembled components' concept is distinct from the previously discussed building envelope concepts, and it can be supplementary to other industrialized or on-site renovation approaches (Table 10.2).

10.5 Outlook for the future

Next to the improvement in energy efficiency, indoor climate and environmental impact that renovation achieves, industrialization has additional benefits, which can also be considered drivers for further development (D'Oca et al., 2018; Hong et al.,

Table 10.2 Overview of the applicable design concepts, linked to reference state-of-the-art projects, and most important degree(s) of industrialization.

Design principle			Reference project	Degree of industrialization
Prefabricated sandwich panels		• Two layers of sheathing boards bonded to either side of a rigid insulation core • Different options for the boards and insulation materials • Possible prefitted windows and finishing	• 2ndSkin: zero-energy apartment renovation via an integrated façade approach (Konstantinou et al., 2017) • Rc Panels (Decorte et al., 2020; Rc Panels, 2021)	Prefabrication Mechanization
Timber-frame panels		• Load-bearing timber frame, sheathing boards, insulation in between studs, moisture and waterproofing foils • Windows incorporated in the framework	• MORE-CONNECT (MORE-CONNECT, 2019) • Transformation Zero (Energiesprong, 2019)	Prefabrication Mechanization
Modular façade		• Exterior finish of building made by separate, prefabricated units (modules) • Modules with different functions • Modules connected directly to existing structure or incorporated in a frame • Standardized interfaces	• MEEFS, (Paiho et al., 2015)	Prefabrication Reproduction

(*Continued*)

Table 10.2 (Continued)

Design principle		Reference project	Degree of industrialization
Prefabricated rainscreen façade	• Outer skin (rainscreen), the air cavity, the substructure and the insulation layer • Outer skin consists of prefabricated, passive and active elements • Based on the principle of ventilated façade, which can be assembled off-site	• BuildHEAT (Avesani et al., 2020) • BRESAER (Aguirre et al., 2018), (Capeluto, 2019)	Prefabrication Reproduction
Preassembled configuration	• Components preassembled and placed in prefabricated constructions • Installed on top or on the side of existing building envelope • Combined with other industrialized or on-site renovation measures	Factory-Zero (Factory Zero, 2020)	Prefabrication Mechanization

MEEFS, Multifunctional energy efficient façade system.

2016; IEA Annex 50, 2012; Jaillon & Poon, 2014; MORE-CONNECT, 2019). The main benefits of industrialized renovation are related to the potential for the following aspects:

- effective strategy for improving the productivity of the construction industry (mass and scaling) to make upscaling possible;
- cost reduction through the economy of scales;
- reduced on-site construction time and disturbance for occupants;
- quality assurance of manufacturing by prefabrication and integration;
- design and engineering efficiency (scanning, simulation and optimizing) is an integral part of the industrial process; and
- environmental and economic benefits related to the reduction of construction waste and material use.

To achieve those benefits of industrialization, the renovation market requires process, marketing and organizational innovation (BPIE, 2016). The following sections elaborate on the aspects that can support the further developments and implementation of industrialized renovation.

10.5.1 Adaptability and circularity

The larger amount of production is at the core of what makes industrialization possible and meaningful (Richard, 2005), which is also in line with the need to upscale renovation to reach the goals for decarbonization of the built environment.

Of course, the building stock is not homogeneous. To be able to apply industrialized renovation in a large number of buildings, there is a need for customized solutions. Using prefabricated envelope elements might sound like a disparity to reach adaptability since mass production components have more or less fixed characteristics (Capeluto, 2019).

The key characteristic of the industrialized components should then be adaptability. Adaptability in manufacturing can be achieved if the whole system is based on the combination of various components and allows for small but significant changes decided during the design stage, which can be implemented during system manufacturing according to demand (Capeluto, 2019). The solution needs to combine a general baseline concept, which can be manufactured in an industrialized way while being able to vary in a rationalized/systematic way.

Moreover, upscaling the renovation also means increasing the material and the energy use for their manufacturing and construction. The construction sector accounts for 38% of the waste generated in the EU, more than any other sector of the economy (European Construction Sector Observatory, 2018). Within the context of decarbonization, renovation approaches need to adopt life cycle thinking that aims not only to be more energy-efficient but also less carbon-intensive over their full life cycle. Applying circularity principles to building renovation will reduce material-related greenhouse gas emissions for buildings (European Commission, 2020).

Modularity and adaptability are key principles for the design of a circular built environment. Off-site construction and modularity reduce the amount of waste produced on-site and enable reuse and repurposing (Arup, 2016). Industrialized construction of the retrofitted components offers advantages to that respect. Modularity and disassembly that are inherent properties in industrialized construction support the application of circular design principles, such as reuse, replace and remanufacture of components and materials (Durmisevic, 2010).

10.5.2 Process optimization

As already discussed, the complexity of holistic application of prefabricated modules in renovation, integrating both the building envelope and building services, is increased. However, it is also essential to achieve high-performance results in energy upgrade. Additional issues that hinder the application of large prefabricated components are logistics limitations, for example, the lack of on-site storage (Jaillon & Poon, 2014). Optimizing and standardizing the process, from design and manufacturing to construction and supply-chain collaboration, can help overcome those problems. Current digital technologies, such as image-based 3D reconstruction (Ying, Lu, Zhou, & Lee, 2018) and BIM (Aldanondo et al., 2014), support this process optimization.

10.5.3 Renovation market

The adoption of industrialized renovation concepts by the building industry requires higher initial costs (Jaillon & Poon, 2014), and investment in equipment for mechanized construction, to reach sufficient capacity (Kamaruddin, Mohammad, & Mahbub, 2016). The lack of those investments hinders the uptake of industrialized renovation.

To overcome such barriers the demand is key to motivating investment in industrialized manufacturing and process optimization needed for the paradigm shift from traditional renovation techniques. As for every product, the market share that is the target should be considered and analysed. Looking at country level, BPIE (2016) has identified the characteristics that result in higher potential for industrialized renovation market integration. Those characteristics are the mature prefab construction market for new constructions, existing building stock in need for renovation, the availability of suitable building typologies for an aggregated prefab construction approach, such as (social) housing, apartment blocks and offices.

10.5.4 Renovation as a product

To increase the market uptake of industrialized renovation, the building industry needs to innovate and look at the renovation not only as the technical challenge but also as a product. Platform- and product-oriented building companies have prerequisites in their company structures and setups that include opportunities to further develop their business models. By following the same paths as other product-oriented industries, industrialized building companies could extend their physical offerings by combining them with services throughout the products' life cycles (Lessing & Brege, 2018), providing the renovation as part of a holistic approach, and combining the technical upgrade with models, such as energy contracting, addresses financial barriers and fragmentation of the supply chain.

Examples of such approach can already be found. In the renovation approach of the Energiesprong (2019), previously discussed in Section 10.4.2, the energy retrofit with industrialized components is combined with a long-term performance guarantee on both the indoor climate and the energy performance. The principle is that the money the residents normally spend on energy bills and maintenance over time is used to finance the retrofit. In this way the total cost of living for the residents remains the same, while their home's quality improves. Similarly, the 2ndSkin project (BIKBouw, 2017) offered zero-energy use of the renovated appartments combined with energy performance contracts. This approach resulted in a viable business case for the housing association, who financed the renovation without increasing the rent after the renovation.

10.6 Conclusion

This chapter discussed industrialized renovation of the building envelope as a way to increase the rate and depth of renovation in the building stock, which are necessary to eliminate carbon emissions from the building sector in the following decades. Industrialization of the retrofits is an effective strategy for improving the productivity of the construction industry, to achieve high performance in existing buildings while minimizing on-site construction time.

Industrialization in the construction, and renovation in particular, has several degrees, from automation and mechanization to robotization and prefabrication. Those degrees are already applied and combined in renovation practice. However, industrialized renovation should go beyond standard practices and aiming to maximize reproduction and off-site construction, through an effective combination of all degrees of industrialization, to target deep renovation at large numbers.

With regards to industrialized renovation of the building envelope, there are different design and construction concepts that can be used. They all aim at high energy performance of the renovated building and incorporation of different technologies that are needed, while making the use of the benefits industrialization for minimizing on-site time, high productivity and quality assurance. Deciding on a concept determines not only the renovation strategy but also the manufacturing and installation techniques and facilities. Therefore it is an important starting point toward industrializing the renovation.

Despite the benefits of industrialized renovation and the several successful state-of-the-art examples, the market uptake is still slow, leaving the potential underexploited. Next to the general barriers to energy renovation, some more specific barriers to industrialized renovation are related to the high initial investment to set-up the production, as well as addressing the adaptability of the concept that is necessary in the heterogenous existing building stock. To overcome those the building industry should focus into more integrated solutions combined with business models, make use of current digital technologies to aim at mass-customization, and collaborate with the demand site and policymakers to target large numbers of buildings to reduce the cost through reproduction.

To achieve that, industrialization of the renovation needs to be made part of the supply chain operation and be prioritized in the strategic decision-making. When successful, thought, industrialized renovation will help overcome some of the challenges of the renovation market such as high costs, lack of capacity, lack of information and fragmentation in the supply chain. In this way the potential of the building stock for decarbonization and sustainability can be realized.

References

Aguirre, I., Azpiazu, A., Lacave, I., Álvarez, I., & Garay, R. (2018). BRESAER. In *Breakthrough solutions for adaptable envelopes in building refurbishment VIII international congress on architectural envelopes*, San Sebastian-Donostia, Spain.

Aldanondo, M., Barco-Santa, A., Vareilles, E., Falcon, M., Gaborit, P., & Zhang, L. (2014). Towards a BIM approach for a high performance renovation of apartment buildings. In S. Fukuda, A. Bernard, B. Gurumoorthy, & A. Bouras (Eds.), *Product life-cycle management for a global market*. Berlin, Heidelberg: Springer.

Artola, I., Rademaekers, K., Williams, R., & Yearwood, J. (2016). *Boosting building renovation: What potential and value for Europe?* http://trinomics.eu/project/building-renovation/.

Arup. (2016). *Circular economy in the built environment* (Arup, Issue). https://www.arup.com/perspectives/publications/research/section/circular-economy-in-the-built-environment.

Astudillo, J., Garcia, M., Sacristan, J., Uranga, N., Leivo, M., Mueller, M., ... De Elgea, A. O. (2018). New biocomposites for innovative construction facades and interior partitions [Article]. *Journal of Façade Design and Engineering*, 6(2), 67−85. Available from https://doi.org/10.7480/jfde.2018.2.2104.

Avesani, S., Andaloro, A., Ilardi, S., Orlandi, M., Terletti, S., & Fedrizzi, R. (2020). Development of an off-site prefabricated rainscreen façade system for building energy retrofitting. *Journal of Façade Design and Engineering*, 8. Available from https://doi.org/10.7480/jfde.2020.2.4830, in press.

Avesani, S., Ilardi, S., Terletti, S., Rodriguez, I., & Fedrizzi, R. (2019). *BuildHeat D3.10a— The active façade kit*. http://www.buildheat.eu/reports/.

BIKBouw. (2017). *Project 2ndSkin®*. Retrieved from https://www.bikbouw.nl/blog/79-pilot-project-2nd-skin. Accessed 21.02.21.

Borodulin, V. Y., & Nizovtsev, M. I. (2021). Modeling heat and moisture transfer of building facades thermally insulated by the panels with ventilated channels. *Journal of Building Engineering*, 40, 102391. Available from https://doi.org/10.1016/j.jobe.2021.102391.

BPIE. (2011). *Europe's buildings under the microscope*. Building Performance Institute Europe. http://www.bpie.eu/eu_buildings_under_microscope.html.

BPIE. (2013). *A guide to developing strategies for building energy renovation: Delivering the energy efficiency directive article 4 Building Performance Institute Europe*. http://www.bpie.eu/renovation_strategy.html.

BPIE. (2016). *Prefabricated systems for deep energy retrofits of residential buildings.* http://bpie.eu/wp-content/uploads/2016/02/Deep-dive-1-Prefab-systems.pdf.

Brand, S. (1994). *How buildings learn: What happens after they're built.* Viking.

Brumana, R., Prisco, M. D., Colombo, M., Marchi, F., Terletti, S., Coeli, F., ... Sonzogni, F. (2016). Aler building in cinisello balsamo (Mi): An example of energy efficient refurbishment with EASEE method. In *Lecture Notes in Civil Engineering, 10,* 467–480.

Bruno, M., & Grecea, D. (2017). Sustainable retrofitting of existing residential buildings: A case study using a rainscreen cladding system. In *International multidisciplinary scientific GeoConference surveying geology and mining ecology management, SGEM.*

Bumanis, K., & Pugovics, K. (2019). Low energy consumption façade pilot project. *Journal of Sustainable Architecture and Civil Engineering, 24*(1), 52–60. Available from https://doi.org/10.5755/j01.sace.24.1.22170.

Bystedt, A., Ostman, L., Knuts, M., Johansson, J., Westerlund, K., & Thorsen, H. (2016). Fast and simple—Cost efficient façade refurbishment. In J. Kurnitski (Ed.), In *Sustainable built environment Tallinn and Helsinki conference Sbe16 Build Green and Renovate Deep* (Vol. 96, pp. 779–787). Elsevier Science Bv. https://doi.org/10.1016/j.egypro.2016.09.140

Callegari, G., Spinelli, A., Bianco, L., Serra, V., & Fantucci, S. (2015). NATURWALL (c)— A solar timber façade system for building refurbishment: Optimization process through in field measurements. In M. Perino (Ed.), In *6th International building physics conference* (Vol. 78, pp. 291–296). Elsevier Science Bv. https://doi.org/10.1016/j.egypro.2015.11.641.

Capeluto, G. (2019). Adaptability in envelope energy retrofits through addition of intelligence features. *Architectural Science Review, 62*(3), 216–229. Available from https://doi.org/10.1080/00038628.2019.1574707.

Chang, T.-S., & Ward, A. C. (1995). Conceptual robustness in simultaneous engineering: A formulation in continuous spaces. *Research in Engineering Design, 7*(2), 67–85. Available from https://doi.org/10.1007/BF01606903.

Commission Recommendation (EU) (2019) *2019/786 of 8 May 2019 on building renovation (notified under document C(2019) 3352) (2019/786).* https://eur-lex.europa.eu/legal-content/GA/TXT/?uri = CELEX:32019H0786.

Cooper, R., Aouad, G., Lee, A., Wu, S., Fleming, A., & Kagioglou, M. (2005). *Process management in design and construction.* Blackwell Publishing Ltd. Available from https://doi.org/10.1002/9780470690758.ch1.

Coupillie, C., Steeman, M., Van den Bossche, N., & Maroy, K. (2017). Evaluating the hygrothermal performance of prefabricated timber frame façade elements used in building renovation. In S. Geving & B. Time (Eds.), In *11th Nordic symposium on building physics* (Vol. 132, pp. 933–938). Elsevier Science Bv. https://doi.org/10.1016/j.egypro.2017.09.727.

D'Oca, S., Ferrante, A., Ferrer, C., Pernetti, R., Gralka, A., Sebastian, R., & Veld, P. O. (2018). Technical, financial, and social barriers and challenges in deep building renovation: Integration of lessons learned from the H2020 cluster projects. *Buildings, 8*(12), 25. Available from https://doi.org/10.3390/buildings8120174, Article 174.

Decorte, Y., Steeman, M., Krämer, U. B., Struck, C., Lange, K., Zander, B., & Haan, A. D. (2020). Upscaling the housing renovation market through far-reaching industrialization. *IOP Conference Series: Earth and Environmental Science, 588,* 032041. Available from https://doi.org/10.1088/1755-1315/588/3/032041.

DIRECTIVE (2018). (2018/844EUU). *On the energy performance of building*. Brussels: The European Parliament And Of The Council. Retrieved from http://data.europa.eu/eli/dir/2018/844/oj.

Du, H., Huang, P., & Jones, P. (2019). Modular façade retrofit with renewable energy technologies: The definition and current status in Europe. *Energy and Buildings*, 205, 109543. Available from https://doi.org/10.1016/j.enbuild.2019.109543.

Durmisevic, E. (2010). *Green design and assembly of buildings and systems, design for disassembly a key to life cycle design of buildings and building products*. VDM Verlag Dr. Müller, https://doi.org/urn:nbn:nl:ui:28−03143149-128b-41ce-bef9-d0877f1d1be9.

EnergieSprong. (2014). *Inspirerende projecten. Platform31*. Retrieved 14−11 from http://energiesprong.nl/blog/category/inspirerende-projecten/.

Energiesprong. (2019). *Energiesprong works!* https://energiesprong.org/publication/.

European Commission. (2019). *The European Green Deal*. Brussels Retrieved from https://ec.europa.eu/info/sites/info/files/european-green-deal-communication_en.pdf.

European Commission. (2020). *A renovation wave for Europe—greening our buildings, creating jobs, improving lives*. Brussels Retrieved from https://ec.europa.eu/commission/presscorner/detail/en/IP_20_1835.

European Construction Sector Observatory. (2018). *Improving energy and resource efficiency*. Brussels: European Commission Retrieved from. Available from https://ec.europa.eu/docsroom/documents/33121.

Factory Zero. (2020). *Factory Zero: De kern van wonen*. Retrieved 11.11 from https://factoryzero.nl/innovatie/.

Ferreira, J., Pinheiro, M. D., & Brito, J. D. (2013). Refurbishment decision support tools review—Energy and life cycle as key aspects to sustainable refurbishment projects. *Energy Policy*, 62, 1453−1460. Available from https://doi.org/10.1016/j.enpol.2013.06.082.

Filippidou, F., & Jimenez Navarro, J. (2019). *Achieving the cost-effective energy transformation of Europe's buildings*. Publications Office of the European Union. Available from https://publications.jrc.ec.europa.eu/repository/handle/JRC117739.

Filippidou, F., Nieboer, N., & Visscher, H. (2016). Energy efficiency measures implemented in the Dutch non-profit housing sector. *Energy and Buildings*, 132, 107−116. Available from https://doi.org/10.1016/j.enbuild.2016.05.095.

Gershenson, J. K., Prasad, G. J., & Zhang, Y. (2003). Product modularity: Definitions and benefits. *Journal of Engineering Design*, 14(3), 295−313. Available from https://doi.org/10.1080/0954482031000091068.

Giebeler, G., Krause, H., Fisch, R., Musso, F., Lenz, B., & Rudolphi, A. (2009). *Refurbishment manual: maintenance, conversions, extentions*. Birkhäuser. Available from http://aleph.tudelft.nl:80/F/?func = direct&doc_number.

Gorse, C., Johnston, D., & Pritchard, M. (2012). *A dictionary of construction. Surveying, and civil engineering*. Oxford: OUP. Available from https://books.google.nl/books?id = KMpXe6ceMSMC.

greenspec. (2013). *Energy-efficient house refurbishment/retrofit*. Retrieved 09/06 from http://www.greenspec.co.uk/.

Hermelink, A., & Müller, A. (2011). *Econimics of deep renovation*. http://www.eurima.org/uploads/ModuleXtender/Publications/51/Economics_of_Deep_Renovation_Ecofys_IX_Study_Design_FINAL_01_02_2011_Web_VERSION.pdf.

Hong, J., Shen, G. Q., Mao, C., Li, Z., & Li, K. (2016). Life-cycle energy analysis of prefabricated building components: An input−output-based hybrid model. *Journal of Cleaner*

Production, *112*, 2198−2207. Available from https://doi.org/10.1016/j.jclepro.2015. 10.030.
IEA Annex 50. (2012). *Prefabricated systems for low energy renovation of residential buildings, project summary report*. http://www.uk.ecbcs.org/Data/publications/EBC_PSR_Annex50.pdf.
Jaillon, L., & Poon, C. S. (2014). Life cycle design and prefabrication in buildings: A review and case studies in Hong Kong. *Automation in Construction*, *39*, 195−202. Available from https://doi.org/10.1016/j.autcon.2013.09.006.
Jensen, P. A., Maslesa, E., Berg, J. B., & Thuesen, C. (2018). Ten questions concerning sustainable building renovation. *Building and Environment*, *143*, 130−137. Available from https://doi.org/10.1016/j.buildenv.2018.06.051.
Kamaruddin, S. S., Mohammad, M. F., & Mahbub, R. (2016). Barriers and Impact of Mechanisation and Automation in Construction to Achieve Better Quality Products. *Procedia - Social and Behavioral Sciences*, *222*, 111−120. Available from https://doi.org/10.1016/j.sbspro.2016.05.197.
Kilaire, A., & Stacey, M. (2017). Design of a prefabricated passive and active double skin façade system for UKK offices. *Journal of Building Engineering*, *12*, 161−170. Available from https://doi.org/10.1016/j.jobe.2017.06.001.
Klein, T. (2013). Integral façade construction. *Towards a new product architecture for curtain walls (Vol. 3) [façade; curtain wall; TUU Delft; future facades; building envelope; product architecture*; Tillmann Klein; Architectural Engineering + Technology]. http://abe.tudelft.nl/article/view/klein.
Knaack, U., Klein, T., Bilow, M., & Auer, T. (2007). *Façades: Principles of construction*. Birkhäuser Basel. https://books.google.nl/books?id = u81G-G4V3aAC.
Konstantinou, T. (2014). *Façade refurbishment toolbox: Supporting the design of residential energy upgrades*, Delft University of Technology. https://books.bk.tudelft.nl/index.php/press/catalog/book/isbn.9789461863379.
Konstantinou, T., Guerra-Santin, O., Azcarate-Aguerre, J., Klein, T., & Silvester, S. (2017). *A zero-energy refurbishment solution for residential apartment buildings by applying an integrated, prefabricated façade module*. PowerSkin, Munich.
Lessing, J., & Brege, S. (2018). Industrialized building companies; Business models: Multiple case study of Swedish and North American Companies. *Journal of Construction Engineering and Management*, *144*(2), 05017019. Available from https://doi.org/10.1061/(ASCE)CO.1943-7862.0001368.
Loebus, S., Ott, S., & Winter, S. (2014). The multifunctional TES-façade joint. In S. Aicher, H. W. Reinhardt, & H. Garrecht (Eds.), *Materials and joints in timber structures*. Dordrecht: Springer.
Ma, Z., Cooper, P., Daly, D., & Ledo, L. (2012). Existing building retrofits: Methodology and state-of-the-art. *Energy and Buildings*, *55*(0), 889−902. Available from https://doi.org/10.1016/j.enbuild.2012.08.018.
Malacarne, G., Monizza, G. P., Ratajczak, J., Krause, D., Benedetti, C., & Matt, D. T. (2016). Prefabricated timber façade for the energy refurbishment of the Italian building stock: The Ri.Fa.Re. project. *Energy Procedia*, *96*, 788−799. Available from https://doi.org/10.1016/j.egypro.2016.09.141.
Mayer, P. (2021). *Whole life costing: Prefabricated structural panels*. Greenspec. Retrieved from http://www.greenspec.co.uk/materials-compared.php. Accessed 09.03.21.
Meijer, F., Straub, A., & Mlecnik, E. (2018). Consultancy centres and pop-ups as local authority policy instruments to stimulate adoption of energy efficiency by homeowners.

Sustainability, *10*(8), 2734. Available from https://www.mdpi.com/2071-1050/10/8/2734.

Moe, K., & Smith, R. E. (2012). *Building systems: Design, technology, and society*. Routledge Chapman & Hall. Available from http://books.google.nl/books?id=1_68cQAACAAJ.

MORE-CONNECT. (2019). *D5.9 Analyses of the total renovation processes in the pilots* https://www.more-connect.eu/wp-content/uploads/2019/07/MORE-CONNECT_WP5_D5.9-Analyses-of-the-total-renovation-processes.pdf.

Ochoa, C. E., & Capeluto, I. G. (2015). Decision methodology for the development of an expert system applied in an adaptable energy retrofit façade system for residential buildings. *Renewable Energy*, *78*, 498−508. Available from https://doi.org/10.1016/j.renene.2015.01.036.

Ochs, F., Siegele, D., Dermentzis, G., & Feist, W. (2015). Prefabricated timber frame facade with integrated active components for minimal invasive renovations. In M. Perino (Ed.), In *6th International building physics conference* (Vol. 78, pp. 61−66). Elsevier Science Bv. https://doi.org/10.1016/j.egypro.2015.11.115.

Pacheco-Torgal, F., Granqvist, C. G., Jelle, B. P., Vanoli, G. P., Bianco, N., & Kurnitski, J. (2017). *Cost-effective energy efficient building retrofitting: Materials, technologies, optimisation and case studies*. Elsevier Science. Available from https://books.google.nl/books?id=aWeyDAAAQBAJ.

Paiho, S., Seppa, I. P., & Jimenez, C. (2015). An energetic analysis of a multifunctional façade system for energy efficient retrofitting of residential buildings in cold climates of Finland and Russia [Article]. *Sustainable Cities and Society*, *15*, 75−85. Available from https://doi.org/10.1016/j.scs.2014.12.005.

Pihelo, P., Kalamees, T., & Kuusk, K. (2017). nZEB renovation with prefabricated modular panels. *Energy Procedia*, *132*, 1006−1011. Available from https://doi.org/10.1016/j.egypro.2017.09.708.

Rc Panels. (2021). *RCC Panels*. Retrieved from https://rcpanels.nl/concept/renovatie/. Accessed 08.03.21.

RIBA. (2020). *Plan of work 2020 overview*. https://www.architecture.com/-/media/GatherContent/Test-resources-page/Additional-Documents/2020RIBAPlanofWorkoverviewpdf.pdf.

Richard, R.-B. (2005). Industrialised building systems: Reproduction before automation and robotics. *Automation in Construction*, *14*(4), 442−451. Available from https://doi.org/10.1016/j.autcon.2004.09.009.

Stroomversnelling. (2014). *Maak de Stroomversnelling mee!*.

Stroomversnelling. (2019). *Nul-op-de-Meter: Prijsontwikkeling 2015−2030*. https://pages.stroomversnelling.nl/rapport-nul-op-de-meter-prijsontwikkeling-2015-2030.

TESEnergyFaçade. (2014). *SmartTES*. Technische Universität München. Retrieved 10−07 from http://www.tesenergyfacade.com/index.php.

Thorpe, D. (2010). *Sustainable Home Refurbishment: The Earthscan Expert Guide to Retrofitting Homes for Efficiency*. Taylor & Francis.

Tsemekidi Tzeiranaki, S., Bertoldi, P., Paci, D., Castellazzi, L., Ribeiro Serrenho, T., Economidou, M., & Zangheri, P. (2020). *Energy consumption and energy efficiency trends in the EU-28, 2000−2018*. In JRC120681. Luxembourg: Publications Office of the European Union.

Tumminia, G., Guarino, F., Longo, S., Ferraro, M., Cellura, M., & Antonucci, V. (2018). Life cycle energy performances and environmental impacts of a prefabricated building module. *Renewable and Sustainable Energy Reviews*, *92*, 272−283. Available from https://doi.org/10.1016/j.rser.2018.04.059.

Ying, H., Lu, Q., Zhou, H., & Lee, S. (2018). A framework for constructing semantic as-is building energy models (BEMs) for existing buildings using digital images. In *ISARC 2018−35th international symposium on automation and robotics in construction and international AEC/FM Hackathon: The future of building things*.

Vertical farming on facades: transforming building skins for urban food security

Abel Tablada[1] and Vesna Kosorić[2,3]
[1]Faculty of Architecture, Technological University of Havana J.A. Echeverría, Havana, Cuba, [2]Balkan Energy AG, Starrkirch-Wil, Switzerland, [3]Daniel Hammer Architekt FH AG, Olten, Switzerland

Abbreviations

CEA	controlled-environment agriculture
BIA	building-integrated agriculture
DLI	daily light integral
DSF	double-skin facade
EPC	energy performance certificates
LCA	life cycle assessment
MCDM	multicriteria decision-making
NFT	nutrient film technique
NUS	National University of Singapore
PF	productive facade
PV	photovoltaics
T² Lab	Tropical Technologies Laboratory
VGS	vertical greenery systems
VF	vertical farming

11.1 Introduction

There are various reasons why producing food on building facades is not only possible but also advantageous for cities and their residents and necessary for the planet. Urban population reached one half of the total world population in 2008. By 2030, this proportion is expected to grow to two-thirds (UN Department of Economic & Social Affairs, 2019). The challenge is so huge that humankind must produce in the next 50 years more food than in the past 10,000 years (Mack, 2014). This problem worsens because of the reduced availability of land area per capita for conventional farming and the area suitable for cultivation. The scarcity of cultivated areas is a result of increasing population, urbanization in former agricultural areas and several climate change impacts such as desertification of land, changes in precipitation patterns, soil composition, flowering and harvesting seasons and increasing frequency and severity of extreme weather events [Kim, 2010; United

States Environmental Protection Agency (EPA), 2017]. Given the shortages of cropland and fresh water, together with the dependence on oil prices and reliance on fossil fuels (Taghizadeh-Hesary, Rasoulinezhad, & Yoshino, 2019) and the impacts of climate change, the era of industrial agriculture delivering cheap limitless food is coming to an end (Cockrall-King, 2012).

On the other hand, cities are getting denser with the growing number of high-rise buildings. While the ratio of vertical versus horizontal surfaces is on the rise, the use of such surfaces is generally nonproductive. Therefore using building skins to produce fresh vegetables and fruits makes more sense than the use of valuable interior space, as it takes advantage of available sunlight and building structure, contributes to indoor and urban climate control and saves indoor spaces for other uses.

In line with the efforts to drastically reduce greenhouse gases emissions in urban areas and in pursuance of the several of the United Nations Development Goals,[1] cities should transform themselves from net consumers to net producers of food and energy. In addition to contributing to food security and resilience at household and neighbourhood levels, farming activities such as those on building facades also improve the quality of life of urban dwellers by improving human health and sociability (Draper & Freedman, 2010; Harris, 2009; Ling & Chiang, 2018), reducing poverty and dependence on large food chains (Akinwolemiwa, Bleil de Souza, De Luca, & Gwilliam, 2018; Gould & Caplow, 2012) and depending on the shading effect of crops, they may also contribute to the reduction of solar gains on facades and the urban heat island effects at city level (Wong, Tan, & Yu et al., 2010).

Adaptable to most common types of facades and building functions such as residential, office and commercial as shown in the several case studies and examples, façade farming is compatible with traditional and advanced technologies such as vertical farming (VF) (Fig. 11.1). Since the availability of solar radiation is the main requirement, its application does not directly depend on income and educational level of potential users. As a decentralized, open-source, small-scale, diverse and scattered farming type, façade farming could contribute to food security,[2] by providing food which is grown, distributed and consumed within urban areas (Cockrall-King, 2012).

The aim of this chapter is to conduct a summary review of studies on VF systems on facades and to highlight their various social, economic and environmental benefits among which food security and the contribution toward a reduction of carbon footprint are probably the most important in our times. This is supported by a

---

[1] Especially Goal 2: End hunger, achieve food security and improved nutrition and promote sustainable agriculture and Goal 11: Make cities and human settlements inclusive, safe, resilient and sustainable. Other goals favoured by the implementation of urban farming could be Goal 10: Reduce inequality within and among countries, Goal 12: Ensure sustainable consumption and production patterns and Goal 13: Take urgent action to combat climate change and its impacts (Department of Economic, & Social Affairs, United Nations, 2015).

[2] 'Food security exists when all people, at all times, have physical and economic access to sufficient, safe and nutritious food that meets their dietary needs and food preferences for an active and healthy life'. (World Food Summit, 1996; Food, & Agriculture Organization of the United Nations, FAO, 2006).

Figure 11.1 Left: Coriander in planters at the HDB public housing in Singapore. *HDB*, Housing and Development Board. Right: Productive façade system at the Tropical Technologies Laboratory (T² Lab).
Source: (Left) Derrick Ng and Eunis Lim (Generation Green). (Right) Abel Tablada.

selection of case studies and examples of actual implementations in diverse latitudes where different technologies are applied to residential, office and commercial buildings. Experimental results in a test bed laboratory are presented in detail as well as the main findings that provide more evidence of the feasibility and benefits of VF on facades as one of the forward-thinking research topics and applications on building skins. The chapter finalizes with a discussion on limitations, future directions of façade farming and main conclusions.

11.2 Urban agriculture: from vacant plots to building-integrated agriculture

The agricultural revolution first occurred around 10,000 years ago in the southeast of the current Turkey, the west of Syria and the Levant in the Middle East. Other agricultural revolutions emerged in an isolated way in other regions all around the world till around 2000 BCE (Harari, 2014). The provision of food in a planned way in contrast to the practice of precedent nomadic hunter-gatherers allowed the creation of more permanent human settlements which then developed into rural villages, towns and cities. Therefore the cultivation of cereals, vegetables and fruits was always in proximity of areas where people lived and in tandem with other nonrural activities (Harari, 2014). With the industrial revolution and the expansion of cities, a clearer land use differentiation occurred between rural and urban areas. The urban expansion continued during the 20th century partly thanks to the development and relative affordability of motorized vehicles in industrialized and emerging economies. The extension of large cities and megacities[3] and the possibility to rapidly transport goods at long distance forced and allowed, respectively, much longer distance between agricultural areas and the places where products were consumed. This distance is known as food miles (Lang, 2006) for the case of food-related products and is one of the causes of current high carbon emissions of the food industry and a threat for food security in a possible future scenario of higher cost of fuel and food chain disruptions due to climate change—related events, especially in developing countries (Szabo, 2016). However, in the majority of small- and middle-size cities, main cultivation areas remained at peri-urban areas that considerably reduce the food miles (Orpiz, Berges, Piorr, & Krikser, 2016).

Since the energy crisis in 1973 and more recently as a consequence of cyclic economic crises and climate change effects, there is an increasing awareness of the need for a shift in the current unsustainable production paradigm, which has incentivized urban farming experiences all around the world for their social and economic benefits (Cockrall-King, 2012; Poulsen, Neff, & Winch, 2017). Urban farming programs in Chicago, Detroit, Havana, London, Paris and many other cities have provided a valuable set of lessons in terms of implementation policies, public participation, practices and technologies (Cockrall-King, 2012; Gorgolewsky, Komisar, & Nasr, 2011; Peña Díaz, 2014; Viljoen, Bohn, & Howe, 2005). A part of this change, especially after 2010, is the move from conventional practices on vacant plots and urban courtyards toward the use of technology and integration of farming systems into buildings.

Building-integrated agriculture (BIA) is an innovative method of food production based on the synergies between buildings and advanced farming systems. Such installations are characterized by recirculating hydroponics, waste heat captured from a building's heating, ventilation and air-conditioning system, photovoltaics (PV), or other forms of renewable energy, rainwater catchment systems and evaporative cooling (Gould & Caplow, 2012).

VF is a variant of BIA aimed at minimizing the use of land and the food miles while taking advantage of urban infrastructure to increase productivity (Despommier, 2010). The use of building facades to produce vegetables (façade farming) is a promising variant of VF systems which combines technological advances and the use of solar energy on already built surfaces. VF can be implemented both indoors and on building envelopes—rooftops and facades. Much of the technology that can be used to produce food on facades has already been developed thanks to the rapid dissemination of the concept and methods of vertical greenery systems (VGS), which is, however, dedicated to growing ornamental plants.

---

[3] According to UN Department of Economic and Social Affairs (2019), a large city has a population between 5 and 10 million inhabitants and a megacity is an urban agglomeration with 10 million inhabitants or more.

11.3 Vertical greenery systems

11.3.1 Definition and classification

VGS are defined as the growing of plants on, up, or against the façade of a building or feature walls (Peck, Callaghan, Kuhn, & Bass, 1999) (Fig. 11.2). According to the growing method, VGS can be classified as 'green facades' or 'living walls' (Perini, Ottelé, Haas, & Raiteri, 2013; Radić, Brković Dodig, & Auer, 2019). Green facades refer to the system where climbers grow from the ground or on planters using the building surface or trellis. Living walls refer to vegetation growing on modular panels or containers. Their basic components are the plants, the system (containers) and the media and the irrigation system. According to Chu (2014), VGS can also be classified as support and carrier types according to the type of plants (nonclimbing, climbing, or hanging), type of support (self-supported, wall, wire, pergolas), medium (substrate—soil or mix, water-hydroponics, or aeroponics), irrigation methods (manual, drip system, channelled) and the type of containers (integrated beds, pots and geotextile felts).

Another term to name the integration of vegetation into buildings is the Skyrise Greenery, which is a solution to leverage on the existing structures of high-rise buildings such as balconies, sky terraces, vertical surfaces and roofs (Tan et al., 2009).

11.3.2 Benefits and drawbacks

In addition to bringing the nature closer to humans (Wong, Tan, & Yok Tan et al., 2010), vertical greenery has considerable functional (Cooper-Marcus & Barnes,

Figure 11.2 Examples of vertical greenery systems (VGS). Left: green façade at Oasia Hotel Downtown by WOHA, Singapore. Centre: living wall at GAIA condominium by UNISEAL, Singapore. Right: sky garden and planters with climbers at Kent Vale II Residences, Singapore.
Souuce: Image credit: Abel Tablada.

1999; Roehr & Laurenz, 2008) and formal-aesthetic values (Sutton, 2014). VGS also has several environmental benefits such as reducing external surface temperature (Pérez, Coma, Sol, & Cabeza, 2017; Wong, Tan, & Yu et al., 2010) and air temperature (Tan et al., 2009), improving air quality and rainwater treatment and retention, enhancing biodiversity and improving acoustic insulation (Wong, Kwang Tan, Tan, Chiang, & Wong, 2010). Economic benefits are related to the added aesthetical value and the reduction of energy use due to insulation and shading provided by plants (Wong, Tan, Tan, & Wong, 2009).

Some disadvantages have been reported in relation to VGS. The most critical are high installation costs, with the green facades being the cheapest and the living walls the most expensive (Perini & Rosasco, 2013). In addition, high operational and maintenance costs, the risk of damaging building walls and the influx of unwanted pests and animals are some other drawbacks of VGS which can also be found in farming on facades.

11.4 Farming on facades

11.4.1 Vertical farming: controlled-environment versus passive agriculture

The concept of VF has been discussed, tested and implemented since 2010 (Despommier, 2010). It is claimed that VF is reshaping agriculture, food production, distribution and consumption thanks to its more efficient and sustainable methods in comparison with traditional approaches (Al-Kodmany, 2018). Three types of VF have been identified (Al-Kodmany, 2018). The most frequent one is a modestly sized urban farm in new and old buildings, including the construction of tall structures with several levels of growing beds, often lined with artificial lights. The second type is found on the rooftops of old and new buildings, atop commercial and residential structures. The third type of VF is that of a newly built multistory building designed specifically to accommodate VF systems. VF on facades can be found on the first and third types. In the last decade, many farms using VF technologies have been installed on rooftops and inside of abandoned buildings (Armanda, Guinée, & Tukker, 2019). Apart from higher yields per unit area, VF has other environmental and economic advantages which make this practice appealing in many cities. However, there is no firm evidence that the third type of VF has been built. Due to the relative recency of this concept, high implementation costs and insufficient life cycle assessments (LCA), VF is still a controversial practice and its success depends on the technological approaches applied and local conditions.

VF using controlled-environment agriculture (CEA) requires artificial lighting and air conditioning systems that increase operational costs and the overall carbon footprint (Beacham, Vickers, & Monaghan, 2019). In a study of a hypothetical high-rise building with indoor VF systems, the simulated carbon footprint (CO_2/kg lettuce) was five times higher than for conventional agriculture during summer conditions (Al-Chalabi, 2015).

Applying VF systems on or close to the building envelope allows the use of sunlight and/or passive thermal controls that considerably reduce energy use and carbon footprint. Rooftop gardens and farming systems on facades are forms of farming that use this principle.

Several studies have measured or simulated the potential sunlight incident on facades to assess the feasibility of growing vegetables on building facades according to orientation and context. The study of Tablada and Zhao (2016) using hypothetical residential buildings in Singapore with different urban densities found that all façade orientations receive enough sunlight for growing leafy vegetables. Song, Tan, and Tan (2018) proved the feasibility of using facades in public housing in Singapore, specifically the building corridors, for vegetable farming. Based on the measured photosynthetically active radiation and the daily light integral (DLI), the study found that vegetables receiving half-day direct insolation can successfully grow. Palliwal, Song, Tan, and Biljecki (2021) applied environmental simulations using 3D city models for the identification of suitable farming sites in residential building facades. Although applied in Singapore, the model is applicable at city levels of any latitude and climate.

11.4.2 Types of vertical farming on facades and technologies

Building facades have the potential to incorporate a large variety of BIA and VF systems. Fig. 11.3 classifies different types of farming technologies and their position on facades. The classification is not exhaustive but tries to cover the existing options of farming on building facades based on literature and actual implementation. VF on facades may include the use of balconies, sky terraces, windows, double-skin façade (DSF) systems and bare walls. Among plant growing systems, horizontal containers can be staked one on top of another or be vertically positioned as in the case of 'green' walls or use vertical cylindrical growth units (Beacham et al., 2019). Each of these typologies and positionings requires specific conditions and can admit a variety of technologies to comply with safety, accessibility, aesthetics and productivity requirements.

Using facades for farming has been an old practice. Pots with flowers, herbs and leafy vegetables placed on balconies and walls can be seen in many buildings all around the world in urban and suburban areas. For families not having a private garden, the use of balconies or windowsills is their only possibility to practice horticulture and farming. Most of these family-running installations use soil-based planting and manual irrigation.

However, the contemporary version of farming on facades may include the use of more advanced VF methods such as hydroponics, aeropionics and aquaponics (Chatterjee, Debnath, & Pal, 2020; Despommier, 2009). Hydroponics refers to the system of tubes or channels where plants' roots are submerged into a circulating nutrient solution by applying nutrient film technique (NFT). The water is continuously drawn back to the nutrient reservoir to keep the correct chemical composition. Aeropionics system is another version of a soilless technique in which a cold nutrient mixture is sprayed onto the plant roots that are suspended in the air (Chatterjee

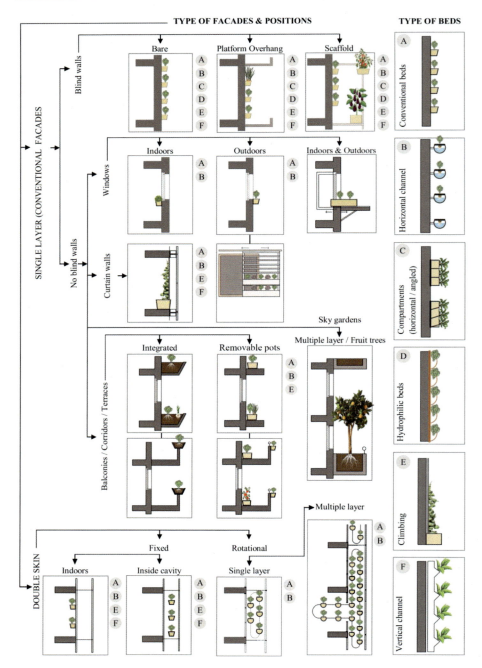

Figure 11.3 Classification of farming facades according to technological systems and position on the façade.
Source: Abel Tablada and Vesna Kosorić.

et al., 2020; Despommier, 2009). Water use is even more controlled and efficiently used in this technique in comparison with conventional hydroponics. However, keeping the nutrient mixture and plant roots cool is a challenge and, therefore, requires more expertise than when applying hydroponics. Aquaponics systems combine fish farming with plant's cultivation—ornamental or vegetable—by creating a synergy between the two reservoirs. Nutrient-rich water from fish waste is provided to plants, while the plants purify the water to be returned to the fish tank.

Due to the complexity of irrigation systems, higher expertise is needed to apply nutrient solutions and such VF systems entailing the risk of plant wilting may not be suitable for residential buildings with amateur operators. However, automated irrigation drip systems can be easily installed and programed for simple VF soil-based systems. For office, commercial and industrial buildings, implementing advanced VF system on facades would represent no risk and the centralized digital management would reduce operational costs and energy. When farming incorporates elements of automation and artificial intelligence, such as light and temperature control, chemical composition of solution and crop growing cycle with the use of software and/or silicon-based hardware, it is classified as digital urban farming (Carolan, 2020).

VF systems may also integrate PV systems for electricity generation and as a shading device. This is the case with the productive façade (PF) system proposed and tested by Tablada and Zhao (2016) and Tablada et al. (2018), described in next sections.

11.4.3 Examples of vertical farming implementation on facades

The most common used crops for façade farming are small leafy vegetables. They are convenient for the ease of access and use of available space as well as productivity. Thanks to their short timeframe from germination to harvest, high profitability and flexibility is achieved in terms of planting regime and crop choice (Beacham et al., 2019). Small fruit trees and grapes can be planted in pots or planters providing at least 31 cm (1 ft) of substrate.

Nowadays, most of the cases of VF on facades are family-running low-tech installations on balconies and open corridors. Thanks to the simplicity and flexibility of installation and low expectations from farmers, such systems have proven their feasibility along time. Cultivating climate-adapted leafy vegetables, herbs and fruits is crucial for this kind of outdoor façade farming. However, in CEA VF systems, the type of crops should be decided according to the artificial environmental conditions created by the system. These conditions may be the typical ones found inside air-conditioned office or residential spaces, or more extreme ones specifically adapted to the type of crops and independent from local climate conditions. Recent research and actual implementation of innovative technologies also show the feasibility of more advanced systems, their advantages and drawbacks.

An interesting proposal was made by an interdisciplinary team led by BrightFarm Systems. They developed a vertically integrated greenhouse to be installed on a curtain wall façade (Fig. 11.4). It consists of a series of horizontal

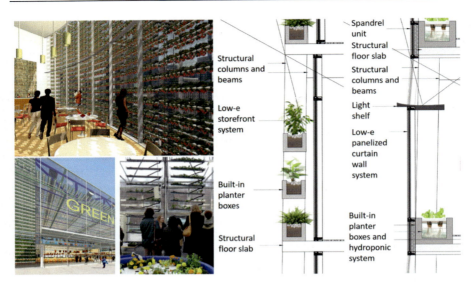

Figure 11.4 Proposed and experimental farming facades. Top left: rendering of a vertical double-skin hydroponic growing system for strawberries. Bottom left: proposal for an agriculture-integrated city food market, Masdar, Abu Dhabi. Bottom centre: detail of vertical integrated greenhouse (VIG) prototype, Science lab/Classroom for PS 84/Jose de Diego School, Brooklyn, New York. Right: sections of proposed VF façade for residential/office space with planter boxes outside and for advanced VF façade with farming area inside. *VF*, Vertical farming.
Source: (Left) Kiss + Cathcart, Architects (Kiss + Cathcart Architects, 2020). (Right) Matharu, J. (2016). Symbiosis in city: How can vertical farming be integrated in a high rise mixed use development? (MArch Thesis). Unitec Institute of Technology, Auckland City.

hydroponic trays (2-m long) suspended on a plant cable lift system connected to a computerized motor for rotation. The entire trip from the bottom level, where seeds are germinated, along the wall before the final return for harvest, takes around 30 days depending on the crop cycle. The trays act as shading devices or blinds allowing also daylight in the interior space (Gould & Caplow, 2012).

Another set of design prototypes of VF on facades for high-rise building were proposed by Matharu (2016). The prototypes responded to site and climatic conditions in Auckland and are adapted to the requirements of residential/office, retail and areas specifically dedicated to VF. The prototypes show a proper integration of building structure, fenestration and other systems with the VF system (Fig. 11.4). However, other aspects related to logistics of the process of cultivation, harvesting, selling and waste disposal as well as the relationship between those activities and office, commercial and residential ones should be further studied before the actual application of the proposed systems in high-rise buildings.

An example of the third type of multistory VF is the proposal from the Swedish company Plantagon International (Plantagon, 2021). They designed a building prototype consisting of a sphere-shape greenhouse with a helix structure in the centre which contains a robot-belt for the automation of farming activities (Al-Kodmany, 2018). The DSF system provides thermal insulation and allows sunlight to reach about 6-m deep. Unfortunately, none of the innovative buildings designed by the company have been built, although several patents related to BIA and small-scale projects have been developed (Plantagon, 2021).

For more traditional residential neighbourhoods where wasted urban spaces can be easily found, the GreenBelly system was created by AVL Studio (Rogers, 2018). The modular system that occupies 35 m^2 of land and is easy to transport, install and remove can produce over 6400 kg of vegetables per year. It contributes to social sustainability by having the potential to provide organic and affordable vegetables for low-income households and by providing education in agriculture practices and healthy eating (GreenBelly Project, 2020). A matter of concern could be the aesthetics of the scaffolding structure and its ability to merge in the image of the city. Care should also be taken in fighting against stormy conditions.

One of the actual implementations of VF on facades is in the atrium of a small hotel in Yogyakarta, Indonesia (Fig. 11.5). A hydroponic system was installed with channels stacked in an inclined way for easier access. Herbs like mint and other spices are cultivated. Originally, the atrium was open for a better light access but was subsequently closed with a semitransparent material due to heavy rain. This resulted in better protection of plants, but with a poorer yield due to the insufficient sunlight, especially on the bottom two floors where no more plants are growing.

Although aquaponics systems require a higher level of knowledge and investment, several studies and actual implementations have been realized on facades. Jenkins, Keeffe, and Hall (2014) and Jenkins (2018) developed a prototype of an aquaponics to be integrated into building facades. In the inner space of a DSF, two layers of soilless growing channel using NFT and a fish tank were installed in a real-scale test bed at an old factory building in Salford, the United Kingdom. For

Figure 11.5 Left: hydroponic system with herbs installed at Greenhost Boutique Hotel, Yogyakarta, 2017. Centre and right: aquaponics developed by Farming Architects, Hanoi.
Source: Abel Tablada.

the design of the prototype, a series of simulations were conducted to assure enough sunlight on each channel and to test the heat transfer reduction of the façade system. The test bed had to be installed indoors, hence the need to test it on an actual building façade. Both technical and social challenges were identified, among them the potential water infiltration and the not yet convincing benefits of such systems. The payback period was estimated to be 7.5 years for the whole façade system considering costs and food prices of Manchester.

A system with similar approach was developed by Farming Architects (2019) for implementation on balconies (Fig. 11.5), and it has been installed on several balconies in Hanoi and other Vietnamese cities. Although originally developed for ornamental purposes, the system can be used to produce vegetables, herbs and edible fishes. Acting as shading devices and as a protection against heavy rains, several planters are placed in a steel grid allowing different configurations and the regulation of daylight.

These short selections of case studies and implementation examples have shown the wide diversity of options in which VF on facades can be applied in terms of budget, locations, climatic conditions and technologies. There is no one single preferable system, rather there should be a preliminary study to adapt the previous proposals to the local requirements and climate, building potentials, and the life cycle of the system.

11.5 Productive facades: a case study

The PF is a concept developed by Tablada et al. (2018) and tested at the Tropical Technologies Laboratory (T^2 Lab) at the National University of Singapore (NUS) (School of Design, & Environment, NUS, 2021) (Fig. 11.6). The façade system integrates PV modules as shading devices and VF systems on balconies and windows. The concept arose in response to the need to utilize existing urban structures and building envelopes exposed to solar radiation as producers of energy and food in addition to their usual functions. The aim was to optimize the design of the two systems to obtain the best performance in terms of energy and food production as well as to contribute to indoor thermal and visual conditions of residential buildings in Singapore.

11.5.1 Design criteria and optimization

Owing to the multiple factors intervening in the final thermal and visual comfort inside buildings with PF systems, a design optimization applying 3D simulation algorithms and a multicriteria decision-making (MCDM)-based process were conducted (Tablada et al., 2018). The development, assessment and design optimization of PF prototypes followed a methodological framework consisting of several phases. In phase 1, the scope, the design concept and the main strategies for the implementation of PFs were defined (Fig. 9.7). In phase 2, preliminary design

Vertical farming on facades: transforming building skins for urban food security 297

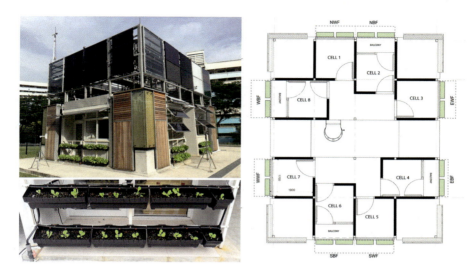

Figure 11.6 Tropical Technologies Laboratory (T^2 Lab at the National University of Singapore (NUS)). Left: east and north facades. Right: floor plan of the T^2 Lab indicating the eight test bed cells and facades.
Source: Adapted from AWP Architects based on lead author's preliminary design. Image Credit: Abel Tablada and Huang Huajing [Tablada, A., Kosorić, V., Huajing, H., Chaplin, I. K., Lau, S. K., Yuan, C., Lau, & S. S. Y. (2018). Design optimisation of productive facades: Integrating photovoltaic and farming systems at the Tropical Technologies Laboratory. *Sustainability 10*(10), 3762. https://doi.org/10.3390/su10103762].

variants for typical façade types in actual residential buildings in Singapore were explored. Two façade categories were defined on each orientation: the façade with a balcony and the façade with a window. The 60-m^2 test bed facility was designed to accommodate eight cells (1.8-m wide, 1.8-m deep, 2.6-m high) on the four cardinal orientations (Fig. 11.6).

In phase 3, the specifications of building integrated PV and VF systems on PFs were defined (Fig. 11.7). The design of each façade would vary depending on the position of planters and the spacing between rows. Also, the position, tilt angle and dimension of PV modules were important design considerations. In all cases the position of planters was on the lowest third of the façade and of PV modules on the top third.

In phase 4, the design variables were planned to be optimized according to five performance indicators. They included (1) food production potential by estimating the DLI which is obtained by converting simulated illumination levels (lx) on top of the planters, (2) potential electricity generation by simulating the incident solar radiation on PV modules, (3) indoor daylight by simulating illumination levels in the interior without the use of artificial lighting, (4) solar heat gain by simulating incident solar radiation and energy flow on the façade and (5) view angles from the interior. Other criteria were also considered prior to the elaboration of the initial design variants such as functional and constructive design parameters.

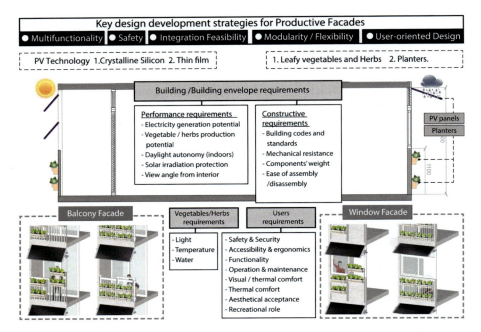

Figure 11.7 Key design development strategies for the productive façade prototypes. *Source:* Vesna Kosorić [Tablada, A., Kosorić, V., Huajing, H., Chaplin, I. K., Lau, S. K., Yuan, C., Lau, & S. S. Y. (2018). Design optimisation of productive facades: Integrating photovoltaic and farming systems at the Tropical Technologies Laboratory. *Sustainability 10*(10), 3762. https://doi.org/10.3390/su10103762].

Computational simulations were then conducted using Rhino-Grasshopper's plug-ins to optimize the design of the eight PF prototypes (Fig. 11.8).

An MCDM was used, in phase 5, to cope with the complexity of the process of selecting the best compromise between the five performance indicators. VIKOR optimization method (Opricovic & Tzeng, 2004, 2007) was applied for its versatility and ease of operation in MS Excel. Based on the simulation results and the weighted-decision matrix VIKOR selected a compromise solution considering the performance indicators, acceptable ranges and targeted performance values.

The final optimal PF prototypes were selected for hypothetical residential buildings in phase 6. Finally, in phases 7 and 8, the final PF prototypes were installed at the T^2 Lab and monitored, respectively.

11.5.2 Optimal productive façade prototypes

Eight PF prototypes were selected, two per façade orientation. Table 11.1 presents the description and values of the five performance indicators per façade prototype. Fig. 11.9 illustrates the configuration of eight optimal PF prototypes and a detail of the planter arrangement on the west façade at the T^2 Lab.

Vertical farming on facades: transforming building skins for urban food security

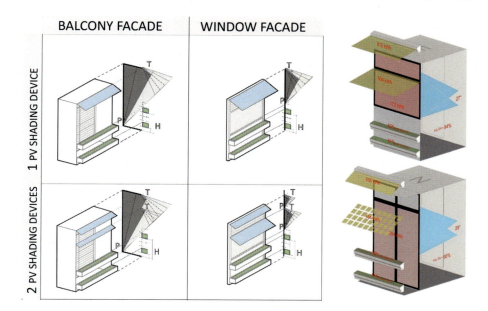

Figure 11.8 Design variants. Left: types of PV configurations and the corresponding start protection and tilt angle. Right: examples of Rhino-Grasshopper models for the assessment of 2135 design variants that are automatically updated according to geometrical inputs. Results are shown and stored after each simulation case. *PV*, Photovoltaics.
Source: Ian K. Chaplin and Huang Huajing [adapted from Tablada, A., Kosorić, V., Huajing, H., Chaplin, I. K., Lau, S. K., Yuan, C., & Lau, S. S. Y. (2018). Design optimisation of productive facades: Integrating photovoltaic and farming systems at the Tropical Technologies Laboratory. *Sustainability 10*(10), 3762. https://doi.org/10.3390/su10103762].

Regarding the number and position of planters, all cases accept only two rows of planters per façade. Having three planters may compromise the amount of sunlight due to smaller spacing and self-shading between the planters. However, given the lower sun altitude on the east and west orientations, the variant of three planters was implemented at one of the facades facing the west at the T^2 Lab. Speaking of the window façade, the position of planters on the lower section of the façade requires the use of additional operable fenestrations for accessibility. Hence, safety elements should also be incorporated in the façade system.

Regarding the arrangement of PV shading devices, two PV modules positioned at the height of the upper level planter is the most preferable variant for most façade types. This configuration allows a larger total surface area of PV cells than with a single module while reducing the obstruction of sunlight toward the planters and achieving the targeted indoor visual comfort.

Both systems can be implemented in higher latitudes providing a similar optimization framework to maximize energy and food production while also complying with indoor thermal and visual performances.

Table 11.1 Optimal cases per façade type and values of performance indicators.

Façade type	Window façade				Balcony façade			
	North	South	East	West	North	South	East	West
Tilt angle of PV-shading (°)	50	50	50	50	50	30	50	50
Top planter height (mm)	700	700	500	500	500	500	500	500
Daylight autonomy interior (%)	95.0	95.2	88.4	83.3	91.6	90.9	89.2	86.8
Energy flow on façade (kW h)	149	149	130	112	194	193	133	123
Irradiance/total PV area (kW h)	1189	1205	1736	1789	1028	1052	1837	1768
Vegetable production (kg/year)	39.2	35.3	42.8	46.5	34.6	28.8	42.6	47.5
View angle (°)	28	28	18	15	37	39	25	24

PV, Photovoltaics.
Source: Adapted from Tablada, A., Kosorić, V., Huajing, H., Chaplin, I. K., Lau, S. K., Yuan, C., Lau, & S. S. Y. (2018). Design optimisation of productive façades: Integrating photovoltaic and farming systems at the Tropical Technologies Laboratory. *Sustainability 10*(10), 3762. https://doi.org/10.3390/su10103762.

Vertical farming on facades: transforming building skins for urban food security

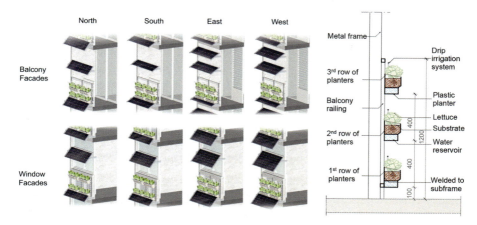

Figure 11.9 Left: artistic impression of the eight optimal PF prototypes. Right: drawing of the location of planters on the balcony railing of the west façade. Original design of planter boxes by UNISEAL. *PF*, Productive façade.
Source: Huang Huajing [adapted from Tablada, A., Kosorić, V., Huajing, H., Chaplin, I. K., Lau, S. K., Yuan, C., Lau, & S. S. Y. (2018). Design optimisation of productive facades: Integrating photovoltaic and farming systems at the Tropical Technologies Laboratory. *Sustainability 10*(10), 3762. https://doi.org/10.3390/su10103762] and Vesna Kosorić.

11.5.3 Test setup and experimental results at the tropical technologies laboratory

The eight PFs at the T² Lab were installed after applying several modifications to the optimal virtual facades. The most important modification was reducing the PV tilt angle to avoid reflections to the neighbouring buildings. Vegetable beds were installed on the heights and separations responding to sunlight accessibility and operational requirements. The position and dimensions of PV modules assure adequate indoor visual and thermal conditions while allowing the required sunlight on the crops. From the list of commonly cultivated leafy vegetables in Singapore, a tropical-adapted variety of lettuce was selected as the most suitable for the experiment considering consumer preferences, short harvest cycles and a moderate light requirement in comparison with other also typical leafy vegetables [DLI > 8 mol/m²/day (Dorais, 2003; Glenn, Cardran, & Thompson, 1984; Schiller, 2017), equivalent to 10,000 lx]. Table 11.2 shows the specifications of the crop, substrate and cultivation actions. Lettuces were cultivated from December 2018 to June 2019, in six rounds.

A rainwater collection system was installed at the T² Lab to ensure water self-sufficiency on top of a symbolic self-sufficiency of energy and food since no person inhabited the lab. Rainwater from the roof and immediate surroundings was filtered and stored in an underground water tank. Electricity generated by PV modules installed on the roof was used to irrigate the crops by a digitally programmed water

Table 11.2 Crop and cultivation specifications.

Vegetable type	Lettuce (*Lactuca sativa*)
Seed germination	Seed trays with one seed per cell with 100% Jiffy Go Bio4 Organic Compost
Seedling transplanting	Day 14
Substrate for planter cells	Cocopeat: Jiffy compost: Perlite = 5:3:2
Total growth period	34 days
Fertilizer application	1 g of 15:15:15 (N:P:K) compound pellet on day 16 and day 26

Source: Provided by Song, S., Department of Biological Sciences, National University of Singapore (NUS).

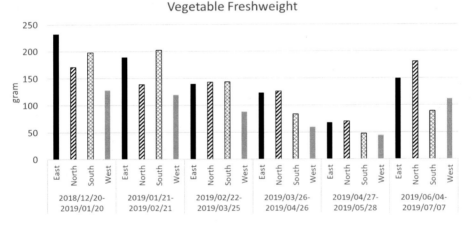

Figure 11.10 Fresh weight of lettuces harvested for six cycles at the Tropical Technologies Laboratory (T^2 Lab), National University of Singapore (NUS). Data provided by Song Shuang and the team from the Department of Biological Sciences, NUS.
Source: Abel Tablada.

pump. A drip irrigation system with micro-spread heads on each plant was used and activated three times per day for 2 minutes.

The experiments using a relatively simple farming system on the two façade categories have provided evidence of the benefits of such implementation on residential building facades in terms of production of fresh vegetables with zero food mile. Fig. 11.10 shows the results of vegetable production for 6-month cultivation. The yield was obtained by weighting the leaves per façade after each harvest and then after drying them.

Comparing the total yield among the facades, east facades have the greatest performance, producing 902-g lettuce. This is explained by the higher direct solar radiation in the morning time in Singapore in comparison with the cloudier afternoon hours. The amount produced represents 55%−103% of the average leafy vegetables' consumption of a 4-member household in Singapore (c. 16 kg/year).

Regarding the generation of electricity, PV modules acting as shading devices produced electricity supply meeting around 45% and 30% of energy demand of a 3-room public housing apartment in Singapore (3345 kW h) (Energy Market Authority, 2018) on east−west and north−south orientations, respectively. East and west facades produce 1.7−3 times more electricity than north and south facades, respectively.

The food and energy produced on the facades of the T^2 Lab, although not satisfying the full demand of a typical apartment in the tropics, do complement other urban farming activities ranging from small private gardens to public intensive farming facilities at district scale. Therefore the cultivation of vegetable and fruits on building facades contributes to food security and urban resilience in an uncertain future scenario of more acute climate change impacts and global health crises.

11.5.4 System's social acceptance

Several studies have indicated the significance of social acceptance as the key factor of success of urban and VF projects (Kalantari, Tahir, Joni, & Aminuldin, 2018; Specht, Siebert, & Thomaier, 2016; Specht, Weith, Swoboda, & Siebert, 2016). Regarding family farming on building facades, Kosorić, Huang, Tablada, Lau, and Tan (2019) conducted a survey among residents living in public housing flats in Singapore. The survey examined potential users' attitudes toward gardening and the PF concept as well as their preferences related to aesthetics, use and maintenance of the new concept aiming to turn building facades into food and electricity producers. The survey results confirmed that the developed concept adequately fits the needs of public housing residents, since the majority of respondents were willing to engage in outdoor gardening activities, preferably on their own floor, rather than on the building roof—specifically in small-scale VF. The costs and safety were identified as the most pressing concerns.

The second survey performed by Tablada, Kosorić, Huang, Lau, and Shabunko (2020) involved professionals and experts in the fields of horticulture, agronomy, PV systems and architecture. While receiving a similar positive response, the survey also obtained a more detailed feedback regarding aesthetic, formal and functional aspects of façade components, including PV modules, planters, safety grills, as well as operation and accessibility of farming systems. The results indicate a certain distrust regarding the aesthetical value of PV modules, coinciding with the problem already pointed out in the relevant literature (Ballif, Perret-Aebi, Lufkin, & Rey, 2018; Heinstein, Ballif, & Perret-Aebi, 2013; Wall et al., 2012). Experts also indicate the potentials for improvement and encourage further development of the system in terms of functionality, maintenance and access to the plants (Tablada et al., 2020).

The insights obtained from the two surveys and measurements could help further the design of PFs, that is, tailor such systems to users' needs and make them comply with productivity, aesthetics and other architectural qualities (Fig. 11.11).

Figure 11.11 Artistic impression of the improved productive façade (PF) design for Housing & Development Board (HDB) residential buildings in Singapore. Left: north orientation; right: west orientation.
Source: Tablada, A., Kosorić, V., Huang, H., Lau, S.S.−Y., & Shabunko, V. (2020). Architectural quality of the productive facades (PFs) integrating photovoltaic (PV) and vertical farming (VF) systems: Survey among experts in Singapore, *Frontiers of Architectural Research*, 9(2), 301−318. https://doi.org/10.1016/j.foar.2019.12.005.

11.6 Limitations and future directions of farming facades

Although BIA and VF are implemented worldwide, with numerous explorations underway, their application on building facades is still in a nascent stage. On the one hand, this may be the result of lower investment cost of VF application inside decommissioned or abandoned buildings or on rooftops in comparison with building façade adaption. On the other hand, it may reflect a relatively low interest of companies and households in implementing farming activities in their buildings or apartments; mostly due to the still low prices of fresh vegetables in the market (with carbon emission externalities not included in the food chain).

Apart from financial risks, several limitations and issues still need to be solved to enable extensive implementation of advanced VF technologies on facades of office and residential buildings. One of the issues is related to potential contamination of crops exposed to road pollution. However, according to Koski (2013) vegetable leaves do not absorb airborne heavy metals and particulates. The crops in proximity of highways can be affected by toxic tyre dust produced by fast running cars, if they are not inside of a greenhouse or other protected installation. Hence, vegetable leaves should always be washed carefully before sale or consumption to remove all particles.

Other issues are related to the control of pests and birds damaging the crops. The solution given in traditional agriculture is the use of local plants with a particular fragrance toward off harmful insects and other pests. Another solution for those cases where planters are exposed to the exterior could be to cover the space where plants grow with a transparent net or plastic which also prevent birds from eating the crops. This exterior protection can also be a solution for strong wind and heavy rain, which are typical conditions of seasonal storms in most parts of the world and pose a concern for potential farmers on building facades.

Regarding maintenance and safety, case-specific measures should be taken according to the type of building. For high-rise buildings, indoor systems behind the glazing façade are recommended. For mid-rise and low-rise buildings, indoor or outdoor VF systems can be implemented. For outdoor VF systems an automated irrigation system is recommended in addition to the incorporation of a fix or movable access structure assuring safety to operators conducting several farming activities from substratum replacement to planting and harvesting.

As in the case of green façade technology, VF on facades would require an evolution of building envelope design to allow an optimal integration of plants, irrigation and maintenance systems that would add aesthetical, environmental, economic and social values to the building. Although 'green' rating systems or energy performance certificates (EPC) such as LEED, BREEAM, BCA Green Mark and others directly or indirectly promote the use of VGS—due to their qualities for insulation, water retention, biodiversity, etc., there is no mention about the use of farming on facades. Therefore building regulations and EPC may incentivize more directly the use of building facades for VF independently if the building institution or user actively participates in farming management.

Further studies and implementation experiences are needed to respond to current reservations and uncertainties stemming from actual problems. Also, education and technology advances are required for higher acceptance and dissemination of VF on facades in office, commercial and residential buildings.

11.7 Conclusion

The VF on facades and the concept of PF in which food and energy-harvesting installations substitute or complement other building envelope elements is a promising design direction in line with the needed deep transformation of conventional agriculture and urban paradigms with high carbon footprint. The main advantage of VF systems on facades is the direct use of solar energy for the photosynthetic process in contrast with the additional energy used in other VF applications inside buildings or in green houses. Moreover, the integration of food and PV systems on facades does not only contribute to energy and food self-sufficiency in buildings but also ameliorates indoor thermal and visual conditions, reducing the overall carbon footprint in the building sector.

Emerging VF studies, experiments and actual applications on building facades have opened a new field of research, proving that producing vegetables on facades is feasible under favourable urban conditions with incident solar radiation according to crop requirements. As with the integration of renewable energy in buildings and urban areas, where a combination of sources and technologies is used, there is not a single sustainable solution to the problem of food supply in cities. A combination of importing certain products from peri-urban areas with the installation of farms on empty plots, of VF systems inside decommissioned industrial or office buildings, or inside new buildings or on rooftops and facades is the most feasible way to achieve food autonomy in urban areas and to reduce the ecological footprint of agriculture.

Experiences and lessons from the application of VGS are important for the success and dissemination of VF on facades. However, further research endeavours, including pilot projects and testing, are needed to overcome financial and operational challenges of VF on facades. Future studies should necessarily include LCA focusing on comprehensive assessment of impacts in terms of all three sustainability dimensions (environmental, economic and social) of façade systems. Further, an extensive multidisciplinary cooperation, including whole systems design approach (Blizzard & Klotz, 2012), could facilitate the development of site- and climate-specific façade prototypes able to demonstrate the benefits of VF on facades. Finally, the synergy of research and practice, together with the work in the domain of social values, could help this promising design concept to take root and become widespread. Implementation of VF on facades through innovative building skin solutions, not only in individual buildings, but more importantly across various scales, including whole communities and cities, could help achieve future urban food security, thus contributing to urban resilience and enhancing urban health and well-being.

Acknowledgements

The research related to the productive facades and their implementation at the T^2 Lab was funded by the City Developments Limited (CDL) (SGH Grant Call/Project 1), Singapore. We also acknowledge the collaboration of Stephen Siu-Yu Lau, Siu-Kit Lau, Yuan Chao, Huang Huajing, Ian Kevin Chaplin and Shi Xuepeng, from the Department of Architecture of the National University of Singapore (NUS), Hugh Tan, Song Shuang and the team from the Department of Biological Sciences, NUS, Veronika Shabunko, Thomas Reindl and the team from the Solar Energy Research Institute of Singapore (SERIS) and to Fadhlina Suhaimi from the Agri-Food and Veterinary Authority (AVA) of Singapore. The farming and irrigation systems were partially financed by UNISEAL.

References

Akinwolemiwa, O. H., Bleil de Souza, C., De Luca, L. M., & Gwilliam, J. (2018). Building community-driven vertical greening systems for people living on less than £1 a day: A case study in Nigeria. *Building and Environment*, *131*, 277−287. Available from https://doi.org/10.1016/j.buildenv.2018.01.022.

Al-Chalabi, M. (2015). Vertical farming: Skyscraper sustainability? *Sustainable Cities and Society*, *18*, 74−77. Available from https://doi.org/10.1016/j.scs.2015.06.003.

Al-Kodmany, K. (2018). The vertical farm: A review of developments and implications for the vertical city. *Buildings*, *8*(2), 24. Available from https://doi.org/10.3390/buildings8020024.

Armanda, D. T., Guinée, J. B., & Tukker, A. (2019). The second green revolution: Innovative urban agriculture's contribution to food security and sustainability—A review. *Global Food Security*, *22*, 13−24. Available from https://doi.org/10.1016/j.gfs.2019.08.002.

Ballif, C., Perret-Aebi, L.-E., Lufkin, S., & Rey, E. (2018). Integrated thinking for photovoltaics in buildings. *Nature Energy*, *3*, 438−442. Available from https://doi.org/10.1038/s41560-018-0176-2.

Beacham, A. M., Vickers, L. H., & Monaghan, J. M. (2019). Vertical farming: A summary of approaches to growing skywards. *The Journal of Horticultural Science and Biotechnology*, *94*(3), 277−283. Available from https://doi.org/10.1080/14620316.2019.1574214.

Blizzard, J. L., & Klotz, L. E. (2012). A framework for sustainable whole systems design. *Design Studies*, *33*(5), 456−479. Available from https://doi.org/10.1016/j.destud.2012.03.001.

Carolan, M. (2020). Urban Farming Is Going High Tech. *Journal of the American Planning Association*, *86*(1), 47−59. Available from https://doi.org/10.1080/01944363.2019.1660205.

Chatterjee, A., Debnath, S., & Pal, H. (2020). Implication of urban agriculture and vertical farming for future sustainability. Chapter. In S. Shekhar, et al. (Eds.), *Urban horticulture—Necessity of the future*. Intechopen. Available from http://doi.org/10.5772/intechopen.91133.

Chu, L. M. (2014). Vertical greening in a subtropical city—System and species selection. In *1st international conference on green wall—Meeting the challenge of a sustainable urban future: The contribution of green walls. 4−5 Sept 2014*, Staffordshire University, Stoke-on-Trent, UK.

Cockrall-King, J. (2012). *Food and the city: Urban agriculture and the new food revolution*. New York: Prometheus Books.

Cooper-Marcus, C., & Barnes, M. (1999). *Healing gardens: Therapeutic benefits and design recommendations*. New York: John Wiley.

Department of Economic and Social Affairs, United Nations. (2015). *Sustainable development goals*. New York: United Nations, http://sdgs.un.org/goals. Accessed 20.01.21.

Despommier, D. (2009). The rise of vertical farms. *Scientific American*, *301*(5), 32−39. Available from https://doi.org/10.1038/scientificamerican1109-80, 80−7.

Despommier, D. (2010). *The vertical farm: Feeding the world in the 21st Century*. New York: Macmillan.

Dorais, M. (2003). The use of supplemental lighting for vegetable crop production: Light intensity, crop response, nutrition, crop management, cultural practices. In *Proceedings of the Canadian Greenhouse Conference, Toronto, ON, Canada, 9−10 October 2003*.

Draper, C., & Freedman, D. (2010). Review and analysis of the benefits, purposes, and motivations associated with community gardening in the United States. *Journal of Community Practice*, *18*(4), 458−492. Available from https://doi.org/10.1080/10705422.2010.519682.

Energy Market Authority (EMA) (2018). https://www.ema.gov.sg/cmsmedia/Publications_and_Statistics/Statistics/8RSU.pdf. Accessed on 19.02.21.

Farming Architects (2019). http://farmingarchitects.com. Accessed 15.11.19.

Food and Agriculture Organization of the United Nations (FAO) (2006). *Policy brief: Food security*. June, Issue 2. http://www.fao.org/fileadmin/templates/faoitaly/documents/pdf/pdf_Food_Security_Cocept_Note.pdf. Accessed 18.02.21.

Glenn, E. P., Cardran, P., & Thompson, T. L. (1984). Seasonal effects of shading on growth of greenhouse lettuce and spinach. *Scientia Horticulatura*, *24*(3−4), 231−239. Available from https://doi.org/10.1016/0304-4238(84)90106-7.

Gorgolewsky, M., Komisar, J., & Nasr, J. (2011). *Carrot city: Creating places for urban agriculture*. New York: The Monacelli Press.

Gould, D., & Caplow, T. (2012). Chapter 8: 'Building-integrated agriculture: A new approach to food production. In F. Zeman (Ed.), *Metropolitan sustainability: Understanding and improving the urban environment*. Cambridge: Woodhead Publishing Limited.

Greenbelly Project (2020). http://www.greenbelly.org/. Accessed 12.06.20.

Harari, Y. N. (2014). *De animales a dioses: Breve historia de la humanidad* (Spanish edition). Barcelona: Penguin Random House Grupo Editorial.

Harris, E. (2009). The role of community gardens in creating healthy communities. *Australian Planner*, *46*(2), 24−27. Available from https://doi.org/10.1080/07293682.2009.9995307.

Heinstein, P., Ballif, C., & Perret-Aebi, L.-E. (2013). Building integrated photovoltaics (BIPV): Review, potentials, barriers and myths. *Green*, *3*(2), 125−156. Available from https://doi.org/10.1515/green-2013-0020.

Jenkins, A. (2018). Building integrated technical food systems (PhD thesis). Queen's University Belfast.

Jenkins, A., Keeffe, G., Hall, N. (2014). Façade farm: Solar mediation through food production. In *Proceedings EuroSun 2014*, Aix-les-Bains.

Kalantari, F., Tahir, O. M., Joni, R. A., & Aminuldin, N. A. (2018). The importance of the public acceptance theory in determining the success of the vertical farming projects. *Management Research and Practice*, *10*(I), 5−16. Available from https://www.ceeol.com/search/article-detail?id = 670355.

Kim, C. G. (2010). *The impact of climate change on the agricultural sector: Implications of the agro-industry for low carbon, green growth strategy and roadmap for the East Asian Region*. https://www.unescap.org/sites/default/files/5.%20The-Impact-of-Climate-Change-on-the-Agricultural-Sector.pdf. Accessed 18.02.21.

Kiss + Cathcart Architects. (2020). http://www.kisscathcart.com/integrated_agriculture.html. Accessed 02.05.20.

Koski, H. (2013). *Guide to urban farming in New York State*. New York: Cornell Small Farms Program. http://nebeginningfarmers.org/publications/farming-guide/.

Kosorić, V., Huang, H., Tablada, A., Lau, S.-K., & Tan, H. T. W. (2019). Survey on the social acceptance of the productive façade concept integrating photovoltaic and farming systems in high-rise public housing blocks in Singapore. *Renewable and Sustainable Energy Reviews*, *111*, 197−214. Available from https://doi.org/10.1016/j.rser.2019.04.056.

Lang, T. (2006). *Locale/global (food miles)* (pp. 94−97). Bra: Slow Food.

Ling, T. Y., & Chiang, Y. C. (2018). Well-being, health and urban coherence-advancing vertical greening approach toward resilience: A design practice consideration. *Journal of Cleaner Production*, *182*, 187−197. Available from https://doi.org/10.1016/j.jclepro.2017.12.207.

Mack, M. (2014). *How can we ensure food security for the future?* World Economic Forum. https://www.weforum.org/agenda/2014/01/can-feed-population-future/. Accessed 28.01.21.

Matharu, J. (2016). Symbiosis in city: How can vertical farming be integrated in a high rise mixed use development? (MArch thesis). Auckland City: Unitec Institute of Technology.

Opricovic, S., & Tzeng, G.-H. (2004). Compromise solution by MCDM methods: A comparative analysis of VIKOR and TOPSIS. *European Journal Operational Research*, *156*(2), 445−455. Available from https://doi.org/10.1016/S0377-2217(03)00020-1.

Opricovic, S., & Tzeng, G.-H. (2007). Extended vikor method in comparison with outranking methods. *European Journal of Operational Research*, *178*(2), 514−529. Available from https://doi.org/10.1016/j.ejor.2006.01.020.

Orpiz, I., Berges, R., Piorr, A., & Krikser, T. (2016). Contributing to food security in urban areas: Differences between urban agriculture and peri-urban agriculture in the Global North. *Agriculture Human Values*, *33*, 341−358. Available from https://doi.org/10.1007/s10460-015-9610-2.

Palliwal, A., Song, S., Tan, H. T. W., & Biljecki, F. (2021). 3D city models for urban farming site identification in buildings. *Computers, Environment and Urban Systems*, *86*, 101584. Available from https://doi.org/10.1016/j.compenvurbsys.2020.101584.

Peck, S. W., Callaghan, C., Kuhn, M. E., & Bass, B. (1999). *Greenbacks from green roofs: Forging a new industry in Canada* (p. 78) CMHC.

Peña Díaz, J. (2014). UPA in Havana: On recent developments. In Viljoen, A., & Bohn, K. (Eds.), *Second nature urban agriculture: Designing productive cities*.

Pérez, G., Coma, J., Sol, S., & Cabeza, L. F. (2017). Green façade for energy savings in buildings: The influence of leaf area index and façade orientation on the shadow effect. *Applied Energy*, *187*, 424−437. Available from https://doi.org/10.1016/j.apenergy.2016.11.055.

Perini, K., Ottelé, M., Haas, E. M., & Raiteri, R. (2013). Vertical greening systems, a process tree for green facades and living walls. *Urban Ecosystems*, *16*(2), 265−277. Available from https://doi.org/10.1007/s11252-012-0262-3.

Perini, K., & Rosasco, P. (2013). Cost−benefit analysis for green facades and living wall systems. *Building and Environment*, *70*, 110−121. Available from https://doi.org/10.1016/j.buildenv.2013.08.012.

Plantagon (2021). http://plantagon.com/. Accessed 19.02.21.

Poulsen, M. N., Neff, R., & Winch, P. J. (2017). The multifunctionality of urban farming: Perceived benefits for neighbourhood improvement. *The International Journal of Justice and Sustainability*, *22*(11), 1411−1427. Available from https://doi.org/10.1080/13549839.2017.1357686.

Radić, M., Brković Dodig, M., & Auer, T. (2019). Green facades and living walls—A review establishing the classification of construction types and mapping the benefits. *Sustainability*, *11*, 4579. Available from https://doi.org/10.3390/su11174579.

Roehr, D. & J. Laurenz. (2008). Living skins: Environmental benefits of green envelopes in the city context. In *Proceeding of Eco Architecture II, WIT Press. Southampton, England* (pp. 149−158).

Rogers, S.A., 2018. Blind Building Facades Become Urban Farms with Scalable Scaffolding System. Available online: https://weburbanist.com/2018/09/17/blind-building-facades-become-urban-farms-with-scalable-scaffolding-system/ (accessed 20.03.2021).

Schiller, L. (2017). *Is my plant getting enough light?* Available online: http://www.ceresgs.com/is-my-plantgetting-enough-light/. Accessed 20.09.17.

School of Design and Environment, NUS. (2021). https://www.sde.nus.edu.sg/arch/research/technologies/tropical-technologies-laboratory/. Accessed 18.02.21.

Song, X. P., Tan, H. T. W., & Tan, P. Y. (2018). Assessment of light adequacy for vertical farming in a tropical city. *Urban Forestry & Urban Greening, 29*, 49−57. Available from https://doi.org/10.1016/j.ufug.2017.11.004.

Specht, K., Weith, T., Swoboda, K., & Siebert, R. (2016). Socially acceptable urban agriculture businesses. *Agronomy for Sustainable Development, 36*, 17. Available from https://doi.org/10.1007/s13593-016-0355-0.

Specht, K., Siebert, R., & Thomaier, S. (2016). Perception and acceptance of agricultural production in and on urban buildings (ZFarming): A qualitative study from Berlin, Germany. *Agriculture and Human Values, 33*(4), 753−769. Available from https://doi.org/10.1007/s10460-015-9658-z.

Sutton, R. (2014). Aesthetics for green roofs and green walls. *Landscape Architecture Program: Faculty Scholarly and Creative Activity, 19*. Available from http://digitalcommons.unl.edu/arch_land_facultyschol/19.

Szabo, S. (2016). Urbanisation and food insecurity risks: Assessing the role of human development. *Oxford Development Studies, 44*(1), 28−48. Available from https://doi.org/10.1080/13600818.2015.1067292.

Tablada, A., & Zhao, X. (2016). Sunlight availability and potential food and energy self-sufficiency in tropical generic residential districts. *Solar Energy, 139*, 757−769. Available from https://doi.org/10.1016/j.solener.2016.10.041.

Tablada, A., Kosorić, V., Huajing, H., Chaplin, I. K., Lau, S. K., Yuan, C., & Lau, S. S. Y. (2018). Design optimisation of productive facades: Integrating photovoltaic and farming systems at the Tropical Technologies Laboratory. *Sustainability, 10*(10), 3762. Available from https://doi.org/10.3390/su10103762.

Tablada, A., Kosorić, V., Huang, H., Lau, S. S. −Y., & Shabunko, V. (2020). Architectural quality of the productive facades (PFs) integrating photovoltaic (PV) and vertical farming (VF) systems: Survey among experts in Singapore. *Frontiers of Architectural research, 9*(2), 301−318. Available from https://doi.org/10.1016/j.foar.2019.12.005.

Taghizadeh-Hesary, F., Rasoulinezhad, E., & Yoshino, N. (2019). Energy and food security: Linkages through price volatility. *Energy Policy, 128*, 796−806. Available from https://doi.org/10.1016/j.enpol.2018.12.043.

Tan, P., Chiang, K., Chan, D., Wong, N., Chen, Y., Tan, A., & Wong, N. (2009). *Vertical greenery for the tropics*. Singapore: National Parks Board, National Parks Board Headquarters.

UN Department of Economic and social Affairs. (2019). *World urbanization prospects: The 2018 revision*. New York: United Nations. https://population.un.org/wup/publications/files/wup2018-highlights.pdf. Accessed 22.04.20.

United States Environmental Protection Agency (EPA) (2017). *Climate change impacts: Climate impacts on agriculture and food supply*. https://19january2017snapshot.epa.gov/climate-impacts/climate-impacts-agriculture-and-food-supply. Accessed 18.02.21.

Viljoen, A., Bohn, K., & Howe, J. (Eds.), (2005). *Continuous productive urban landscapes: Designing urban agriculture for sustainable cities* (1st ed.). Oxford: Architectural Press.

Wall, M., Probst, M. C. C., Roecker, C., Dubois, M. C., Horvat, M., Jorgensen, O. B., & Kappel, K. (2012). Achieving solar energy in architecture—IEA SHC Task 41. *Energy Procedia, 30*. Available from https://doi.org/10.1016/j.egypro.2012.11.138, 1250−1250.

Wong, N. H., Kwang Tan, A. Y., Tan, P. Y., Chiang, K., & Wong, N. C. (2010). Acoustics evaluation of vertical greenery systems for building walls. *Building and Environment, 45*(2), 411−420. Available from https://doi.org/10.1016/j.buildenv.2009.06.017.

Wong, N. H., Tan, A., Yok Tan, P., Sia, A., Chung, A., & Wong, N. (2010). Perception studies of vertical greenery systems in Singapore. *Journal of Urban Planning and Development*, *136*(4), 330–338. Available from https://doi.org/10.1061/(ASCE)UP.1943-5444.0000034.

Wong, N. H., Tan, A. Y. K., Tan, P. Y., & Wong, N. C. (2009). Energy simulation of vertical greenery systems. *Energy and Buildings*, *41*(12), 1401–1408. Available from https://doi.org/10.1016/j.enbuild.2009.08.010.

Wong, N. H., Tan, A. Y. K., Yu, C., Sekar, K., Tan, P. Y., Chan, D., ... Wong, N. C. (2010). Thermal evaluation of vertical greenery systems for building walls. *Building and Environment*, *45*(3), 663–672. Available from https://doi.org/10.1016/j.buildenv.2009.08.005.

Interactive media facades—research prototypes, application areas and future directions

Martin Tomitsch[1,2]
[1]School of Architecture, Design and Planning, The University of Sydney, Sydney, NSW, Australia, [2]Central Academy of Fine Arts, Beijing Visual Art Innovation Institute, Beijing, P.R. China

12.1 Introduction

Media facades have become a key feature of cities across the globe, animating the urban skyline at night. Their pervasiveness has been spurred by technological progress and declining costs of lighting technologies. Their success is underpinned by architectural and urban design principles that equally consider the elements of place, people and technology (Foth, Choi, & Satchell, 2011). Indeed, one of the first interactive media facades, the Blinkenlights project (Haeusler, 2009, p. 108), was launched as a socio-technological experiment. The project was initiated by the Chaos Computer Club in Berlin, Germany, in 2001. Using the windows of an empty high-rise building at Alexanderplatz, the building's façade was converted into an 18-pixel wide and 8-pixel tall display that was capable of showing animated low-resolution graphics (Fig. 12.1). The installation allowed people to play the famous computer game Pong on the façade by calling a number and consequently interacting with the keys on their mobile phone to control the paddle shown on the façade. Using a web portal, people could also create their own animation to be displayed on the building's façade.

Today more advanced networking and sensing technologies allow for more complex forms of interaction with media facades of all shapes, types and scales. At the same time, the media façade technology industry has recognized the value of interactive features as a way to innovate their products and to create more engaging media façade installations. However, adding interactivity is a challenging process as there is no standardized approach for what form these interactive features would take. While innovations in noninteractive media facades are largely driven by technological challenges—for example, how to increase the resolution by being able to address a higher number of light-emitting diodes (LEDs) with the same number of controllers—adding interactivity requires a more multidisciplinary approach that goes beyond technological considerations. Getting the interaction design right is challenging as it requires a deep understanding of the sociocultural context, which is unique to the media façade's location (Vande Moere & Wouters, 2012). While it is increasingly more feasible and affordable to add interactive features, the success

Rethinking Building Skins. DOI: https://doi.org/10.1016/B978-0-12-822477-9.00019-X
© 2022 Elsevier Inc. All rights reserved.

Figure 12.1 The Blinkenlight project installed at the 'Haus des Lehrers' in Berlin, Germany, used 144 lights to turn the building's façade into a giant interactive low-resolution media screen. *Source:* Courtesy: Tim Pritlove via Wikimedia.

of interactive media facades relies on whether people can meaningfully engage with those interactive features.

This chapter introduces different approaches for creating dynamic building skins that can respond to external stimuli from within the building's environment. These approaches are demonstrated through media façade projects spanning from technologies that allow passersby to directly interact with content displayed on the façade to buildings that adapt their visual appearance based on activities taking place inside the building. Through an analysis of state-of-the-art projects and research studies, the chapter offers a practical as well as a conceptual contribution. As a practical contribution, the chapter presents a new framework to guide the design and implementation of interactive media facades. As a conceptual contribution, the chapter outlines criticalities, limitations, challenges and opportunities for future research and practice in this domain.

The remainder of the chapter is organized as follows. First, it provides an account of the evolution of media facades, offering definitions of terms used in this domain. The second section discusses recent innovations in practice and research. The third section introduces interaction concepts and technologies. The fourth section outlines approaches for designing and evaluating interactive systems and how these approaches can be applied to interactive media facades. The fifth section presents a framework to offer guidance on the implementation of interactive media façade projects. The chapter ends with a reflection of future directions for this domain.

12.2 Interactive architecture, urban screens and media facades

The concept of interactive architecture provides a rich foundation for the field of interactive media facades. Predating media facades, interactive architecture can be traced back to 1958 and the interactive installations displayed as part of pavilions for the Brussels Expo. These early explorations of creating architectural environments that react to people and their interactions used information technologies as a building material to create real-time changes through means such as light, sound or even moving elements of the building. To date, interactive architecture installations are often explorative and implemented for a predetermined time span, for example, as part of art or architecture festivals.

Urban screens are related to media facades but distinctive through the way they are integrated into the built environment. As defined in previous publications (Tomitsch, McArthur, Haeusler, & Foth, 2015; Tscherteu & Tomitsch, 2011), urban screens typically take the form of mid-to-large-scale screens that can either be freestanding or attached to a building façade. Urban screens that appear as freestanding, independent architectonic elements take on the single purpose of communicating media content (McQuire, Martin, & Niederer, 2009). These kinds of urban screens are typically used in urban environments for displaying advertizing, wayfinding or other types of informational messages.

Media facades, as a concept, first emerged in Richard Roger and Renzo Piano's competition entry for the Centre Pompidou in Paris in 1971 (Silver, 1994). The competition entry included a giant media screen displaying cultural and political news as well as electronic messages about events at the centre (Haeusler, 2009, p. 21). Due to the technological complexity of the proposed concept given the state of technology at the time, the envisioned media façade installation was never realized. Subsequent advancements in the production of LED technologies and their uptake by experimental architecture firms as a way to transform buildings at night saw a surge of media façade installations during the first decade of the 21st century. Because of the costly installation and operation of media facades, they initially found their application on flagship buildings for large corporations, such as banks and insurance companies, before becoming more widespread and extending their application to shopping centres and cultural centres (Haeusler, Tomitsch, & Tscherteu, 2012; Haeusler, 2009). Strategically, developers and organizations invested in media facades as a way to promote their brand and to project innovation. Architecturally, media facades enable the aesthetic transformation of a built structure at night. A significant step in the evolution of media facades and their relevance, use and acceptance in architectural practice was marked by their inclusion in the catalogue of the 14th Venice Architecture Biennale, which listed media facades as 1 of 12 influential types of building facades (Zaera-Polo, Trüby, Koolhaas, & Boom, 2014).

Technologically, media facades are more closely integrated with the building compared to urban screens and follow the typology of the building, creating a new hybrid structure (Haeusler, 2009; Kronhagel, 2010). Media facades commonly use individually controllable LEDs, which form the pixels of the larger media façade screen. However, other technologies have been successfully used for turning a building façade into a media screen, such as installing lights behind windows (Fig. 12.1), using a grid of fluorescent bulbs as seen in the iconic Kunsthaus Graz in Austria (Haeusler, 2009, p. 99) or integrating physical actuators into the façade to create a three-dimensional image (Fig. 12.2). Media façade displays tend to

Figure 12.2 The MegaFaces media façade is an example of a kinetic media façade that uses physical actuators to represent three-dimensional images.
Source: Reproduced with permission from Asif Khan, Courtesy: Asif Khan, Photographer: Hufton + Crow.

feature a lower resolution compared to urban screens, meaning that the individual pixels (eg, represented by LEDs) are further apart from each other. The distance between pixels is referred to as pixel pitch and determines the type of content that can be displayed and the distance from which the content is readable. Thus media facades can be described as 'digital public screens with arbitrary form factors and of arbitrary resolution' (Gehring & Wiethoff, 2014).

Costs associated with installation and operation (which both increase with the number of pixels used in the media façade) are determining factors when it comes to deciding on the resolution of a media façade. However, opting for a low resolution may also be a deliberate choice to create an aesthetic effect. The unusual shape and low resolution make it more difficult to display text or images compared to urban screens (Offenhuber & Seitinger, 2014). As a consequence, media facades often feature abstract content, for example, to show light patterns (Fig. 12.3), or to visually support or manipulate the perceived geometry of the building at night (Fig. 12.4). Local regulations also impact both the implementation and content of media facades. For example, in some cities, media facades need to be a minimum distance above ground to control their potential impact on traffic; in some cases, traffic authorities also impose restrictions regarding the frame rate for videos or animated content displayed on the façade.

While content is key in urban screens, content in media facades is sometimes (falsely) perceived to play a less critical role during the design phase of a building or its façade. This can lead to installations, where the content creation becomes an afterthought, potentially resulting in unforeseen costs due to the need to develop and regularly update the displayed content. Creating customized content is complex as it needs to respond to the irregular shape of the media façade, making it difficult to simply transfer existing content (Gehring, Hartz, Löchtefeld, & Krüger, 2013). Seen as a media surface, media facades consist of architecture, media and art and

Figure 12.3 The low-resolution media façade installed at the Ars Electronica Centre in Linz, Austria, activates the building at night through abstract light patterns that wrap the building in a dynamic digital layer.
Source: Courtesy: Ars Electronica.

Figure 12.4 The Uniqa tower in Vienna, Austria, features a wide-meshed grid of LEDs that transforms the perceived physical building structure at night—generating a 'dynamic form of architecture' (Haeusler, 2009, p. 137). *LED*, Light-emitting diode.
Source: Reproduced with permission from Mader Stublic Wiermann.

as such draw on the vocabulary of media communication (Gasparini, 2017). Furthermore, the content needs to be able to visually work on the resolution offered through the façade and take into account views from close-up as well as from a distance (Haeusler, 2009). Several cities in China have extended the content to span multiple buildings, turning the cityscape into a massive screen to display, for example, images that are connected to local cultural events.

As another example for connecting multiple media facades, the Times Square Moment media project displays public artwork across multiple screens embedded in the façade of New York Times Square's buildings every night just before midnight (Haeusler et al., 2012, p. 142). The Times Square Moment initiative demonstrates how media screens can be used beyond advertizing, reducing visual pollution while creating a spectacle moment that has the potential to increase footfall and contributing cultural value. Indeed, there is a level of corporate responsibility that comes with turning an entire building into a large-scale media screen. To ensure that the displayed content supports this corporate responsibility and considers the impact of the media façade on its urban environment, it may be useful to determine content rules and to set up a curatorial or editorial board, as was the case for the media façade installed at the Bayer AG corporate office in Leverkusen, Germany (Kronhagel, 2010; Scully & Mayze, 2018).

Following the definitions outlined in this section, rectangular high-resolution media facades, such as those found at the iconic New York Times Square, represent urban screens rather than media facades. However, the distinction is somewhat blurred given that the primary purpose of these large-scale, high-resolution media screens is to communicate content, in many cases in the form of advertising (a characteristic of urban screens), while the technological layer is closely integrated with the physical building façade (a characteristic of media facades). This observation demonstrates that not every project will fall into one category or the other. If the underlying rationale behind installing a media screen on a building façade is known, the following distinguishing classification can be used. If the media surface has a function in addition to that of being a display, it can be considered a media façade. If that is not the case, it should be considered an urban screen. Based on the examples provided thus far, additional functions beyond being a display include altering the social dynamics of an urban space (eg, Blinkenlights), connecting people (eg, MegaFaces) and changing the perceived physical architecture of a building (eg, Uniqa tower). As the Times Square Moment shows, this classification may change over time, with the media surface transitioning from acting as an urban screen to becoming a media façade and vice versa.

In more practical terms the distinction between media facades and urban screens is useful as it offers input on the design, implementation and curation of media surfaces—both in terms of how the installation can be technically approached and determining its purpose, which drives decisions about aspects such as content and interaction.

12.3 Innovations in interactive media facades in practice and research

Innovations in LED manufacturing have led to new forms of media facades over the last two decades. The domain has seen rapid development and uptake of media facades as a building component since the first experiments through projects, such as Blinkenlights, almost 20 years ago. However, a review of projects across the globe (Haeusler et al., 2012) and projects nominated as finalists for the biennial Media Architecture Awards between 2010 and 2018 shows that much of this progress is in relation to the technology and form of media facades. For example, this includes the integration of solar panels (eg, in the GreenPix project in Beijing, China, Haeusler, 2009, p. 166), the use of transparent media surfaces (eg, in the Klubhaus St. Pauli in Hamburg, Germany, Fig. 12.5), and the appropriation of mechanical elements to create a kinetic façade for visualizing three-dimensional images (eg, in the MegaFaces installation in Sochi, Russia, Fig. 12.2). Comparatively only few media facades have been implemented and documented that make use of interactive features. This section reviews some of those key projects, linking them to transformative research studies involving interactive media facades.

One of the early examples of a commercial large-scale media façade that uses sensors to control the content displayed on the façade in real time is the National Football Stadium, in Lima, Peru. The building, which was opened in 2011, features a permanent LED-based media façade, which reflects noise levels inside the stadium (Haeusler et al., 2012, p. 135). A network of customized sound level sensors deployed along the roofline of the stadium captures the crowd's noise levels during games. These data are then mapped in real time to colour patterns using a custom-developed software algorithm, performing a series of comparative mathematical calculations to create the mood state of the crowd inside the stadium. The system was designed to capture the atmosphere during sports events, communicating it to the wider city through its dynamic building skin.

The MegaFaces media façade (Fig. 12.2), which was installed for the 2014 Winter Olympics in Sochi, Russia, allowed people to directly alter its content by

Figure 12.5 The Klubhaus St. Pauli in Hamburg, Germany, combines different media façade technologies, including a transparent LED solution that transforms the glass elevator into a media display. *LED*, Light-emitting diode.
Source: Courtesy: ONLYGLASS, Photographer: Christian O. Bruch.

sending photos to the system. Using photo booths located inside the building and across Russia, people could take selfies, which were then displayed on the media façade. A custom-developed software algorithm translated the selfie images into three-dimensional large-scale portraits. The media facades spanned 18 by 8 m and consisted of 10,477 telescopic actuators arranged in a trigonal grid (Hespanhol, Haeusler, Tomitsch, & Tscherteu, 2017, p. 195). Each actuator was able to extend by up to 2 m, creating a kinetic building skin. LEDs were used to further enhance the visual imagery displayed on the building façade.

Over the last decade, several research studies investigated different techniques for enabling people to manipulate or control the content displayed on media facades. In many cases, these studies involved the creation of temporary installations—either creating a temporary media façade or appropriating an existing media façade. One of the first accounts of an interactive media façade in the academic literature is the Aarhus by Light installation (Brynskov et al., 2009). The installation allowed people to control digital avatars displayed on a 180 square metre LED screen mounted behind the glass façade of the Concert Hall Aarhus in Denmark. Three cameras were used to cover three interaction zones mapped to three parts of the façade.

Using the existing media façade of the Galeria de Arte Digital in São Paulo, Brazil, the Smart Citizen Sentiment Dashboard research project (Behrens, Valkanova, gen. Schieck, & Brumby, 2014) involved a large-scale visualization displayed on the façade. The media façade spanned three sides of the building and consisted of approximately 26,000 LED clusters. Using the Smart Citizen Sentiment Dashboard interface, people could vote on five different issues, regarding the environment, mobility, security, housing and public space, through a dial and by swiping a radio-frequency identification card, such as their public transport card, to select one of the issues. The aggregated data were then displayed on the media façade to visualize the general mood of the city and to highlight the points that might need attention from the public authorities.

A research project, carried out by the University of Munich and the German Research Centre for Artificial Intelligence, investigated new forms of interaction with low-resolution media facades using mobile phones. The project team prototyped their proposed solution for the Ars Electronica Center media façade in Linz, Austria (Fig. 12.3). The custom-developed mobile application allowed people to directly interact through live video on their mobile phones with the media façade (Boring et al., 2011). Transforming the media façade into a giant collaborative canvas, people could, for example, collaboratively solve a puzzle or paint on a media façade (Fig. 12.6). The Ars Electronica Centre also features a permanent touch screen terminal located at the nearby riverside, allowing passersby to 'play the façade' by drawing light patterns and connecting it to their mobile music player.

The In The Air Tonight project (Colangelo, 2020, p. 124), developed by art collective Public Visualization Studio, used the permanent media façade of the Ryerson Image Centre in Toronto, Canada. The low-resolution façade consists of 1400 full colour LED fixtures mounted on custom brackets behind

Figure 12.6 A research application developed to evaluate how people could collaboratively interact with a media façade through their mobile phones.
Source: Reproduced with permission from Alexander Wiethoff, Photographer: Martin Knobel.

Figure 12.7 The In The Air Tonight project transformed a permanent media façade to raise awareness about homelessness.
Source: Reproduced with permission from Dave Colangelo/Public Visualization Studio, Photographer: Maggie Chan.

rectangular translucent glass panels backlit by two fixtures each (Hespanhol et al., 2017, p. 111). The media installation responded to Twitter posts with the hashtag 'homelessness', which would trigger a red pulse on the façade (Fig. 12.7). The installation glowed white when a donation was made to a local organization helping people with lived experience of homelessness transition into meaningful employment and stable housing. In between tweets and donations, a blue wave-like animation was displayed, which changed following the wind speed and direction.

These examples from commercial projects, research studies and art installations demonstrate how interactivity can be added in various forms to enhance existing media facades, either on a temporary or permanent basis.

Motivated by technological progress, research and artistic explorations, media façade technology has transcended the boundaries of building facades. Indeed, recent innovative approaches are found in smaller scale experiments and applications. For example, the TetraBIN project (Tomitsch, 2018, p. 122), which was first exhibited as part of the Vivid Sydney light festival in 2014, uses LEDs to transform the exterior surface of a common city bin (Fig. 12.8). By turning the act of disposing rubbish into a playful experience, the project aimed to encourage positive behaviour change and to reduce littering in cities. Another example for using media façade technology in an urban-furniture scale intervention can be found in a new set of public toilets installed in two parks in Tokyo, Japan. The design uses sensor and LED technology to reimagine public toilets (Wong & Enjoji, 2020). The toilet cubicles are made of translucent coloured glass that turns opaque when locked from the inside. The see-through design addresses concerns about cleanliness and allows people to see whether anyone is inside the toilet before entering. Additionally, LEDs are used to illuminate the park at night, improving safety in the area at night and creating an aesthetic effect that aims to transform the perception of public toilets.

Figure 12.8 Exploratory projects, like TetraBIN, demonstrate the applicability of interactive media façade technology to smaller scale urban applications—transforming experiences and activities in cities. TetraBIN uses LEDs and sensors to turn the act of disposing rubbish into a gamified experience. *LED*, Light-emitting diode.
Source: Courtesy Steven Bai, Sam Johnson and Martin Tomitsch, Photographer: Martin Tomitsch.

12.4 Interaction concepts, technologies and applications

Media facades interact with their surroundings in different ways. Gasparini (2014) categorizes these interactions into three groups: the interaction of media facades with the landscape (shaping them as landmarks), their interaction with the city (contributing to the image of a city) and their interaction with human beings (creating immersive environments where 'the passer-by becomes an active part of the new urban performance'). The design of media facades needs to consider all of these interactions.

12.4.1 Interaction concepts

To further segment the interaction between media facades and human beings, Giannetti's (2004) three types of interactivity in media installations—active, reactive and interactive—offer a useful template for determining the type of interactivity in a media façade. These types also correlate with the types of behaviour for media facades introduced by Brynskov, Dalsgaard, and Halskov (2013), which are dynamic, reactive and interactive. The first type, active or dynamic systems, refers to applications where the content, such as a video or animation, is prerecorded. There is some form of control in terms of starting the replay of the content, but once the content is running it is not responding to any input from the user or the environment (except for pausing or stopping the replay).

The second type, reactive systems, refers to applications that reflect external stimuli, such as user behaviour or environmental conditions. This can be seen, for example, in the Peru National Football Stadium, where noise levels within the stadium are used as implicit input to generate the content displayed on the building's media façade. Reactive (or responsive) installations have successfully been used in media architecture projects across research (eg, Hespanhol & Tomitsch, 2012) and practice (Bullivant, 2006).

The third type, interactive systems, refers to systems that allow people to explicitly influence the system's output or to add new information in the case of more complex systems. In the case of media façade installations, this explicit input can, for example, be realized via gestures, voice control or smartphone applications. As demonstrated in the In The Air Tonight project (Fig. 12.7), the different types can also be combined, for example, by triggering content changes based on explicit input from smartphones as well as implicit input triggered by changes in the environment.

12.4.2 Technologies

Technologies used for adding interactivity to media facades depend on the specific application, use case and situation. For example, many experimental projects have used Microsoft's Kinect sensor, a depth camera, which provides depth information and identification of people and their limbs. The Solstice LAMP installation, exhibited during the Vivid Sydney light festival in 2013, used two Kinect sensors to capture the movement of people in front of a skyscraper, which was then transformed into a musical score and animations projected onto the building's façade (Hespanhol, Tomitsch,

Bown, & Young, 2014). Given the temporary nature and large scale of the building, two laser projectors were installed to activate the façade. The Kinect, like all sensors, comes with its limitations. For example, it does not work during daylight and has a limited range of up to about 6 m. As machine learning algorithms are becoming more advanced, it is increasingly possible to capture the poses and motions of passersby with standard image cameras combined with toolkits, such as PhysCap (Shimada, Golyanik, Xu, & Theobalt, 2020). Using camera-based sensors allows for all kinds of interactions, including the use of a physical lighter to trigger a firework animation on a media façade (Hoggenmueller & Wiethoff, 2014).

Interactive installations may also rely on custom-developed mobile applications running on the user's phone (eg, the project developed for the Ars Electronica Center media façade) or custom-developed kiosk-like input mechanisms (eg, the Smart Citizen Sentiment Dashboard). Sensors for reactive installations include microphones (eg, used in Peru's National Football Stadium), motion sensors and so on. For applications responding to environmental data, such as wind conditions (eg, the In The Air Tonight installation) or the local weather, it may not even be necessary to install site-specific sensors as much of this information is available via application programming interfaces.

The different kinds of technologies are summarized in Table 12.1. The choice of sensing technology depends on many factors, including the type of interactivity and

Table 12.1 Technologies used in interactive media façade installations.

Technology	Types and examples
Sensing	Camera (Aarhus by Light, MegaFaces) Depth camera (Solstice LAMP) Noise level metres (Peru National Football Stadium) RFID (Smart Citizen Sentiment Dashboard) Touch screen (Ars Electronica Center) Voice input (Megaphone)
Communication	Local (Peru National Football Stadium) Remote (In The Air Tonight) Flexible (Ars Electronica Centre)
Interface	Active (Ars Electronica Centre, Blinkenlights, MegaFaces, Megaphone) Passive (Peru National Football Stadium) Hybrid (In The Air Tonight)
Data processing	Direct mapping (MegaFaces) Direct control (Aarhus by Light, Blinkenlights, Ars Electronica Centre) Advanced data analytics (Peru National Football Stadium, Smart Citizen Sentiment Dashboard)

RFID, Radio-frequency identification.

environmental conditions. Communication technologies can be implemented using a local approach, where people interact with the media façade while being in close proximity or a remote approach, where people interact over a distance or from a different location altogether. Interface technologies can be active, passive or supporting both. An active interface technology requires a user to perform an explicit input action; a passive interface technology implicitly responds to people's behaviour or other trigger events. Data processing can be in the form of one-to-one mapping, where an input event directly changes the content of a media façade, direct control, where the user input controls an element on the façade, or advanced data analytics to translate user input into a visual output, which is the most abstract form of data processing.

12.4.3 Application areas

Interactivity extends the application areas of media facades, which predominately have been used to create landmarks and to shape the image of a city (Gasparini, 2014), and for corporations and organizations to project an image of being innovative (Haeusler et al., 2012). Four future directions for emerging application areas enabled by interactive features are outlined next.

Raising awareness

Given the large scale of media facades, they offer great opportunities for drawing people's attention to specific topics, which may include exposing people to narratives they may not initially be interested in, thus breaking down filter bubbles (Ergin & Fatah gen. Schieck, 2018; Foth, Tomitsch, Forlano, Haeusler, & Satchell, 2016). Communicating messages on media facades is more difficult compared to urban screens due to their low resolution. However, adding interactivity can be an effective way to engage people through additional channels, and that way, communicating messages on a personal as well as through a publicly amplified channel. For example, the In The Air Tonight project achieves this by linking the media façade visualization to occurrences of a specific hashtag on Twitter to raise awareness about homelessness. This application area is particularly relevant during times of crises that affect urban communities, as was the case during the Covid-19 pandemic. Media facades can serve as a way to amplify messages, as demonstrated through the artwork Pulso, displayed during the SP_Urban Arte Conecta festival in São Paulo, Brazil (Fig. 12.9).

Playful encounters

Media facades can engage passersby in short playful experiences, to improve the hedonic quality of life in cities (Nijholt, 2017, 2020). Solstice LAMP invited people working in a corporate building to enter a moment of play while walking through the digitally augmented forecourt of the building. TetraBIN, representing a small-scale media architectural urban interface, demonstrates how gamification can be used to positively influence people's behaviour.

Figure 12.9 The artwork Pulso projected messages and images onto various buildings in São Paulo, Brazil, as part of the SP_Urban Arte Conecta festival, raising awareness about the importance for people to stay at home and to take care of each other as the city was hit by the Covid-19 pandemic.
Source: Reproduced with permission from Luke Hespanhol, Videographer: ZaniZ Alberto Zanella/Juliana Cretella.

Participatory civics

Enabling people to manipulate the content of a media façade can be an effective mechanism for engaging citizens in discussions of civic topics. For example, the Megaphone project, a temporary art installation in Montreal, Canada, provided citizens with a stage to express their opinions (Hespanhol et al., 2017, p. 119). A software system analysed the voice input, selecting keywords to be displayed on the building façade (Tomitsch, 2018, p. 18). Similarly, the abovementioned Smart Citizen Sentiment Dashboard enabled people to vote on civic topics that they felt needed to be brought to the attention of the city authorities, and the City Bug Report installation in Aarhus, Denmark, used a media façade to amplify issues logged by citizens via an online portal (Korsgaard & Brynskov, 2014).

Digital placemaking

Media facades offer opportunities to contribute to the emerging field of digital placemaking, which applies digital technologies to connect people with each other and with the places they share (Tomitsch, 2016). For example, the Peru National Football Stadium connects people inside the stadium with people all over the city as the façade is visible from various vantage points. The Solstice LAMP installation activated an otherwise mundane public space, not only transforming it into a playful space but also encouraging social interactions between passersby while making those interactions visible to people within the wider precinct (Hespanhol et al., 2014).

12.5 Approaches for designing and evaluating interactive media facades

The design of interactive media facades should draw on but needs to go beyond established architectural design practices. With the introduction of information technologies as a new building element, it is necessary to adopt new methods for designing and implementing digital solutions within an architectural context. Here we can turn to the fields of human−computer interaction and interaction design, which have long focused on the various ways of developing information technologies and the interactions between people and those technologies. Originally this movement started from the field of ergonomics and the first efforts focused primarily on solutions for the office as a workplace, but over the last few decades, the focus has broadened to include other contexts, such as the urban environment.

A particular challenge for interactive media facades is their situatedness within a dynamic and complex context, which consists of 'multiple co-existing and intricately intertwined situations' — requiring a 'more extensive or differently planned initial exploration than in other design domains' (Dalsgaard & Halskov, 2010). Dalsgaard (2017) identified three challenges that need to be tackled when designing an interactive media façade installation. The first challenge is to find appropriate input (eg, an array of noise level metres as used in Peru's National Football Stadium) and output (ie, what type of LEDs or other forms of actuators are installed to transform the building façade into a media surface) technologies. The second challenge involves two parts. The first part is to develop the interaction model that allows people to interact with the architecture. The second part relates to the creation of relevant content that supports that interactive experience (eg, enabling spectators inside the stadium to implicitly manipulate the visual appearance of the building's façade and translating the recorded noise levels into visual patterns). The third challenge is bringing the input−output technologies, the interaction model and the content together in a unified whole. This challenge requires the careful consideration of how the interactive media façade fits into the sociocultural context in a meaningful way. The application areas identified in the previous section offer a useful starting point for defining in what way the interactive media façade meaningfully enhances the urban experience.

Previous work has presented tools and approaches for tackling these challenges (Dalsgaard, Halskov, & Wiethoff, 2016) and highlighted some of the techniques across the fields of architecture, interaction design and industrial design for prototyping interactive urban applications (Tomitsch, 2018) (Fig. 12.10). A further key consideration for designing permanent interactive media facades is their ability to activate an urban space beyond the short-term 'spectacle' moment (Tomitsch, 2016) and to create strategies for meaningful long-term engagement (Fatah gen. Schieck, 2017). The long-term engagement with and success of interactive media facades is more difficult to assess. Research in this area has identified and documented methods for evaluating the interactive elements of a media façade, to ensure that it is intuitive and achieves the initial objectives (Dalsgaard et al., 2016). For example, this includes the application of user experience evaluation methods, such as valence markers for measuring emotions and conducting laddering interviews with people after they have interacted with the installation (Wiethoff & Gehring, 2012).

The success of commercial media facades can be assessed by contrasting the installation costs against revenue generated from tenant's rental rates, advertizing income or royalties from event programming (Wouters, Hunt, Dziemidowicz, Hiscock, & Vetere, 2018). However, the return on investment is difficult if not impossible to assess for a media façade, if its purpose is to activate a space or to create a landmark, offering cultural or societal benefits instead of economic value.

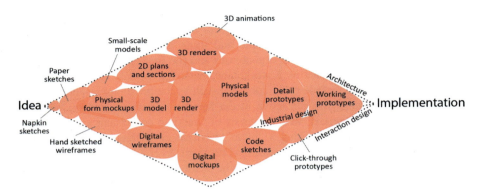

Figure 12.10 Prototyping techniques used in architecture, interaction design and industrial design (Tomitsch, 2018). Ideally, these techniques are carried out in parallel to consider the spatial, physical and digital elements that define the experience of interactive media facades. *Source:* Courtesy Martin Tomitsch.

The impact of such media façade installations is less tangible and hence more difficult to capture. For example, the Kunsthaus Graz was featured worldwide in architectural magazines and other media channels due to its innovative dynamic façade, contributing to the museum becoming a well-known iconic building and a destination for locals and tourist alike. Wouters et al. (2018) suggest a shift from calculating the return on investment to assessing the return on objectives for media facades that fulfil cultural, societal or academic purposes. In other words, they encourage to assess the success of project goals rather than cost recovery. A similar business model has been proposed for digital signage that goes beyond advertizing use cases (Schaeffler, 2012).

12.6   A framework for interactive media façade design

The framework introduced in this section builds on and extends previously published studies and reviews of media facades. For example, Tscherteu (2008) identified 11 properties of media facades: display technology, image properties (including resolution and pixel pitch), integration, permanency, dimensionality, translucency, sustainability, matching of content with the building (ie, aesthetic integration), interaction, socio-urbanistic properties and artistic quality. Drawing on a series of experimental and explorative research studies, Dalsgaard and Halskov (2010) identified eight challenges for media façade design: interfaces, physical integration, robustness, content, stakeholders, situation, social relations and emerging use. Biskjaer et al. (2014) introduced the use of simple design space schemas to identify and explore the aspects that need to be considered when designing an interactive media façade, such as display, location, situation, interaction, content, purpose and experience. Halskov and Fischel (2019) extended this to define the pixel characteristics used in a media façade and their aspects, such as their shape, pattern, spacing and colour range.

Further contributing to this domain but expanding the realm to the overarching field of media architecture, Hoggenmueller, Wiethoff, Vande Moere, and Tomitsch (2018) proposed five media architecture principles based on previous work and their own research studies: physical integration of display technology with the underlying architecture, material aesthetics of the medium (ie, the media surface), communicative and informative aspects of media architecture, contextual aspects and social interactions (ie, active and passive engagement of and between passersby). Specific to interactive features in media facades, Hespanhol and Tomitsch (2015) published strategies for intuitive interaction in public urban spaces, which included strategies for providing feedback on people's interaction with urban interfaces, such as media facades (Fig. 12.11). Based on their previous work in this area, the authors identified three broad categories of urban interfaces: performative, allotted and responsive. These categories extend Giannetti's (2004) types of interactivity and influence the technologies and content used in an interactive media façade.

The framework presented in this section consolidates some of these considerations from previous studies and approaches discussed in the previous section to guide the design and implementation of interactive media facades (Table 12.2). It offers guidance on selecting an application area, an interaction concept, interaction

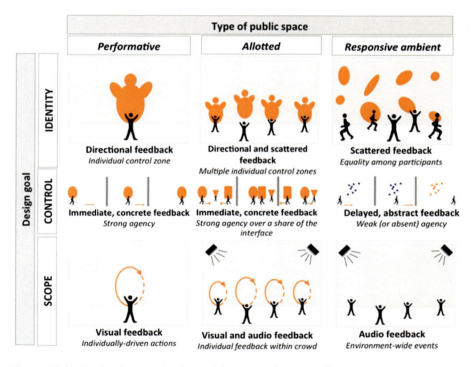

Figure 12.11 Feedback strategies for intuitive interaction in public spaces.
Source: Reproduced with permission from Luke Hespanhol.

Table 12.2 A framework for designing interactive media facades.

Component	Options	Source
Application area	Raising awareness Playful encounters Participatory civics Digital placemaking	Behrens et al. (2013), Ergin and Fatah gen. Schieck (2018), Foth et al. (2016), Hespanhol et al. (2014), Korsgaard and Brynskov (2014), Nijholt (2017, 2020), Tomitsch (2016)
Interaction type	Active Reactive (responsive) Interactive Performative Allotted	Giannetti (2004), Hespanhol and Tomitsch (2015)
Interaction technology	Personal mobile devices Mobile augmented reality Kiosk with a touch screen Kiosk with physical input controls Image cameras Infrared cameras Noise level sensors Motion sensors Proximity sensors APIs	Behrens et al. (2014), Boring et al. (2011), Hespanhol et al. (2014), Korsgaard and Brynskov (2014), Moloney, Globa, Wang, Khoo, and Tokede (2019), Parker (2017), Tomitsch (2018)
Output technology	LEDs Physical actuators Projectors	Behrens et al. (2014), Hespanhol et al. (2014), Hespanhol, et al. (2017), Korsgaard and Brynskov (2014)
Content	Text Images Video Abstract graphics Prerecorded Live Generative	Haeusler (2009), Halskov and Fischel (2019), Sade (2014)
Feedback strategies	Directional feedback Scattered feedback Immediate concrete feedback Delayed abstract feedback Visual feedback Audio feedback	Hespanhol and Tomitsch (2015)

Options within each of the components can be combined and are not limited to the specific examples included in the table. *APIs*, Application programming interfaces; *LED*, light-emitting diodes.

technologies, output technologies, content and feedback strategies. Given that each media façade and its location is unique and, thus, comes with specific opportunities and challenges, the framework is represented in a way that allows for a fluid construction of a large range of combinations that make up the specific interaction model chosen for the intervention. All of the components included in the framework must be considered in light of the unique urban context. Vande Moere and Wouters (2012) suggest that in addition to the content, media facades need to carefully take into account socio-demographic (environment) and architectural (carrier) perspectives. Through an analysis of four media façade and urban screen installations, they demonstrate that failing to successfully address those perspectives can result in vandalism against the installation or the installation simply being turned off. Projects can draw on and apply human-centred design methods used in the design of digital technologies and applications (eg, Tomitsch et al., 2021) to understand how to address these perspectives for a specific context.

The primary purpose of the framework is its use during the design and implementation phase of a project. However, it can also serve as a structured approach for considering the life of a media façade, and whether and how any of the components may need to change after the façade has been launched. This ensures that the building owner is not limited to a particular content and type of interaction for the entire lifetime of the façade. Physical interaction technologies, such as sensors installed in the façade or around the building, may be difficult to change, but if set up with an adequate server and communication infrastructure (cf. Table 12.1), interactive features can be added, for example, through mobile applications. For the same reason, output technologies need to be carefully selected, considering whether the technology will lock the façade into a particular type of content. For example, the pixel grid used in the Kunsthaus Graz media façade allows for different kinds of video content to be displayed (eg, linked to exhibitions), yet its abstract visual appearance and aesthetic ensure that it is not perceived as a large television screen (Haeusler, 2009, p. 227). The lifecycle model for designing interactive urban applications (Tomitsch, 2018, p. 206) can serve as a conceptual reminder that the design of a media façade does not necessarily end with its launch (Fig. 12.12). Ongoing updates and modifications may include either the underlying technology or the content, or both.

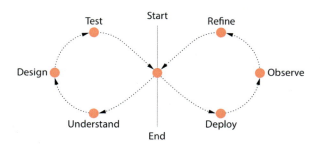

Figure 12.12 The lifecycle of interactive urban applications (Tomitsch, 2018).

12.7 Conclusion

Digital technologies are pushing existing notions of building skins and, as demonstrated specifically through this chapter, enable the transformation of the building into a media surface that can dynamically communicate ambient information, respond to activities happening inside or nearby and allow people to directly manipulate the content. Through adding interactivity, the building skin becomes a digitally enhanced interface allowing for new interactions between the building, its environment and the people within and around the building.

However, interactivity must not be added to media facades for its own sake. The key to getting interaction in media facades right is taking a human-centred approach, which involves an iterative process of understanding, designing, prototyping and deploying (Tomitsch, 2018). In addition to considering the technology (for input and output) and the spatial aspects, each interactive media façade project needs to start with a deep understanding of the existing sociocultural context. The process of developing the conceptual model of the interactive features is critical to the success of the end product. The framework presented in this chapter offers guidance for selecting the elements that shape how people will experience the media façade.

For manufacturers of media façade technology to enter the domain of interactive media facades, a conceptual shift is required. Rather than purely focusing on manufacturing and installing products, they need to develop capability in the design of human-centred technology solutions. This requires a new kind of knowledge and, ideally, experience with designing interactive technologies. As the history and evolution of media facades demonstrate, each project is unique and slightly different due to the spatial configuration of the building envelope. Similarly, interactive features will significantly differ from project to project. They might not only use different types of sensors but also different kinds of conceptual models for engaging passersby depending on the building's spatial and sociocultural context.

Future innovation in this field should consider themes that have recently emerged in research studies. For example, this includes the negative impact of light emitted from the building's façade, adding to the light pollution caused by cities (Zielinska-Dabkowska & Xavia, 2019). In extreme cases, such as New York Times Square, this may lead to situations where an urban environment is brighter at night than during the day (Ergin & Fatah gen. Schieck, 2018). Scholars have also raised concerns about the impact of light pollution on urban wildlife, advocating for a more-than-human approach to media architecture (Foth & Caldwell, 2018). This notion can guide the design of building facades to specifically cater for wildlife. This can be seen, for example, in a proposal for the façade of a new high-rise building in New York City. The concept uses biomaterials to create a habitat for the Monarch butterfly, which is at the cusp of distinction due to urban developments (Joachim, Aiolova, & Terreform ONE, 2020, p. 71). Media façade elements are integrated into the bio façade to raise awareness about wildlife extinction by amplifying live images of caterpillars and butterflies nesting inside the façade (Fig. 12.13).

Interactive media facades—research prototypes, application areas and future directions 333

Figure 12.13 A concept for a hybrid media façade that uses a vertical meadow to provide a natural habitat for the Monarch butterfly and integrates media façade elements to provide magnified live views of the caterpillars and butterflies inside the façade.
Source: Reproduced with permission from Maria Aiolova/Terreform.

Innovations can also emerge from connecting interactive media facades with other façade concepts discussed in this book, for example, by integrating solar panels to power the LEDs at night (Haeusler, 2009, p. 166). Recent research has highlighted the potential of combining kinetic sunscreens and low-resolution media facades in the form of hybrid environmental-media facades (Moloney et al., 2019). With robotic technology advancing, new opportunities are also emerging to create innovative media façade systems that represent information through physical self-moving elements (Hoggenmueller, Hespanhol, Wiethoff, & Tomitsch, 2020).

The field is entering a new era, as many of the technological challenges have been solved and costs of media façade technologies are continuing to decline. However, other challenges remain that require careful consideration to ensure media facades are developed in a way that improves urban life, for example, by drawing on digital placemaking principles. These challenges require attention and input from multiple disciplines across academia and industry, to ensure that interactive media facades contribute to addressing the global issues of our time, including extreme climates, political unrest and the health of urban communities.

Acknowledgements

Thank you to Marius Hoggenmueller for providing feedback and input on an early draft version of this chapter. I would also like to thank the anonymous reviewers for providing excellent suggestions that helped with strengthening the contribution that this chapter makes. Some parts of sections are adapted from my book Making Cities Smarter—Designing Interactive Urban Applications (published by Jovis).

References

Behrens, M., Fatah gen Schieck, A., Kostopoulou, E., North, S., Motta, W., Ye, L. & Schnadelbach, H. (2013). Exploring the effect of spatial layout on mediated urban interactions. In *Proceedings of the 2nd ACM international symposium on pervasive displays* (pp. 79–84).

Behrens, M., Valkanova, N., gen. Schieck, A. F. & Brumby, D. P. (2014). Smart citizen sentiment dashboard: A case study into media architectural interfaces. In *Proceedings of the international symposium on pervasive displays* (pp. 19–24).

Boring, S., Gehring, S., Wiethoff, A., Blöckner, A. M., Schöning, J. & Butz, A. (2011). Multi-user interaction on media facades through live video on mobile devices. In *Proceedings of the SIGCHI conference on human factors in computing systems* (pp. 2721–2724).

Brynskov, M., Dalsgaard, P. & Halskov, K. (2013). Understanding media architecture (better): One space, three cases. In *Proceedings of the CHI 2013 workshop on interactive city lighting, Paris, France*.

Brynskov, M., Dalsgaard, P., Ebsen, T., Fritsch, J., Halskov, K., & Nielsen, R. (2009). Staging urban interactions with media facades. IFIP conference on human-computer interaction (pp. 154–167). Berlin, Heidelberg: Springer, August.

Bullivant, L. (2006). *Responsive environments: Architecture, art and design*. Victoria & Albert Museum.

Colangelo, D. (2020). *The building as screen—A history, theory, and practice of massive media*. Amsterdam University Press.

Dalsgaard, P. & Halskov, K. (2010). Designing urban media facades: Cases and challenges. In *Proceedings of the SIGCHI conference on human factors in computing systems* (pp. 2277–2286), ACM.

Dalsgaard, P. (2017). Tools and approaches to support collaborative digital placemaking in media architecture. In L. Hespanhol, H. M. Haeusler, M. Tomitsch, & G. Tscherteu (Eds.), *Media architecture compendium: Digital placemaking* (2017, pp. 127–129). Avedition.

Dalsgaard, P., Halskov, K. & Wiethoff, A. (2016). Designing media architecture: Tools and approaches for addressing the main design challenges. In *Proceedings of the 2016 CHI conference on human factors in computing systems* (pp. 2562–2573).

Ergin, E. & Fatah gen. Schieck, A. (2018). Times square in the era of post-truth politics. In *Proceedings of the 4th media architecture biennale conference* (pp. 11–18).

Fatah gen. Schieck, A. (2017). Designing conditions for digital placemaking: Embodied, performative, and participatory. In L. Hespanhol, H. M. Haeusler, M. Tomitsch, & G. Tscherteu (Eds.), *Media architecture compendium: Digital placemaking* (2017, pp. 127–129). Avedition.

Foth, M. & Caldwell, G. A. (2018), November. More-than-human media architecture. In *Proceedings of the 4th media architecture biennale conference* (pp. 66−75).

Foth, M., Choi, J. H. J. & Satchell, C. (2011). Urban informatics. In *Proceedings of the ACM 2011 conference on computer supported cooperative work* (pp. 1−8).

Foth, M., Tomitsch, M., Forlano, L., Haeusler, M. H. & Satchell, C. (2016). Citizens breaking out of filter bubbles: Urban screens as civic media. In *Proceedings of the 5th ACM international symposium on pervasive displays* (pp. 140−147).

Gasparini, K. (2014). Media facades and the immersive environments. *International Journal of Architectural Theory, 19*(33), 251−261.

Gasparini, K. (2017). Media-surface design for urban regeneration: The role of colour and light for public space usability. *Journal of the International Colour Association (2017), 17*, 38−49.

Gehring, S., & Wiethoff, A. (2014). Interaction with media facades. *Informatik-Spektrum, 37*(5), 474−482.

Gehring, S., Hartz, E., Löchtefeld, M. & Krüger, A. (2013). The media façade toolkit: Prototyping and simulating interaction with media facades. In *Proceedings of the 2013 ACM international joint conference on pervasive and ubiquitous computing* (pp. 763−772).

Giannetti, C. (2004). *Ästhetik des Digitalen: Ein intermediärer Beitrag zu Wissenschaft, Medien- und Kunstsystemen*. Springer.

Haeusler, M. H. (2009). *Media facades: History, technology, content*. Avedition.

Haeusler, M. H., Tomitsch, M., & Tscherteu, G. (2012). *New media facades: A global survey*. Avedition.

Halskov, K., & Fischel, A. (2019). The design space of media architecture displays. *Interactions, 26*(6), 60−63.

Hespanhol, L. & Tomitsch, M. (2012). Designing for collective participation with media installations in public spaces. In *Proceedings of the 4th media architecture biennale conference: Participation* (pp. 33−42).

Hespanhol, L., & Tomitsch, M. (2015). Strategies for intuitive interaction in public urban spaces. *Interacting with Computers, 27*(3), 311−326.

Hespanhol, L., Haeusler, H. M., Tomitsch, M., & Tscherteu, G. (2017). *Media architecture compendium: Digital placemaking*. Avedition.

Hespanhol, L., Tomitsch, M., Bown, O. & Young, M. (2014). Using embodied audio-visual interaction to promote social encounters around large media facades. In *Proceedings of the 2014 conference on designing interactive systems* (pp. 945−954).

Hoggenmueller, M. & Wiethoff, A. (2014). LightSet: Enabling urban prototyping of interactive media facades. In *Proceedings of the 2014 conference on designing interactive systems* (pp. 925−934).

Hoggenmueller, M., Hespanhol, L., Wiethoff, A., & Tomitsch, M. (2020). Self-moving robots and pulverised urban displays: Status quo, taxonomy, and challenges in emerging pervasive display research. *Personal and Ubiquitous Computing*, 1−17.

Hoggenmueller, M., Wiethoff, A., Vande Moere, A. & Tomitsch, M. (2018). A media architecture approach to designing shared displays for residential internet-of-things devices. In *Proceedings of the 4th media architecture biennale conference* (pp. 106−117).

Joachim, M., Aiolova, M., & Terreform ONE. (2020). *Design with life: Biotech architecture and resilient cities*. Actar.

Korsgaard, H. & Brynskov, M. (2014), November. City bug report: Urban prototyping as participatory process and practice. In *Proceedings of the 2nd media architecture biennale conference: World cities* (pp. 21−29).

Kronhagel, C. (2010). *Mediatektur: The design of medially augmented spaces*. Springer.

McQuire, S., Martin, M., & Niederer, S. (Eds.), (2009). *Urban screens reader* (Vol. 5). Amsterdam: Institute of Network Cultures.

Moloney, J., Globa, A., Wang, R., Khoo, C. K., & Tokede, O. (2019). Hybrid environmental-media facades: Rationale and feasibility. *Architectural Engineering and Design Management, 15*(5), 313–333.

Nijholt, A. (2017). Towards playful and playable cities. In A. Nijholt (Ed.), *Playable cities: The city as a digital playground*. Springer.

Nijholt, A. (2020). Playful introduction on 'Making Smart Cities More Playable. In A. Nijholt (Ed.), *Making smart cities more playable: Exploring playable cities*. Springer.

Offenhuber, D. & Seitinger, S. (2014). Over the rainbow: Information design for low-resolution urban displays. In *Proceedings of the 2nd media architecture biennale conference: World cities* (pp. 40–47).

Parker, C. (2017). Augmenting public spaces with virtual content. In L. Hespanhol, H. M. Haeusler, M. Tomitsch, & G. Tscherteu (Eds.), *Media architecture compendium: Digital placemaking* (2017, pp. 133–135). Avedition.

Sade, G. (2014). Aesthetics of urban media facades. In *Proceedings of the 2nd media architecture biennale conference: World cities* (pp. 59–68).

Schaeffler, J. (2012). *Digital signage: Software, networks, advertising, and displays: A primer for understanding the business*. CRC Press.

Scully, M. & Mayze, S. (2018). Media facades: When buildings perform. In *Proceedings of the 4th media architecture biennale conference* (pp. 19–27).

Shimada, S., Golyanik, V., Xu, W., & Theobalt, C. (2020). PhysCap: Physically plausible monocular 3D motion capture in real time. *arXiv, 2008*, 08880.

Silver, N. (1994). *The making of Beaubourg—A building biography of the Centre Pompidou*. Cambridge: MIT Press.

Tomitsch, M. (2016). Communities, spectacles and infrastructures: Three approaches to digital placemaking. In S. Pop, T. Toft, N. Calvillo, & M. Wright (Eds.), *What urban media art can do*. Stuttgart: Avedition.

Tomitsch, M. (2018). *Making cities smarter: Designing interactive urban applications*. Jovis.

Tomitsch, M., Borthwick, M., Ahmadpour, N., Cooper, C., Frawley, J., Hepburn, L. A., ... Wrigley, C. (2021). *Design. Think. Make. Break. Repeat. A handbook of methods* (revised edition). Amsterdam: BIS Publishers.

Tomitsch, M., McArthur, I., Haeusler, M. H., & Foth, M. (2015). The role of digital screens in urban life: New opportunities for placemaking. In M. Foth, M. Brynskov, & T. Ojala (Eds.), *Citizen's right to the digital city: Urban interfaces, activism, and placemaking*. Singapore: Springer.

Tscherteu, G. & Tomitsch, M. (2011). Designing urban media environments as cultural spaces. *CHI 2011 workshop on large urban displays in public life* (p. 4), May 2011, Vancouver, Canada.

Tscherteu, G. (2008). *Media facades exhibition companion*. Berlin: Media Architecture Institute Retrieved from. Available from https://www.mediaarchitecture.org/wp-content/uploads/sites/4/2008/11/media_facades_exhibition_companion.pdf.

Vande Moere, A. & Wouters, N. (2012). The role of context in media architecture. In *Proceedings of the 2012 international symposium on pervasive displays* (pp. 1–6).

Wiethoff, A. & Gehring, S. (2012), June. Designing interaction with media facades: A case study. In *Proceedings of the designing interactive systems conference* (pp. 308–317).

Wong, M. H. & Enjoji, K. (2020). Tokyo's latest attraction: Transparent public toilets. *CNN*, 19th August 2020. Retrieved from: https://edition.cnn.com/travel/article/tokyo-toilet-project-transparent-toilets/.

Wouters, N., Hunt, T., Dziemidowicz, O., Hiscock, R. & Vetere, F. (2018). Media architecture in knowledge and innovation districts: Designing a canvas for research, culture and collaboration. In *Proceedings of the 4th media architecture biennale conference* (pp. 35−44).

Zaera-Polo, A., Trüby, S., Koolhaas, R., & Boom, I. (2014). In 14. International Architecture Exhibition, la Biennale di Venezia (Ed.), *Elements of Architecture #7*. Façade. Marsilio.

Zielinska-Dabkowska, K. M., & Xavia, K. (2019). Global approaches to reduce light pollution from media architecture and non-static, self-luminous LED displays for mixed-use urban developments. *Sustainability*, *11*(12), 3446.

Part B

Process Innovation

The building envelope: failing to understand complexity in tall building design

José L. Torero

Civil, Environmental and Geomatic Engineering, University College London, London, United Kingdom

13.1 Introduction

The Grenfell Tower fire (June 14, 2017) is the most tragic example of a long list of similar failures that have recurrently filled the news headlines with horrific images of tall buildings fully engulfed in flames. These failures have been systematically occurring in the last 20 years and have continued to occur after the Grenfell Tower fire. These have become so common, and our capacity to prevent them so limited, that many already consider them an unavoidable feature of modern architecture. Currently, governments, at a global scale, are spending billions in massive retrofit operations; nevertheless, there is no confidence on any of these actions. Therefore these retrofit operations are being accompanied by a constant, and ever more draconian, set of regulatory changes. The real question to be asked remains how did we reach this situation?

Construction outcomes are a balance between drivers and constraints. Drivers represent society's aspirations, while constraints enforce society's responsibilities. Only a healthy balance between drivers and constraints can enable science to foster equity and deliver social objectives. Regulation is a complex process that embeds social, economic and technological aspects and that is enabled by competency, enforcement and accountability.

A fire safety strategy is the means by which fire safety is delivered by regulations (Torero, 2013, 2019). The fire safety strategy includes components that are directly introduced to support the fire safety strategy (ie, detection, alarm, signalling and fire proofing) and the adequate fire performance of other multifunction systems (ie, combustibility, egress distances, occupancy, compartmentalization and structural behaviour).

Social objectives are translated into functionality requirements to set targets for regulation. As such concepts like as low as reasonably practical, maximum acceptable consequences are devised to describe society's aspirations (Cadena et al., 2019). These concepts can be described in a deterministic manner (eg, people will not interact with smoke in a fire) or in a probabilistic manner (eg, the probability of death shall not exceed 10^{-6}).

Once the targets are understood and articulated, it is normally the case that before defining the implementation, several assumptions need to be made. These assumptions can be physical, social, contextual, economic, etc. In the case of fire safety, many of these assumptions are applied globally but some are specific to a local context. Within the global assumptions a first key assumptions in fire safety is that there will be only one fire and that this fire will have a probability of unity; in other words, it is accepted that a fire will always occur (Fire Safety Engineering, 2020). The implications of this assumption are that every building has to include fire safety provisions and that arson has to be treated as a crime. Given that fire does not focus on probabilities, the main objective of fire safety is the management of consequences (Cadena et al., 2019). A social assumption common to fire safety is that the problem is deemed to be of sufficient complexity that the users (eg, occupants of buildings, passengers in trains) are not required to have any knowledge in matters of fire safety, their role is simply to follow a directed evacuation process, and, in some cases, to inform the fire brigades. A contextual assumption that varies among countries is the presumption of competency of the professionals involved (eg, architect, fire safety engineer and constructor). Different countries will require different competency frameworks (Fire Safety Engineering, 2020).

On the basis of the objectives and assumptions, methodologies can be built that will enable to adequately deliver the objectives. Many different approaches can be followed, with prescriptive and performance-based solutions being the most common when applied to buildings and infrastructure. Prescriptive and performance-based solutions are constructed on the basis of the same objectives and physical principles but can differ on assumptions. The presumption of competency is an assumption that tends to vary significantly between these two approaches.

Prescriptive solutions describe problems that have already been explicitly solved. Therefore they inevitably include a classification. A classification provides the description that bounds a typology for which a solution has already been proven to deliver the desired outcomes (Fire Safety Engineering, 2020). As such, any building that falls within he characteristics of the classification can be resolved using a predefined solution without the need for demonstration. The key to prescriptive solutions is an appropriate classification scheme (eg, high-rise buildings, industrial facilities, medical facilities and public spaces) and a competent body of professionals that can assess if a building falls within the bounds of the classification.

The development of classifications and prescriptive solutions is an essential aspect of professional practice. Thus those developing classification and prescriptive solution framework are expected to have a higher standard of competency than those who apply such framework.

Performance-based solutions do not require a classification and are primarily based on professional competency. The tools to be used to demonstrate that the objectives have been met and their application are the prerogative of the professional.

The process of the assessment of any of such solutions remains the responsibility of the state. A feedback loop is essential to confirm the performance of a system. Prescriptive solutions require the confirmation that the solution has been implemented correctly but also that the building is within the bounds of the classification. The latter tends to be the most complex aspect and the one that requires the highest competency. Meeting a classification requires a deep and detailed understanding of all the assumptions embedded in the classification. Furthermore, those providing confirmation of compliance have also to be well versed in the assumptions embedded in the regulatory framework and in the manner the fire safety strategy has been resolved (Fire Safety Engineering, 2020). Given that the focus is on the solution, the process of approval is a simple review, nevertheless, the presumption of competency remains high for the regulator.

Confirmation of performance is much more complex in performance-based solutions; this is not only associated with the presumption of competency but also with a process that requires not only the assessment of the solution but also the assessment of the adequacy of the methodology used. The existence of multiple solutions and multiple tools results in a presumption of competency that is high for all parties and, therefore, the dialogue can be very complex and involve many stakeholders.

The product of the regulatory framework is the building or infrastructure. Therefore the inspection of the final product is an integral part of the process. Inspections also carry a presumption of competency for the inspector that is intimately related to the process used, the complexity of the building (and of the solution) and the multiple functionalities of the systems used.

Fire safety is structured not only by components that are designed specifically to support fire safety strategy but also by multifunctional systems that have to perform in a manner that supports the fire safety strategy. Therefore competency is very difficult to define. Not only the fire safety professional needs to understand the fire performance of all systems but also other professionals delivering other performance requirements (eg, insulation, acoustics, structural behaviour and architecture) need to be sufficiently familiarized with the fire safety implications of all the decisions that are being made. Therefore the presumption of competency involves fire safety professionals, inspectors, as well as other professionals involved in the delivery of the building.

For a building to satisfy the needs of drivers and constraints, it is essential that all aspects are treated in a balanced manner. Investment on drivers leads to new technologies, increased complexity and the development of competency around the drivers. If constraints are not given an equal treatment, and a balanced investment is not guaranteed, then the increased complexity of the multifunctional systems will not be followed by an increased competency of those delivering the constraints. This will render those professionals focused on constraints incompetent.

This brings us to the final aspect of a regulatory framework, which is enforcement. Enforcement has to be a transparent and powerful process that enables to guide society in the direction where the social objectives are met. Enforcement is, therefore, also based on the presumption that there is sufficient competency to enable adequate judgment.

The case of reduction of energy consumption (driver) and façade systems as an integral component of tall building fire safety (constraint) is a clear case, where a potential conflict of performance could arise. Stringent energy management requirements have resulted in the introduction of materials and systems that have the capacity to actively promote the rapid spread of a fire (Grenfell Tower Inquiry, 2019).

A massive investment imbalance (between driver and constraint) by governments globally have rendered every sector that has a function in the fire safety strategy (eg, manufacturers, engineers, regulators and fire brigades) incompetent (Shergold & Weir, 2018; Grenfell Tower Inquiry, 2019; Hackitt). The presumption of competency embedded in the regulatory framework is no longer satisfied. This chapter explores the role of the building envelope in the implementation of fire safety and details the knowledge base necessary to be able to design and implement a building envelope that fulfills the requirements and assumptions of the fire safety strategy.

13.2 The building envelope and the fire safety strategy for high-rise buildings

The design of a fire safety strategy for high-rise buildings requires careful consideration because the safe use of a high-rise building is a complex problem (Torero, 2009; Cowlard et al., 2013). The fire safety strategy is linked to how the building is defined or classified. The definition of a building that is to be classified as a high-rise building is also complex. Regulations many times propose simple definitions of a high-rise building only on the basis of height (eg, Todd, 2018); nevertheless, numerous assumptions hide behind the classification. These assumptions, together with the many protective measures implemented, allow these buildings to be used in a safe manner. The assumptions and protective measures will vary depending on the specific characteristics and use of a building. This section provides a brief summary extracted from the study of Torero (2018) of how measures and assumptions interact in the context of the design of a building envelope.

Conceptually, the fire safety strategy for a high-rise building recognizes that the main characteristic that defines a high-rise building is a convergence of time scales. In a high-rise building, people will take significant time to evacuate (several minutes); therefore the time to egress is of the same order of magnitude as the time for failure or the time required for the fire and rescue services intervention (Cowlard et al., 2013). Time for failure could be defined in many ways, such as attainment of conditions that are untenable, structural failure, etc. In buildings that are not classified as high-rise, egress times are generally very short compared to all other characteristic times; therefore occupants are not expected to interact with firefighting operations or with the different potential modes of failure. It is clear that this will only be the case, if the fire safety strategy works appropriately during the fire event.

Given this convergence of time scales, there is insufficient time to evacuate everyone and, therefore, a high-rise building requires the existence of safe areas within the building. These safe areas are intended to assure the well-being of occupants, while the fire grows and while countermeasures and fire fighter operations are in progress. Furthermore, in the case of vulnerable people, these safe areas will serve to provide protection until rescue is achieved (Todd, 2018).

The most common safe areas are the stairwells. Stairwells are intended to remain isolated from the event during the duration of the fire, guaranteeing the egress process. There is no limit to the time where stairwells are to remain safe. To maintain the stairs as safe areas during the fire, these have to be constructed such that the fire is prevented from damaging the enclosure (ie, walls and doors). Furthermore, redundancies are necessary for all safety systems; therefore supplemental protection can be introduced to prevent smoke from entering the stairs. Typical approaches are ventilated lobbies that create a buffer between areas with combustible materials and the stairs, or increasing the pressure within the stair, thus ensuring a flow of air from the stair to the lobby (as opposed to smoke from the lobby to the stair) (Cowlard et al., 2013).

Also, it is important for safety systems to have redundancies; therefore having more than one means of egress is highly desirable. Nevertheless, it is recognized that emergency stairs can occupy a significant fraction of the surface area of a high-rise building, challenging its functionality. Limiting the number of stairs, therefore, might be necessary. In this case, other forms of redundancy might be introduced. A common form of redundancy is to prevent the fire or smoke from escaping the sector of the building where the fire originated. This is achieved by means of barriers that block the progression of a fire out of a sector. Egress, in this case, can be contained to the high-risk sectors of the building and the rest of the occupants will remain in place. All other sectors of the building are deemed safe. Firefighting operations will proceed with occupants in the building; therefore provisions have to be made to account for firefighter—occupant interactions. These provisions are in part designed into the building but also relate to firefighting operations. This strategy is generally named 'stay put' or 'defend-in-place'.

Buildings will bound these sectors by means of barriers that are qualified as 'fire resistant'. Fire resistance is a standard term used to describe the performance of structural components in a standard test. This test is recognized internationally (BS 476, ISO 834, ASTM-E 119, etc.) and it provides a standard and severe thermal exposure to the structural element by means of a furnace. Typically, for residential units, each unit represents a sector; therefore the perimeter of the unit needs to meet 'fire-resistance' requirements. Certain boundaries of the sector can be deemed more important than others; therefore a higher 'fire resistance' might be required. It is typical to recognize that global structural integrity and containment of a fire to the floor of origin are of critical importance for high-rise buildings; therefore floor slabs and main structural elements generally will have higher fire-resistance requirements than doors or nonload bearing partition walls. This is of critical importance when discussing the building envelope because the building envelope is both part of the floor system and the external wall.

It is recognized that certain components of these sectors cannot behave as barriers to the same extent as walls or floor slabs. Windows are some of these necessary components. Windows will incorporate glazing and could be potentially open; nevertheless, provisions are still necessary to prevent a fire from entering the adjacent sectors.

Figure 13.1 Trump Tower fire, April 8, 2018: (A) the fire during burning showing a fully developed postflashover fire and (B) the aftermath showing the extent of damage of the glazing.

Fig. 13.1 shows an external photograph of the Trump Tower fire in New York (April 8, 2018). Fig. 13.1A shows the magnitude of the fire and Fig. 13.1B the external glazing after the event. The figure shows that the fires did not progress to the sectors above or adjacent to the sector of fire origin. In this case the compartmentalization provisions performed adequately.

It is important to note that robustness and redundancies are paramount for high-rise buildings because for these buildings the evolution of a fire, as it scales-up, can be extremely complex (Cowlard et al., 2013), and the behaviour of occupants over such long timescales is highly uncertain. Furthermore, the performance of many of the fire safety systems and that of the fire brigades is stretched. Given the characteristics of high-rise buildings, building regulations will, therefore, stipulate robust solutions and many redundancies (Todd, 2018).

Sprinklers can be used as a means to control the rate of growth of a fire. While sprinklers can potentially extinguish the fire of origin, thus eliminating the problem, responsible design cannot assume that, due to the presence of sprinklers, a fire event that challenges the lives of tenants will not occur. Sprinklers will only reduce the probability of a life-threatening fire and, therefore, can be included as supplemental protection but do not supersede other elements of the strategy.

The previous description of a fire safety strategy for high-rise buildings defines the role of the building envelope in matters pertaining fire safety. The building envelope has to contain or delay the progression of fire spread in a manner that is consistent with the egress strategy used for emergency evacuation of the high-rise building. If the evacuation strategy is defined as a 'stay-put' strategy then the building envelope has to contain the fire in the unit of origin following the same principles of fire resistance used for the most strained structural component, that is, the floor slab. If the evacuation strategy is 'full and simultaneous evacuation' then the building envelope cannot allow fire spread to exceed a spread rate that results in compromising the occupants before they reach a safe sector of the building (ie, the stairs). If the evacuation strategy is 'full but staged evacuation' then the same principles apply; nevertheless, the delays induced by the evacuation sequencing need to be included when assessing the permissible rates of fire spread. It is important to note that firefighting operations have also to be consistent with the fire safety strategy, thus firefighting protocols have an expectation of how big a fire can be upon arrival of the fire brigade. This size is rarely defined in an explicit manner and is generally considered bigger than a fire that hampers evacuation.

13.3 Assumptions embedded in the fire safety strategy

A fire safety strategy relies on many assumptions. Some of these assumptions are associated with the design and implementation process, others with maintenance and many with adequate performance. Having explained the rationale behind the fire safety strategy, it is important to establish some of the key assumptions that enable the performance of these measures:

- That the means to establish performance of all fire safety systems can deliver a performance assessment that is sufficiently accurate and robust. Thus performance is assumed to be as intended.
- That all components of the fire safety strategy are designed with the intention that they can be built such that the prescribed performance of the design is achieved. The most common means to validate this assumption is through precompletion inspections and commissioning of the different systems. It is assumed that these are appropriate and they occur where necessary.
- That all components of the fire safety strategy are built such that they can be appropriately maintained. Provisions for inspection and maintenance with adequate means to manage repairs and improvements are assumed. It is also assumed that these provisions and means are consistent with the complexity of the systems implemented.
- That all professionals involved in the design, building, commissioning, inspection and maintenance have the competency necessary to perform their duties for the specific systems being addressed (Fire Safety Engineering, 2020). Thus it is possible for the user to rely on the outcome of all professional assessments which have been made.
- A fundamental performance assumption, key to all high-rise buildings, is the containment of the fire within the unit of origin, for a 'stay-put' egress strategy and the reduction of fire spread to a rate that does not hamper evacuation, for all strategies that enable full evacuation.
- In the case of fires progressing inward, there are no provisions to prevent the inward spread of fires. Therefore in the event of vertical or horizontal flame spread, there is no structured barrier that will prevent the involvement of further units through fires spreading inwards.
- Design for 'burn-out' is a concept implicit in the definition of required fire resistance. Structural components should be capable of withstanding the heat generated by a fire until all the fuel has been consumed. Therefore the energy delivered by a fire until burnout should be less than the energy delivered by the fire-resistance furnace until the time associated with the required 'fire resistance'. Other factors can enhance the required fire resistance such as safety factors that serve as multipliers used to manage situations with different levels of risk.

13.4 Fundamental performance principles for the building envelope

13.4.1 Performance principles

The fundamental performance requirements for any building envelope are defined by the fire safety strategy and by the assumptions embedded within this strategy. The performance requirements can be defined in three groups:

1. Fire ingress: A fire occurring in any adjacent building shall not generate a fire within the interior or exterior of the building. Thus the building envelope has to provide sufficient protection for the interior of a building. The protective capacity of a building envelope will accept that open windows represent an unsurmountable problem but will mitigate the risk of ignition by requiring that separation within buildings is sufficient to reduce the incident heat flux to values that are not deemed a threat to combustible materials.
2. Compartmentalization: Any internal fire shall be compartmentalized in a manner that it enables egress in accordance to the fire safety strategy. Requirements of compartmentalization are defined in terms of fire resistance and have to conform with all fire-resistance assumptions in the fire safety strategy as well as with all assumptions embedded in the definition of fire resistance.
3. Fire spread: A fire is expected to exit through glazing but shall not spread vertically or horizontally at any rate that could compromise egress as defined by the fire safety strategy. Fire spread has no simple definition when it comes to complex building envelopes because it involves complex concepts such as flammability (or 'reaction to fire'), encapsulation and mechanical response. It is generally assessed in terms of a combination of large-scale tests (Agarwal, Wang, & Dorofeev, 2020; BS 8414−2, 2007; National Fire Protection Association, 2012; Standards Australia, 2016), small-scale flammability tests (European Committee for Standardization, 2009, 2010; International Organization for Standardization, 2010; International Organization for Standardization, 2010) and competent professional interpretation of the test evidence (Colwell and Baker, 2013).

A building envelope will have to satisfy specific performance requirements in all three areas to enable the fire safety strategy to operate in an appropriate manner. Given the importance of each performance requirement, it is essential to understand all concepts involved.

13.4.2 Fire ingress

External fires (ie, requirement no. 1) are less onerous than internal fires because energy is allowed to dissipate upward because of buoyancy. A smoke plume will carry energy away from the fire and energy can only be transferred to adjacent buildings by means of radiation. While radiation is a complex phenomenon, it can be approximated by means of simple, yet conservative, analysis (Drysdale, 2011). Minimum required separation distances between buildings are then embedded in building regulations (eg, Torero (2016)) allowing to estimate the magnitude of the incident heat flux. Ignition of most combustible materials does not occur below a critical heat flux for ignition (The Building Regulations, 2010) enabling the use of a characteristic critical heat flux for ignition as the criterion for the design of the building envelope. A commonly accepted approach is, therefore, that the building envelope shall be capable of resisting a heat flux of 12.6 kW/m^2 (Torero, 2016). While this is not the only value quoted in building codes globally, all values utilized by all building codes are within ± 20% of this value.

13.4.3 Compartmentalization

Compartmentalization provisions have to withstand internal fires (ie, requirement no. 2). Internal fires retain the majority of the energy within the compartment and, thus, achieve conditions that are much more onerous. The thermal solicitation of walls, floor and ceiling slabs can reach heat-fluxes greater than 200 kW/m^2 (Drysdale, 2011), and direct flame impingement by localized fires has been measured to exceed 120 kW/m^2 (Agarwal et al., 2020). While these characteristic values could be used for the purpose of designing walls, ceiling and floor slabs, the thermal solicitation imposed by an internal fire has been standardized to what is known as the 'standard temperature vs time curve' [eg, ASTM-E-119, ISO-834, BS-476 (Parts 20−24), Buchanan (2002)]. This evolution of the temperature within the compartment is deemed as a worst case fire scenario and, therefore, demonstration of adequate performance of a structural element when exposed to the standard fire is deemed as an appropriate assessment. The test duration is defined as a time greater than the time required for complete consumption of the fuel within the compartment. Therefore fire-resistance requirements are defined as a testing time. Structural elements have very clear failure criteria (Buchanan, 2002) but this is not the case for the building envelope. The building envelope is traditionally a lightweight system attached to the structure at specific points by means of brackets. The system structure is commonly made of lightweight metals such as aluminium that have high thermal conductivities thus will heat in bulk. In contrast, concrete slabs (or any modern alternative such as composite systems, Cross Laminated Timber or hybrid slabs) have much lower thermal conductivities and will heat producing in-depth temperature gradients. The behaviour of materials that heat in bulk is dominated by thermal expansion, while those that enable temperature gradients tend to lead to thermally induced curvature. The interactions between the primary structure and the building envelope are, therefore, dominated by relative motion between the components of the overall system (building + building envelope). Standard furnaces used for fire-resistance testing cannot characterize this behaviour; therefore the failure criteria that enable the designer to ascertain that the building envelope provides the necessary protection from internal fires are ill-defined.

A common misconception is that by filling the gaps formed between the building envelope and the perimeter of the slabs and walls with fire stopping material (ie, cavity barriers), the system can be considered a single element (Fig. 13.2A). This works in the case of penetrations where the wall or slab surrounding the pipe provides full restraint to the filler (Fig. 13.2B). In the case of the building envelope, the structure of the system is fully unrestrained in the outward direction and, therefore, is free to deform and disassemble (Fig. 13.2A).

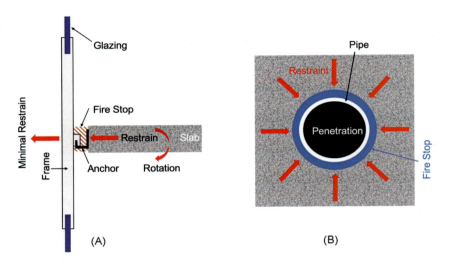

Figure 13.2 Penetration fillings (A) for a façade system (eg, cavity barrier) and (B) for a penetration (eg, pipe).

13.4.4 Fire spread

An adequate control of fire spread (ie, requirement no. 3) is probably the most complex of all three requirements. Here, a brief description of the controlling mechanisms will be provided to highlight what are the specific parameters that control flame spread.

Flame spread is a process by which the heat from a flame increases the temperature of a combustible material to a point where this material starts producing combustible gases. Once combustible gases are produced and they mix with air, they produce a combustible mixture. The flame serves as an ignition source that ignites the combustible gases, thus spreading the flame to a different location.

A fire is formed by a burning material and a flame established in the gas above the burning material. Fig. 13.3 presents a schematic of a generic fire transferring heat in all directions. The region in red will receive heat in sufficient quantities to enable the surface of the material to heat up to a temperature where the combustible material starts degrading and producing combustible gases. This region will also be sufficiently close to the fire so that the flame can ignite the combustible mixture (this process is called 'piloted-ignition'). The fire will, therefore, spread into this region. As the fire spreads, this red region will increase in size, the area of material burning will increase, and consequently, the fire size will also increase. It is not absolutely necessary for the fire to be close enough to ignite the combustible gases, because the combustible gases can heat up until the ignition of a flame is achieved simply by them absorbing the heat radiated by the flame (this process is called 'auto-ignition'). While this is physically possible, fire spread by autoignition is difficult and not very common. In Fig. 13.2 the region where autoignition can occur is schematized by the area between the red and the brown ellipses. The region

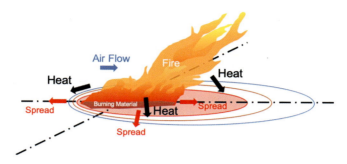

Figure 13.3 The fire spread process.

between the blue and the brown ellipses is receiving enough heat to produce combustible gases but it is too far for the combustible gases to be ignited by the flame or for autoignition to occur. Thus the fire will not spread into that region until the fire grows, heat transfer is enhanced and the flame gets closer to this region.

The larger the heat supplied [heat flux from the flame: $\dot{q}''_f(kW/m^2)$] and the larger the area through which the heat enters the combustible material [area heated by the flame: A_S (m^2)], the faster the spread. So, in the case depicted in Fig. 13.3, at the front end of the fire, the heat is being carried away from the combustible material (ie, heat goes forward and airflow goes backward) and, therefore, the heated region and the heat supplied are small. This process is normally referred to as opposed flame spread. In contrast, behind the fire, air flow and heat go in the same direction and, therefore, the heat supplied and the area heated will be large. This process is normally referred to as forward flame spread. On the sides, heat transfer conditions are somewhere in-between opposed and forward; thus the fire will most likely spread at a rate bounded by opposed and forward flame spread rates. This process will normally be referred to as lateral fire spread. When discussing vertical fire spread, upward fire spread is of the forward type and downward fire spread is of the opposed type; thus fires will spread rapidly upward and slowly downward because the airflow induced by buoyancy will move upward. Lateral fire spread is generally very close to downward fire spread because the flow induced by buoyancy is very strong taking away very fast the heat transferred laterally.

The explanation provided earlier is based on a single fire spreading. In many instances the heat flux originating from the flame ($\dot{q}''_f$) can be assisted by heat coming from other flames or hot smoke. This is very common in compartments where the accumulated hot smoke can be comparable to the heat flux from the flame. Another example where the flame is assisted by a significant heat flux are cavities with multiple combustible materials. If more than one combustible material is burning within a cavity then the flame from one material will assist the spread of a flame over the surface of another material. Defining the potential assisting heat flux is a complex process that requires very detailed understanding of the fire dynamics adjacent to the flame; nevertheless, here this will be simply referred to as an external heat flux [$\dot{q}''_e(kW/m^2)$].

Flame spread over combustible materials tends to be a continuous process by which the flame creeps over the surface of the material (Fig. 13.4A). Nevertheless, flame spread does not have to be continuous; there can be gaps with no combustible materials that can, therefore, not produce combustible gases (Fig. 13.4B). In this case the flame will only spread if the fire can heat the closest combustible material until it starts producing combustible gases and the flame is close enough that it can still act as an ignition source. This is the scenario depicted in Fig. 13.3B. If the non-combustible material extends beyond the heated area (Fig. 13.4C), the flame will not be able to spread. There are many details and nuances to these processes that make their assessment very complex.

Two conditions emerge that need to be characterized, the first corresponds to the 'stay-put' evacuation strategy that requires 'no' fire spread and the second the fire spread velocity that is consistent with 'full evacuation' approaches. The former is an extinction condition, while the latter requires to establish how fast the combustible materials can ignite.

Figures 13.3 and 13.4 show idealized approaches to fire spread, while Fig. 13.5 (Torero, 2018) shows the true complexity of a typical cladding system that uses an aluminium composite panel rainscreen. This is given as an example of a possible configuration, but there are many other different approaches with similar levels of complexity. Given the description provided earlier, at the core of flame spread are the material (or system) characteristics that lead to its decomposition into combustible gases, the transfer of heat from the flame to the combustible material and then the transfer of heat so that the ignition of the combustible gases produced. Therefore to establish the capability of a system (or a material) to allow for fire spread, it is essential to quantify these two forms of heat transfer in the context of the systems involved. This is an extremely complex process that currently exceeds the knowledge, tools and competency of great majority of fire safety professionals.

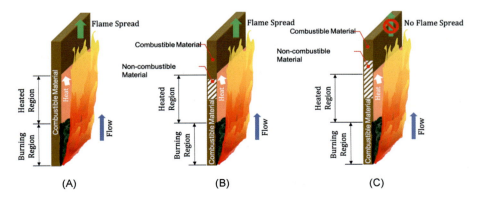

Figure 13.4 Schematic of vertical flame spread showing three different conditions: (A) flame spread over a continuous combustible material, (B) flame spread over a discontinuous combustible material and (C) arrested flame spread by means of discontinuities in the combustible material.

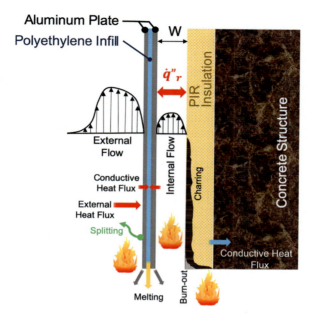

Figure 13.5 Typical layout of a conventional cladding system using an aluminium composite panel rainscreen and showing the full complexity of fire spread.

Proper performance assessment of requirement no. 3 for a typical modern building envelope is of such complexity that has rendered the whole industry incompetent.

13.5 Major misconceptions

As explained in the previous section, requirement no. 1 can be properly characterized by very simple means. Requirement no. 2 already exhibits complexity beyond existing standardized testing and common misconceptions regarding the application of fire resistance prevail in current design and implementation practices. Nevertheless, analysis of the performance of the system (from a structural perspective) still remains possible by means of existing detailed tools (Buchanan, 2002). In contrast, requirement no. 3 presents the greatest challenges and complexity. The current approach has fully ignored the fundamental principles of behaviour and, therefore, is filled with misconceptions.

Current building regulations and firefighting practices globally do not have provisions for a scenario where vertical and horizontal flame spread occurs. An analysis describing this limitation for the United Kingdom is given by Todd (2018) but this analysis is consistent with other jurisdictions. Thus from a regulatory perspective, the assumption of 'no' vertical and horizontal flame spread is implicitly made. Instead, most building regulations introduce language that is intended to recognize the hazard associated with external fire spread, in particular when combustible materials are present. For example, Section 12.5 of the UK building regulations—Approved Document B (ADB) (The Building Regulations, 2010): 'External Wall Construction. The external envelope of a building should not provide a medium for fire spread if it is likely to be a risk to health and safety. The use of combustible materials in the cladding system and extensive cavities may represent such a risk in tall buildings'. But, when the matter is addressed as in ADB functional requirements, the functional requirement B4 indicates: 'External fire spread: (1) The external walls of the building shall adequately resist the spread of fire over the walls and from one building to another, having regard to height, use and position of the building'. The language used in B4 introduces the ambiguity of 'adequately' and only requires that external walls shall 'resist the spread of fire over the walls and from one building to another'.

While codes commonly introduce ambiguity to allow for flexibility (which is often necessary for design purposes), in this case it hides a misunderstanding of the significance of fire spread on the fire safety strategy. Furthermore, most codes globally mix two fundamentally different issues: fire spreading from another building (requirement no. 1), with fire initiated within the building and spreading to the external walls (requirement no. 3). In the case of ADB it indicates: 'The aim is to ensure that the building is separated from the boundary by at least half the distance at which the total thermal radiation intensity received from all unprotected areas in the wall would be 12.6 kW/m^2 (in still air)'. The value of 12.6 kW/m^2, therefore, defines the thermal load expected on external walls (and glazing) for the scenario of fire spreading from an adjacent building. This is of fundamental importance because it defines the requirement that all external components need to meet to prevent fires igniting or entering a compartment from the outside.

In the case of a fire initiated within a building (ie, issued from a compartment fire spill plume) and spreading on an external wall (including arson scenarios such as waste bin fires), the expected heat fluxes on the external wall can reach values in excess of 120 kW/m^2 (Agarwal et al., 2020). This is not only an order of magnitude greater in intensity, but it is also a fire, temporal evolution of which will be very different.

When it comes to the parameters controlling fire spread building regulations never specify [$\dot{q}''_e$ or $\dot{q}''_f$(kW/m^2)] instead talk about combustibility in terms of ignition and heat release rates. This stems from the confusion between requirements nos. 1 and 3. In the case of requirement no. 1, the external heat flux $\dot{q}''_e \leq 12.6$(kW/m^2) therefore many materials will not ignite or spread a flame under these conditions; therefore it is relevant to discuss this external heat flux value as a potential ignition threshold. Material classifications will separate materials in groups, such as combustible, limited combustibility and noncombustible. Thus in that context, the provision is logical. But, in the case of a 120 kW/m^2 'external source', only completely inert materials such as metals or ceramics will not ignite or sustain spread (materials classified as noncombustible). All other classifications become irrelevant and so the provisions for low heat release rates.

The alternative path contemplated by codes is the use of large-scale tests that are intended to address system performance. While complex systems need to be addressed at the system level, before a test is developed there has to be a clear understanding of all the phenomena involved in the process, the performance of which the test is intended to address (ie, fire spread or 'no' fire spread). All tests currently used (Agarwal, Wang, & Dorofeev, 2020; BS 8414−2, 2007; National Fire Protection Association, 2012; Standards Australia, 2016) were not developed for systems of the complexity of current building envelopes and, therefore, do not interrogate key phenomena. For example, none of those tests contemplate heating from an 'external source' of a magnitude of 120 kW/m^2 (Agarwal et al., 2020) or use diagnostics that enable to gather the key information necessary to assess fire spread. Diagnostics included in these tests are very limited (Agarwal, Wang, & Dorofeev, 2020; BS 8414−2, 2007; National Fire Protection Association, 2012; Standards Australia, 2016) and none of them even establish fire spread rates. A key limitation of these tests is the failure criteria utilized. In all of these tests a threshold flame height is established as the failure criteria without acknowledging that there is no relationship between this specific threshold value and fire spread rates.

As a result of the awareness raised by the Grenfell Tower fire, many decisions have been made that range from increasing provisions for fire safety (eg, addition of sprinklers, improving detection and alarm) to introducing bans to many products. These decisions might reflect that the need for immediate action nevertheless is not based in an increased understanding of the problem or in an improvement of any of our performance assessment metrics. Furthermore, none of these changes are aimed at improvement the competency of all those professionals involved in establishing the influence of a façade design on the fire safety strategy.

It is clear, from the information provided in this section, that fire performance assessment for building envelopes needs to be entirely revisited. In this process the aim is to deliver a more integrated, multifunctional design where fire safety performance is seen as another driver for a better building envelope. Nevertheless, a hurdle that needs to be resolved first is that of attaining the appropriate competency of all parties involved in this multifunctional design process (Fire Safety Engineering, 2020).

13.6 Incompatibilities

It might appear, from the narrative of this section, that the strong drive toward sustainability in construction (eg, the reduction of energy consumption, reduction of embodied carbon) is in direct contradiction with fire safety. Unfortunately, this is currently the case. This is not because these drivers are naturally incompatible with fire safety but because of the manner in which fire safety is addressed. Fire safety is addressed as a necessary constraint; therefore product or system development is achieved and completed on the basis of the demands dictated by the drivers and only once it is optimized and finalized, all that is left is to demonstrate that the constraints have been met. Thus fire safety is not part of the research and development that leads to innovation, or part of the building envelope design process in an application, but simply a final verification. In as such, resources to deliver fire safety are minimal and the aim becomes the simplest and fastest process to achieve performance validation with minimal change. This is not conducive to adequate testing protocols but to over simplistic (ie, fast and cheap) approaches. Years of this approach to building envelope development have led to products, new systems and the ubiquitous presence of combustible materials that have delivered exceptional functionality in many areas but unfortunately have also delivered an extraordinary challenge to the fire safety management of high-rise buildings. It is unfortunate, but currently there is no concerted research effort anywhere in the world that aims to address these issues and, therefore, it is of extreme importance that this situation changes.

13.7 Conclusion

The fire performance of building envelopes has been addressed in the context of fire safety and the fire safety strategy that is intended to deliver adequate levels of safety. This analysis shows that decades of efforts in support of new building envelopes have led to the development of complex multifunction systems that, while being highly performant in many of their functions, have completely ignored fire safety. The result is an unstoppable sequence of very large failures that have claimed many lives and will continue to occur until fire performance of these systems is fully revisited. Currently, society does not have the technical means or competency to address how modern building envelopes support or negatively affect the fire safety strategy. It is, therefore, necessary to develop a better understanding of the fire performance of building envelopes while simultaneously reducing their complexity. Society might, therefore, need to accept, as new knowledge in fire safety emerges, less performance in other building envelope functions (eg, the reduction of energy consumption). But most importantly, it is in competency where the investment is necessary and this will only be enabled by an understanding that fire safety cannot be treated as a constraint but a driver for design.

References

Agarwal, G., Wang, Y., & Dorofeev, S. (2020). Fire performance evaluation of cladding wall assemblies using the 16-ft high parallel panel test method of ANSI/FM 4880. *Fire and Materials*, 1−15.
Buchanan, A. H. (2002). *Structural design for fire safety*. John Wiley and Sons Ltd.
Cadena, J. E., Hidalgo, J. P., Maluk, C., Lange, D., Torero, J. L., & Osorio, A. F. (2019). Overcoming risk assessment limitations for potential fires in a multi-occupancy building. *Chemical Engineering Transactions*, 77, 463−468.
Colwell, S., & Baker, T. (2013). *BR 135—Fire performance of external thermal insulation for walls of multi-storey buildings* (3rd Ed.). Watford: IHS BRE Press.
Cowlard, A., Bittern, A., Abecassis-Empis, C., & Torero, J. L. (2013). Some considerations for the fire safe design of tall buildings. *International Journal of High-Rise Buildings*, 2(1).

Drysdale, D. (2011). *Chapter 1—Fire science and combustion. Introduction to fire dynamics* (3rd ed.). John Wiley and Sons.

BS 8414−2:2015 + A1:2007. (2015). Fire performance of external cladding systems. *Test method for non-loadbearing external cladding systems fixed to and supported by a structural steel frame*, BS1.

European Committee for Standardization. (2009). '*EN 13501−1:2007 + A1:2009 Fire classification of construction products and building elements—Part 1: Classification using data from reaction to fire tests*', Brussels, Belgium.

European Committee for Standardization. (2010). *EN 13823. Reaction to fire tests for building products—Building products excluding floorings exposed to the thermal attack by a single. burning, item*, 1−104.

Fire Safety Engineering. (2020). *The Warren Centre for Advanced Engineering*. The University of Sydney.

Grenfell Tower Inquiry. (2019). Phase 1 report, report of the public inquiry into the fire at Grenfell Tower on 14 June 2017, Chairman: The Rt. Hon Sir Martin Moore-Bick, October 2019.

Hackitt, J., *Building a safer future − Independent review of building regulations and fire safety: Final report, Crown Copyright*, 2018.

International Organization for Standardization, (2010) *EN ISO 1716:2010 Reaction to fire tests for products—Determination of the gross heat of combustion (calorific value)*, Geneva, Switzerland, 2010.

International Organization for Standardization, (2010) *EN ISO 1182:2010 Reaction to fire tests for products—Non-combustibility test*, Geneva, Switzerland, 2010.

National Fire Protection Association. (2012). '*NFPA 285 standard fire test method for evaluation of fire propagation characteristics of exterior non-load-bearing wall assemblies containing combustible components*', 1−30.

Shergold, P., & Weir, B. (2018). *Building confidence: Improving the effectiveness of compliance and enforcement systems for the building and construction industry across Australia,*. Canberra: Australian Government—Department of Industry, Science, Energy and Resources.

Standards Australia. (2016). '*AS5113:2016 + A1 Classification of external walls of buildings based on reaction-to-fire performance*', 1−37.

The Building Regulations 2010, *Fire safety, approved document B—Volume 2, Buildings other than dwelling houses*, HM Government, 2013.

Todd, C. (2018). Legislation, guidance and enforcing authorities relevant to fire safety measures at Grenfell Tower, *Report for the Grenfell Public Inquiry*, 2018.

Torero, J. L. (2009). *The risk imposed by fire to high rise buildings, introduction, fire safety in high rise buildings* (pp. 1−16). VDM Publishing.

Torero, J. L. (2013). Scaling-up fire. *Proceedings of the Combustion Institute, 34*(1), 99−124.

Torero, J. L. (2016). *Flaming ignition of solid fuels. SFPE handbook of fire protection engineering* (pp. 633−661). New York: Springer.

Torero, J.L. (2018). 'Grenfell Tower inquiry: Phase 1 report', *Report of the Public Inquiry into the Fire at Grenfell Tower on 14 June 2017*, October 21, 2018.

Torero, J. L. (2019). Fire safety of historical buildings: Principles and methodological approach. *International Journal of Architectural Heritage, 13*(7), 926−940.

Resilience by design: building facades for tomorrow   14

Mic Patterson[1,2]
[1]Façade Tectonics Institute, Newington, CT, United States, [2]University of Southern California, Los Angeles, CA, United States

14.1 Introduction

The preamble of the climate crisis is playing out right now all over the world with a huge cost in lives (human and otherwise) and dollars, and challenging the resilience of buildings and urban habitat. The global pandemic of 2020 laid bare a profound lack of resilience in many aspects of contemporary societies and the built environment.

> It there's one lesson to be learned from the pandemic, it's the benefits of flexibility. In a few months, we've converted parking spaces into cafés, restaurants into food pantries, closets into broadcast centers, parks into hospitals, hotels into homeless shelters, porches into concert stages, and laptops into schools. . . .the metropolis is not a rigid machine but a text we can rewrite when we need to *(Davidson, 2020)*.

But much of this adaptation has not been easy, and many facilities and systems have proven resistant to change. Anticipating the needs and designing for greater adaptability in buildings and urban habitat in response to an uncertain future—a future where the metropolis needs to be even less of a rigid machine and more of a text easily rewritten as needed—is imperative.

Designing for the future is nothing less than a paradigm shift from our ongoing practice of designing based on historical data. Designing for uncertainty and change alters fundamental design criteria and brings new considerations and constraints to the forefront.

Organizations like the Façade Tectonics Institute (2021) are increasingly being joined by others in recognizing the building skin as

1. the lynchpin to the attainment of critical goals for carbon emission reductions from the building sector;
2. an integral influence in indoor air quality and related health, wellness and productivity considerations; and
3. a determining factor in the resilience of buildings and urban habitat.

In Section 14.2, we will discuss the links between resilient buildings, urban habitat and the building skin. There is more to resilience in the façade system than simply the use of laminated glass to prevent façade breach from windborne debris. The urban resilience dialogue as dominated by attention to the impacts of extreme weather events will be considered, but we will explore beyond these short-term shocks to the deeper long-term stresses that can strip resilience from the urban fabric and, again, we will find direct links to the façade system.

Durability, and its manifestations as service life, is explored in Section 14.3 as a resilience attribute. The causal forces of obsolescence that limit service life in buildings and their façade systems are also reviewed. The discussion links durability to adaptability, which takes us back to abovementioned Davidson's observations, and the recognition of adaptability as a primary resilience attribute. Maintenance and renovation are explored as strategies to extend service life.

Experiences with the renovation of early curtain wall systems have revealed a brittleness to these systems resulting from the failure to anticipate and accommodate the need for future retrofit in their initial design. In Section 14.4, these paired attributes of durability and adaptability suggest new paradigms for thinking about extended or even perpetual service life in buildings and façade systems, the opportunity of designing for the ages, and a resilient future. Finally, select strategies to enhance resilience reflecting these considerations are suggested.

14.2 Bending strength

> The bamboo that bends is stronger than the oak that resists.—*Japanese Proverb*

Resilience is a concept that has been adopted by and adapted to a wide variety of disciplines ranging through the sciences to business management. The variant most relevant to buildings is urban resilience. The dialogue around urban resilience addresses the façade system only in the most obvious fashion, but there is more to it and we will peel back another layer of the resilience onion. We will adopt the following definition for this investigation.

> Resilience is the capacity to adapt to changing conditions and to maintain or regain functionality and vitality in the face of stress or disturbance. It is the capacity to bounce back after a disturbance or interruption of some sort *(Resilient Design Institute, nd)*.

The 'capacity to adapt to changing conditions' will be a focus of the resilience discussion here, one that takes us well beyond the usual focus on the impact of extreme weather events and natural disasters.

14.2.1 Resilience and the building skin

Typical considerations of urban resilience address what occurs during and following a natural disaster and involve risk assessment, hazard mitigation and recovery. Life-safety and property risk are a predominant focus. The Insurance Institute for Business & Home Safety (IBHS) (IBHS, 2013) categorizes property risk as earthquake, flood, extreme temperatures, hail, high winds, hurricane, lightning, tornado and wildfire, resilience considerations in buildings that all link directly or indirectly to the building skin.

Disruptions to the power grid are a recognized resilience threat that can result not only from any of these property risks but also from other causes like system deterioration and acts of terrorism and social discord. Façade-related life-safety issues not only concern envelope breach and falling windborne debris but also the life-safety threat that can arise from the extreme hot and cold internal temperatures that can occur during a power disruption (Urban Green, 2014). These concerns have given rise to the consideration of thermal resilience and the important, more recent notion of passive survivability (Kesik and O'brien, 2019; Leigh et al., 2014; Wilson, 2006).

These comprise the common focus of façade-related resilience issues. They are primarily addressed through engineered systems and materials, for example, curtain walls that accommodate seismic movements and provide better insulation performance, operable windows, heat-treated and laminated glazing materials, fire-resistant glazing and panel materials. But this is not a complete picture.

There always seems to be another layer of the onion, and we need to be ever more diligent in peeling back the next layer. The next layer of the resilience onion reveals the consideration of durability—not typically a part of the resilience dialogue—which, in turn, reveals related layers of adaptability, repairability, upgradability, maintainability and reusability, what I have referred to as the (dis)-abilities (Patterson, 2019) because they are often vulnerabilities in contemporary curtain wall systems. Embracing the (dis)-abilities in façade system design holds the potential to bring positive disruptive change to façade technology as discussed in Section 14.4.

14.2.2 Shocks and stresses

Resilience is often discussed in terms of shocks and stresses; shocks being high intensity short duration impacts like hurricane or blast events, whereas stresses, like droughts and economic recessions, are distributed over a longer timescale. Shocks command the bulk of media attention and dominate the resilience dialogue, as well as the resilience planning of cities. Shocks can lead to stresses, however, as when a hurricane or earthquake results in extensive damage that may take years to repair, drain public and private financial resources, and have a lasting impact on the social fabric of a community. Stresses can also produce shocks, as when aging or damaged infrastructure leads to electrical grid malfunctions that cause wildfires. The 2020 global pandemic was a long-term stress that manifested resilience vulnerabilities across the built environment from elderly care facilities to hospitals and airports. The shocks are familiar from the popular resilience dialogue. Here we will explore the stresses induced over a longer timeframe by forces of obsolescence to see how they may inform a deeper understanding of resilience.

14.2.3 Stresses link resilience to durability

Durability is a resilience issue: 'Strategies that increase durability enhance resilience'. (Resilient Design Institute, nd.) Aging building stock and other infrastructure systems can become 'brittle', compromising resilience (Lovins and Lovins, 1982). The forces of obsolescence are creeping stressors that can play out over longer timeframes than shocks and result in unplanned or premature service life termination.

The consideration of durability in buildings and their façade systems yields the question: how long should buildings (or their façade systems) last? A literature review reveals a lack of convergence on this issue (Patterson, 2017, 150). Opinions on an appropriate lifespan for buildings range predominantly from 30 to 60 years with 100 years being the rarer stretch. Dr. Antony Wood (2015), executive director of the Council for Tall buildings and Urban Habitat, strongly objects and points to the importance of both durability and adaptability:

> It's patently ridiculous that we talk about buildings have[Ing] a design life of only 50 to 100 years. We should be designing for the ages, as there is very little practical experience in dismantling tall buildings—not to mention [it being] destructive to the environment and a waste of embodied energy—and modifications can be prohibitively expensive. We have to start designing and building for a future we cannot fully anticipate. Durability is important, but adaptability is perhaps more so. Facades are the first line of defense in this cause.

I concur; buildings and their façade systems should last until we are done with them. Especially when the massive commitment of resources required for a large commercial or multiunit residential building is considered. When we begin thinking in terms of 'designing for the ages', durability and adaptability pop to the foreground and alter the context for considerations of resilience.

14.3 Embodied resilience: durability and adaptability

Durability and adaptability are not generally included among urban resilience considerations. The growing awareness of embodied carbon provides a lesson on the importance of durability and service life—doubling the service life of an assembly approximately halves the life cycle embodied carbon footprint of that assembly (Wilson, 2005). Durability in buildings and their major systems also has resilience implications. Kesik and O'brien (2019) note that, 'durability is a prerequisite for resilience in buildings'.

The termination of service life in buildings and façade systems can have many causal forces but is often rooted in a lack of adaptive capacity: the ability of a system to accommodate change in response to a threat of obsolescence. There are many good reasons for extending the service life of buildings and façade systems, but to do so we need to understand their existential threats.

14.3.1 Why buildings die: the forces of obsolescence

> Every building is potentially immortal, but very few last half the life of a human.—
> Stewart Brand, 1994.

If we want to extend the service life of buildings and their façade systems, it is important to understand why they die. A common assumption is that they just wear out, but degradation is seldom the cause of a building's demise, especially with large commercial buildings. Table 14.1 lists factors that affect service life.

Table 14.1 Factors that affect service life (Athena Institute, 2006, 9–16; Straube & Burnett, 2005, 37–42; ISO 2000).

Service environment	Site conditions Climate and microclimate Intended use Indoor environment
Design	Material quality Design and detailing quality Design complexity Unconventional design or material use Fabrication and installation quality
Service quality	Minimum parameters Maintenance planning
Social, technological and economic obsolescence	Obsolescence and social factors Urban development plans and policies Changing needs Heritage value Adaptive capacity Life cycle cost

More often it is some other variant, like technical or functional obsolescence, that renders a building unfit for its intended purpose. Obsolescence is a phenomenon well studied in the realm of consumer products, where the notion of 'planned obsolescence' was pioneered in the mid-20th century. It has been far less studied with respect to buildings and façade systems and there is likely much to be learned in doing so. Silva et al. (2016) list forms of obsolescence as categories of threat to service life that may apply to buildings or facades. They specifically include envelope failure, but the forms range from physical deterioration and economics to social, legal and aesthetic obsolescence. We want to design and construct the façade system in a manner intended to anticipate and neutralize these forces of obsolescence to prevent premature service life termination.

14.3.2 Planned cycles of maintenance/renovation as a strategy to extend service life

The longer service life is extended the higher the probability of these forces of obsolescence becoming an issue. If we want a façade system that is going to last indefinitely, it is going to involve some cyclical maintenance and very likely renovation at some intervals during its life cycle. Under the umbrella of maintenance, let us include the following:

1. repair: fixing any damage that occurs from any cause;
2. refit: returning the standard of service of an assembly to a predetermined target level on or before it reaches a baseline minimum standard; and
3. retrofit: installing new components, materials or finishes that raise the original standard of service of the assembly above its original value when new.

Renovation is an intervention that may include any or all of these maintenance procedures and, in addition, may involve modification to the appearance or functionality of a building through its façade system. Maintenance activities are primarily intended to treat physical deterioration. Renovation interventions address potential functional–technological and aesthetic forces of obsolescence.

If cycles of maintenance and partial renovations are anticipated and planned, they can be designed for, with the façade system design facilitating these procedures to maximize their efficiency and minimize resource requirements for their execution. Systems can also be designed to minimize damage from the shocks discussed earlier, and to facilitate their repair, thereby enhancing resilience.

Contemporary curtain wall systems are often designed with no consideration of maintenance (beyond an assumed annual cleaning) and are even sold as zero-maintenance systems, this catering to the desires of the building owner. What this really means, however, is that the systems are not designed to be maintained, repaired, refit or retrofit; at the end of their service life, there are essentially no viable options but to strip them from the building to be replaced with a new one. With an aluminium and glass curtain wall, much of the aluminium may be recycled, but the glass is often sent to landfill or crushed and downcycled as a filler material. They are certainly not designed to persist 'until we are done with them'.

14.3.3 Adaptive capacity: durability links adaptability to resilience

Curtain wall technology to meet the various carbon emission reduction goals (Architecture 2030, World Green Building Council; Intergovernmental Panel on Climate Change) does not yet exist, meaning that not just the older buildings, but the many newer tall curtain wall buildings constructed through the early decades of the 21st century will likely require some level of retrofit to meet these goals. Nonetheless, research thus far has yet to reveal a single verified example of a curtain wall system designed to accommodate maintenance and future renovation (Patterson et al., 2012, 213).

Contemporary unitized curtain wall systems may prove even more challenging to renovate than their stick-built antecedents. If the failure to anticipate the future need for renovation is indeed the reason why more practical and efficient façade intervention options are unavailable—leading to complete façade replacement as the dominant renovation, a strategy both costly and disruptive—have we then all been busy building tomorrow's problems?

A deeper study of resilience reveals considerations that may, at first, seem counterintuitive. For example, the energetic pursuit of increasing efficiency often proves problematic: as systems evolve toward increasing efficiency, they typically lose redundancy—overefficiency can strip resilience. Fisher (2013, loc 306–393) uses the 2007 I-35W bridge collapse in Minneapolis as an example of this.

Another lesson from the resilience playbook is the downside of interconnectedness or interdependence; systems that become overly interdependent are linked like dominoes and the failure of a single component can lead to system collapse or catastrophic failure. Complexity is yet another factor that can strip system resilience (Zolli, 2012). The modular prefabricated unitized curtain wall systems, characteristic of façade systems today, are considerably more efficient to manufacture and install than the older stick systems. They are also more interconnected, interdependent and complex; the panel components are bonded and fastened together in a manner that makes disassembly a challenge, and substantive in-service maintenance or repair impossible.

Upgrading the glass on a system-wide basis is rendered impractical, as it is typically bonded in place, with the bond even inaccessible on some system designs. Bonding new glass lites into the system under field conditions would be impractical in any case. Removing a single module from a unitized system without damaging it or adjacent units is generally impossible because the units interlock with adjacent units on all four sides; they are installed in sequence and not intended for removal. Fig. 14.1 is a rendering of a typical stack joint section (horizontal intersection between units) showing the penetration of the lower unit into the upper, tongue-and-groove fashion.

The tongue of the lower extrusion is free to slide up and down within the groove of the upper extrusion to accommodate relative movement between the adjacent

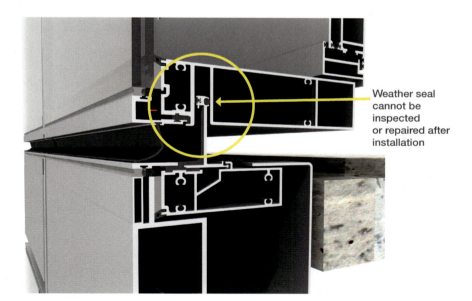

Figure 14.1 Typical stack (horizontal) joint characteristic of a unitized curtain wall system. *Source:* Courtesy Enclos Corp.

units, one of the advantages of a unitized curtain wall system. At the top of the tongue is the business end of the unit; the gaskets that act as the air and moisture barrier. The prevention of water and air migration across the weather barrier is the most fundamental and important function of the façade system. The detail represented in Fig. 14.1 is obviously not designed for maintenance. Once installed, the gasket seals are no longer available for inspection, repair or replacement, despite being subject to wear as they move within the groove.

These gaskets are the 'weak link' in the durability of the system (along with the sealants in an insulated glass unit, if present) and should be easily accessible for inspection, repair and replacement, if the potential for extended service life is to be realized. The majority materials in a typical unitized glass and metal curtain wall system are float glass and extruded aluminium, both with extraordinary service life potential: indefinitely for the glass and hundreds of years for the extrusions if properly maintained. A best practice is design to realize the full service-life potential of the components and materials that comprise an assembly (or at least so that they can be efficiently reclaimed for reuse or recycling). Any compromise to the service life potential represents wasted durability.

We can recognize the challenges to maintaining, repairing, refitting and retrofitting contemporary curtain wall systems—the actions required to extend their service life—as resulting from a lack of adaptability in the design of these systems.

14.3.4 Adaptability as a vaccine for the forces of obsolescence

The capacity of a system or assembly to endure is easily compromised by a lack of adaptability. This phenomenon became evident in the building stock during the 2020 pandemic; some buildings proved difficult to adapt to the new and unanticipated requirements of use initiated by the public health response to the pandemic. Future unforeseen conditions resulting from accelerating social evolution and as promised by the climate crisis over coming centuries can only be accounted for with strategies that maximize adaptive capacity in buildings and their major systems.

Adaptive capacity is another concept linked to resilience and borrowed from the social and ecological sciences and adopted by the business community (PricewaterhouseCoopers, 2012), where the relationships between efficiency, redundancy, adaptability and resilience have been aggressively explored. The challenges of maintaining and renovating curtain wall systems discussed earlier highlight the lack of adaptive capacity in the design of the systems, a flaw that leaves them vulnerable to the forces of obsolescence. As accelerating technological and social change is further amplified by various manifestations of climate change, resource scarcity, future public health challenges, and more, how can we possibly anticipate how buildings might be used and experienced in upcoming decades?

A rational strategy is to design buildings and façade systems with as much adaptive capacity as possible. In recognition that net-zero façade technology is not yet available and that the product suppliers are steadily producing higher performing solutions, the building industry has begun to use the term *net-zero-ready*, meaning systems designed to accommodate future retrofit of these new products as they become available (Coughlin, 2016; Patterson, Vaglio & Noble, 2014).

Failing to provide adequate adaptive capacity in the façade systems we build today to accommodate the needs of future generations, regardless of how uncertain these needs may be to us now, holds the potential of burdening those generations with buildings unfit for the needs of the time.

14.4 Resilience by design: designing for the ages and a resilient future

Understanding resilience as a design problem—and the exercise of resilience thinking—is informed by an examination of the existing building stock and characteristics of how they have aged. Existing curtain wall buildings dating back to the mid-20th century provide a useful example.

Despite the favourable economy of the 2010s, many of the tall curtain wall buildings in urban centres like New York City, buildings approaching 50–60 years and more in age, remained badly in need of façade renovation. Cost and disruption to ongoing building operations proved effective barriers to these needed renovations. Given the spectre of accelerating climate change and the new perspective of a global pandemic, it is not difficult to imagine future generations burdened with an aging and increasingly obsolete building stock and without the economic wherewithal to do anything about it; an urban habitat stripped of resilience.

We can conceivably design our way out of this scenario, but the design and construction practices that anticipate and act as remedy need to happen now. They include significantly extending the potential service life of buildings and their façade systems by enhancing their durability and adaptive capacity.

Research into façade renovation practices on early curtain wall buildings confirmed that the façade systems were most often replaced rather than renovated because of the lack of viable alternative options (Martinez et al., 2015) and that lesser interventions, for example, replacing the glass and refinishing aluminium mullions, could prove to be more costly than complete replacement (Patterson, 2017, 332). The research concluded that the system designs failed to anticipate and facilitate the need for future renovation (Patterson, 2017, 201).

There are serious problems with the replacement of a curtain wall system on an existing building, particularly a tall building in an urban environment. It is wasteful from a material/carbon standpoint, it is very costly to the building owner, and most problematic, it is highly disruptive to building occupants. Few building owners can afford the lost lease revenues resulting from this strategy. So, most building renovations involving façade replacement remain occupied throughout the process, with tenants being shuffled around within the building and exposed to construction noise, dust and other inconveniences that may pose both a health and productivity risk. The owner is exposed to liability for this risk, bringing to the fore strategies for mitigating the disruption to ongoing building operations during the renovation process.

The cost and disruption of façade replacement are effective barriers to façade renovation on tall curtain wall buildings; building owners are reluctant to include the façade system in a building renovation, instead favouring a less costly strategy like lighting and mechanical systems upgrades. Most of early tall curtain wall buildings are single-glazed without any form of thermal break and were never thermally or acoustically efficient. Delaying the renovation of these façade systems continues the exposure of building occupants to substandard interior environments, with negative comfort, health, wellness and productivity impacts, with the potential to erode the psychophysiological resilience of occupants. Furthermore, when façade system replacement becomes unavoidable, these barriers may force the building owner to consider building demolition and replacement as a viable alternative to building renovation, resulting in unfortunate embodied carbon impacts.

Durability and adaptability are powerful attributes for buildings and façade systems intended to last 'until we are done with them'. These twin attributes enhance resilience by minimizing the potential for unplanned obsolescence. Planned cycles of maintenance and partial renovation can be envisioned as a strategy to perpetually extend the service life of buildings and their major systems. Assemblies can be designed for ease of disassembly, with component parts readily accessed, inspected, repaired or replaced, any material or finish easily renewed.

Service life in this context becomes an entirely different paradigm; an assembly is in uninterrupted service over decades, and even centuries, although at some point perhaps not a single original component of the assembly remains. The assembly is conceivably replaced many times over, but incrementally and while remaining in service. When we are eventually done with it, the assembly is disassembled and its component parts reused, repurposed or if nothing else, recycled. The challenges of façade replacement—high cost, major disruption to ongoing building operations, occupant impacts on health and productivity, the associated risk to the building owner, and the threat these pose to building obsolescence—are avoided.

So, why has not this already happened?

14.4.1 Can we afford resilience?

I hear this question often when I lecture on aspects of resilience and sustainability. As always, the costs of preventative measures must be weighed against potential benefits, and the costs and liabilities resulting from inaction must also be considered. Life cycle cost analysis is the necessary strategy for this assessment yet is seldom employed. The value of risk mitigation must also be included in the assessment and is, again, rarely practiced. In the face of an escalating climate crisis posing an unprecedented threat of property damage and an existential threat to humanity itself, the better question may be can we afford to ignore these emerging conditions? Can we afford not to anticipate and design for these conditions by providing optimally adaptive systems? Nonetheless, the barriers to adoption of such practices persist.

The challenges of durable and adaptable systems, noted long ago by Lynch (1958), include

1. a potential need for additional resources in the form of design, fabrication and installation time, and added materiality;
2. a potential for added complexity;
3. a potential for increased first cost (in both carbon and dollars) and the prospect of ongoing maintenance and renovation costs (even if life cycle costing analysis evidences a reduction in life cycle cost); and
4. an ongoing and continuous tension between efficiency in the present and adaptability in the future.

Consider a simple conceptual example: a unitized curtain wall system design that utilizes a cassette strategy to facilitate the future changeout of the glazing panels (Fig. 14.2). A minimal frame around the glazing panel that functions to easily fix and release the panel in the façade system could facilitate the removal of existing glass and its substitution with a higher performing product during off hours so as not to disturb occupants. This type of feature is a challenge to accomplish without the addition of at least some additional system complexity in the form of more parts and more time that propagate through the delivery process, which inevitably translate into an incrementally higher cost. Very clever design solutions are required, solutions that aspire to cost reductions over conventional practices while providing enhanced durability and adaptability; in other words, a ripe but challenging area for innovation.

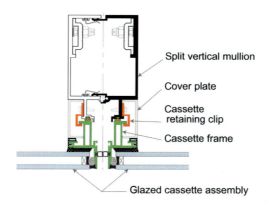

Figure 14.2 Section, vertical split mullion: simplified conceptual detail of a cassette glazing system that accommodates installation and removal of prefabricated cassette from inside the building.
Source: Courtesy Enclos Corp.

The cassette strategy suggested here may be a relatively moderate, incremental step toward greater façade system adaptability and durability. More potentially disruptive and effective concepts might be borrowed from other industries; take the computer and IT industry, for example.

14.4.2 The future: plug-and-play and hackable systems

The notion of a plug-and-play façade system makes sense from the standpoint of an easily adaptable modular prefabricated curtain wall system where not just the vision glass, but opaque panels and even entire units could easily be swapped out, plugged in and fully integrated with lighting, sensors, controllers, shading devices, etc., through an automated building management system. If standard façade grids were shared between buildings, units could even be interchanged or reused elsewhere if a building were to become obsolete for whatever reason.

This concept is certainly not new but one that we are far from in practice with the ongoing trend of increasing geometric complexity in bespoke façade system design. We could also consider 'hackability' as a resilience attribute in façade system design that enhanced the adaptive capacity of the system. A hackable system would enhance the ease with which future building owners and users could morph the system in response to altered interior or exterior conditions; changing conditions of environment, users and use. The hack could potentially apply to the performative or aesthetic attributes of the façade system or to both. Adopting such design considerations and goals would shape a very different building landscape.

Contemporary unitized curtain wall systems and the best practices by which they are currently designed and delivered appear themselves to be obsolete, or close to it. Plug-and-play and hackable system concepts begin to suggest *next* practices for façade system design and delivery.

14.5 Strategies to enhance resilience

The following includes three categories of resilience strategies starting with resilience planning, a prerequisite for resilient design; followed by thermal resilience, a passive adaptive response to extreme temperatures; and finally, by the primary focus of this exercise, durability and adaptability.

Note that these are basic strategies and require the development of robust and detailed tactical plans for execution. Additional information on resilience and the building skin may be found among the resources on the Façade Tectonics Institute website.

14.5.1 Resilience planning

Setting resilience-specific goals early in the design process aids in assuring resilient outcomes. The IBHS recommends developing code-plus goals over the minimum life-safety standards included in model codes like the International Building Code. Developers in hurricane prone coastal areas may chose, for example, to adopt the South Florida Building Code's higher standards for wind velocities and missile-impact resistance even if they are not required for the building's location. Goals and strategies to extend façade system service life should be an integral part of resilience planning.

1. Develop a written resilience plan that specifically addresses the façade system.
2. Resilience considerations are highly site and climate sensitive; base assessments on specific site and projected (not historic) climate and microclimate data. Building type and use are also important factors.
3. Develop performance goals on the basis of projected data anticipating future climate change extreme weather impacts.
4. Assess the impact of extended disruptions to the power grid.
5. Develop code-plus goals for building resilience early in the design process as part of a basis-of-design between the owner and design team.
6. Consider code-plus goals around considerations of extreme winds, temperatures, floods, fires and seismic events, as well as the potential for security threats.
7. Develop strategies to minimize damage from extreme events and plan for emergency repair and replacement of damage prone materials and components.
8. Develop strategies to enhance system durability and adaptability that can avoid premature obsolescence and extend service life.
9. Establish a specific design service life for the façade system that harmonizes with a specified design service life for the building (see following strategy on 'adaptive resilience').
10. Develop a façade system maintenance and renovation plan in support of an extended design service life.

14.5.2 Thermal resilience: consider extreme temperatures and assess passive survivability as a function of climate and exposure

The cold snap in Texas during the winter of 2021 that left nearly 70% of the state's population without power for an extended period has provided the latest example of the life-threatening conditions that can result from extreme weather events. Extreme temperatures are not the only thing that threatens the power grid. California utility companies now routinely cut power to the grid during high wind events to prevent accidental wildfires sparked by power lines. The combination of high wind and high outdoor temperatures can quickly translate to life-threatening indoor temperatures, especially among elderly populations. The National Resource Defense Council projects nearly 30,000 heat-related deaths per year in the United States by 2100 at current CO_2 levels (Constible, 2017). For new building construction and existing building renovation:

1. What happens when the lights go out—at minimum, model the building interior under blackout conditions using predicted temperature extremes over a 7-day period, one cold period and one hot period. Kesik and O'brien (2019) provide thermal resilience design methodology and modelling conventions.

2. Assembly u-factor—glass area and type, air infiltration/exfiltration, thermal mass and insulation values in walls and roof are primary determinants of interior temperature conditions during extreme exterior temperatures; use these inputs to model indoor temperatures.
3. Exterior shading—heat gain is a critical problem during extreme high temperature events. Sunshades and high-performance glazings that reflect heat help minimize heat gain to the interior.
4. Ventilation—operable windows are vital in providing ventilation during blackouts over periods of high temperatures.
5. Window-to-wall-ratio: triple and quad glazings add insulative value to glass area, but significantly underperform compared to highly insulated opaque wall assemblies.
6. Fire resistance—carefully evaluate the efficacy of code minimums as a function of building location and exposure; consider code-plus goals in high-risk areas.
7. Façade commissioning—verify air tightness, operability and other relevant performance metrics of the as-built building envelope through a commissioning process.

14.5.3 Adaptive resilience: durability and adaptive capacity as resilience attributes

Kesik and O'brien (2019) note that, 'durability is a prerequisite for resilience in buildings'. In Section 14.3, we linked durability to adaptability. The considerations discussed in Section 14.4 provide context for a very different kind of thinking about façade system design; one that embraces the constraints imposed by considerations of durability and adaptability to leverage a departure from contemporary practice and mere incremental improvement; one that opens the door on the potential for perpetual service life and 'designing for the ages'.

1. Embeds—although typically designed and provided by the façade contractor, embeds should be considered part of the building structure and designed for a minimum 300-year lifespan that anticipates predicted climate change, amplified wind velocities and expanded hurricane zones. Oversizing embeds is a miniscule added construction cost and may ultimately make the difference between a building's renovation or demolition.
2. Avoid complexity—simplicity is a resilience attribute; complexity strips resilience. Geometric, material and assembly simplicity enhance maintenance, repair, replacement and renovation requirements throughout the systems lifespan.
3. Expose weak links—design sealants and gaskets to be easily accessible for inspection, repair and replacement and insulated glass units for easy removal and replacement.
4. Enhance redundancy—the pursuit of efficiency can strip resilience. Consider redundant strategies for priority system functionality like air and vapour barriers and ventilation.
5. Design for disassembly—in addition to facilitating maintenance, repair and replacement requirements and renovation interventions throughout the façade system's service life, the ability to disassemble the system provides options for efficient end-of-life disposition, including optimum potential for adaptive reuse. Avoid bonded assemblies.
6. Plug-and-play—design vision glass and opaque façade panels for ease of installation and removal from inside the building. This can facilitate glazing product upgrades and aesthetic enhancements to the building façade through a building's service life.
7. Design for the ages—consider the generational legacy of the building and design for your great-grandchildren's great grandchildren!

8. Resilience in ruins—even the eventuality of our buildings as ruins should be considered (Dale & Burrell, 2011). Tribes, if not civilizations, can be rebuilt from ruins, with building sites conceivably providing shelter and a head start on habitable dwellings, or at least a concentration of materials with the potential for use in building construction.

14.6 Conclusion

Resilience is a critically important sustainability attribute, yet its consideration in the sustainability dialogue remains incomplete. The dominant focus on the shocks produced by extreme weather events and natural disasters shadows deeper vulnerabilities resulting from long-term stresses that fail to register at the level of shocks. The risk factors common to resilience—earthquake, flood, extreme temperatures, hail, high winds, hurricane, lightning, tornado and wildfire—have direct and indirect linkages to the façade system. We have explicitly linked the less common consideration of durability to resilience, explored the forces of obsolescence in buildings and their façade systems, and further linked adaptability to durability.

Contemporary façade system designs uniformly fail to anticipate and facilitate the need for future maintenance and renovation as a means to extend service life. The resulting lack of adaptive capacity in contemporary façade system design exposes these façade systems and the buildings they clad to the threat of premature obsolescence. Building tomorrow's problems today could saddle future generations with costly and disruptive façade renovations, or worse, with façade systems and buildings unfit for purpose. These twin considerations of durability and adaptability, when fully considered, embraced and applied in façade system design, hold the potential to significantly advance the resilience and sustainability of buildings and urban habitat.

References

Athena Institute (2006). *Service life considerations in relation to green building rating systems: An exploratory study*. Athena Sustainable Materials Institute. <http://www.athenasmi.org/wp-content/uploads/2012/01/Service_Life_Expl_Study_Report.pdf>. Accessed 10.05.21.

Constible, J. (2017). *Killer summer heat: Paris Agreement compliance could avert hundreds of thousands of needless deaths in America's cities*. National Resource Defense Council. <https://www.nrdc.org/sites/default/files/killer-summer-heat-paris-agreement-compliance-ib.pdf>. Accessed 04.03.21.

Coughlin, B. (2016). *Building the future: Net zero & net zero ready*. In *Presentation to 2016 public sector climate action leadership symposium*. RDH Building Science. <https://www2.gov.bc.ca/assets/gov/environment/climate-change/cng/symposium/2016/23_building_the_future_net_zero_and_net_ready_-_brittany_coughlin.pdf>. Accessed 12.01.21.

Dale, K., & Burrell, G. (2011). Disturbing structure: Reading the ruins. *Culture and Organization*, *17*(2), 107–121, Taylor & Francis online. Available from https://www.tandfonline.com/doi/abs/10.1080/14759551.2011.544888.

Davidson, J. (2020). The 15-minute city: Can New York be more like Paris? In the Intelligencer. *New York Magazine*. Available from https://nymag.com/intelligencer/2020/07/the-15-minute-city-can-new-york-be-more-like-paris.html, July 17, 2020; Accessed 20.10.20.

Façade Tectonics Institute. (2021). *Art, science and technology of the building skin.* <https://www.facadetectonics.org/about/mission-vision>. Accessed 05.05.21.

Fisher, T. (2013). *Designing to avoid disaster: The nature of fracture-critical design* (Kindle ed.). New York: Routledge.

IBHS (2013). *Risks*. The Insurance Institute for Business & Home Safety. <https://www.disastersafety.org/>. Accessed 26.01.15.

Kesik, T., & O'brien, L. (2019). *Thermal resilience design guide*. University of Toronto. <https://pbs.daniels.utoronto.ca/faculty/kesik_t/PBS/Kesik-Resources/Thermal-Resilience-Guide-v1.0-May2019.pdf>. Accessed 08.05.21.

Leigh, R., Unger, R., Scheib, C., Kleinberg, J. & Druelinger J. (2014). *Baby it's cold inside. Urban green*. United States Green Building Council. <https://www.urbangreencouncil.org/sites/default/files/2013_baby_its_cold_inside_report.pdf>. Accessed 08.05.21.

Lovins, A. B., & Lovins, L. H. (1982). *Brittle power*. Brick House Publishing Company.

Lynch, K. (1958). Environmental adaptability. *Journal of the American Institute of Planners*, *24*(1), 16−24. Available from https://doi.org/10.1080/01944365808978262.

Martinez, A., Patterson, M., Carlson, A., & Noble, D. (2015). Fundamentals in façade retrofit practice. *Procedia Engineering*, *118*, 934−941.

Patterson, M., Martinez A., Vaglio J., & Noble D. (2012). New skins for skyscrapers: Anticipating façade retrofit. Council for tall buildings and urban habitat. In Proceeedings fot the *ninth world congress*, Shanghai.

Patterson, M., Vaglio, J., & Noble, D. (2014). Incremental façade retrofits: Curtainwall technology as a strategy to step existing buildings toward zero net energy. *Energy Procedia*, *57*, 3150−3159.

Patterson, M. (2017). *Supple skins: A methodology and framework for considering façade system resilience. From Skin fit and retrofit: Challenging the sustainability of curtainwall practice in tall buildings* (Ph.D. dissertation). University of Southern California. <http://digitallibrary.usc.edu/cdm/compoundobject/collection/p15799coll40/id/457628/rec/16>.

Patterson, M. (2019). Materiality and embodied carbon considerations in contemporary curtainwall systems. In *Proceedings of the PowerSkin conference*. 17 January; Munich.

PricewaterhouseCoopers (2012). Prospering in an era of uncertainty: The case for resilience. *Second report in a series on risk and resilience by PwC in association with the University of Oxford*. <http://pwc.blogs.com/files/the-case-for-resilience1.pdf>. Accessed 21.05.15.

Resilient Design Institute (nd). *Resilient design*. <https://www.resilientdesign.org/resilient-design/>. Accessed 10.05.21.

Resilient Design Institute (nd.). *The resilient design principles*. <https://www.resilientdesign.org/the-resilient-design-principles/>. Accessed 10.05.21.

Silva, A., de Brito, J., & Gaspar, P. L. (2016). *Methodologies for service life prediction of buildings: With a focus on façade claddings*. Springer.

Urban Green (2014). *Baby it's cold inside*. New York: Urban Green Council. <https://www.urbangreencouncil.org/sites/default/files/2013_baby_its_cold_inside_report.pdf>. Accessed 20.10.21.

Straube, J. F., & Burnett, E. F. P. (2005). *Building science for building enclosures*. Building Science Press.

Wilson, A. (2006). Passive survivability: A new design criterion for buildings. *Environmental Building News*, *15*(5). Available from https://www.buildinggreen.com/feature/passive-survivability-new-design-criterion-buildings, Accessed 18 October 2020.

Wilson, A. (2005). Durability: A key component of green building. *Environmental Building News*. <https://www.buildinggreen.com/feature/durability-key-component-green-building>. Accessed 20.10.20.

Wood, A. (2015). Skyscraper expert Antony Wood calls for a facades revolution. *The Architect's Newspaper*. Available from http://blog.archpaper.com/2015/03/skyscraper-expert-antony-wood-calls-facades-revolution/, 18 March 2015; Accessed 19.10.20.

Zolli, A. (2012). Want to build resilience? Kill the complexity. *Harvard Business Review*. Available from https://hbr.org/2012/09/want-to-build-resilience-kill-the-complexity, September 26, 2012; Accessed 20.10.20.

Inverse design for advanced building envelope materials, systems and operation

Roel C.G.M. Loonen[1], Samuel de Vries[1] and Francesco Goia[2]
[1]Department of the Built Environment, Eindhoven University of Technology, TU/e, Eindhoven, The Netherlands, [2]Department of Architecture and Technology, NTNU Norwegian University of Science and Technology, Trondheim, Norway

Abbreviations

BL	baseline
BPS	building performance simulation
CABS	climate adaptive building shell
DGPs	daylight glare probability simplified
HSA	horizontal shading angle
PB	performance based
PVSD	photovoltaic integrated shading devices
RB-BL	roller blind baseline
RB-PB	roller blind performance based
VB-BL	vertical blind baseline
VB-PB	vertical blind performance based
VF	view fraction
VSA	Vertical shading angle
WWR	window-to-wall ratio

15.1 Introduction

Technological developments and novel design approaches have been prime drivers of the continuous evolution of building skins required to meet increasingly strict performance requirements across different domains (Jelle et al., 2012). Several research and development efforts are targeting improved building envelope materials, components and systems (Delmastro et al., 2019) to achieve advanced building skins capable of providing occupant-centric health, well-being and productivity (Steinemann, Wargocki, & Rismanchi, 2017) while also minimizing the use of energy to ensure high indoor environmental quality.

Advanced building skins can balance competing performance aspects (e.g., those related to energy efficiency, thermal comfort, daylighting, indoor air quality) by using a combination of advanced material properties, advanced components and advanced integrated control strategies (Taveres-Cachat, Favoino, Loonen, & Goia, 2021).

Because of their intrinsic multidomain nature, the design of high-performance advanced building skins is a challenging task. Existing design guidelines and rules-of-thumb typically provide only limited guidance since they disregard the dynamic interactions and trade-offs in multiple physical domains (Attia et al., 2018). In addition, due to the complexity of advanced building skins, it is important that advanced controls and operational principles are already considered during the design stage. Similar challenges also play a role in the product development of novel building envelope materials, components and systems.

The planning and operation of advanced building envelopes, especially in comparison to conventional building envelope solutions, is supported by and follows the general trend in building design of the transition from a prescriptive approach to a performance-based approach (Bianco et al., 2018; Boswell, Hoffman, Selkowitz, & Patterson, 2021). In the prescriptive approach, minimum requirements for the behaviour of each subsystem of the construction are described under a limited set of representative boundary conditions in detailed codes and regulations, and the designer is required to select the technical solutions that meet the given specifications. Conversely, the performance-based approach emphasizes the global, final behaviour of the entire system, which needs to meet a series of high-level performance criteria without a predetermined list of solutions, methods or specifications. This approach allows designers to freely explore a large design option space and focus on the holistic building system performance rather than on the characteristics of individual subsystems.

Exploiting a performance-based approach involves more time, resources and complexity than the simpler prescriptive route, and this challenge in the construction industry is particularly visible when considering the speed of development and innovation that characterizes this sector (Kesidou & Sorrell, 2018). The prevailing approach for innovations in the construction industry is through incremental improvements and experiential learning using trial-and-error, which is characteristic of the prescriptive approach, even in cases where the performance-based design approach is applied. The most conventional (direct) approach to solving design problems is indeed that of taking one or more solutions from a certain range of known cases and then calculating the performance effects of one or more particular configurations of a material, geometry or process based on the knowledge of the set of physical laws that describe the problem. Such a procedure can be iterated until the ultimate performance goal is eventually met.

In this chapter, we forward the concept of inverse design as a driver for the development and innovation of the building skin, for new design and implementation, for the development of new materials and systems and for the operation of advanced building envelopes. Our aim is to outline the opportunities presented by inverse design for high-performance building skins. To do this, we rely primarily on our first-hand experience of using this approach for research and innovation of building envelopes. After a brief definition of inverse design, we present some examples of different applications of inverse design targeting a range of problems in the design of high-performance building skins. Advancements in building performance simulation (BPS) in the last decade have enabled the implementation of inverse approaches. In this chapter, we briefly pinpoint the key technical simulation methods that can be employed to apply the concept of inverse design. Further, we give a short demonstration of inverse design using a particular case where we investigated the combined optimization of fixed and dynamic solar shading elements while targeting both optimal daylighting exploitation and high visual comfort. Finally, we conclude the chapter with an outlook of future possibilities and trends in inverse design, not limited to the field of façade engineering, but potentially relevant for the entire built environment.

15.2 What is inverse design?

Inverse design, also known by the term 'inverse problem', is a collection of relatively new research approaches that are growing in popularity in natural sciences and engineering. In general terms, in the case of an inverse design problem, the particular configuration of a material, geometry or process is determined as the result of a targeted search activity (Fig. 15.1). While the techniques adopted to perform such a search can differ, what makes the idea of the inverse approach unique is that it formulates the functional requirements for a given design as an optimization problem. This means that in an inverse approach, a parameter search is systematically carried out, in an automated way, until a design solution is found that meets the specified objectives in the best way possible. Through such structured design space explorations, the trade-offs between various performance aspects can be analysed in a systematic way, thus fostering not only a better understanding of the underlying, unexplored, relationships between properties and their effects but also driving the development of new solutions that may not have been explored so far.

Among the first applications, the concept of inverse design has found fertile ground in solid-state chemistry since the 1990s (Zunger, 2018). The exploration of

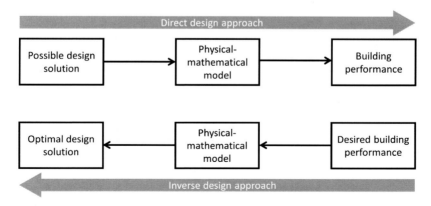

Figure 15.1 Schematic representation of inverse design in comparison to direct design approach.
Source: Original illustration by the authors.

different atomic arrangements in search of a configuration with the desired electronic and optical properties is limited by the vast number of possible combinations, making a simple trial-and-error approach unlikely to be successful. Theoretical methods have, therefore, been developed to address the problem of finding the atomic structure of a complex and multicomponent system that has the target electronic-structure properties. The idea of using inverse design has expanded since then and found widespread applications, for example, in a broad range of materials targeting energy applications such as lithium-ion batteries, hydrogen production and storage materials, superconductors, photovoltaics and thermoelectric materials (Jain, Shin, & Persson, 2016).

Different engineering disciplines have also made significant use of the concept of inverse design since the late 1980s, especially for mechanical applications (Dulikravich, 1988), targeting primarily structural and thermal problems for both industrial and civil applications. In the field of construction, inverse design has been adopted for form-finding problems aimed at structural design (Nagy, Zhao, & Benjamin, 2018) and energy efficiency (Kämpf & Robinson, 2010), as well as for optimizing heating, ventilation and air-conditioning configurations (Wright, Loosemore, & Farmani, 2002; Zhai, Xue, & Chen, 2014). More recently, inverse design has also been applied to the development and study of advanced building envelope solutions.

15.3 A flavour of applications of inverse design for building envelope research

Inverse design may be based on different methods and techniques and address different domains in building envelope research. The concept of inverse design is scalable across different levels of the elements that constitute the building skin, as it can be applied to search for basic properties of building materials, the integration and balance of the different materials in a technology or system, the control of dynamic properties for components and systems, and, in general, for the overall building envelope design.

In this section, we aim at demonstrating the flexibility and broad applicability of such an approach by describing five activities we have carried out in recent years. The reason to focus on our own activities rather than on a more extensive literature review lies in the desire to exemplify and reflect on the adopted methods in cases in which we have first-hand experience. The particular examples were selected as they cover different levels or parts of the building skin, and at the same time they show the different methods that can be adopted in inverse design.

15.3.1 The material level

An example of the use of inverse design at the building material level is shown in a study to define the optimal retroreflectivity properties of surfaces to reduce urban overheating effects (Manni, Lobaccaro, Goia, & Nicolini, 2018). Retroreflectivity is the ability of a particular class of material coatings to reflect a beam of incoming solar radiation back along its incident direction. This feature has been proposed to reduce the so-called urban heat island effect, but we proposed that a more advanced utilization of this characteristic could occur if the retroreflective behaviour had a strong angular dependency. The reasoning behind this is the desirability to reflect unwanted irradiance (eg, in the warm months) and to collect useful irradiance (eg, in the cold months), hence playing with the geometry of solar radiation in relation to the climatic conditions.

In this study, the relation between the 'useful' irradiation and the geometrical features of solar irradiation in different climates was, therefore, explored to identify the optimal angular properties required to activate an angular-dependent retroreflective material. By systematically analysing different surfaces and orientations in nine climatic locations ranging from Oulu (Finland) to Doha (Qatar), we found that the selective angular properties of an ideal retroreflective material should be in the angular interval between 25° and 55° for vertical surfaces. This research also developed best practices related to the application of retroreflective materials and the activation of their selective angular properties in different climate zones (Fig. 15.2). Because of the nature of the problem (optical properties that are only geometry-dependent; one single domain analysed), we could use a relatively simple strategy to perform inverse design based on the parametric analysis of solar irradiation incident on urban surfaces (ie, façade, roof and paving), performed with validated solar dynamic simulation tools based on raytracing techniques (Radiance).

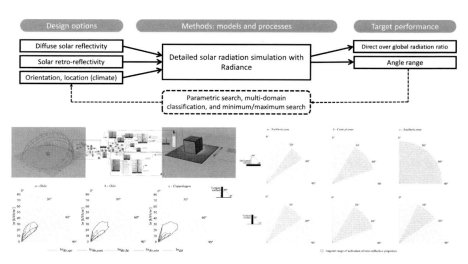

Figure 15.2 Summary of the results from the study by Manni et al. (2018) on the identification of the optimal ranges for the activation of the retroreflective property in an ideal material for surface treatment targeting the reduction of the urban heat island effect.
Source: Illustration derived from Manni, M., Lobaccaro, G., Goia, F. & Nicolini, A. (2018). An inverse approach to identify selective angular properties of retro-reflective materials for urban heat island mitigation. Solar Energy *176*, 194–210.

Much of the inverse design was performed as an automated data analysis task that led us to reconstruct the best angular ranges given the performance criteria that we had set in the investigation (the minimal angle range that collects the highest value for the ratio between direct irradiation and global irradiation). This application of the idea of inverse design can be seen as a sort of 'entry-level' in terms of both scale (a material property) and complexity of the simulation framework (parametric analysis on a single simulation engine). However, it shows the potential of using inverse design, even with a relatively simple method, to drive the development of materials for high-performance envelopes while also providing a first understanding of the theoretical limits in the effectiveness of using retroreflective materials.

Another case of inverse design for material property research is presented in a study to assist the decision-making during the product development of new switchable window types (Loonen, Singaravel, Trčka, Cóstola, & Hensen, 2014). The key features of these glazing solutions are their dynamic optical (luminous and solar) properties, which have a direct influence on both energy performance and indoor environmental quality of a building. Depending on the glazing's state (from a low-transparent dark state to high transparency in the bright state), the positive and negative sides of incoming solar radiation can be tuned over time. From a theoretical perspective, a large dynamic switching range has the highest potential for reducing energy use and maximizing occupant satisfaction. However, from the viewpoint of material development, it is not realistic to aim at very large ranges of dynamicity with very high-/low transparency states. The goal of this study was, therefore, to identify high-potential solutions and set priorities to support the development of the switchable glazing product based on integrated comfort and energy performance considerations.

By controlling the switchable glazing with a strategy based on indoor daylighting illuminance, which was found to be the control strategy that resulted in an improvement of all of the four performance criteria identified (low-energy use for lighting, heating and cooling; high degree of daylight utilization; low occurrence daylight glare risk; high levels of thermal comfort), it was shown that it is more beneficial (at least for the set of boundary conditions in the study) to tune window specifications in response to the requirements under different design scenarios instead of striving for the largest switching range possible. The study demonstrated that the role of spectral selectivity for different states is crucial and can help address the multicriteria nature of solar shading control, and that having high visible transmittance in the bright state or lower solar transmittance in the dark state is the key to obtaining higher performance, depending on the type of building where the switchable glazing is installed. This suggested that for short-term product development goals, these divergent requirements could, therefore, be met in the form of two different variants of the same product family.

This research combined BPS with sensitivity analysis and structured parametric studies to provide multiscale, multidisciplinary information about the performance of different product variants. In this inverse design process, a more complex, high-resolution, coupled simulation strategy was adopted (Fig. 15.3). Daylight simulations were first conducted in a preprocessing stage for all window states

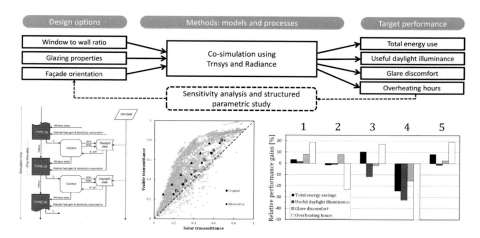

Figure 15.3 Summary of the results from the study by Loonen et al. (2014) on the identification of the optimal properties for a switchable glazing.
Source: Illustration derived from Loonen, R. C. G. M., Singaravel, S., Trčka, M., Cóstola, D. & Hensen, J. L. M. (2014). Simulation-based support for product development of innovative building envelope components. Automation in Construction *45*, 86–95.

independently with a Radiance-based tool, and these data were then supplied to an integrated building energy simulation tool (TRNSYS), which selected the right data during runtime corresponding to the operational logic in the window controller.

15.3.2 The building technology level

Moving from material level to technology level, a recent example of an inverse design problem is the design of an external fixed louvre system coated with a photovoltaic active layer, which is known by the name PVSD (photovoltaic integrated shading devices) (Taveres-Cachat & Goia, 2021; Taveres-Cachat, Lobaccaro, Goia, & Chaudhary, 2019). The aim of this investigation was to optimize the design of fixed PVSDs based on multicriteria performance requirements that tackled the thermal, electric and lighting energy quantities, and daylighting exploitation. From a more general perspective, the results presented in these studies aimed to contribute to the understanding of the extent to which competing solar energy uses can be balanced though optimized multifunctional building envelope technologies, as well as investigating how problem formulation in building design optimization impacts the obtained results.

The variables in the optimization were the number of louvre-blades as well as their individual tilt angle and position along the vertical axis. This allowed the introduction of a higher degree of freedom compared to standard external louvre solutions. Different objective functions were tested, and the most complex case consisted of three objectives: the total net energy demand, the energy converted by the photovoltaic material, and the daylighting level in the zone measured as the continuous daylight autonomy. The results highlighted that configurations with smaller louvre counts were preferable for the specific case study (a conclusion that could appear counterintuitive due to the lower surface of the system covered by photovoltaic material) and that optimization increased the performance of the PVSD compared to the reference case. When it comes to the role of the size of the solution space and how it is searched, it was shown that increasing the size of the solution space through greater degrees of freedom in terms of design variables led to better designs compared to both a full factorial parametric analysis and an optimized but more rigid model, regardless of the nature and number of objectives (Fig. 15.4).

In this study, a comprehensive cosimulation infrastructure was used, based on different thermal (EnergyPlus) and lighting (Radiance) engines, and simple photovoltaic models directly implemented in the overall simulation workflow. To carry out the inverse design approach, this was coupled to a tool for multiobjective evolutionary optimization that incorporates the SPEA-2 and HypE algorithm for the automated search in the design space. This implementation benefited from the progress in BPS that occurred in the last years, showing how current state-of-the-art workflows for inverse design can address a varied family of problems, including shape and form optimization using parametric-based modelling coupled to simulation engines for energy calculations.

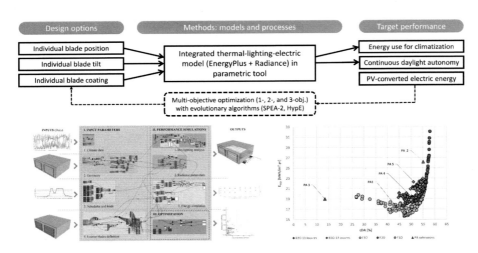

Figure 15.4 Summary of the results from the study by Cachat et al. (2019), and Cachat and Goia (2021) on the identification of the underlying principles for optimized PVSD design.
PVSD, Photovoltaic integrated shading devices.
Source: Illustration derived from Taveres-Cachat, E., Lobaccaro, G., Goia, F., & Chaudhary, G. (2019). A methodology to improve the performance of PV integrated shading devices using multi-objective optimization. Applied Energy, *247*, 731–744 and Taveres-Cachat, E., & Goia, F. (2021). Exploring the impact of problem formulation in numerical optimization: A case study of the design of PV integrated shading systems. Building and Environment, *188*, 107422.

15.3.3 The façade system level

Scaling up the analysis from a technological element to an entire façade module, inverse design has been used to understand the optimal configuration of a façade in terms of window-to-wall ratio (WWR) for office buildings characterized by best-available technologies. In the context of the two studies that addressed this investigation (Goia, Haase, & Perino, 2013; Goia, 2016), the optimal WWR value is defined as the one that minimizes, on an annual basis, the sum of the energy use for heating, cooling and lighting. The optimization of a façade configuration becomes complicated when a comprehensive approach to reduce the total energy need of the building is used because nonlinear relationships are often disclosed between different energy uses linked to façade features. This occurs to a greater extent when the façade module, as the one investigated in this study, is equipped with a dynamic shading system, and the question of how to control such devices becomes fully integrated with the overall performance assessment.

Optimal ranges for the WWR for each of the main orientations were found in four different locations (mid-latitude European region from 35° to 60°N latitude) from temperate to continental climates (Fig. 15.5). The results of the investigation showed that although there is a numerical optimal WWR in each climate and orientation, most of the ideal values for the transparent percentage can be found in a

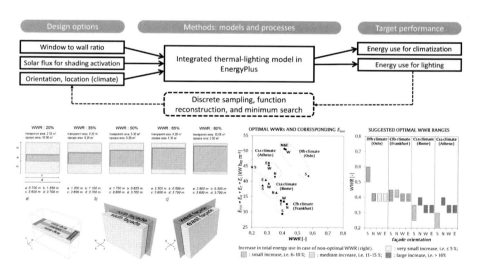

Figure 15.5 Summary of the results from the study by Goia et al. (2013) and Goia (2016) on the identification of the optimal WWR for office buildings in different climatic locations in Europe. *WWR*, Window-to-wall ratio.
Source: Illustration derived from Goia, F., Haase, M., & Perino, M. (2013). Optimizing the configuration of a façade module for office buildings by means of integrated thermal and lighting simulations in a total energy perspective. Applied Energy *108*, 515−527 and Goia, F. (2016). Search for the optimal window-to-wall ratio in office buildings in different European climates and the implications on total energy saving potential. Solar Energy *132*, 467−492.

relatively narrow range between 30% and 45%. Only the south-oriented facades in very cold or very warm climates should have transparent percentage values outside this range (higher in colder climates and lower in warmer climates) to ensure the overall optimal performance. In this study, it was also demonstrated that the total energy use may increase in the range of 5%–25% by adopting the worst WWR configuration compared to the optimal WWR. Furthermore, the control logic and threshold values for the activation of the shading device (an interpane venetian blind) were shown to be dependent on the orientation and the WWR. This finding highlighted how rules-of-thumb (that are independent on the window's dimension and orientation) for shading activation are not suitable to address optimal performance, but an inverse design approach can tackle the multidomain complexity of the building skin, including determining the best control strategies.

In this study the inverse design problem was addressed by means of integrated thermal and lighting simulations carried out in EnergyPlus, coupled with a systematic parametric search and data-postprocessing that recreates a continuous function for the problem variable starting from a relatively small number of simulations. The findings of this investigation supported the development of an integrated, modular façade system (a multifunctional façade module) capable of dynamically interacting with the other building services to reduce the building energy use and maximize the indoor comfort conditions (Favoino, Goia, Perino, & Serra, 2016).

The analysis of an ideal adaptable façade element (Kasinalis, Loonen, Cóstola, & Hensen, 2014), a so-called climate adaptive building shell (CABS), is presented to conclude this exemplificatory overview of cases of inverse design. Through enhanced, dynamic interaction with the environment, the CABS seeks to reduce the energy use for heating, cooling and lighting while maintaining high levels of indoor environmental quality. This study focused on long-term CABS, which means a solution with seasonal façade adaptation capability. The fundamental idea was that compared to short-term adaptability, long-term CABS were expected to be more feasible as they were seen as more likely to be built as low-cost add-on solutions with less challenging technologies and simpler controls. Despite the promising perspective, little information was available to quantify the relationship between seasonal adaptable façade principles and (potential) building performance improvements.

The study investigated the impact of the variability of six CABS' parameters (opaque wall density, specific heat capacity, thermal conductivity, external opaque surface absorptance, WWR and glazing type, equipped with an optimally controlled shading device—the latter within a range of seven solutions). The performance objectives were the thermal zone's primary total energy use and the number of overheating hours during occupied hours based on adaptive thermal comfort theory. Visual comfort requirements were indirectly considered through the operation of artificial lighting and blinds, which were controlled with the aim of maximizing daylight availability while preventing direct sunlight on the work plane (hence avoiding glare risk).

The finding of this study showed that a south facing office with a monthly adaptive façade in the Netherlands can achieve up to 15% energy savings and significantly improved thermal comfort conditions compared to the best performing nonadaptive façade (Fig. 15.6). The view to the outside was also improved in the

Inverse design for advanced building envelope materials, systems and operation 387

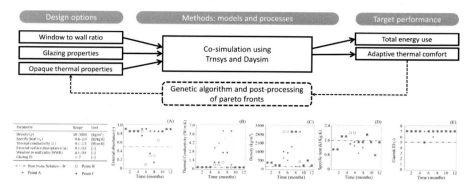

Figure 15.6 Summary of the results from the study by Kasinalis et al. (2014) on the identification of variability of six key CABS' parameters (opaque wall density, specific heat capacity, thermal conductivity, external opaque surface absorptance, WWR and glazing type—equipped with an optimally controlled shading device) to obtain the best energy and luminous performance in the Dutch climate. *CABS*, Climate adaptive building shell; *WWR*, window-to-wall ratio.
Source: Illustration derived from Kasinalis, C., Loonen, R. C. G. M., Cóstola, D., & Hensen, J. L. M. (2014). Framework for assessing the performance potential of seasonally adaptable facades using multiobjective optimization. Energy and Buildings 79, 106–113.

case of the adaptive façade as higher values of WWR were achieved in most of the year. In general, the design characteristics of such a CABS showed a strong relationship between some of the variables and the seasonal boundary conditions.

This inverse design study was conducted using dynamic BPSs with the multizone building modelling tool TRNSYS. The energy simulations in TRNSYS were coupled with the outcomes of daylight simulations in the Radiance-based DAYSIM tool, which were conducted in a preprocessing stage for all the daylight opening configurations. The long-term adaptation was investigated using 12 separate analyses based on bi-objective optimization carried out with NSGA-II, where each month of the year was addressed by a different set of simulations and optimization, and the annual performance was then obtained by the combination of monthly results. The conclusions of this study, in addition to demonstrating a framework for designing adaptive building skins, highlighted some possible avenues for product development, showing how inverse design can also be adopted in the context of research and development of dynamic envelope systems, where it demonstrates great potential to support disruptive innovation (Fig. 15.6).

15.4 Methods and digital tools for inverse design

As illustrated in the previous section, the methods and tools used to carry out inverse design can differ according to the complexity and the aim of the problem. In general, mathematical terms, an inverse design problem can be approached with

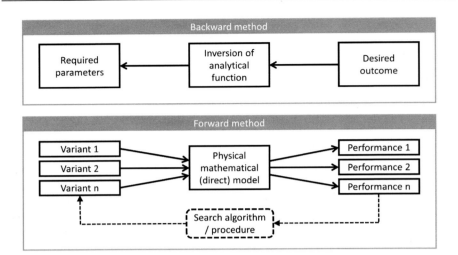

Figure 15.7 Direct and indirect function, backward methods and forward methods.
Source: Original illustration by the authors.

an analytical solution by the direct inversion of the original function (Fig. 15.7). In inverse design problems, the methods based on the identification of the inverse function and its solutions are commonly known as backward methods. However, such an approach is seldom possible in many building physics models due to the nature of the physical–mathematical descriptions adopted, which does not allow an analytical solution for the direct (or the inverse) problem.

When finding the inverse function is not possible, or when finding the inverse function does not lead to significant advantages in terms of computational efficiency or reliability of the results, forward methods are instead employed. The term 'forward methods' covers a large range of different approaches that share the common aim of exploring the entire solution domain through sampling with the use of the direct function. In other words, we can describe the forward method as the intelligent use of the direct function to search for the (sometimes extremely large) domain of possible alternatives to determine those that represent the best options in light of the preset performance requirements. The complexity of the solution domain and the complexity of the physical–mathematical modelling are the two main factors that should drive the modeller toward deciding on one or another approach for the solution of the inverse problem.

A key aspect in forward methods is the 'efficiency' of the search. The systematic, homogenous exploration of the entire space domain (which we can call the full factorial design) is often impossible because of computational and time limitations. The full factorial design can, however, be suited for simple design problems with, for example, just one variable and a single domain. This option becomes more viable considering the increasing availability of computational resources and can offer the modeller the safety of a comprehensive analysis.

More often, the exploration of the entire domain is achieved, using a sensibly lower amount of simulations, through different strategies that aim to ensure the same accuracy in the searched performance as would be found by running the full factorial design.

Several recent developments in digital support tools have cleared the way for the application of inverse design to promote design and operation of advanced building envelopes.

Computational *BPS* models are at the centre of inverse design approaches for building envelope systems. The heat and mass balances that are solved in such first-principle models can be used to understand cause-and-effect relationships that emerge from the dynamic interactions between occupants, the building skin and structure, energy systems and the environment. Particularly, the advent of multidomain modelling approaches and cosimulation methods facilitates integrated consideration of multiple physical phenomena that influence building skin performance (Taveres-Cachat et al., 2021). Moreover, increasing possibilities to model the performance of adaptive and/or advanced building envelope features in state-of-the-art BPS tools has extended the range of possibilities for inverse design of advanced façade systems.

Inverse design furthermore relies on the reconciliation of multiple competing performance requirements. Advances in *multicriteria decision making* are nowadays allowing for discovery of a refined trade-off solution among multiple candidate solutions. Within the building performance domain, we can, for example, benefit from Pareto analysis, multiattribute utility theory and risk-informed decision-making protocols (Hopfe, Augenbroe, & Hensen, 2013; Kotireddy, Loonen, Hoes, & Hensen, 2019).

Optimization algorithms constitute another important element for inverse design methods. The exploration of the design option space becomes efficient if it combines the strategic exploitation of promising design options with the search for novel combinations to avoid becoming stuck in local optima. Novel optimization algorithms have been designed to do just that while simultaneously also minimizing computational cost. Among the most employed optimization algorithms are metaheuristics (eg, evolutionary algorithms) and model-based optimizers, where it is important to note that the choice of the algorithm should be matched with the nature of the objective function and the type of variables (continuous vs discrete) that are included in the optimization problem.

Long simulation times are one of the possible caveats of inverse design on the basis of BPS models. Among the mitigation measures that are currently being investigated in this context are the use of *surrogate models* based on statistical techniques or artificial intelligence. The aim of such models is to approximate simulation outcomes at a fraction of the simulation time. Advances in computational science and machine learning are likely to represent factors that will leverage a widespread use of surrogate model optimization in the near future.

15.5 A demonstration of inverse design: combining static and dynamic solar shading

15.5.1 Background

This study intends to show how inverse design can represent a powerful tool for early design phase identification of façade configurations and solar shading solutions that exhibit optimal performance trade-offs. The study furthermore shows ways of ensuring optimal control rules or sequences for operating dynamic solar shading devices, while targeting the different performance requirements.

Building design commonly follows a stepped process where façade design features like static shading devices are specified in early design stages, whereas dynamic solar shading systems and their controls are selected and refined at a later design stage. Façade design optimization studies, therefore, commonly exclude dynamic solar shading devices or assume simple control strategies. The hypothesis we investigate in this study is that this sequential design approach leads to less beneficial outcomes than a process in which dynamic solar shading controls and specifically building-aware control strategies, are considered an integral part of early-stage façade design. We expect that tackling this problem through an inverse design approach will lead us to high-performance solutions that would likely be overlooked in conventional design processes, or even in more advanced ones based on stepped design optimization.

The question of whether or not combinations of static and dynamic solar shading devices can be designed and controlled to exploit their unique geometrical features (leading to more effective daylight utilization) will be explored in a case study that focuses on a south facing office space and that looks for optimal trade-offs between visual comfort, daylighting and view performance. Two combinations of static and dynamic types of solar shading devices are evaluated, and each combination is aimed at utilizing the strengths of one of the two types of shading systems in blocking direct sunlight at different solar positions.

In the first configuration of shading devices, vertical static fins on the exterior of the building are combined with an automated interior roller blind system (Fig. 15.8, left). At solar positions with a low solar altitude (α) and a high solar azimuth relative to the window surface normal (γ^*), blocking direct sunlight requires positioning a roller blind such that it largely covers the window, thereby negatively affecting the admission of daylight and views. Vertical fins, however, can block direct sunlight from this direction while preserving a more unobstructed view of the sky, even when the fins are modestly dimensioned.

In the second configuration, horizontally oriented shading elements on the exterior of the building are combined with an automated interior vertical blind system (Fig. 15.8, right). At solar positions characterized by a low γ^* (eg, at solar noon for the south-facing façade), blocking direct sunlight requires that the blinds are fully closed. At these sun positions, an overhang or horizontal louvre is more effective at obstructing direct sunlight while preserving a large unobstructed view of the sky.

To test if including building-aware control strategies in façade design optimization leads to different façade design choices, and possibly more modestly dimensioned static shading devices, different designs are also tested for both the horizontal overhang and the vertical fins.

15.5.2 Methodology

This study employs a forward inverse design method for optimizing both the design of the static solar shading devices as well as for choosing optimal control actuations

Inverse design for advanced building envelope materials, systems and operation

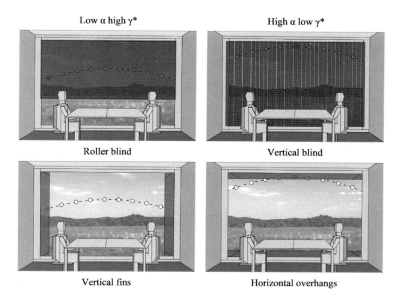

Figure 15.8 Combinations of dynamic (top) and static (bottom) solar shading devices. The graph illustrates how the weaknesses of the dynamic solar shading devices in blocking direct sunlight at particular sun positions could potentially be mitigated by using static shading devices.
Source: Original illustration by the authors.

for the dynamic shading systems. The full solution domain is homogenously sampled in both cases, which is facilitated by efficient matrix-based daylight and view simulation methods. The simulation workflow that we use in this study gives an example of how inverse design can be applied to static design features on the system level as well as to control strategies for dynamic operation of advanced building skins.

Performance goals and indicators

The performance aspects of interest in this study are visual comfort, daylighting and view to the outdoors. Among these aspects, visual comfort is given the highest priority. Visual comfort is operationalized as a lack of discomfort and quantified using daylight glare probability simplified (Wienold, 2010), abbreviated as DGPs. DGPs is assessed for the four seating positions and viewing directions, and at each timestep the maximum of the four values is used ($DGPs_{mx4p}$). To assess glare performance across a year, the $DGPs_{mx4p}$ value that is exceeded for at least 5% of the year is taken ($DGPs_{mx4p;95\%}$). The goal in this study is to achieve a high degree of glare protection where daylighting conditions lead to 'perceptible' glare for less than 95% of the occupied time ($DGPs_{mx4p;95\%} \leq 0.35$).

Daylighting performance is assessed using spatial daylight autonomy $sDA_{300lx/50\%}$ (Heschong et al., 2012). To evaluate instantaneous daylighting performance, the daylit area fraction (D_{300lx}) is used. This indicator gives the percentage of floor area that receives at least 300 lux at a point in time. The goal in this case study is to maximize daylighting performance. This is operationalized as a requirement to achieve at least good ($sDA_{300lx/50\%} \geq 55\%$) and preferably excellent daylighting performance ($sDA_{300lx/50\%} \geq 75\%$).

The view toward the outside is assessed in terms of view quantity, and an isovist field approach (Benedikt, 1979; Turan, Reinhart, & Kocher, 2019) is used to assess the part of the exterior that is visible to occupants. At each timestep the view fraction (VF) is assessed, which is defined as the cone of human vision that is not obstructed by the façade or shading devices. The VF is assessed for the four occupant positions, and at each timestep the average VF is used. Performance across a year is assessed using a time-weighted VF $VF_{av4p;tw}$ where each timestep is given an equal weight.

Overview of the simulation workflow

Fig. 15.9 shows a graphical overview of the applied simulation methods. Vertical illuminance for predicting glare at the four occupant positions, and the interior daylight illuminance distribution across the grid of sensor points are simulated using the Radiance three-phase method (McNeil & Lee, 2013) and the Honeybee [+] visual programming interface plug-in for Rhinoceros 3D. The optical behaviour of the combined glazing and shading system is described using subsystem simulations with LBNL-Window. The VF is simulated through raytracing using the Ladybug plug-in (Roudsari, Pak, & Smith, 2013). Indoor illuminance and VF are simulated for each design (d), each timestep (t), and each discrete position of the dynamic shading systems, referred to as the shade state (ShSt). This way, matrices are created that describe indoor daylighting conditions ($E_{ShSt,t}$) and effects on view ($VF_{ShSt,t}$) across the complete control space. The performance effects of the different control strategies are found by computing the ShSt at each timestep (t) and referencing the corresponding subset of the indoor illuminance and view matrices.

Façade design elements: static solar shading devices

Within the inverse design approach, the different static solar shading elements are described in a generic way by specifying the geometric relationships, but without determining the exact shape and dimensions. For simple rectilinear shading devices, this can be done using a set of angles indicating the part of the sky that they obstruct (Lyons, Curcija, & Hetzel, 2017; Whitsett & Fajkus, 2018). A horizontally oriented overhang, or a set of horizontal louvers, can be described using a vertical shading angle (VSA). This angle is defined in this study as the angle between the furthest edge of shading device and the vertical façade surface (Fig. 15.10).

Inverse design for advanced building envelope materials, systems and operation 393

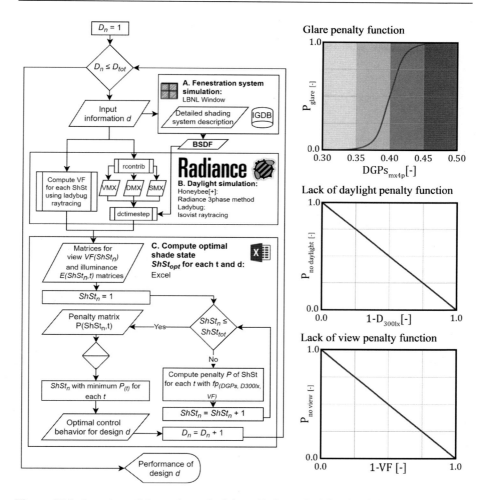

Figure 15.9 Overview of the study methodology. Left: method for simulating design (*d*) and control performance and the exhaustive search method for deriving optimal building-aware control behaviour (C). Right: penalty functions that are used to define penalty scores for each ShSt at each timestep (*t*). *ShSt*, Shading state.
Source: Original illustration by the authors.

Likewise, vertically oriented shading fins can be described using the horizontal shading angle (HSA).

Table 15.1 gives an overview of the different vertical fin and horizontal overhang designs that are investigated. In these alternatives a single overhang and vertical fin are used to achieve a particular VSA or HSA. It should be noted that this approach leads to very deep shading devices that would not be feasible in practice. Each VSA or HSA alternative, however, can be achieved with a variable number of

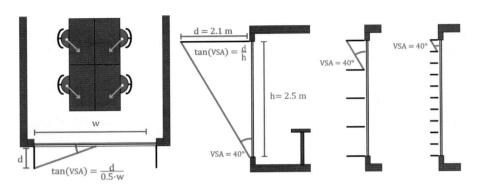

Figure 15.10 Left: the HSA and the four user positions and viewing directions that are used for the glare and view performance assessments. Right: The VSA and three horizontal louvre designs that are all described using the same VSA and lead to similar degrees of solar exposure. *HSA*, Horizontal shading angle; *VSA*, vertical shading angle.
Source: Original illustration by the authors.

Table 15.1 Investigated horizontal overhang and vertical fin designs described in terms of their vertical shading angle (VSA) and horizontal shading angle (HSA).

Vertical fin design alternatives									
HSA	0°	10°	20°	30°	40°	50°	60°	70°	80°
Fin depth (m)	0.00	0.40	0.82	1.30	1.89	2.68	3.90	6.18	12.76
Horizontal overhang design alternatives									
VSA	0°	10°	20°	30°	40°	50°	60°		
Overhang depth (m)	0.00	0.44	0.91	1.44	2.09	2.97	4.31		

shading devices of a smaller depth that offer equivalent degrees of solar exposure (Fig. 15.3). A series of sensitivity analyses, which can be found in Bodde, Vries de, Loonen, and Hensen (2020), were executed to validate the assumption that varying static shading designs with the same VSA can be considered equivalent in terms of their daylighting and visual comfort performance.

Façade design elements: the baseline dynamic solar shading strategies

For the roller blinds a conventional up–down control strategy is used as a baseline (RB-BL) that fully raises or lowers the shade in response to an exterior horizontal irradiance threshold of 200 W/m^2. For the vertical blind a sun-tracking control strategy is used (VB-BL) that aligns the surface normal of the blinds with γ^* (Table 15.2).

Table 15.2 Baseline control strategies in pseudo code for the vertical blind system (left) and the roller blind (right) system.

VB-BL: Vertical blind baseline		RB-BL: Roller blind baseline	
If $\gamma^* \leq WN_\gamma$	Else	If $I_h \geq 200$ W/m²	Else
bRA = 90 + γ^*	bRA = 90 − γ^*	Fully lower shade	Fully raise shade

Table 15.3 Penalty weights.

Function	W_{glare}	$W_{nondaylit}$	$W_{no\ view}$
P1	0.330	0.330	0.330
P2	0.440	0.560	0.000
P3	0.560	0.330	0.110
P4	0.670	0.330	0.000
P5	0.949	0.050	0.001

Façade design elements: dynamic solar shading control strategy

In the building-aware control alternatives, a performance-based controller (RB-PB and VB-PB) is used to select control actuations by comparing the effects that all possible control actions would have on weighted performance indicators (step C in Fig. 15.9). H-D$_{300lx}$, DGP$_{smx4p}$ and VF are computed for each design, timestep and discrete dynamic solar shading position. The three performance indicators are combined into a single penalty score that is to be minimized using Eq. 15.1 and the weights reported in Table 15.3. A complicating factor in inverse design of shading controls is that performance indicators, such as sDA$_{300lx/50\%}$ and DGP$_{smx4p;95\%}$, measure performance across a period of time and cannot directly be translated into functions and weights that describe performance trade-offs at individual points in time. Five different sets of weights are, therefore, used to represent different priorities in making performance trade-offs, and the most beneficial control weights for each case are selected afterward based on the desired performance outcomes.

$$P_{tot(t)} = w_{glare} \cdot P_{glare(t)} + w_{nondaylit} \cdot P_{nondaylit(t)} + w_{no\ view} \cdot P_{no\ view(t)} \quad (15.1)$$

15.5.3 Results

Façade design option 1: roller shade with vertical fins

Fig. 15.11 (left) shows daylighting performance (sDA$_{300lx/50\%}$) and visual comfort (DGP$_{smx4p;95\%}$) for the roller blind strategies combined with exterior vertical fin designs with a varying HSA. Alternatives using the baseline (RB-BL) strategies are labelled with a diamond and the cases where the performance-based (RB-PB)

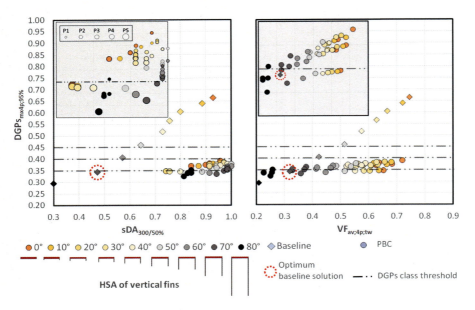

Figure 15.11 Glare in relation to daylighting performance (left), and glare in relation to view performance (right) for the simulated combinations of different exterior shading fin designs and interior roller blind control strategies.
Source: Original illustration by the authors.

controller is used are labelled with a circle. It is found that glare and daylighting performance are very sensitive to the chosen HSA of the vertical fin. Additionally, if a designer was to optimize the vertical fin design assuming only conventional strategies, only the 70° and 80° HSA alternatives would fulfil the visual comfort requirement and the 70° HSA design would be chosen for its superior daylighting performance.

The cases where the performance-based controller is used are less sensitive to the design of the fins. The RB-PB alternatives are clustered close to the utopia point and the control variations that meet the visual comfort requirements all offer very good daylighting performance (73%–98% $sDA_{300lx/50\%}$) regardless of which vertical fin design is chosen. Significant differences among the different HSA alternatives can still be observed, however, and the 70° HSA design is Pareto optimal. For this design alternative, going from the RB-BL strategy to the building-aware control strategy offers significant improvements in daylighting performance (55% higher $sDA_{300lx/50\%}$).

For the RB-PB strategies, increasing the vertical fin depth has the counterintuitive effect of increasing daylighting performance in many cases. The improvements in daylighting performance are the result of the PB strategy raising the roller blind further at instances where the vertical fins prevent direct sunlight from reaching occupants. Both control strategies lead to the same Pareto optimal solution. However, the RB-PB strategy leads to a range of vertical fin solutions being near

Pareto optimal (50°−70° HSA), whereas only a single design option fulfils the design goals and requirements using the RB-BL strategy.

The increased near-Pareto optimal design space with the performance-based controller potentially offers benefits in terms of other performance aspects. Fig. 15.11 (right) shows the extent to which occupants have a view to the outdoors ($VF_{av4p;tw}$) with the different control strategies and vertical fin designs. The graph shows that $VF_{av4p;tw}$ is highly sensitive to both the control strategy and the HSA of the vertical fins.

The 70° HSA fin design that was selected when only the RB-BL strategy was considered shows very poor view performance (32% $VF_{av4p;tw}$). In contrast the range of fin designs that was identified as having near-optimal daylighting and visual comfort performance for the RB-PB strategy offers more beneficial view performance (35%−52% $VF_{av4p;tw}$). Overall, the 50° HSA vertical fin design offers beneficial trade-offs when all performance aspects are considered (0.34 $DGPs_{mx4p;95\%}$, 88% $sDA_{300lx/50\%}$ and 51% $VF_{av4p;tw}$).

Façade design option 2: vertical blinds with horizontal overhang

Fig. 15.12 shows daylighting performance ($sDA_{300lx/50\%}$) and view performance ($VF_{av4p;tw}$) in relation to visual discomfort ($DGP_{smx4p;95\%}$) for the vertical blind

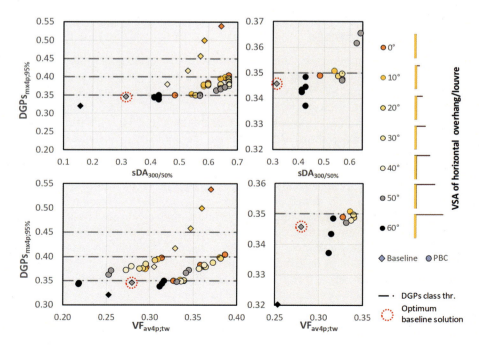

Figure 15.12 Glare, daylighting and view performance for the simulated combinations of different exterior horizontal overhang designs and interior vertical blind control strategies. *Source:* Original illustration by the authors.

strategies combined with exterior overhang designs with a varying HSA. As with the roller blind strategies, the performance-based (VB-PB) control alternatives are clustered more closely to the utopia point for daylighting performance and visual discomfort. In the case of the vertical blinds, however, the VB-PB controller does not guarantee good daylighting performance for all VSA designs and the 60° VSA design leads to poor performance.

Again, only two options fulfil the visual comfort requirement with the RB-BL strategy and a single option can be identified as optimal (50° VSA). When the horizontal overhang design is optimized assuming the VB-PB controller, a range of near-Pareto optimal solutions (20°−50° VSA) is found. Additionally, the performance-based controller offers significant improvements in the daylighting performance of all shading fin designs compared to the RB-BL (eg, 25% higher $sDA_{300lx/50\%}$ for 50° VSA).

As with the roller blinds, the overhang design (50° VSA), which is identified as optimal when only the conventional VB-BL strategy is considered, offers very poor view performance (28% $VF_{av4p;tw}$), whereas the range of designs that is found when the VB-PB controller (20°−50° VSA) offers more beneficial view performance (36%−39% $VF_{av4p;tw}$). The effect of the depth of the overhang on $VF_{av4p;tw}$ is less strong than with the vertical fins and all alternatives in the 20°−50° VSA range offer similar glare, daylighting and view performance.

15.5.4 Lessons learned

In both shading configurations, different optimal static shading designs could be identified when a simple rule-based dynamic solar shading control strategy was assumed (70° HSA and 50° VSA) and in the cases where a performance-based and building-aware control strategy was used (50° HSA and 20° VSA). Generally, the performance-based control approach caused daylighting and glare performance to be less sensitive to the external shading design. In both the investigated cases, a wider range of near-optimal solutions was found for the performance-based controls (50°−70° HSA and 20°−50° VSA) than for the simple rule-based controls.

The results illustrate the advantages of considering performance-based building-aware controls in the early design phase. Applying a stepped-optimization approach, where the façade design is first optimized assuming simple rule-based controls and advanced building-aware strategies are implemented at a later stage, does not lead to façade design outcomes with poor performance. Including a building-aware control approach in the early-stage design optimization, however, does present the opportunity of having greater design freedom. When shading controls are inversely designed together with the façade design, this leads to a larger part of the design space offering near-optimal performance and hence to a larger range of suitable façade design options. This design freedom can be exploited to improve other performance aspects that were not investigated in this study such as the architectural expression of the façade, investment and operational costs, energy performance and thermal comfort.

The results also offer insight into the potential of inverse design of solar shading control strategies for specific building applications. For the façade design alternatives that were identified using the simple BL strategy (70° HSA and 50° VSA), switching to the PB control strategy that exploits façade features led to significant performance improvements (25−55 higher $sDA_{300lx/50\%}$ and 3%−10% more $VF_{av4p;tw}$ with compatible glare performance).

This case study gives an example of how the inverse design process can be used to identify optimal façade design features and derive optimal control behaviour for dynamic building skins. The presented simulation workflow illustrates how performance indicators and design features can be formulated such that they can be made operational within an inverse design framework. Additionally, the workflow shows how the multicriteria nature of design and control of advanced building skins can be addressed within the inverse design process.

15.6 Conclusion

This chapter has introduced the background principles and highlighted the merits of inverse design for supporting the design and operation of advanced building envelopes. At the heart of inverse design is a targeted search activity that aims to find the design solution that meets the specified objectives in the best way possible. Through a description of five illustrative examples at the material, technology and system levels, the various possibilities were demonstrated in a practical context. By showing an in-depth example of inverse design for optimizing the combination of static and dynamic solar shading system, the added value of inverse design was further demonstrated with a focus on informed decision-making among multiple conflicting performance criteria in an early design context.

We expect that inverse design approaches will play a more prominent role in future product development and design of advanced building envelopes. New developments in multidomain (co)simulation workflows and user-friendly interfaces will make it easier for practitioners to add such design and analysis methods to their repertoire. Likewise, the increased application of cloud computing and advances in data-driven surrogate modelling methods will eliminate constraints related to computation time or the inability to explore the entire design option space. We anticipate a particularly large potential for inverse design for combined application of material development advanced control of dynamic components, as this is an area where traditional methods offer limited support. Lastly, by taking advantage of developments in generative design and performance-driven form-finding studies, it is foreseen that engineering and design disciplines will mutually strengthen each other through increased integration of inverse design approaches.

Acknowledgement

We would like to acknowledge all coauthors of the cases presented in Section 15.3 for the fruitful research collaborations over the years: Andrea Nicolini, Charalampos Kasinalis, Daniel

Cóstola, Ellika Taveres-Cachat, Gabriele Lobaccaro, Gaurav Chaudhary, Jan Hensen, Marco Perino, Marija Trčka, Matthias Haase, Mattia Manni and Sundaravelpandian Singaravel. We also would like to express our special gratitude to Kim Bodde who has extensively collaborated with us for the case presented in Section 5 in the framework of her MSc thesis.

References

Attia, S., Bilir, S., Safy, T., Struck, C., Loonen, R., & Goia, F. (2018). Current trends and future challenges in the performance assessment of adaptive façade systems. *Energy and Buildings*, *179*, 165–182.

Benedikt, M. L. (1979). To take hold of space: Isovists and isovist fields. *Environment and Planning B: Planning and design*, *6*(1), 47–65.

Bianco, L., Cascone, Y., Avesani, S., Vullo, P., Bejat, T., Koenders, S., ... Favoino, F. (2018). Towards new metrics for the characterisation of the dynamic performance of adaptive façade systems. *Journal of Façade Design and Engineering*, *6*(3), 175–196.

Bodde, K., Vries de, S. B., Loonen, R. C. G. M., & Hensen, J. L. M. (2020). Coupled design optimization of façade design and automated shading control for improving visual comfort in office buildings (Master thesis). Eindhoven University of Technology, Eindhoven.

Boswell, K., Hoffman, S., Selkowitz, S., & Patterson, M. (2021). Skin metrics: The wicked problem of façade system assessment. *Journal of Façade Design and Engineering*, *9*(1), 131–146.

Delmastro, C., Abergel, T., Dulac, J., Lane, K., Janoska, P., & Prag, A. (2019). *Perspective from the clean energy transition—The critical role of buildings*. Paris: IEA.

Dulikravich, G. S. (1988). Special issue on inverse design and optimization in engineering: Introduction. *Applied Mechanics Reviews*, *41*(6), 216–216.

Favoino, F., Goia, F., Perino, M., & Serra, V. (2016). Experimental analysis of the energy performance of an ACTive, RESponsive and Solar (ACTRESS) façade module. *Solar Energy*, *133*, 226–248.

Goia, F. (2016). Search for the optimal window-to-wall ratio in office buildings in different European climates and the implications on total energy saving potential. *Solar Energy*, *132*, 467–492.

Goia, F., Haase, M., & Perino, M. (2013). Optimizing the configuration of a façade module for office buildings by means of integrated thermal and lighting simulations in a total energy perspective. *Applied Energy*, *108*, 515–527.

Heschong, L., Wymelenberg van den, K., Andersen, M., Digert, N., Fernandes, L., Keller, A., ... Mosher, B. (2012). *Approved method: IES spatial daylight autonomy (sDA) and annual sunlight exposure (ASE)*. IES-Illuminating Engineering Society.

Hopfe, C. J., Augenbroe, G. L., & Hensen, J. L. (2013). Multi-criteria decision making under uncertainty in building performance assessment. *Building and Environment*, *69*, 81–90.

Jain, A., Shin, Y., & Persson, K. A. (2016). Computational predictions of energy materials using density functional theory. *Nature Reviews Materials*, *1*(1), 15004.

Jelle, B. P., Hynd, A., Gustavsen, A., Arasteh, D., Goudey, H., & Hart, R. (2012). Fenestration of today and tomorrow: A state-of-the-art review and future research opportunities. *Solar Energy Materials and Solar Cells*, *96*, 1–28.

Kämpf, J. H., & Robinson, D. (2010). Optimisation of building form for solar energy utilisation using constrained evolutionary algorithms. *Energy and Buildings*, *42*(6), 807–814.

Kasinalis, C., Loonen, R. C. G. M., Cóstola, D., & Hensen, J. L. M. (2014). Framework for assessing the performance potential of seasonally adaptable facades using multi-objective optimization. *Energy and Buildings, 79*, 106–113.

Kesidou, S. L., & Sorrell, S. (2018). Low-carbon innovation in non-domestic buildings: The importance of supply chain integration. *Energy Research & Social Science, 45*, 195–213.

Kotireddy, R., Loonen, R., Hoes, P. J., & Hensen, J. L. (2019). Building performance robustness assessment: Comparative study and demonstration using scenario analysis. *Energy and Buildings, 202*, 109362.

Loonen, R. C. G. M., Singaravel, S., Trčka, M., Cóstola, D., & Hensen, J. L. M. (2014). Simulation-based support for product development of innovative building envelope components. *Automation in Construction, 45*, 86–95.

Lyons, P., Curcija, C., & Hetzel, J. (2017). Fenestration. *Handbook—Fundamentals* (pp. 1–68). Atlanta, GA: ASHRAE.

Manni, M., Lobaccaro, G., Goia, F., & Nicolini, A. (2018). An inverse approach to identify selective angular properties of retro-reflective materials for urban heat island mitigation. *Solar Energy, 176*, 194–210.

McNeil, A., & Lee, E. S. (2013). A validation of the radiance three-phase simulation method for modelling annual daylight performance of optically complex fenestration systems. *Journal of Building Performance Simulation, 6*(1), 24–37.

Nagy, D., Zhao, D., & Benjamin, D. (2018). Nature-based hybrid computational geometry system for optimizing component structure. In K. De Rycke, C. Gengnagel, O. Baverel, J. Burry, C. Mueller, M. M. Nguyen, P. Rahm, & M. R. Thomsen (Eds.), *Humanizing digital reality: Design modelling symposium Paris 2017* (pp. 167–176). Singapore: Springer.

Roudsari, M. S., Pak, M. & Smith, A. (2013). Ladybug: A parametric environmental plugin for grasshopper to help designers create an environmentally-conscious design. In: *Proceedings of the 13th International IBPSA conference, Lyon, France* (pp. 3128–3135).

Steinemann, A., Wargocki, P., & Rismanchi, B. (2017). Ten questions concerning green buildings and indoor air quality. *Building and Environment, 112*, 351–358.

Taveres-Cachat, E., Favoino, F., Loonen, R. C. G. M., & Goia, F. (2021). Ten questions concerning co-simulation for performance prediction of advanced building envelopes. *Building and Environment, 191*, 107570.

Taveres-Cachat, E., & Goia, F. (2021). Exploring the impact of problem formulation in numerical optimization: A case study of the design of PV integrated shading systems. *Building and Environment, 188*, 107422.

Taveres-Cachat, E., Lobaccaro, G., Goia, F., & Chaudhary, G. (2019). A methodology to improve the performance of PV integrated shading devices using multi-objective optimization. *Applied Energy, 247*, 731–744.

Turan, I., Reinhart, C. & Kocher, M. (2019). Evaluating spatially-distributed views in open plan work spaces. In: *Proceedings of building simulation* (pp. 1098–1105), Rome 2019: IBPSA.

Whitsett, D., & Fajkus, M. (2018). *Architectural science and the sun: The poetics and pragmatics of solar design.* New York: Routledge.

Wienold, J. (2010). *Daylight glare in offices.* Freiburg: Fraunhofer Institute for Solar Energy Systems ISE.

Wright, J. A., Loosemore, H. A., & Farmani, R. (2002). Optimization of building thermal design and control by multi-criterion genetic algorithm. *Energy and Buildings, 34*(9), 959–972.

Zhai, Z. J., Xue, Y., & Chen, Q. (2014). Inverse design methods for indoor ventilation systems using CFD-based multi-objective genetic algorithm. *Building Simulation*, *7*(6), 661–669.

Zunger, A. (2018). Inverse design in search of materials with target functionalities. *Nature Reviews Chemistry*, *2*(4), 0121.

Towards automated design: knowledge-based engineering in facades

Jacopo Montali[1], Michele Sauchelli[2] and Mauro Overend[3]
[1]Algorixon Srl, Parma, Italy, [2]WSP UK Ltd, London, United Kingdom, [3]Department of Architectural Engineering + Technology, Chair of Structural Design & Mechanics, Delft University of Technology, Delft, The Netherlands

Abbreviations

AEC	architecture, engineering and construction
KBE	knowledge-based engineering
PM	product model
UML	unified modelling language
BIM	building information modelling
IFC	industry foundation classes
API	application programming interface
GUI	graphical user interface
MOKA	methodologies and tools oriented to knowledge-based engineering applications
PCC	precast concrete cladding

16.1 Introduction

16.1.1 Context

The process of designing any product is characterized by a series of tasks which, by their own intrinsic nature, require the concurrent evaluation of many design possibilities (Ashby, 2000). Only a few of these are eventually selected after assessing them against a set of requirements such as expected performance, manufacturability/buildability, cost or even aesthetics. The route to generating and selecting such viable design possibilities is often impeded by the lack of proper design/manufacturing knowledge, especially at early design stages when crucial decisions are made. As a rule-of-thumb, it is commonly agreed that 80% of the overall costs of a product are implicitly committed in the initial 20% of the design phase of that product (Cavalieri, Maccarrone, & Pinto, 2004). This means that poorly informed choices made early in the design stage can result in disproportionately higher final costs. Therefore it is essential to bring knowledge from downstream stages and apply it in upstream processes, for example, using knowledge about the subsequent manufacturing or procurement processes in the early design stages. Knowledge-

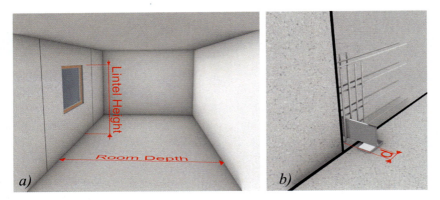

Figure 16.1 Two examples of required early incorporation of design knowledge in facades to design (A) for performance (window position) and (B) for manufacturability (corbel bracket position).

based engineering (KBE), the management and digitalization of engineering knowledge into ready-to-use software applications, provides a means to address this challenge. This chapter describes how early design stages influence cost and environmental performance of the final product and shows the basic components of a KBE application and its development process, along with a representative example.

As products to be designed and built, building facades are no exception to this process. For example, Fig. 16.1 illustrates two cases a façade designer may encounter when designing for performance or for manufacturability—the example is based on a precast façade system for a residential building, commonly used in the United Kingdom. In Fig. 16.1A the window's global dimensions and position in residential buildings will strongly influence the amount of daylight in the room. In this case, adopting the heuristic[1] rule 'lintel height > 40% room depth' will tend to reduce design changes at later stages of design when the design cannot easily be modified. Fig. 16.1B shows the detail of a cast-in steel billet supporting a prefabricated concrete façade panel. In this case the early determination of the billet location with respect to the panel edge (distance d in figure) strongly influences how easily reinforcement bars can be installed.

Both are simple examples of how upstream knowledge utilized in the early stages to reduce the risks of onerous late changes. In the first example, altering window size or location at later stages would require redesign, changing glass properties to increment transparency may come at a cost, and increasing the room's inside reflectance of surfaces will significantly affect the design of the interior spaces. In

---

[1] The term heuristic (from Greek *heurískō*: I discover) refers to a form of knowledge that is achieved by experience, rather than from first principles or more explicit forms. For that reason, it is also referred to as 'implicit knowledge'. Despite the authors provide a formula via rule-of-thumb in the previous example, collecting and reporting this type of knowledge can often be challenging.

the precast concrete panels example, late-stage alterations to the billet bracket, if at all possible, will come at a relatively high cost, due to potential clashes with other more relevant structural or architectural elements and the need for costly bespoke rebar design.

In façade design, many requirements from different disciplines generate a series of interdependencies that constrain the scope of each separately considered requirement. For the previous two examples, if the project was a residential building with precast façade panels, the size of the windows would be influenced by minimum daylight levels and its position by the required clearance for the support location, among other things.

Managing design knowledge and its early and concurrent incorporation along the design process is one of the major challenges in façade design (Kassem & Mitchell, 2015; Montali, Overend, Pelken, & Sauchelli, 2017a; Voss, Jin, & Overend, 2013). The process is characterized by increasing levels of detail, starting from a façade concept as a rationalized set of bidimensional surfaces (Henriksen, Lo, & Knaack, 2016), to the specific choice of materials and buildup at detailed design stage to produce a better estimate of the expected performances and costs, until the technical design stage where the design solution is detailed for production. Each step commits to a specific set of choices that can be difficult and costly to reverse as design progresses. A schematic view of this process is shown in Fig. 16.2 (Montali et al., 2017a). Most of the time designers are assisted by façade consultants, who bring design experience and knowledge on different façade systems. However, design

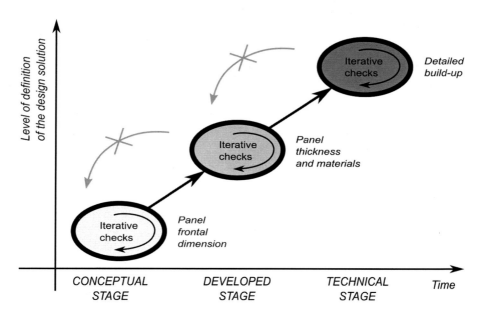

Figure 16.2 Schematic view of the façade design process (Montali et al., 2017a).

iterations may still be required to incorporate manufacturing constraints arising from the individual manufacturer's capabilities.

This challenge has been recognized in various countries, to the extent that new design and procurement processes have recently been devised. For example, the RIBA Plan of Work (RIBA Plan of Work, 2020) in the United Kingdom has been recently updated by introducing new procurement routes, to anticipate the level of information at early stages and to align the design and build process with the national sustainability and digital agenda. Another example is the Singapore Construction Industry Transformation Map (Construction Industry Transformation Map, 2017), introducing specific actions in the design process requiring the implementation of manufacturing and installation constraints during the early stages of design. These new forms of procurement rely also on information[2] exchange via the Industry Foundation Class (IFC) data schema, which sets the ground for building information modelling (BIM) (Alfieri, Seghezzi, Sauchelli, Di Giuda, & Masera, 2020). Despite these new procurement approaches, the time spent to retrieve useful information and the implementation of design knowledge still remains a challenge (Aram, Eastman, & Sacks, 2014; Belsky, Sacks, & Brilakis, 2016). Furthermore, these new processes require the active involvement of manufacturers/specialist contractors at early stages.

Advances in computer-aided design in the last decade, fuelled by digitalization and a substantial maturity of modern programming languages (e.g., Python, C#, Ruby...), have made it possible to develop systems to automatically reuse design and manufacturing knowledge to support product design. Other industries, such as maritime and aerospace, have developed digital tools [for a complete review, see (Montali et al., 2017a)] to manage design and manufacturing knowledge at early stages in a holistic and automatic manner and to ensure that the developed design concept faithfully represents the expected final product. This approach is referred to as Knowledge Based Engineering (KBE). The main objective of KBE is to automate repetitive and interdisciplinary knowledge and integrate it into one unique software platform (Cooper & La Rocca, 2007; La Rocca, 2012). A comprehensive review of what these two industries have developed in terms of KBE development is provided in the study of Montali et al. (2017a). Some examples of application of KBE in facades design can be found in some early researches on precast concrete cladding (PCC) cladding and timber walls (Day, Gasparri, & Aitchison, 2019; Montali, Sauchelli, Jin, & Overend, 2019).

16.1.2 KBE to optimize performance and cost

The fundamental idea behind KBE applications is twofold: (1) retrieving knowledge about the design and manufacture of a specific product and (2) implementing this knowledge into a highly parametric software model, usually supported by 3D

---

[2] The fundamental difference between information and knowledge is that the former can be defined as 'data put into a context', whereas knowledge is the ability to produce useful information, although there is no full agreement on the definition of the latter.

representations, to integrate design disciplines. The model, commonly referred to as 'product model' (PM) (La Rocca, 2012), is based on design and manufacturing knowledge and informs the users about their design choices automatically, by interrogating a database of rules and constraints. The implemented knowledge is reusable, that is, suitable for performing routine tasks that would otherwise be delegated to manual, time-consuming and error-prone activities.

The major benefit of using KBE applications is that design times are shortened, yet the spectrum of all possibilities is explored thoroughly to identify the domain of feasible design solutions. From this the near-optimal design solutions may be selected. The whole process can also be achieved automatically, for example, alongside optimization algorithms to minimize one or more measurable function, known as objective functions. More formally, if $z(x) = [z_1(x), \ldots, z_p(x)] \in R^p$ is a vector of objective functions calculated against a specific configuration of the PM described by the vector $x = (x_1, \ldots, x_i, \ldots, x_n) \in R^n$, an optimization problem would respect the following:

$$\min z(x) \tag{16.1}$$

while

$$g_i(x) \leq 0 \quad i = 1, \ldots, m \tag{16.2}$$

$$x_j \geq 0 \quad j = 1, \ldots, n \tag{16.3}$$

If $p > 1$, the problem is multiobjective, otherwise it is single-objective. KBE, apart from automating the calculation of a series of p objective functions $z_i(x)$, places the emphasis on constraining appropriately the problem with the largest amount of m knowledge-based functions $g_i(x)$. An example that uses this approach is provided in this chapter.

Fig. 16.3 shows the benefits in terms of cost and time when using KBE to automate a portion of the design process. As mentioned briefly earlier, the design process presents an intrinsic discrepancy between *actual* and *committed* costs since the value of present choices includes the cost of the design task (actual costs) as well as the downstream (ie, manufacturing and construction) costs of executing that particular choice. A more holistic design approach includes future costs that these choices entail, such as the cost of not achieving the expected performance or the cost of an exceedingly complex manufacturing process (committed costs). This discrepancy between actual and committed costs is most pronounced in the early stages of design (approximately equal to 20%) and it reaches zero at the end of the construction process, when actual and committed costs converge. In this context, KBE allows designers to commit earlier in the process to optimized manufacturable/buildable design solutions, with the resulting benefit of reducing design times and overall costs.

Despite the evident benefits of KBE, there are a series of barriers and shortfalls. The main shortcomings are the initial development costs of a KBE system; in

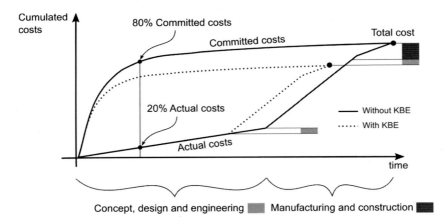

Figure 16.3 Qualitative diagram for actual and committed costs versus time in product development. As a rule-of-thumb, when 20% of actual costs have been produced, nearly 80% of the total costs have been committed, even if unknown. KBE has the potential to shorten design time and total costs by committing early in the process to manufacturable/buildable design solutions. *KBE*, Knowledge-based engineering.
Source: Adapted from Cavalieri, S., Maccarrone, P., & Pinto, R. (2004). Parametric vs. neural network models for the estimation of production costs: A case study in the automotive industry, *International Journal of Production Economics, 91,* 165–177.

particular, knowledge collection, knowledge maintenance (change/updating of the knowledge database), software implementation challenges. A clear business case for setting up a KBE approach is, therefore, required. Benefits and challenges will be further discussed by means of a 'real-world' example of a KBE application.

16.2 Constructing a KBE application

The benefits of capturing and using knowledge through a KBE approach in the design process are clear, but how can this be achieved? The first KBE applications were developed around 30 years ago in various sectors such as aerospace, automotive and shipbuilding. The process of constructing a KBE application consists of a series of steps to collect and store knowledge leading to the final implementation into a software application. Specific methodologies were developed for this purpose, such as MOKA (Methodologies and tools Oriented to Knowledge-based engineering Applications) (Stokes, 2001) or, more recently, Knowledge Nurture for Optimal Multidisciplinary Analysis and Design (KNOMAD) (Curran, Verhagen, & Van Tooren, 2010). The software implementation and the relevant principles were also formalized recently (La Rocca, 2012). These approaches, both on the procedural and software implementation side, consider a KBE application as an independent software platform based on specific programming languages (La Rocca, 2012) (eg, Intelligent Computer Aided Design (ICAD), General-purpose Declarative

Language (GDL) and Adaptive Modeling Language (AML)). These programming languages form the basis for a KBE 'system', a general-purpose software platform used to build a highly parametric software representation of a specific product to be designed (PM). The PM developed via a KBE system is defined as a KBE 'application'.

The previous approaches were conceived for design teams appointed to work on medium- to long-term design projects of products intended for mass manufacture (eg, airplanes, cars). Conversely, the architecture, engineering and construction (AEC) sector is characterized by the fragmentation of design responsibilities and tasks requiring the integration into existing platforms for one-of-a-kind projects. This poses a challenge when integrating existing KBE systems into current AEC-specific software platforms. Moreover, using existing AEC-specific software platforms favors the integration of KBE into BIM workflows via the IFC schema. This integration is fueled by recent development efforts from the major software vendors in the AEC sector to access the software functionalities via increasingly sophisticated API (application programming interface), alongside the traditional GUI (graphical user interface). The APIs permit automatic software control via widely used programming languages, such as C# and Python, as opposed to GUIs requiring manual mouse- and keyboard-based approaches.

In an attempt to streamline the creation of the KBE application in the façade sector, Montali, Overend, Pelken, and Sauchelli (2017b) proposed a four-step iterative process to transition from knowledge captured from real-world experienced personnel that is often in a natural language (low level of formality), up until the creation of the PM, where the knowledge is stored in the form of programming code (high level of formality). Fig. 16.4 illustrates the proposed process. The four steps are described in studies of Montali et al. (2019, 2017b).

1. *Knowledge capture*: The aim of the first step is to elicit and collect the type of knowledge that is available and its impact in terms of benefits for the company. If a specific design aspect is impossible to collect, due to the lack of analyses/experts and, at the same time, it is not relevant for the final delivery of the product, no implementation is needed. For those aspects that are required but not available (eg, cost information, constraints for specific manufacturing machines/operations), further studies might be needed. Unstructured interviews with domain experts provide a sense of the major gaps in the design and manufacturing process and how to approach them. The interviewee should be informed of the future opportunities arising from the development of such applications to maximize his/her contribution. Semistructured interviews can be then conducted to retrieve knowledge more systematically, once the problem has been set and the business case for developing the application has been defined. Document-based research of documents already produced by the company is also useful to retrieve knowledge and information that would otherwise require excessive effort to be used repetitively by humans (eg, large PDF documents that contain guidelines and technical datasheets). The availability of such documents varies from one company to another. A standard methodology for capturing knowledge is illustrated by Milton (2007).

2. *Knowledge base*: The next step structures the knowledge collected in step 1 by selectively sorting, storing and linking it into a knowledge base, a structured repository where knowledge is easily accessible. The creation process of a knowledge base consists of the

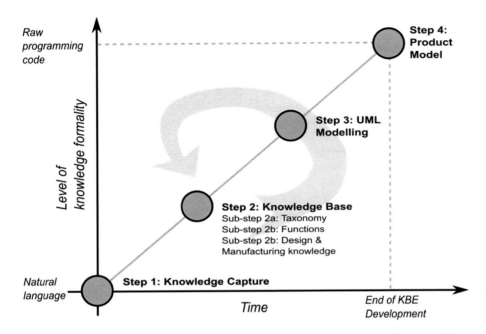

Figure 16.4 Knowledge formalization process of the proposed methodology, from natural language to raw programming code (Montali et al., 2019).

analysis and categorization of the knowledge related to the design and manufacture of the product under question. The process of creating the knowledge base requires the identification of the fundamental units representing knowledge (knowledge units). The ICARE—illustration, constraint, activity, rule and entity—forms (Stokes, 2001) are standard tables representing a type of unit of knowledge which can be used to create a knowledge repository. Table 16.1 shows the most common forms of knowledge. Knowledge is, represented in tables and stored into these standard forms, which are then cross-referenced (eg, through hyperlinks, if forms are developed in HTML), thus resulting in a network of interlinked knowledge units. An example is shown in Fig. 16.5 where an 'Entity' form is referenced to a 'Rule' form. Graphical representations of the network help visualize the overall network and the correlation between different concepts. The knowledge base is then validated against the opinion of domain experts that help correct or extend it. Step 2 also includes the analysis of the product architecture to define the product's taxonomy, product component's functions and associated design and manufacturing knowledge as explained in the study of Montali et al. (2019) and as adapted from the study of Klein (2013).

3. *Unified modeling language (UML) modeling*: The next step after knowledge collection and its structuring in the KB is the implementation into a more formal (ie, lower level) language. UML (OMG, 2009) is used to model each knowledge unit through an object-oriented programming (OOP) approach, where each physical product component and function (ie, 'entity' ICARE forms) are represented by a class. Through OOP, it is also possible to model the engineering rules ('rule' ICARE forms) and constraints ('constraint' ICARE form) by specifying the function. UML captures all the features characterizing

Towards automated design: knowledge-based engineering in facades 411

Table 16.1 Methodologies and tools oriented to knowledge-based engineering applications illustration, constraint, activity, rule and entity forms.

Form	Represented knowledge
Illustration	Experience on past projects
Constraint	Physical/geometrical limits on product/processes
Activity	Single step in design and manufacturing activity
Rule	Design/manufacturing engineering rule
Entity	Physical entity: 'Entity-Structure'
	Function: 'Entity-Function'
	Change in state of a product: 'Entity-Behavior'

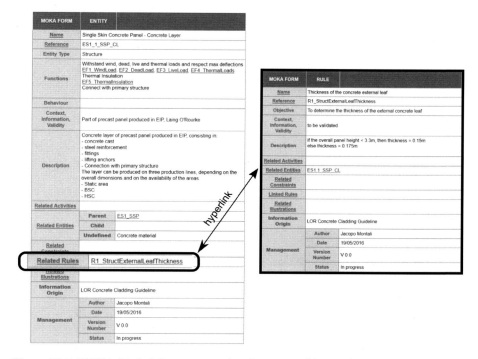

Figure 16.5 MOKA 'Entity' form representing the structural layer of a precast concrete single-skin panel, linking to a 'Rule' form containing a simplified engineering rule for dimensioning the concrete thickness (Montali et al., 2017b). *MOKA*, Methodologies and tools oriented to knowledge-based engineering applications.

OOP in terms of interrelationship between classes, such as inheritance, association, composition and aggregation. The taxonomy of the product is, therefore, created: Fig. 16.6 shows a typical 'composition' link between the product and its subcomponents, represented by a black diamond, describing the 'contains' relationship between physical

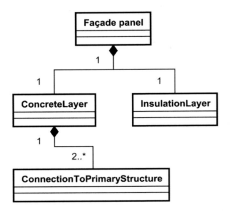

Figure 16.6 Simplified UML diagram showing the taxonomy of a façade product. Each box corresponds to an 'Entity' MOKA form (Montali et al., 2017b). *MOKA*, Methodologies and tools oriented to knowledge-based engineering applications; *UML*, unified modeling language.

entities. Once the taxonomy has been defined, the design and manufacturing knowledge is included into the taxonomy to form a lower level ontological framework of the product.

4. *Product model (PM)*: The PM is then translated into a programming code, based on the software architecture specified by the UML diagram. The overall process (steps 1–4) is iterative, where new knowledge is included or replaces outdated concepts. The development of a software architecture that facilitates modifications is, therefore, desirable. Object-orientation, in this sense, allows the creation of custom libraries of standard objects with associated knowledge that can be reused whenever a new tool for a new product is created. For instance, the insulation material of a single-skin precast concrete panel is identical to that used for a loadbearing, precast concrete sandwich panel in terms of intrinsic properties such as thermal resistance and material cost.

16.3 An example of KBE in facades

This section provides an example of a KBE application that was constructed for the design of PCC façade panels, a technology largely used in the United Kingdom for high-rise residential developments. The KBE is then implemented in a case study of a real-world residential building. For more details, refer to the study of Montali et al. (2019).

The purpose of the KBE digital tool in this example is to assist designers in the early stages of the process by giving valuable real-time feedback on the performance of the façade (ie, manufacturability, transportability, U-value, environmental impact and daylight factor) which would otherwise become available much later in the process when most of the design decisions have already been taken.

The case study consists of a recently built residential building in London. The tower is a 36-storey building clad with precast, single-leaf concrete panels. The prefabricated panels include precast concrete, insulation and glazing elements. The

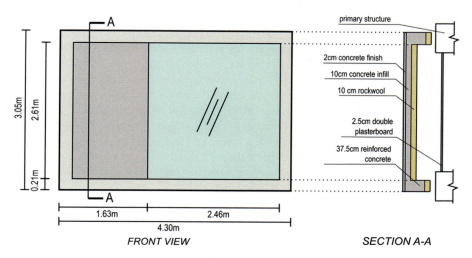

Figure 16.7 Main frontal dimensions (left) and buildup (right) of the investigated panel (Montali et al., 2019).

total area of the façade is 3532 m². Once the component was installed, the dry lining, vapour control layer and plasterboard were applied onsite. Fig. 16.7 shows the panel's main frontal dimensions, position with respect to the primary structure and buildup. The south-east façade is considered in this chapter.

The panels were manufactured at the Explore Industrial Park, a manufacturing facility located in Steetley, Nottinghamshire (the United Kingdom) part of the Laing O'Rourke group (Laing O'Rourke, 2021). The facility provides production lines with different degrees of automation for different types of products ranging from low automation for unique products to high automation for mass production. The analysed panels in this chapter were manufactured in the 'Bespoke Carousel System' (BSC), which consists of a semiautomated line. In the BSC the panels are manufactured on steel tables that are conveyed through a series of stations were specific operations (eg, mould lay-up, reinforcement and fitting installation, casting, panel turning for demoulding) are performed. Each station presents some manufacturing constraints and rules that affect the design of the precast panel (eg, maximum panel weight to be lifted by single gantry or tandem cranes, maximum panel dimensions due to steel tables geometrical limitations).

The database used for this study comprises six types of insulation boards with different thicknesses, three types of windows (low-, medium- and high performance), three types of jointing materials (mastic seal, cast-in drainage channel and expanding tape air seal). A knowledge-based rule governs the combination of multiple insulations (up to two) based on different criteria such as sustainability, potential installation risk from the contractor and condensation risk. Data on embodied carbon was taken from the ICE V2.0 database (Hammond & Jones, 2008).

16.3.1 Step 1: knowledge capture

The first step consisted in collecting the knowledge from relevant personnel within the company: experts in the manufacturing division of the company giving advice on the constructability issues arising at late-stages, or people working at earlier stages on the thermal design of the panel. All useful knowledge was then stored and used later to build the PM. To begin with, a series of semistructured interviews were conducted. To facilitate the process of knowledge collection, the interviewees were shown the latest version of the developed tool and asked to provide comments. Once the feedback about the tool was collected, the discussion moved toward adding more design and manufacturing rules/constraints to the model (eg, more detailed knowledge on limitations of manufacturing machines).

16.3.2 Step 2: knowledge base

The knowledge base creation first investigates the taxonomy of the product by considering the fundamental components that constitute the panel. The taxonomy is characterized by a relationship between the overall product and its constituents [in accordance with Klein's 'product levels' (Klein, 2013)] of the type 'contains'. The taxonomy for the precast, single leaf panel is shown in Fig. 16.8, in which each element is associated with a corresponding 'Entity-Structure' ICARE form that stores information about their upper and lower level constituents. Grey boxes represent the 'leaves' of the diagram, which were then assigned a function in step b. An 'Entity-Function' form was created to store a description of the function.

Then, the product's functions were linked with the physical components. This is represented diagrammatically in the directional force-directed layout shown in Fig. 16.9. The meaning of the directed arrow depends on the start and end elements: if an arrow points to an 'Entity-structure' element (*dark circle*) from an 'Entity-function' element, this signifies that the link will be of the type 'function associated with the physical element'. Conversely, two 'Entity-structure' elements connected to another share a part-whole relationship ('has a').

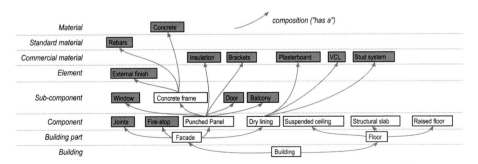

Figure 16.8 Taxonomy of the PM of precast concrete single-skin panel, based on the classification scheme proposed (product levels) by Klein (2013). Grey boxes represent the 'leaves' of the tree (Montali et al, 2019). *PM*, Product model.

Towards automated design: knowledge-based engineering in facades 415

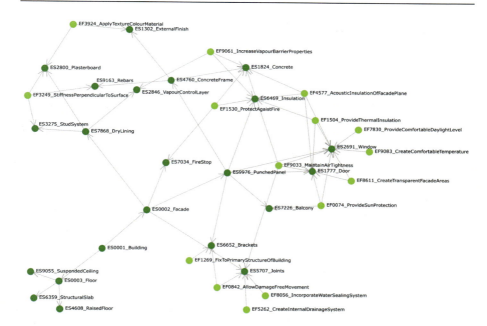

Figure 16.9 Taxonomy of the PM of precast concrete single-skin panel (*dark dots*) and associated functions (*light dots*) in a force-directed layout generated in D3.js (Montali et al., 2019). *PM*, Product model.

At this stage the ontological framework of the PM created includes information about the product breakdown and the associated functions. The knowledge about rules and constraints associated with the design and manufacture of the product have yet to be linked. This is achieved by creating the remaining 'Illustration', 'Rule', 'Constraint', and 'Activity' forms and by linking them with the relevant 'Entity-structure' and 'Entity-functions' forms created in the preceding steps (Fig. 16.10).

For example, rule RU0190_Daylighting refers to minimum daylight in the room, whereas rule RU2114 refers to the location of the steel billet (with $D = 200$ mm), both presented at the beginning of the chapter.

The final product is a network of interrelated concepts, creating semantic links between features for defining the product architecture. These features are physical components and their functions, or design and manufacturing criteria under the form of rules and constraints. Given the large number of links between knowledge units, the final knowledge base was represented by hierarchical edge bundling, to reduce the 'visual clutter when dealing with large numbers of adjacency edges' (Holten, 2006). Fig. 16.11 shows the diagram generated through the JavaScript library D3.js (Bostock, Ogievetsky, & Heer, 2011).

The resulting KB works as follows. From the hierarchical edge bundling, the user can hover on specific elements such as rules, constraints, description of a physical component, or its functions. The diagram is interactive, that is, it highlights all the links and interrelated elements to that specific element. If the user selects a specific

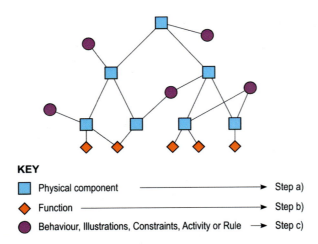

Figure 16.10 Graphical representation of the three substeps to build the knowledge base.

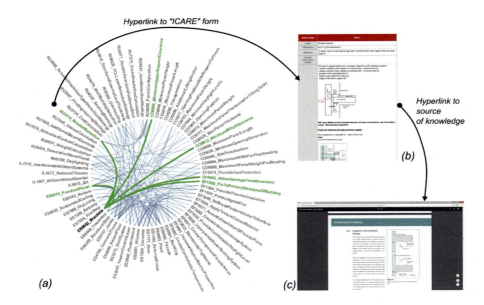

Figure 16.11 Knowledge base in the form of a 'hierarchical edge bundling' (Holten, 2006) and links to more detailed descriptions of the underlying knowledge related to the selection of the supporting brackets for precast concrete single leaf panel: (A) overarching view of the links with other elements of knowledge, (B) MOKA 'Rule' form containing the logic and (C) original source of knowledge (Montali et al., 2019). *MOKA*, Methodologies and tools oriented to knowledge-based engineering applications.

element, a hyperlink redirects to a webpage containing the MOKA form describing the element in question. The form contains further links to the sources of knowledge. In the example shown in Fig. 16.11A, the user hovers on the rule 'RU2114_BracketSelection'. Clicking on the hyperlink opens a webpage containing the logic behind the selection of the appropriate support bracket for the precast panel (Fig. 16.11B). The form also contains a field (Information origin) with a hyperlink to a specific page of a PDF document containing the original source of knowledge (Fig. 16.11C). In this way, it is possible to achieve different levels of granularity of the relevant information/knowledge, from the highest level possible (the hierarchical edge bundle), to the most detailed description (the original PDF guideline document).

16.3.3 Step 3: UML modeling—class diagramming

The definition of the fundamental components of knowledge and their storage into appropriate forms is followed by a UML class diagram to represent the product architecture for the subsequent implementation. Fig. 16.12 shows the diagram generated for the type of façade panel in this case study, in which each class represents a physical component. Functions (eg, thermal) and properties (eg, weight) are assigned via interfaces that are implemented by the classes. In some cases, interfaces were not assigned to certain elements since they have a negligible effect on the performance of the panel (eg, vapour control layer on weight).

16.3.4 Step 4: product model—digital tool implementation

The last step consists of the implementation of the PM into a usable digital tool. The tool shown here constitutes an evolution from an earlier version (Montali,

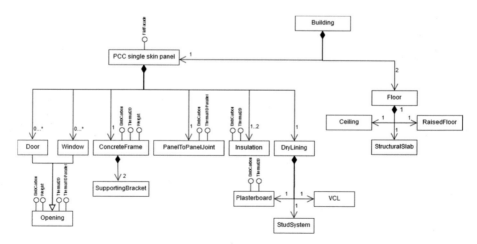

Figure 16.12 UML diagram representation of the product architecture (Montali et al., 2019). *UML*, Unified modeling language.

Overend, Pelken, & Sauchelli, 2017c), after several iterations of the process shown in Fig. 16.4. The chosen platform is Rhinoceros 5 (McNeel & Associates, 2016) and the tool was under the form of a series of Grasshopper's custom components written in C# representing the PM (Fig. 16.13). Once the user has drawn the surface representing the overall façade, a specific Grasshopper definition is associated to the surface. Fig. 16.13A shows the window to configure the panel in terms of buildup, the type of jointing solutions, the external finish type, as well as other properties such as the thickness of the concrete layer (which can be automatically determined based on the rule described in substep 2c) or the thickness of the air layer. All configurations are selected from a prebuilt set of solutions, which embed knowledge about the preferred design and manufacturing practices from the manufacturer. Fig. 16.13B shows that a series of performance indices are automatically calculated based on the selected configuration, such as U-value, daylight factor, embodied carbon emissions, panel weight and total panel thickness. Fig. 16.13C contains the KB that is updated automatically as the user configures the PM.

The tool offers to the users the opportunity to interrogate the 'Knowledge Wheel' which is the knowledge base that summarizes all the rules that the designed façade

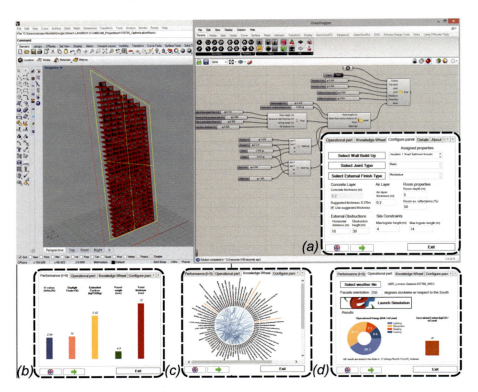

Figure 16.13 Digital tool's GUI for panel buildup configuration (A), performance analysis (B), compliance to constraints (C) and operational performance via EnergyPlus (D) (Montali et al., 2019). *GUI*, Graphical user interface.

panel should comply with and the subsequent breach of any of them with the current design. The resulting advantage is that users are informed of the design choices' implications and they can enhance the knowledge of the façade system they are designing. For example, if a 'hard constraint' is violated, the corresponding element will turn red; if the broken constraint is 'soft', the text will turn orange. Furthermore, by clicking on the coloured text the user will be able to open the MOKA forms (see earlier) that constitute the knowledge source. In this way the users are informed about the consequences of their design choices. Fig. 16.13D shows that it is also possible to determine an early-stage estimate of the expected operational energy or carbon by running a dynamic, single-zone energy simulation at run time via a link to building energy simulation software, EnergyPlus (Crawley, Pedersen, Lawrie, & Winkelmann, 2000), based on the solution that is currently configured by the user.

16.4 Results

The knowledge base and the digital tools serve as configuration tools to understand trade-offs between design criteria. The following approach seeks for an optimized solution that takes into account for the optimal trade-off between performance and number of violated design and manufacturing constraints. The objective functions chosen in this instance are operational carbon emissions (measured in kgCO$_2$/y m^2 of floor area) and embodied carbon emissions (measures in kgCO$_2$/kg of panel weight). The optimization, therefore, aims to identify manufacturable façade design solutions that minimize both the operational carbon and the embodied carbon by automatically testing all the possible design configurations.

The operational energy was determined computationally by means of a dynamic building energy simulation in EnergyPlus (v8.7). This involved creating a single-zone model with adiabatic surfaces except for the façade under investigation. In this model the width of the zone corresponds to the width of the panel, which does not necessarily correspond to the room width. For this reason the analysis should be seen as conducted over the area of influence of the façade, rather than for a specific room. A 'Building Area Method' as per ASHRAE 90.1 (ANSI/ASHRAE/IES Standard 90.1–2016, 2016) was, therefore, followed, in which internal gains are given for generic end uses, rather than for specific space types (eg, office vs open office or single office). This approach is particularly suitable for early-stage conceptual stages, where the internal distribution of spaces is poorly defined.

Analyses were run on a Dell Inspiron with 8GB RAM and processor Intel Core i7–3630 QM, 2.40 GHz. The optimization was run three times by generating 150, 1500, 15,000 unique solutions for the façade by varying the values of the PM parameters $x = (x_1, \ldots, x_i, \ldots, x_n) \in R^n$ from a Gaussian distribution of zero mean and variance σ, that is, $x_i \sim N(x_i|0, \sigma) = x_i + \sigma \cdot N(0, 1)$. The variance σ was chosen so that there is a 95% chance of having $|\Delta x_i| \leq 10\%$. Discrete variables were instead sampled from a uniform distribution. Calculation times were 20 minutes, 2 hours and 8 hours, respectively. The number of discarded analyses due to unfeasible geometries (eg, window outline overlapping panel outline) was equal to 46, 473 and 4722, respectively.

The results were also compared with those obtained from a nondominated sorting genetic algorithm II (NSGA-II) in Matlab, which represents the benchmark for the analyses that were run. The prototype whole-life value optimization tool for façade design model (Jin & Overend, 2013) was adapted to take into account the variables and objectives in this study. While the database of materials was incorporated in the NSGA-II, design knowledge from the knowledge base was not included due to confidentiality reasons, thus leading to solutions that are optimized for performance but complex or impossible to build. For the implementation of the genetic algorithm, a convergence test was carried out for different population sizes and numbers of generations. A population size of 1000 and number of generations of 50 were selected to ensure that a close approximation of the real Pareto Front can be obtained. The crossover probability was set to 70% in the algorithm. Analyses were run on a Windows with 8GB RAM and processor Intel Core i7-4650 U, 1.70 GHz. The total simulation time is 32 hours for the GA optimization.

Fig. 16.14 shows the results from the optimization, where the dark circle represents the original solution adopted in the real-world building. Solutions associated with very low U-values (darker colour) do not constitute optimal trade-offs between embodied and operational energy: given the relatively large window-to-wall area of this study (circa 40%), the optimal solutions instead correspond to an intermediate level of specification of the window. The window spec increase corresponds also to a lower U-value. The mid-spec configuration determines lower cooling loads during the summer season of the London climate, if compared to the configuration with the lowest U-value window. The incidence of the window type also determined two separate groups of solutions, one corresponding to the lower performance window to the right, and one associated with the remaining two (mid- and high-performance) window types to the left of the diagram. The radii of the solutions (ie, inverse of number of violated design and manufacturing constraints) do not follow

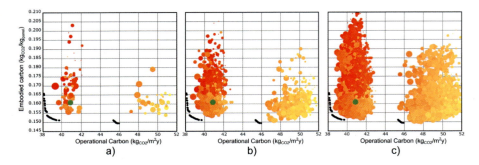

Figure 16.14 Results from the optimization of the case study. Analyses for 150 (A), 1500 (B) and 15,000 (C) design variants with respect to the original design, represented by the dark circle. The colour scale refers to reducing levels of overall U-value (red = low U-value; yellow = high U-value). Black points correspond to the values obtained from the GA optimization (Montali et al., 2019). *GA*, Genetic algorithm.

a specific trend, but the interactive visualization technique allows the user to browse through each solution individually.

The digital application allows quick design checks over multiple possible design solutions to understand expected performance and manufacturability, thus reducing the time spent over repetitive and manual tasks. When combined with an optimization algorithm, the KBE application has permitted to explore a broad range of feasible (ie, manufacturable) solutions and to improve the originally conceived design by applying small geometrical and material variations (eg, window size increased along the two dimensions, overall panel geometry). In this way the architect's original intent is retained while reducing the carbon footprint of the façade. Conversely, the use of an optimization algorithm without the concurrent incorporation of useful knowledge (like the GA algorithm used here) may lead to solutions not viable from a design and manufacturing standpoint.

As already anticipated in the introduction, the process of creating such applications presents some challenges. For the case study, retrieving useful design and manufacturing information/knowledge was time-consuming due to the interdisciplinary nature of the problem, requiring the input from people from different divisions in the company. Once the knowledge was translated into code, changes to hard-coded heuristic rules required the creation of a new version of the tool and the subsequent installation on the user's personal computer. Future work is required to deal with change management and version release more smoothly.

16.5 Conclusion

The AEC industry is striving for improvements in the way it operates, with a strong need of advanced design and construction processes to reduce the overall environmental impact and improve quality and efficiency. The façade, with its significant impact on construction costs (up to 30% of the total) and on the whole building performance, is pivotal in this transformation.

Various tools and processes have been developed to improve façade design and construction processes. However, it is still a complex process that involves the creativity of the designer to be combined with the engineering and manufacturing knowledge from the supply chain. In this context few, if any, tools bridge the gap between design and construction processes. KBE has been successfully used to integrate engineering design and manufacturing knowledge in other industries and is deemed suitable for the AEC industry and for the façade sector.

The process presented in this chapter shows in practical terms how a digital knowledge-based application can be achieved. It consists of four principal steps, from the knowledge collection stage to the subsequent knowledge implementation into a usable digital application. This process should be seen as iterative, with revisions that modify existing or incorporate new knowledge.

A knowledge-rich application that can incorporate various design and manufacturing criteria about the design of a specific product can be used to improve the design, test ideas early in the process and increase efficiency. The presence of a GUI allows the user to manipulate the design solution to achieve the required targets of performance and manufacturability. If specific performance targets are set, automatic configuration via optimization algorithms is possible as shown in the previous example. More sophisticated techniques, such as machine learning algorithms, can be used to allow custom-built intelligent agents to learn how design and manufacturing criteria affect the final design of the product.

The challenges in implementing KBE across the whole façade industry are mainly related to capturing, storing and updating of knowledge in a structured way, as well as the ability to deal with a product, underlying design and manufacturing knowledge of which change over time. Future work will provide tools to collect and store continuously evolving knowledge and the automatic implementation into hard-coded rules. In this way the majority of the challenges encountered when developing such applications will be addressed, leaving the users to reap the benefits of KBE applications.

References

Alfieri, E., Seghezzi, E., Sauchelli, M., Di Giuda, G. M., & Masera, G. (2020). A BIM-based approach for DfMA in building construction: Framework and first results on an Italian case study. *Architectural Engineering and Design Management, 16*(4), 247–269.

ANSI/ASHRAE/IES Standard 90.1-2016, (2016). *Energy standard for buildings except low-rise residential buildings.*

Aram, S., Eastman, C., & Sacks, R., (2014). A knowledge-based framework for quantity take-off and cost estimation in the AEC industry using BIM. In *ISARC. Proceedings of the international symposium on automation and robotics in construction* (Vol. 31, ii, p. 1).

Ashby, M. F. (2000). Multi-objective optimization in material design and selection. *Acta Materialia, 48,* 359–369.

Belsky, M., Sacks, R., & Brilakis, I. (2016). Semantic enrichment for building information modeling. *Computer-Aided Civil and Infrastructure Engineering, 31*(4), 261–274.

Bostock, M., Ogievetsky, V., & Heer, J. (2011). D3: Data-driven documents. *IEEE transactions on visualization & computer graphics,* 2301–2309.

Cavalieri, S., Maccarrone, P., & Pinto, R. (2004). Parametric vs. neural network models for the estimation of production costs: A case study in the automotive industry. *International Journal of Production Economics, 91,* 165–177.

Construction Industry Transformation Map. (2017) [Online]. Available: https://www1.bca.gov.sg/buildsg/construction-industry-transformation-map-ITM.

Cooper, D. & La Rocca, G., (2007). Knowledge-based techniques for developing engineering applications in the 21st century. In *7th AIAA Aviat. Technol. Integr. Oper. Conf., no.* (pp. 1–22) July, 2007.

Crawley, D. B., Pedersen, C. O., Lawrie, L. K., & Winkelmann, F. C. (2000). EnergyPlus: Energy simulation program. *ASHRAE Journal, 42,* 49–56.

Curran, R., Verhagen, W. J. C., & Van Tooren M. J. L., (2010). The KNOMAD methodology for integration of multi-disciplinary engineering knowledge within aerospace production. In *48th AIAA Aerosp. Sci. Meet. Incl. New Horizons Forum Aerosp. Expo., no.* (pp. 1–16) January, 2010.

Day, G., Gasparri, E., & Aitchison, M. (2019). Knowledge-based design in industrialised house building: A case-study for prefabricated timber walls. In F. Bianconi, & M. Filippucci (Eds.), *Digital wood design: Innovative techniques of representation in architectural design* (pp. 989–1016). Cham: Springer International Publishing.

Hammond, G. P., & Jones, C. I. (2008). Embodied energy and carbon in construction materials. *Proceedings of the Institution of Civil Engineers—Energy, 161*.

Henriksen, T., Lo, S., & Knaack, U. (2016). The impact of a new mould system as part of a novel manufacturing process for complex geometry thin-walled GFRC. *Architectural Engineering and Design Management, 12*(3), 231–249.

Holten, D. (2006). Hierarchical edge bundles: Visualization of adjacency relations in hierarchical data. *IEEE Transactions on Visualization and Computer Graphics, 12*(5), 741–748.

Jin, Q., & Overend, M. (2013). A prototype whole-life value optimization tool for façade design. *Journal of Building Performance Simulation, 1493* (July).

Kassem, M., & Mitchell, D. (2015). Bridging the gap between selection decisions of façade systems at the early design phase: Issues, challenges and solutions. *Journal of Façade Design and Engineering, 3*(2), 165–183.

Laing O'Rourke (2021). Website. [Online]. Available: http://www.laingorourke.com/.

La Rocca, G. (2012). Knowledge based engineering: Between AI and CAD. Review of a language based technology to support engineering design. *Advanced Engineering Informatics, 26*(2), 159–179.

Klein, T. (2013). *Integral façade construction—Towards a new product architecture for curtain walls*.

McNeel & Associates R. (2016). Rhinoceros webpage. [Online]. Available: https://www.rhino3d.com/.

Milton, N. R. (2007). *Knowledge acquisition in practice: A step-by-step guide* (1st ed.). London: Springer-Verlag.

Montali, J., Overend, M., Pelken, P. M., & Sauchelli, M. (2017a). Knowledge-based engineering in the design for manufacture of prefabricated facades: Current gaps and future trends. *Architectural Engineering and Design Management, 14*, 1–17.

Montali, J., Overend, M., Pelken, P. M., & Sauchelli, M. (2017b). Towards facades as make-to-order products—the role of knowledge-based-engineering to support design. *Journal of Façade Design and Engineering, 5*(2), 101–112.

Montali, J., Overend, M., Pelken, P. M., & Sauchelli, M., (2017c). Knowledge-based engineering applications for supporting the design of precast concrete façade panels. In *ICED17: International conference of engineering design*.

Montali, J., Sauchelli, M., Jin, Q., & Overend, M. (2019). Knowledge-rich optimisation of prefabricated facades to support conceptual design. *Automation in Construction, 5*.

OMG (2009), *OMG unified modeling language (OMG UML), Superstructure*, no.

RIBA Plan of Work, (2020). [Online]. Available: https://www.architecture.com/knowledge-and-resources/resources-landing-page/riba-plan-of-work.

Stokes, M. (2001). *Managing engineering knowledge—MOKA: Methodology for knowledge based engineering applications*. American Society of Mechanical Engineers.

Voss, E., Jin, Q., & Overend, M. (2013). A BPMN-based process map for the design and construction of facades. *Journal of Façade Design and Engineering, 1*, 17–29.

Additive manufacturing in skin systems: trends and future perspectives

17

Roberto Naboni[1,2] *and Nebojša Jakica*[2,3]
[1]CREATE, Odense, Denmark, [2]Section for Civil and Architectural Engineering, University of Southern Denmark (SDU), Odense, Denmark, [3]FACETS lab, Odense, Denmark

Abbreviations

AM	additive manufacturing
C3DP	construction 3D printing
FDM	fused deposition modelling
WAAM	wire and arc additive manufacturing
ME	material extrusion
BJ	binder jetting
G3DP	glass 3D printing
TRL	technology readiness level

17.1 Introduction

Additive manufacturing (AM) is one of the most disruptive technologies of our time (Attaran, 2017). Starting from the first experiences with rapid prototyping in the 1980s, passing through initial industrial applications at the beginning of the century (Wohlers & Gornet, 2014), this technology finds to date growing applications in a large number of sectors, including aerospace (Najmon, Raeisi, & Tovar, 2019), automotive (Delic & Eyers, 2020), medical replacements (Li, Pisignano, Zhao, & Xue, 2020), defence (Busachi et al., 2017), manufacturing and architecture (Camacho et al., 2018).

Construction 3D printing (C3DP) is considered an established and continuously expanding research field, which has evolved rapidly over the last decade. Novel applications are constantly implemented, including 3D-printed houses, walls, structures, pedestrian bridges, urban furniture, villages and so forth. Among the various applications, the use of AM is of great interest when it comes to building skins, façade design and enveloping solutions in general, where C3DP has the potential to extend functional integration (Strauss, 2013) and enhance design freedom (Fig. 17.1). Functional challenges, a multitude of performances, building codes and liabilities, economic aspects and market acceptance are some of the difficulties identified for using AM in façade construction (Strauss & Knaack, 2016).

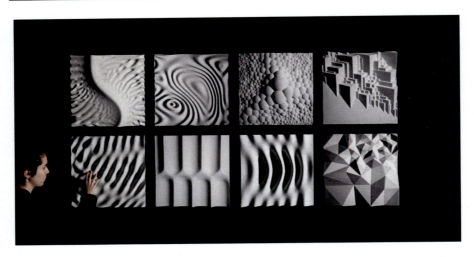

Figure 17.1 High-resolution concrete panels, with topological and geometrical complexity enabled by the use of FDM formworks at CREATE—University of Southern Denmark. *FDM*, Fused deposition modelling.

Considering these limitations that refer to the current industrial practice, the adoption of AM for façade construction implies a conceptual and production shift.

The implementation of AM in facades construction offers a number of technical and contextual advantages. Advanced design workflows can be implemented with 3D printing through a direct link that empowers data-driven computational design. This provides a prominent edge to exploring complex geometrical solutions, which can be manufactured thanks to extended design freedom. In some instances the inherent characteristics of 3D printing allow the fabrication of objects with topological and geometrical features that would be impossible to fabricate otherwise (Naboni & Breseghello, 2018), unless with the use of inflexible purpose-specific tooling (Camacho et al., 2018). The digitally led flexibility of AM propagates the design into the realm of material optimization and the opportunity to strategically use the material (Vantyghem, Boel, De Corte, & Steeman, 2018), introducing new concepts, such as functionally graded material. This involves a variation in the distribution of a single material across an object, such as mapping specific performance targets (Oxman, Keating, & Tsai, 2011) or dosing the ratio between different materials along a component's volume to achieve specific properties (Oxman, 2011).

In comparison with other techniques, AM does not require moulds and formworks, therefore contributing to reducing sacrificial and waste material. Moreover, it allows product customization and complex design without a sharp increase in costs than standard manufacturing (Naboni & Paoletti, 2015). The 3D printing supports on-demand, individualized production for small- and medium-size production, which fits the needs of constructions, where the final product (ie, the building) is always different. From a market perspective, AM has the potential to shorten the supply chain and enable localized production, bringing the production closer to the final consumer (Thomas & Gilbert, 2014), offering solutions that mitigate the decreased availability of skilled workforce through automation (Karimi, Taylor, Goodrum, & Srinivasan, 2016) and contribute to the reduction of labour costs by 50%−80% (Aslam, Altynay, Suvash, & Jong, 2020).

C3DP has a clear potential to enhance digitalization, customization, functional integration, material efficiency and building skins' environmental impact. Research and innovative practices are looking into it with increasing interest to overcome the many constraints that standard manufacturing imposes on the design of facades. The chapter navigates the reader through the latest advancements and discusses future opportunities and challenges. An overview of the existing techniques is provided, with an insight into construction experiments and material developments. Subsequently, the chapter focuses on the applications of these processes for building skins and façade components. Lastly, the industry's current and future role of additive processes is discussed, focusing on challenges and foreseeable opportunities.

17.2 AM techniques for construction 3D printing

Techniques and materials for C3DP have been developing at a fast pace in the last 5 years, which contributes to rendering AM more and more affordable globally. Advancements to technology and materials are both undertaken by research institutions and private companies prevalently based in Europe and North America. In 2018 the market of AM in construction encountered 65 companies, ranging from prototyping services for architectural and engineering firms to the 3D printing of structures and entire buildings (de Laubier, Wunder, Witthöft, & Rothballer, 2018).

From a technical perspective, AM is an umbrella term for several technologies utilized to print three-dimensional objects or components with the progressive allocation of material, following instructions obtained from digital models. An overview of the available processes can be found in the study of Gibson, Rosen, and Stucker (2010). Most of the employed AM techniques have their origin in small-scale industrial applications. Scaling up such techniques for use in architecture and constructions implies inherent challenges. First, architectural systems are typical of a large scale. Second, building components for facades, among others, require complex assemblies to address multiple performance requirements. Third, in traditional design and construction, a multitude of materials are used to ensure specific functional requirements and performance levels. Lastly, prefab components or in situ construction are inevitably coming together on the building site, a complex, dynamic and noncontrolled production environment that is difficult to automate with the current technology conceived for factory scenarios. In the last decade, extensive research and practice-led investigations have been conducted to develop AM methods suitable for construction purposes (Paolini, Kollmannsberger, & Rank, 2019). Innovative technologies are under continuous development and refinement, becoming available to the first commercial projects, and slowly filling the gap between research and market expectations.

Table 17.1 Overview of the available techniques and materials currently available for construction 3D printing, following the classification ISO/ASTM 52900.

ISO/ASTM 52900:2015 classification	Technique	Materials in constructions	Invention	Use	On-site
ME	FDM	Thermoplastics	1988	Direct Indirect	No
	SME	Concrete Mortar Clay Earth	1995	Direct Indirect	Yes
BJ	BJ	Polymer–sand composites	2005	Direct	Yes
	Particle-bed 3D printing	Cementitious materials Polymer–sand composites Polymers	1993	Indirect	No
Directed energy deposition	WAAM	Metals and alloys	1920	Direct	No
Powder bed fusion	DMLS/SLM MJF	Metals and alloys polymers	1995	Direct	Yes

BJ, Binder jetting; *DMLS*, direct metal laser sintering; *FDM*, fused deposition modelling; *ME*, material extrusion; *MJF*, multijet fusion; *SLM*, selective laser melting; *SME*, semifluid material extrusion; *WAAM*, wire and arc additive manufacturing.

According to the international standard on Additive Manufacturing ISO/ASTM 52900 (ISO/ASTM 52900.2015, 2015), additive processes are grouped into seven technological categories, specifically referring to the machine interaction with the processed material. Most of them also find applications in façade construction, specifically material extrusion (ME), binder jetting (BJ), directed energy deposition and powder bed fusion (Table 17.1).

Several techniques have been systematically developed within this classification until reaching sufficient maturity for experimental to commercial application. The main processes and materials are briefly described in the following paragraphs, including notions of their historical development.

17.2.1 Fused deposition modelling

Fused deposition modelling (FDM) or fused filament fabrication is a process based on the extrusion of a heated thermoplastic feedstock through a nozzle tip to deposit layers onto a platform to build parts layer by layer (Masood, 2014). The technique

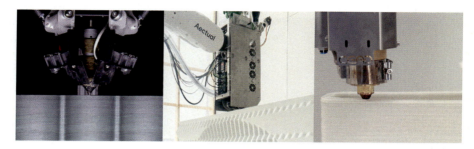

Figure 17.2 Examples of FDM printing. From Left, WASP 3MT at the CREATE Lab (SDU); Aectual (2020) Robotic Printing; CEAD (2020) fibre Reinforced printing. *FDM*, Fused deposition modelling.

was first introduced by Stratasys in 1988 (Stratasys Inc, 1992) and had since become the most common and inexpensive printing technique for many applications. FDM is the most used technique in 3D printing. Large enclosed printers have been developed by BigRep (BigRep, 2020), WASP (2020) and Cincinnati Incorporated under the name of Big Area Additive Manufacturing (Roschli et al., 2019) to manufacture at a sufficient scale. More recently, FDM printing has been implemented into robotic arms by companies that target more explicitly the field of construction, such as Aectual, which is looking into bio-based plastic (Aectual, 2020), AI Build (2020), and CEAD, which developed a system for structural elements with large-scale long fibre−reinforced polymer extrusion (CEAD, 2020). FDM printing was also extended to non-layered cellular structures, as seen from Branch Technology (Branch, 2020). All these systems can be utilized for direct 3D printing of façade elements, mostly with non-structural applications, temporary structures (Naboni, Kunic, Breseghello, & Paoletti, 2017a; Naboni, Kunic, & Breseghello, 2020) or nondirect manufacturing of moulding and formworks to concrete and other elements (Naboni & Breseghello, 2018, 2020) (Fig. 17.2). The last application includes a method that prints quickly with wax to build large recyclable formwork volumes, which are subsequently computer numerically controlled milled for high-resolution results (Gardiner & Janssen, 2014).

17.2.2 Semifluid material extrusion

Semifluid material extrusion (SME) includes techniques in which fresh material in a semifluid state such as mortar, concrete, clay, earth-based compounds and mud are pumped into a nozzle which deposits a filament along a defined path. After extrusion, once the material sets and forms a strong bond between layers. This technique was first patented by Khosnevis (1995) and subsequently experimented in 2004 with the invention of Contour Crafting (Khoshnevis, 2004). Over the last decade, more than 30 research and industry groups investigated this technique applied to concrete-like extrusion, which is already applied at a large scale. However, the technology still presents material, process and design challenges for achieving comparable results to ordinary cast concrete structure (Buswell, Leal de Silva, Jones, & Dirrenberger, 2018). Different strategies for integrating reinforcements have been

Figure 17.3 Semifluid material extrusion (SME). From left, COBOD (2020) BOD2 Gantry printer; WASP Crane Printer for Clay; Ceramic production by Studio RAP (2020).

investigated in recent works (Asprone et al., 2018; Mechtcherine et al., 2018). A number of companies are currently providing equipment for large-scale 3D concrete printing (3DCP), which is by volume the most utilized material in the market of additive construction (Lee et al., 2019). The issue of large-scale automation has been tackled with different approaches, including large gantry systems (COBOD, 2020); stationary robots (XTree, 2020), sliding (Vertico, 2020) and movable robotic arms (CyBe, 2020); and movable machines for 3D printing. An alternative material group used in semifluid extrusion includes clay/earth/mud/adobe. Examples of large-scale direct 3D printing with clay have been developed by WASP (CRANE, 2020) and IAAC (Terraperforma, 2017) for on-site construction. Ceramic façade components have been printed at a smaller scale and higher resolution by StudioRAP (Studio RAP, 2019) (Fig. 17.3).

17.2.3 Particle-bed 3D printing and binder jetting

Particle-bed 3D printing is a process that consists of the application of successive layers of dry material in particles, and the subsequent deposition of a fluid-binding material on the layer of particles through a printing head or a nozzle, to form two-dimensional patterns, which are staked to build 3D artifacts (Ziaee & Crane, 2018). The process was introduced with the BJ technique at Massachusetts Institute of Technology (MIT) in 1993 to realize ceramic components (Sachs, Cima, Williams, Brancazio, & Cornie, 1992). Pegna (1997) proposed for the first time the use for producing concrete-like elements. Dini (2006) introduced the technique for building applications (Fig. 17.4, left). In reference to construction purposes, the main techniques are direct printing with cementitious material and nondirect printing of a polymer—sand composite (Lowke et al., 2018). Particle-bed 3D printing represents a viable alternative for highly complex geometries, shape and topology optimization and ornamental possibilities (Deckard, 1986) (Fig. 17.4, right).

17.2.4 Wire and arc additive manufacturing

Wire and arc additive manufacturing (WAAM) is a direct energy deposition process that uses metal wire feedstock and an electric arc as the energy source to print

Additive manufacturing in skin systems: trends and future perspectives 431

Figure 17.4 Examples of particle-bed printing. Left, D-shape printer (Dini, 2006); right, ETH binder jetting (DBT, 2020).

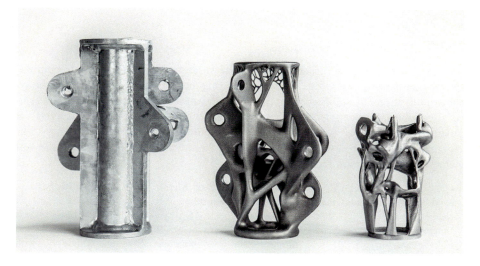

Figure 17.5 Left MX3D (2020) printing with WAAM; right, SLS printing by EOS for ARUP. *SLS*, Selective laser sintering © Davidfotografie; *WAAM*, wire and arc additive manufacturing.

objects of medium-to-large size incrementally (Derekar, 2018). It is typically implemented over a robotic arm, building upon a substrate material that can be cut afterward. WAAM operates with a wide range of metals, including stainless steel, nickel-based alloys, titanium alloys and aluminium alloys. This technology derives from standard welding technology, and its conceptual origin can be traced back to a patent from 1925 (Baker, 1925). However, systematic developments as an AM technique were made during the 1990s, once associated with computer control. The technique is particularly suited for construction applications as it enables a relatively quick production for complex geometries and has an unlimited theoretical size, yet with limitations in dimensional accuracy and surface finishing. The work of Joris Laarman and MX3D has successfully demonstrated the applicability of such a technique to a large scale, including load-bearing steel structures and envelopes, through the use of a fabrication system based on robotic arms (Buchanan & Gardner, 2019) (Fig. 17.5, left).

17.2.5 Directed energy deposition

Directed energy deposition is an AM process in which focused thermal energy from an energy source such as a laser/electron beam is used to fuse materials by melting as they are deposited. The technique's origin is from Deckard, Beaman, and Darrah (1986), who invented selective laser sintering. An evolution of this technique, direct metal laser sintering, was then commercially introduced in 1995 by EOS (2014). This technique fundamentally supports the high-resolution printing of metal objects and has its use in load-bearing complex steel joints since 2015 (Galjaard, Hofman, Perry, & Ren, 2015) (Fig. 17.5, right). Multijet fusion printing, developed by HP, combines the selective application of heat, printing powder, fusing and detailing agents to achieve higher mechanical performance and detail of the printed parts (HP, 2021).

17.2.6 Emergent techniques

C3DP is a highly active research field where research and developments continuously expand the possibilities of manufacturing architectural systems. Currently, many applications in building skins are under development. These are alternative to the already established methods described earlier and are natively conceived for construction purposes, expanding the range of material and performance possibilities currently available on the market. Among them, it is worth mentioning the following for future uses in facades.

- Glass 3D printing (G3DP) was introduced in 2015 by Micron3DP (2015) and MIT (Klein et al., 2015), utilizing fine glass powder and extruding it in a liquid form at very high temperature.
- Shotcrete 3D printing is developed at TU Braunschweig (Lindemann et al., 2018) and ETH (Taha et al., 2019) to produce reinforced concrete structures additively, with enhanced geometric and mechanic freedom compared to existing techniques.
- Mud-spraying drones have been experimented in various projects by MuDD architects (Chaltiel, Bravo, Veenendaal, & Sayers, 2020) to perform clay coating on-site and at inaccessible locations.
- Robotic Wood Printing with continuous wood filament, obtained from solid willow withies, is currently experimented at the University of Kassel (Ochs, Akbar & Eversmann, 2020) to produce freeform wood structures without subtractive methods.

17.3 Review of AM projects for skin systems

There is now extensive literature on the general topic of C3DP, but specific studies on AM in the façade field are still scarce. Kragh (2020) identifies AM as one of the core Industry 4.0 technologies where bespoke façade components and tuned material properties may yield advantages in terms of constructability while potentially providing optimal design solutions for a given set of circumstances. A total of 113 projects are collected from 2012 to date, which have engaged with the design and production of additive facades, exploring unconventional concepts and implementing different materials in research, development and commercial projects.

This study analyses 56 works in particular, including construction-ready technology, early research experiments and successfully realized prototypes. The selection was operated with a combination of criteria, based on the potential for functional and aesthetic qualities, the practicality of the solutions, market potential and the AM technique's appropriateness to address specific performative aspects. The case studies are classified according to their typology of application as facades components in walls and envelopes, cladding, shading and curtain walls. They are grouped by the manufacturing technology used; the material employed in a *direct* production technique or an *indirect* process requiring further operations; the year of realization to provide a temporal perspective to the analysis; and an indication of their technology readiness level (TRL), deducted from a literature study. The following paragraphs describe the projects individually, regarding their potential integration within the contemporary façade value chain; the design features enabled by the use of AM, in terms of complexity, aesthetic and functional features; the construction, functional and performance requirements that are met or yet to be improved for prospective applications.

17.3.1 Walls and envelopes

This section presents solid skin elements such as walls (Table 17.2) and envelopes (Table 17.3), with structural and load-bearing functions. The main difference between walls and envelopes lies in the structural loads they respond to. Walls are vertical and subjected to compression only, while envelopes might also be inclined, like roofs, and, therefore, are subjected to bending forces as well. The most frequent technology used in additive construction, more specifically for façade walls, is 3DCP. This technology is typically used to produce vertical wall elements with an infill structure to provide structural strength during the printing process and afterward. The technique makes hollow cavities, which can either be filled in situ with reinforced concrete, left as air cavities, or otherwise filled with a foam-expanding material to provide thermal insulation. Applications of this technique can be found in various projects (Apis Cor, 2019; ICON, New Story, & ECHALE, 2019) (Fig. 17.6, left), including a 290-panels façade by Besix 3D (BESIX 3D, 2020). In cases with air cavities, the effectiveness of thermal insulation depends on compartmentalization. Small closed cavities are more effective than big and tall cavities where the air pressure differential creates a stack effect that creates a convective airflow and concentrates heat in the upper part of the cavity. As a result, such walls have poor heterogeneous performance. Due to the particular tectonic character of 3D printed concrete walls, cavity walls are often not plastered or protected on both the interior and exterior sides. A temporary pavilion built in Milan (CLS Architetti & Arup, 2018) was used to demonstrate the effective employability of 3D printing for houses, with specific attention to exterior and interior surface aesthetics, an approach that can also be found in the Office of the Future (Siam Research & Innovation Company SRI, 2017). The drawback of such an approach can be seen in

Table 17.2 Overview of the additive manufacturing (AM) technologies and materials for wall structures.

Citation	Technology	Material (AM) + postprocess (no AM)	Year	TRL
Le et al. (2012)	SME	Concrete	2012	4–5
CLS Architetti and Arup (2018)	SME	Concrete	2018	4–5
Apis Cor (2019)	SME	Concrete	2019	8–9
ICON, New Story, and ECHALE (2019)	SME	Concrete	2019	8–9
AlOthman et al. (2019)	SME	Clay (indirect) + concrete (postprocess)	2019	4–5
BESIX 3D (2020)	SME	Concrete	2020	8–9
Kokon (2020)	SME	Concrete	2020	1–3
Gramazio Kohler Research (2014)	FDM	Thermoplastics (indirect) + concrete (postprocess)	2014	4–5
Branch Technology (2015)	FDM	Fiber-reinforced polymers (indirect) + polyurethane foam (postprocess)	2015	4–5
Mirreco (2015)	FDM	Polymer/hemp	2015	8–9
Lab3D (2015)	FDM	Thermoplastics	2015	4–5
Ai Build (2017)	FDM	Thermoplastics (indirect) + concrete (postprocess)	2017	8–9
NOWlab (2018)	FDM	Thermoplastics (indirect) + concrete (postprocess)	2018	4–5
Taha et al. (2020)	Shotcrete	Concrete	2020	4–5
Dubor et al. (2018)	SME	Earth/mud	2018	4–5
WASP (2018)	SME	Earth/mud	2018	6–7
Chiou and Christofer (2020)	SME	Earth/mud	2019	4–5
WASP & MC A—Mario Cucinella Architects (2019)	SME	Earth/mud	2019	6–7
IAAC (2018)	SME	Clay	2018	4–5
Shi et al. (2019)	SME	Clay	2019	4–5

(*Continued*)

Table 17.2 (Continued)

Citation	Technology	Material (AM) + postprocess (no AM)	Year	TRL
Aguilar et al. (2020)	SME	Clay	2020	4–5
Lange et al. (2019)	SME	Ceramic	2019	4–5
Emerging Objects (2015b)	DMLS/SLM	Ceramic	2015	4–5
Keating et al. (2017)	ME	Polymer/polyurethane	2017	4–5

DMLS, Direct metal laser sintering; *FDM*, fused deposition modelling; *ME*, material extrusion; *SME*, semifluid material extrusion; *TRL*, technology readiness level.

Table 17.3 Overview of the additive manufacturing (AM) technologies and materials for building envelopes.

Citation	Technology	Material (AM) + postprocess (no AM)	Year	TRL
(Killa Design. 2016)	SME	Concrete	2016	8–9
Siam Research and Innovation Company (SRI) (2017)	SME	Concrete	2017	4–5
(3dprintedhouse 2019)	SME	Concrete	2019	8–9
SOM (2016)	FDM	Polymer	2016	8–9
Branch Technology (2016)	FDM	Fiber-reinforced polymers (indirect) + polyurethane foam (postprocess)	2016	6–7
Emerging Objects (2018)	FDM	Bio-based thermoplastic/corn	2018	4–5
Naboni et al. (2019)	FDM	Thermoplastics	2019	4–5
Navillus Woodworks (2019)	MJF	Thermoplastics (indirect) + concrete (postprocess)	2019	4–5

FDM, fused deposition modelling; *SME*, semifluid material extrusion; *TRL*, technology readiness level.

maintenance and durability as those surfaces are rough and porous and, therefore, collect dirt that cannot be cleaned. Projects where external walls are printed in components can be found in the art installation in Millennium Park (Navillus Woodworks, 2019); and Urban Cabin (Emerging Objects, 2018). The 3D polymer building for off-grid living (SOM, 2016), Trabeculae Pavilion (Naboni, Kunic, Breseghello, & Paoletti, 2017b; Naboni, Breseghello, & Kunic, 2019), and Curve Appeal (Branch Technology, 2016) go further in exploring how the technology can

Figure 17.6 Left, Apis Cor (2019); middle, WASP (2018); right, Branch Technology (2015).

be used for complex and freeform envelopes. Printing components of the façade helps in handling and logistics as small components can be printed off-site on small printers and do not require expensive cranes and machinery setups to position them in place.

The list of case studies includes refurbishment projects (Kokon, 2020), where AM is used for railings on balconies. However, this construction method produces thermal bridges due to the direct connection of boundary layers with the infill structures and the absence of an insulation layer. A print-in-place technique for thermal insulation layer that could be potentially used with multihead extrusion or as formwork for in situ concrete is presented in the study of Keating, Leland, Cai, and Oxman (2017). Alternatively, naturally sourced materials like hemp may also create walls with better thermal properties than concrete but with a reduced structural capacity (Mirreco, 2015).

Examples where earth-based materials are additively manufactured using SME include: the 3D printed house Gaia, in which natural ventilation and thermoacoustic insulation are integrated (WASP, 2018) (Fig. 17.6, middle); TECLA—Eco-housing, experimenting with dome-like structures (WASP & MCA—Mario Cucinella Architects, 2019); Digital Adobe, in which modularity and surface configurations are explored (IAAC, 2018); Terraperforma (Dubor, Cabay, Chronis, & Thomsen, 2018); and (Chiou & Christofer, 2020) focus on openings and in particular aesthetics, privacy and daylighting. These projects use various mixtures with natural materials sourced from the surrounding area to reduce the environmental footprint and lower costs. However, all of them are characterized by limited freedom for nonvertical free-form surfaces due to the poor rheological behaviour as well as structural and layer bonding strength of earth-based materials. Furthermore, similarly to exposed concrete, exposed and unfinished earth-based materials have reduced durability and poor maintenance. Due to the low resistance to rain drain, the exterior surface can be washed away and, over time, impose risk to the structural stability. Another drawback is an earth-mix design that allows for fast casting and sufficient strength in the dry state. As a consequence, the compromise in low production speed limits the implementation rate.

For more complex surfaces and free-form nonvertical extrusion, cementitious composites, clays and other ceramic materials are of particular relevance due to their improved young modulus and yield stress properties compared to simple earth-based materials. However, the inherent material instabilities of clay extrusion, especially at fast rates, might be considered problematic for precise fabrication. Nevertheless, such property was investigated in an architectural masonry façade system (Shi, Cho, Taylor, & Correa, 2019). Similarly, one case study attempts to reconcile an imprecise material process with an automated workflow to create customized bricks inspired by traditional Chinese architecture (Lange, Holohan, & Kehne, 2019). Clay material may also be suitable to explore complex material behaviour for façade components with photocatalytic and climatic properties such as particle filtration and evaporative cooling (Aguilar, Borunda, & Pardal, 2020; Emerging Objects, 2015b).

Thermoplastics can also be used in façade walls, especially for high-resolution printing that allows for the integration of construction, piping, window frames and good thermal insulation via air cavity pockets in infill structures (Lab3D, 2015). However, thermoplastics do not have structural and thermal stability to be considered for permanent and multistorey buildings.

AM may also be used indirectly as formwork for in situ cast concrete. The Smart Concrete Wall project uses additively fabricated moulds to create a concrete hexagonal grid with responsive integrated lighting (NOWlab, 2018). Formworks can also take the form of cellular lattices produced via nonlayered FDM from a variety of composites according to structural and morphological demands (Ai Build, 2017). These formworks may be recycled and reused. A similar technique is employed for a structure that combines formworking and reinforcement of geometrically complex elements produced with either shotcrete or cast concrete with low viscosity applied within printed cellular lattices. Depending on the material, various load-bearing capacities may be achieved (AlOthman, Im, Jung, & Bechthold, 2019; Gramazio Kohler Research, 2014). The process is also used for insulation layers when combined with spray-foam (Branch Technology, 2015) (Fig. 17.6, right). Alternatively, to increase efficiency and reduce material, a slender, permeable formwork and reinforcement mesh can be used (Hack et al., 2017).

Furthermore, one project shows a high potential for surface treatments using AeroCrete robotic spraying (Taha, Walzer, & Ruangjun, 2020). However, composites used for the cellular lattice meshes have certain limits in structural strength that are inferior to steel. Therefore further investigation and testing is needed to characterize structural behaviour before using this method in facades subjected to bending and tension.

17.3.2 Cladding

Unlike façade walls, additive cladding applications are nonload bearing interfaces designed as rainscreens and protective shields for structural and insulation components (Table 17.4). Therefore visual and aesthetic features are also included in the list of performances. In the project Seed Stitch a cladding surface is built from several ceramic elements with a textile-like texture, obtained by nonlinear toolpath design, which generates occasional 'stitch' drops with unique features for each element (Rael & San Fratello, 2017) (Fig. 17.7, left). Consequently, every tile becomes

Table 17.4 Overview of the additive manufacturing (AM) technologies and materials for cladding systems.

Citation	Technology	Material (AM) + postprocess (no AM)	Year	TRL
Emerging Objects (2015b,c)	SME	Wood-cement composite/ recycled waste products	2015	4—5
Rael and San Fratello (2017)	SME	Ceramic/thermoplastic	2017	4—5
Studio RAP (2019)	SME	Ceramic	2019	8—9
TrigonArt (2013)	DMLS/SLM	Sand	2013	8—9
DUS Architects and Duivenbode (2016)	FDM	Bio-based plastic	2016	4—5
Yu and Yuexiu Group (2018)	FDM	Thermoplastics	2018	4—5
Aectual (2019a, 2019b)	FDM	Bio-based plastic/linseed	2019	8—9
Terreform ONE and BASF (2019)	FDM	Silicone (indirect) + concrete (postprocess)	2019	4—5
Engholt and Pigram (2019)	FDM	Thermoplastics (indirect) + concrete (postprocess)	2019	4—5
Branch Technology (2020)	FDM	Fiber-reinforced polymers (indirect) + polyurethane foam (postprocess)	2020	8—9
Naboni and Breseghello (2020)	FDM	Thermoplastics (indirect) + concrete (postprocess)	2020	6—7

DMLS, Direct metal laser sintering; *FDM*, fused deposition modelling; *SLM*, selective laser melting; *SME*, semifluid material extrusion.

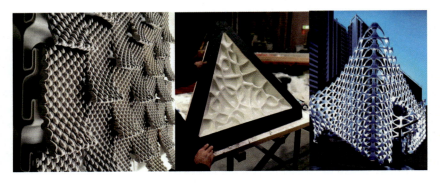

Figure 17.7 Left, Rael and San Fratello (2017); middle, Engholt and Pigram (2019); right, Yu and Yuexiu Group (2018).

a unique piece due to the controlled randomness of 'stitches', producing a unique aesthetic out of deliberate production flaws. However, the project lacks consideration of maintenance and durability as there are no indications of postprocessing that would protect the clay from weathering. In the project Tailored Flexibility, a fabric formwork is combined with plastic FDM to achieve a high-level material manifestation through a low-complexity fabrication setup (Engholt & Pigram, 2019) (Fig. 17.7, middle). Esthetic possibilities with this method are endless, while setup remains simple.

Similarly, a complex surface articulation and high-resolution concrete texturing can also be obtained using an additive formwork (Naboni & Breseghello, 2019). New Delft Blue's project relies on ceramic postprocessing techniques on clay 3D printing to create highly reflective coloured ceramic tiles inspired by Delft pottery's world-famous decorative qualities (Studio RAP, 2019). The project has excellent potential as it considers previously mentioned aspects like maintenance and durability. However, it remains unclear whether this process is intended for commercial uses and mass customization in other uses. In a less expressive project proposal for a new Tennessee Valley Federal Credit Union building, Branch Technology used subtle customization of floor-to-floor cladding panels (Branch Technology, 2020). Although relatively modest innovation regarding cladding panel system design, the project uses AM to produce custom moulds. Besides aesthetics, additive moulds may be reused much more times than wooden ones and recycled. Therefore this approach also presents a better environmental alternative to wooden moulds.

A different approach using mass-produced panels by Aectual demonstrate the potential for commercialization of AM façade panels with excellent surface quality and durability (Aectual, 2019a,b). The company's offer presents one of the rare cases where the product is optimized for mass production. No further details are provided to demonstrate innovation value chain and production advantages. Alternatively, nonstandard functions can be integrated into additive facades to embrace social interaction (DUS Architects & Duivenbode, 2016) or to enable plant growth tailored to specific species (Emerging Objects, 2015c; Terreform ONE & BASF, 2019). Such biophilic designs tend to evoke connectivity with nature and well-being. Nevertheless, the risk of maintenance malfunctioning due to the watering shortage or excessive solar radiation are not presented. In the case of damage, such designs might be very short-lived, expensive and wretched looking. Therefore special attention needs to be dedicated to addressing these issues.

Some projects like Arachne promote an understanding of cladding beyond the functional level toward the sculptural domain (Yu & Yuexiu Group, 2018) (Fig. 17.7, right). However, such an approach has limited functional impact and will likely be adopted marginally in future. On the other hand, the Liebknecht portal utilized 3D printing to produce an exact copy of a cultural heritage building (TrigonArt, 2013). Although focusing mainly on aesthetic and ornamentation, the project still builds on top of cultural heritage refurbishment and has a much larger market and potential.

17.3.3 Shading

A domain with one great potential for AM façade applications is shading. Shading devices can take many different forms and positions concerning a specific façade design. Moreover, aesthetic possibilities are endless, and performance demands are relatively limited to the g-value and the amount of daylight the shading device allows to pass through. Only a limited number of projects to date focused on 3D-printed shading devices (Table 17.5). The list includes exterior (Grassi, Lupica Spagnolo, & Paoletti, 2019), embedded (Louter et al., 2018) (Fig. 17.8, left) and interior (Meibodi, Kladeftira, Kyttas, & Dillenburger, 2019) (Fig. 17.8, middle) applications. Similarly to the traditionally produced shading systems, additively manufactured ones share the same advantages and disadvantages. Exterior shading is better for mitigating solar heat gains before entering the building interior, while exposure to weathering makes it less durable and more costly to maintain. Interior shading is less effective for sun protection but better in terms of control and access. Also, since there is no direct exposure to weathering, maintenance and durability are much easier and longer, respectively. Embedded shading seems to be the best

Table 17.5 Overview of the additive manufacturing (AM) technologies and materials for shading systems.

Citation	Technology	Material (AM) + postprocess (no AM)	Year	TRL
Meibodi et al. (2019)	DMLS/SLM	Sand (indirect) + metal (postprocess)	2018	4—5
Tenpierik et al. (2018)	FDM	Aerogel/phase change material	2018	4—5
Louter et al. (2018)	FDM	FDM	2019	4—5
Grassi et al. (2019)	FDM	Thermoplastics	2019	6—7

DMLS, Direct metal laser sintering; *FDM*, fused deposition modelling; *SLM*, selective laser melting.

Figure 17.8 Left, Louter et al. (2018); middle, Meibodi et al. (2019); right, Tenpierik et al. (2018).

blend of advantages in solar control and protection from weathering. When static, this approach can even structurally activate the core and reduce the need for glass lamination, especially in big and jumbo-size glass panels (Jakica and Kragh, 2019). Multifunctional concepts like the Trombe wall (Tenpierik, Turrin, Wattez, Cosmatu, & Tsafou, 2018) (Fig. 17.8, right) represent a promising way forward and demonstrate AM applications' versatility. However, for reaching full potential, the innovation will need to focus on finding the right material composition to support the additive process and yet have a high thermal mass.

17.3.4 Curtain walls

Previous studies have shown that AM of metal curtain walls has limited potential compared to the high efficiency of mass-produced frame profile extrusion or extremely controlled processes such as glass and glass coating. Therefore most of the current market-oriented applications focus on nodes (Attaran, 2017) (Fig. 17.9, left), knots (NextStudio, 2020) (Fig. 17.9, middle) and brackets (Altair, Materialise, & Renishaw, 2016) (Table 17.6) as these elements require a high level of customization that is not easily achievable with traditional manufacturing processes. Other conceptual projects explore the AM potential for curtain wall applications by integrating furniture (Taseva, Eftekhar, Kwon, Leschok, & Dillenburger, 2020), electrical components (Kwon, Eichenhofer, Kyttas, & Dillenburger, 2019) and thermal regulation systems (Sarakinioti et al., 2018) (Fig. 17.9, right). However, it is unclear how these projects will evolve when approaching higher TRLs. Furthermore, since thermoplastic facades are known for their thermal instability, yellowing, low scratch resistance and short life span, extensive testing and material improvement are needed to be considered for permanent façade applications. A range of testing (Mungenast, 2019) showed ways to improve the TRL of current concepts. On the other hand, for applications with concrete, AM has found its way to improve the efficiency and durability of formworks by replacing wood with thermoplastics (CookFox Architects, Two Trees, & Gate Precast Company, 2018).

Figure 17.9 Left, Strauss and Knaack (2016); middle, NextStudio (2020); right, Mungenast (2019).

Table 17.6 Overview of the additive manufacturing (AM) technologies and materials for curtain walls.

Citation	Technology	Material (AM) + postprocess (no AM)	Year	TRL
CookFox Architects, Two Trees, and Gate Precast Company (2018)	FDM	Thermoplastics (indirect) + concrete (postprocess)	2018	4–5
Sarakinioti et al. (2018)	FDM	Thermoplastics	2018	4–5
(Kwon, Eichenhofer, Kyttas, & Dillenburger, 2019)	FDM	Thermoplastics/ electrical	2019	4–5
Mungenast (2019)	FDM	Thermoplastics	2019	6–7
Taseva et al. (2020)	FDM	Thermoplastics	2020	4–5
Strauss and Knaack (2016)	DMLS/SLM	Metal	2016	4–5
Altair, Materialise, and Renishaw (2016)	DMLS/SLM	Metal	2016	4–5
NextStudio (2020)	DMLS/SLM	Metal	2020	4–5

DMLS, Direct metal laser sintering; *FDM*, fused deposition modelling; *SLM*, selective laser melting.

17.4 Conclusion

This chapter provided an overview of the current status of AM in the field of architecture and constructions. It focused on the possibilities and implications for the design and manufacturing of facades. The available methods for building skins with various sizes, materials and resolution have been discussed. The analysis highlights an evolving field, developed rapidly in the last 10 years. An increasing number of subtechniques are tuned for construction purposes, including several commercially ready solutions. Techniques based on ME provide the highest potential for the architectural field. They rely on relatively fast and economic processes that employ conventional construction materials such as concrete and clay. Metals often employed in building skins can be 3D printed with a range of sizes and resolutions, available for fine detailing and more significant components. However, both approaches are demanding in terms of material costs. Polymer-based 3D printing is very mature and affordable but has limited applications in constructions because of the limited resistance to high temperature and fire. G3DP is still experimental and does not deliver the expected optical properties of glass in a façade yet.

Most applications are still in the phase of research and prototyping. Approximately two-thirds of the project case studies are at the TRL 4–5. About one-third of the case studies can be considered innovation projects, positioning themselves in the higher end of the TRL scale. This shows that the technology is moving up toward actual construction. However, some constraints are still reducing their marketability.

Concerning available technology, improvements need to be adopted to compare to serial industrial production. Exact reproducibility among different parts is still difficult to achieve when looking at dimensional, functional and mechanical performance. Most of the techniques are still lacking the needed speed or scalability for large batches of production. About material properties, 3D-printed materials are not matching the functional and aesthetic qualities of standard façade systems. The need for specific material standards, which refer to printed elements as a particular class of materials, is a needed step to realistically catch the material behaviour toward the use in advanced applications—where mechanical, geometrical and material properties can be tailored to a specific design case.

Producing an entire building façade is doable with specific techniques but requires significant economic and development efforts. More specifically, technological solutions that can provide high-resolution and competitive finishing at a large scale are still lacking, without impacting prohibitively on the cost. In this sense, interdisciplinary research is still needed to make AM a technically reliable and economically viable option (Camacho et al., 2018). Research developments in manufacturing technology and parallelization of production can ensure product customization at a lower price. At the same time, additively manufactured components will require dedicated material testing standards. Finally, from the point of view of building technology, more design efforts are needed in developing façade systems that can benefit truly from the advantages of just-in-time bespoke production and functional integration. Lastly, AM processes and automation within the construction site are still open topics that need more research.

The diffusion of C3DP in the market would also demand that a complete digitalization of the supply chain takes effect, involving new competencies that are only now introduced in educational programs. Several companies are working with software implementation for the easy management of 3D data to production. However, the workflows are still new to the supply chain. It is foreseeable that if these techniques have gained more presence in practice in a few years, the overall cost efficiency will match or overcome conventional building products for a specific range of applications where advanced customization plays a fundamental role. The need for reduced material usage, minimization of carbon emissions, lightweight and structurally efficient construction and bespoke design for building retrofit are opening new opportunities and market niches. As one of the most prominent digitally led production technologies, AM has a clear potential to make a difference by supporting advanced design that can respond to the upcoming needs of the future building envelopes.

References

Aectual. (2019a). *Aectual façade GFRC panels*. Retrieved December 2, 2020, from <https://www.aectual.com/architectural-products/façade-cladding/variants/gfrc-panels/intro>.

Aectual. (2019b). *Circular façade panels*. Retrieved December 1, 2020, from <https://www.aectual.com/architectural-products/façade-cladding/variants/circular-facades/intro>.

Aectual. (2020). <https://aectual.com> Accessed 22.10.20.

Aguilar, P., Borunda, L., & Pardal, C. (2020). Additive manufacturing of variable-density ceramics, photocatalytic and filtering slats. In L. C. Werner & D. Koering (Eds.), *Anthropologic—Architecture and fabrication in the cognitive age—Proceedings of the 38th international online conference on education and research in computer aided architectural design in Europe, Berlin, Germany, 16th—17th September 2020, Vol. 1*. (Vol. 1, pp. 97—106). eCAADe (Education and Research in Computer Aided Architectural Design in Europe).

Ai Build. (2017). *Concrete formwork*. Retrieved December 2, 2020, from <https://ai-build.com/concreteformwork.html>.

AI Build. (2020). <http://www.ai-build.com> Accessed 24.10.20.

AlOthman, S., Im, H. C., Jung, F., & Bechthold, M. (2019). Spatial print trajectory. In J. Willmann, P. Block, M. Hutter, K. Byrne, & T. Schork (Eds.), *Robotic fabrication in architecture, art and design* 2018 (pp. 167—180). <https://doi.org/10.1007/978-3-319-92294-2_13>.

Altair, Materialise, & Renishaw. (2016). *Spider bracket structure*. Retrieved December 2, 2020, from <https://parametrichouse.com/spider-bracket-structure/>.

Apis Cor. (2019). *3D-printed building in Dubai*. <https://www.dezeen.com/2019/12/22/apis-cor-worlds-largest-3d-printed-building-dubai/?li_source = LI&li_medium = bottom_block_1> Accessed 29.09.20.

Aslam, H., Altynay, Z., Suvash, C. P., & Jong, R. K. (2020). A review of 3D printing in construction and its impact on the labor market, sustainability, MDPI. *Open Access Journal*, *12*(20), 1—21.

Asprone, D., Menna, C., Bos, F. P., Salet, T. A. M., Mata-Falcón, J., & Kaufmann, W. (2018). Rethinking reinforcement for digital fabrication with concrete. *Cement and Concrete Research*, *112*, 111—121.

Attaran, M. (2017). Additive manufacturing: The most promising technology to alter the supply chain and logistics. *Journal of Service Science and Management*, *10*(3), 189—206. Available from https://doi.org/10.4236/jssm.2017.103017.

Baker, R. (1925). *Method of making decorative articles, patent*.

BESIX 3D. (2020). *Concrete façade*. Retrieved December 1, 2020, from <https://www.six-construct.com/en/news/besix-3d-prints-largest-concrete-façade-in-the-world>.

BigRep. (2020). *Large format 3D printers*. <https://www.bigrep.com> Accessed 19.10.20.

Branch Technology. (2015). *Modular wall system*. Retrieved December 2, 2020, from <https://www.architectmagazine.com/technology/this-architect-designed-wall-system-has-a-3d-printed-core_o?fbclid = IwAR08DJLJbzhOtQr_eutZb5kwGIl12YET3zjtdyHJ0jwJNrReDSkVhyHpmaU>.

Branch Technology. (2016). *Curve appeal*. Retrieved December 2, 2020, from <https://www.3dprintingmedia.network/branch-technologies-c-fab-3d-process-can-take-us-mars/>.

Branch Technology. (2020). *Tennessee valley federal credit union*. Retrieved December 2, 2020, from <https://www.branch.technology/projects-1/2019/2tvfcu>.

Branch Technology. (2020). <http://www.branch.technology> Accessed 28.10.20.

Buchanan, C., & Gardner, L. (2019). Metal 3D printing in construction: A review of methods, research, applications, opportunities and challenges. *Engineering Structures*, *180*, 332—348.

Busachi, A., Erkoyuncu, J., Colegrove, P., Martina, F., Watts, C., & Drake, R. (2017). A review of Additive Manufacturing technology and cost estimation techniques for the defence sector. *CIRP Journal of Manufacturing Science and Technology*, *19*, 117−128.

Buswell, R. A., Leal de Silva, W. R., Jones, S. Z., & Dirrenberger, J. (2018). 3D printing using concrete extrusion: A roadmap for research. *Cement and Concrete Research*, *112*, 37−49.

Camacho, D. D., Clayton, P., O'Brien, W. J., Seepersad, C., Juenger, M., Ferron, R., & Salamone, S. (2018). Applications of additive manufacturing in the construction industry—A forward-looking review. *Automation in Construction*, *89*, 110−119.

Chaltiel, S., Bravo, M., Veenendaal, D., & Sayers, G. (2020). Drone spraying on light formwork for mud shells. In B. Sheil, M. R. Thomsen, M. Tamke, & S. Hanna (Eds.), *Design transactions: Rethinking information modelling for a new material age* (pp. 150−157). UCL Press.

CEAD. (2020). <http://www.cead-am.com> Accessed 23.10.20.

Chiou, J., & Christofer, B. (2020). *Openings in 3D printing with earth*. Retrieved December 1, 2020, from <http://www.iaacblog.com/programs/openings-3d-printing-earth/?fbclid=IwAR0P3imMIVcTP83rVE98xvXO07oOPIOdjn_u0gKFpqFs0E9nn_1NxTK3YVk>.

CLS Architetti, & Arup. (2018). *3D printed house in Milan*. Retrieved December 1, 2020, from <https://www.dezeen.com/2018/04/20/cls-architetti-arup-use-portable-robot-3d-print-house-milan/>

COBOD. (2020). *Modular 3D construction printers. 3D printed buildings*. <https://www.cobod.com> Accessed 13.11.20.

CookFox Architects, Two Trees, & Gate Precast Company. (2018). *The concrete façade of a former sugar factory*. Retrieved December 2, 2020, from <https://redshift.autodesk.com/3d-printed-concrete-mold/>.

CRANE WASP. (2020). https://www.3dwasp.com/en/3d-printer-house-crane-wasp/ Accessed in December 2020.

CyBe. (2020). CyBe Construction, *We redefine construction*. <https://www.cybe.eu>. Accessed 17.11.20.

DBT. (2020). <https://dbt.arch.ethz.ch/project/smart-slab/>. Accessed in December 2020.

de Laubier, R., Wunder, M., Witthöft, S., & Rothballer, C. (2018). *Will 3D printing remodel the construction industry?* The Boston Consulting Group.

Deckard, C. R. (1986). *Method and apparatus for producing parts by selective sintering, patent*.

Deckard, C. R., Beaman, J. J., & Darrah, J. F. (1986). *Method for selective laser sintering with layerwise cross-scanning. Filed 17.10.1986 and assigned 13.10.1992*.

Delic, M., & Eyers, D. R. (2020). The effect of additive manufacturing adoption on supply chain flexibility and performance: An empirical analysis from the automotive industry. *International Journal of Production Economics*, *228*.

Derekar, K. S. (2018). A review of wire arc additive manufacturing and advances in wire arc additive manufacturing of aluminium. *Materials Science and Technology*, *34*(8), 895−916.

Dini, E. (2006). *Method and device for building automatically conglomerate structures, patent*.

Dubor, A., Cabay, E., & Chronis, A. (2018). Energy efficient design for 3D printed earth architecture. In K. De Rycke, C. Gengnagel, O. Baverel, J. Burry, C. Mueller, M. M. Nguyen, & ... M. R. Thomsen (Eds.), *Humanising digital reality* (pp. 383−393). Singapore: Springer. Available from https://doi.org/10.1007/978-981-10-6611-5_33.

DUS architects, & Duivenbode, O.van. (2016). *3D printed façade for EU building*. Retrieved December 1, 2020, from <https://divisare.com/projects/307190-dus-architects-ossip-van-duivenbode-3d-printed-façade-for-eu-building-amsterdam>.

Emerging Objects. (2015b). *Cool brick*. Retrieved December 1, 2020, from <http://emergingobjects.com/project/cool-brick/>.

Emerging Objects. (2015c). *Planter tile in cement*. Retrieved December 1, 2020, from <http://emergingobjects.com/project/planter-tile-in-cement/>.

Emerging Objects. (2018). *Urban cabin*. Retrieved December 1, 2020, from <https://www.treehugger.com/d-printed-seed-stitch-tile-emerging-objects-4856200>.

Engholt, J., & Pigram, D. (2019). Tailored flexibility reinforcing concrete fabric formwork with 3D printed plastics. In M. H. Haeusler, M. A. Schnabel, & T. Fukuda (Eds.), *Intelligent & informed, proceedings of the 24th international conference of the association for Computer-Aided Architectural Design Research in Asia (CAADRIA) 2019*. The Association for Computer-Aided Architectural Design Research in Asia (CAADRIA).

EOS. (2014). GmbH—taking the laser lead. *Metal Powder Report, 69*(3), 24−27.

Galjaard, S., Hofman, S., Perry, N., & Ren, S. (2015). *Optimizing structural building elements in metal by using additive manufacturing*. In: Proceedings of the International Association for Shell and Spatial Structures (IASS) Symposium 2015. The International Association for Shell and Spatial Structures, Amsterdam.

Gardiner, J. B., & Janssen, S. R. (2014). FreeFab: Development of a construction-scale robotic formwork 3D printer (2014). In W. McGee, & M. Ponce de Leon (Eds.), *Robotic fabrication in architecture, art and design 2014*. Springer International Publishing. Available from https://doi.org/10.1007/978-3-319-04663-1_9.

Gibson, I., Rosen, D. V., & Stucker, B. (2010). *Additive manufacturing technologies. Rapid prototyping to direct digital manufacturing*. New York, Heidelberg, Dordrecht, London: Springer.

Gramazio Kohler Research. (2014). *Mesh mould*. Retrieved December 2, 2020, from <https://gramaziokohler.arch.ethz.ch/web/e/forschung/221.html>.

Grassi, G., Lupica Spagnolo, S., & Paoletti, I. (2019). Fabrication and durability testing of a 3D printed façade for desert climates. *Additive Manufacturing, 28*(September 2018), 439−444. Available from https://doi.org/10.1016/j.addma.2019.05.023.

Hack, N., Wangler, T., Mata-Falcon, J., Dörfler, K., Kumar, N., Walzer, A. N., ... Kohler, M. (2017). Mesh mould: An on site, robotically fabricated, functional formwork, In *Second concrete innovation conference (2nd CIC), paper no. 19*. Tromsø, Norway (2017).

HP Multi Jet Fusion Technology—*Technical white paper*. (2021). <https://reinvent.hp.com/us-en-3dprint-wp-technical>. Accessed 21.02.21.

https://3dprintedhouse.nl/en/project-info/project-milestone/, (2019). Accessed 1 December 2020.

https://www.archdaily.com/875642/office-of-the-future-killa-design, (2016). Accessed 1 December 2020.

IAAC. (2018). *Digital adobe*. Retrieved December 1, 2020, from <https://iaac.net/project/digital-adobe/>.

ICON, New Story, & ECHALE. (2019). *3D-printed community*. Retrieved December 1, 2020, from <https://www.iconbuild.com/updates/icon-new-story-echale-unveil-first-homes-in-3d-printed-community>

ISO/ASTM 52900.2015. (2015). *Additive manufacturing—general principles terminology*. Geneva: International Organization for Standardization.

Jakica, N., & Kragh, M. K. (2019). Activating optical behaviour of cellular lattices in glass sandwich facades. In *Proceedings of the advanced building skins conference 2019* (pp. 292−299).

Karimi, H., Taylor, T. R. B., Goodrum, P. M., & Srinivasan, C. (2016). Quantitative analysis of the impact of craft worker availability on construction project safety performance. *Construction Innovation*, *16*(3), 307−322.

Keating, S. J., Leland, J. C., Cai, L., & Oxman, N. (2017). Toward site-specific and self-sufficient robotic fabrication on architectural scales. *Science Robotics*, *2*(5). Available from https://doi.org/10.1126/scirobotics.aam8986.

Khoshnevis, B. (2004). Automated construction by contour crafting—related robotics and information technologies. *Automation in Construction*, *13*(1), 5−19.

Khosnevis, B. (1995). Additive fabrication apparatus and method. *United States 5529471A. Filed 03.02.1992 and assigned 25.06.1995.*

Klein, J., Stern, M., Franchin, G., Kayser, M. A. R., Inamura, C., Dave, S. H., ... Oxman, N. (2015). Additive manufacturing of optically transparent glass. *3D printing and additive manufacturing*, *2*(3), 92−105.

Kokon. (2020). *Hendrik Baskeweg*. Retrieved December 4, 2020, from <https://www.kokon.nl/en/projects/hendrik-baskeweg>.

Kragh, M. K. (2020). Façade Engineering 4.0 façade engineering for the Fourth Industrial Revolution. In D. Noble, K. Kensek, & S. Das (Eds.), *Face Time 2020: Better buildings through better skins conference proceedings* (Vol. 1, pp. 203−208). Los Angeles, CA: Tectonic Press.

Kwon, H., Eichenhofer, M., Kyttas, T., & Dillenburger, B. (2019). Digital composites: Robotic 3D printing of continuous carbon fibre-reinforced plastics for functionally-graded [1] building components. In J. Willmann, P. Block, M. Hutter, K. Byrne, & T. Schork (Eds.), *Robotic fabrication in architecture, art and design 2018* (pp. 363−376). Cham: Springer. Available from https://doi.org/10.1007/978-3-319-92294-2_28.

Lab3D. (2015). *3D printed façade*. Retrieved December 1, 2020, from <https://archello.com/project/3d-printed-façade>.

Kwon, H., Eichenhofer, M., Kyttas, T., & Dillenburger, B. (2019). Digital composites: Robotic 3D printing of continuous carbon fiber-reinforced plastics for functionally-graded building components. *Robotic Fabrication in Architecture, Art and Design 2018*. Available from https://doi.org/10.1007/978-3-319-92294-2_28.

Lange, C. J., Holohan, D., & Kehne, H. (2019). Ceramic constellation I robotically printed brick specials. In J. Willmann, P. Block, M. Hutter, K. Byrne, & T. Schork (Eds.), *Robotic fabrication in architecture, art and design 2018* (pp. 434−446). Cham: Springer. Available from https://doi.org/10.1007/978-3-319-92294-2_33.

Le, T., Austin, S. A., Lim, S., Buswell, R. A., Gibb, A. G. F., & Thorpe, T. (2012). Mix design and fresh properties for high-performance printing concrete. *Materials and Structures*, *45*(8), 1221−1232. Available from https://doi.org/10.1617/s11527-012-9828-z.

Lee, D., Kim, H., Sim, J., Lee, D., Cho, H., & Hong, D. (2019). Trends in 3D printing technology for construction automation using text mining. *International Journal of Precision Engineering and Manufacturing*, *20*(5), 871−882. Available from https://doi.org/10.1007/s12541-019-00117-w.

Li, C., Pisignano, D., Zhao, Y., & Xue, J. (2020). Advances in medical applications of additive manufacturing. *Engineering*, *6*(11), 1222−1231.

Lindemann, H., Gerbers, R., Ibrahim, S., Dietrich, F., Herrmann, E., Dröder, K., ...Kloft, H. (2018). Development of a shotcrete 3D-printing (SC3DP) technology for additive manufacturing of reinforced freeform concrete structures. In: T. Wangler, R.J. Flatt (Eds), *First RILEM international conference on concrete and digital fabrication—Digital concrete 2018*. DC 2018. RILEM Bookseries, 19. Springer, Cham.

Louter, C., Akilo, M. A., Miri, B., Neeskens, T., Ribeiro Silveira, R., Topcu., ... O'Callaghan, J. (2018). Adaptive and composite thin glass concepts for architectural applications. *Heron, 63*(1–2), 199–218.

Lowke, D., Dini, E., Perrot, A., Weger, D., Gehlen, C., & Dillenburger, B. (2018). Particle-bed 3D printing in concrete construction—Possibilities and challenges. *Cement and Concrete Research, 112*(August), 50–65. Available from https://doi.org/10.1016/j.cemconres.2018.05.018.

Masood, S. H. (2014). Introduction to advances in additive manufacturing and tooling. In S. Hashmi, C. J. Van Tyne, G. F. Batalha, & B. Yilbas (Eds.), *Comprehensive materials processing*. Oxford: Elsevier.

Mechtcherine, V., Nerella, V. N., Ogura, H., Grafe, J., Spaniol, E., Hertel, M., & Füssel, U. (2018). Alternative reinforcements for digital concrete construction, DC 2018. RILEM Bookseries, 19 In T. Wangler, & R. J. Flatt (Eds.), *First RILEM international conference on concrete and digital fabrication—Digital concrete 2018* (pp. 167–175). Cham: Springer.

Meibodi, M. A., Kladeftira, M., Kyttas, T., & Dillenburger, B. (2019). Bespoke cast façade. In *Acadia 19: ubiquity and autonomy, proceedings of the 39th Annual Conference of the Association for Computer Aided Design in Architecture (ACADIA)* (pp. 100–109). The University of Texas at Austin School of Architecture, Austin.

Micron3DP. (2015). *Breakthrough in 3D printing glass!* <https://micron-eme.com/blogs/news/34473924-breakthrough-in-3d-printing-glass> Accessed December 2020.

Mirreco. (2015). *Hemp houses*. Retrieved December 1, 2020, from <https://returntonow.net/2020/02/26/hemp-houses-are-being-3d-printed-in-australia/>.

Mungenast, M. B. (2019). *3D-printed future façade*. Technische Universität München.

MX3D. (2020). <https://mx3d.com/industries/infrastructure/mx3d-bridge/>, Accessed in December 2020.

Naboni, R., & Breseghello, L. (2019). Additive formwork for concrete shell constructions. In: C. Lázaro, K.U. Bletzinger, and E. Oñate (Eds), *Form and force IASS symposium 2019 conference proceedings* (pp. 87–94).

Naboni, R., & Breseghello, L. (2018). Fused deposition modelling formworks for complex concrete constructions, *SIGraDi 2018—Proceedings of the 22nd Conference of the Iberoamerican Society of Digital Graphics, São Carlos, Brazil*—November 07–09, 2018, 5(1) (pp. 700–707), https://doi.org/10.5151/sigradi2018-1648.

Naboni, R., & Breseghello, L. (2020). High-resolution additive formwork for building-scale concrete panels. In: F. P. Bos, S. S. Lucas, R. J. M. Wolfs, & T. A. M. Salet (Eds.), *Second RILEM international conference on concrete and digital fabrication—Digital concrete 2020*. DC 2020. RILEM Bookseries, 28. Springer, Cham.

Naboni, R., Breseghello, L., & Kunic, A. (2019). Multi-scale design and fabrication of the Trabeculae Pavilion. *Additive Manufacturing, 27*, 305–317. Available from https://doi.org/10.1016/j.addma.2019.03.005.

Naboni, R., Kunic, A., & Breseghello, L. (2020). Computational design, engineering and manufacturing of a material-efficient 3D printed lattice structure. *International Journal of Architectural Computing, 18*(4), 404–423.

Naboni, R., Kunic, A., Breseghello, L., & Paoletti, I. (2017a). Load-responsive cellular envelopes with additive manufacturing. *Journal of façade Design and Engineering, 5*(1), 37–49. Available from https://doi.org/10.7480/jfde.2017.1.1427.

Naboni, R., Kunic, A., Breseghello, L., & Paoletti, I. (2017b). Cellular lattice-based envelopes with additive manufacturing. In T. Auer, U. Knaack, & J. Schneider (Eds.), *Powerskin conference 2017 proceedings*. TU Delft Open TU Delft/Faculty of Architecture and the Built Environment.

Naboni, R., & Paoletti, I. (2015). *Advanced customization in architectural design and construction.* Springer.

Najmon, J. C., Raeisi, S., & Tovar, A. (2019). Review of additive manufacturing technologies and applications in the aerospace industry. In F. Froes, & R. Boyer (Eds.), *Additive manufacturing for the aerospace industry* (pp. 7–31). Elsevier.

Navillus Woodworks. (2019). *Art installation in Millennium Park.* <https://www.fastradius.com/case-studies/navillus/> Accessed 24.11.20.

NextStudio. (2020). *3D-printed façade knot.* Retrieved December 2, 2020, from <https://www.imagine-computation.com/post/3d-printed-façade-knot-at-nextstudio>.

NOWlab. (2018). *The smart concrete wall.* Retrieved December 2, 2020, from <https://additivenews.com/nowlab-partners-creates-3d-printing-illuminating-architecture-smart-concrete-wall/>.

Ochs, J., Akbar, Z., & Eversmann, P. (2020). Additive manufacturing with solid wood: Continuous robotic laying of multiple wicker filaments through micro lamination. In (1st (ed.)). A. Maciel (Ed.), *Design computation input/output* (2020). London: Design Computation.

Oxman, N. (2011). Variable property rapid prototyping. *Virtual and Physical Prototyping, 6*(1), 3–31.

Oxman, N., Keating, S., & Tsai, E. (2011). Functionally graded rapid prototyping. In Paulo Jorge da Silva Bártolo (Ed.), *Innovative developments in virtual and physical prototyping.* CRC Press.

Paolini, A., Kollmannsberger, S., & Rank, E. (2019). Additive manufacturing in construction: A review on processes, applications, and digital planning methods. *Additive Manufacturing, 30*(September), 100894. Available from https://doi.org/10.1016/j.addma.2019.100894.

Pegna, J. (1997). Exploratory investigation of solid freeform construction. *Automation in Construction, 5*(5), 427–437.

Rael, R., & San Fratello, V. (2017). Clay bodies: Crafting the future with 3D printing. *Architectural Design, 87*(6), 92–97. Available from https://doi.org/10.1002/ad.2243.

Roschli, A., Gaul, K. T., Boulger, A. M., Post, B. K., Chesser, P. C., Love, L. J., ... Borish, M. (2019). Designing for big area additive manufacturing. *Additive Manufacturing, 25,* 275–285.

Sachs, E., Cima, M., Williams, P., Brancazio, D., & Cornie, J. (1992). Three-dimensional printing: rapid tooling and prototypes directly from a CAD model. *Journal of Manufacturing Science and Engineering, 114,* 481–488.

Sarakinioti, M. V., Konstantinou, T., Turrin, M., Tenpierik, M., Loonen, R., De Klijn-Chevalerias, M. L., & Knaack, U. (2018). Development and prototyping of an integrated 3D-printed façade for thermal regulation in complex geometries. *Journal of façade Design and Engineering, 6*(2), 29–40. Available from https://doi.org/10.7480/jfde.2018.2.2081.

Shi, J., Cho, Y., Taylor, M., & Correa, D. (2019). Guiding instability a craft-based approach for modular 3D clay printed masonry screen units. In J. P. Sousa, G. C. Henriques, & J. P. Xavier (Eds.), *eCAADe SIGraDi 2019 architecture in the age of the 4th industrial revolution* Vol. 1 (pp. 477–484). https://doi.org/10.5151/proceedings-ecaadesigradi2019_522.

Siam Research and Innovation Company (SRI). (2017). *Triple S.* Retrieved December 1, 2020, from <https://www.archdaily.com/887403/3d-printing-fuses-thai-craftsmanship-to-create-habitable-concrete-structures?ad_medium = gallery>.

SOM. (2016). *3D-printed building designed for off-grid living*. Retrieved December 1, 2020, from <https://www.dezeen.com/2016/01/25/additive-manufacturing-integrated-energy-3d-printed-structure-som/>.

Strauss, H. (2013). *AM envelope: The potential of additive manufacturing for façade constructions. Architecture and the built environment*.

Strauss, H., & Knaack, U. (2016). Additive manufacturing for future facades: The potential of 3D printed parts for the building envelope. *Journal of façade Design and Engineering, 3*(3–4), 225–235.

Stratasys Inc. (1992). Apparatus and method for creating three-dimensional objects. *United States 5121329A, filed 30.10.1989 and issued 09.06.1992*.

Studio RAP, *New delft blue* (2019), https://studiorap.nl/#/poortmeesters, Accessed 01.12.20.

Taha, N., Walzer, A. N., & Ruangjun, J. (2020). Robotic AeroCrete A novel robotic spraying and surface treatment technology for the production of slender reinforced concrete elements. In *Proceedings of the ECAADe 37/SIGraDi 23 Matter—DIGITAL production and robotics 2*, Vol. 3 (pp. 245–256). <https://doi.org/10.5151/proceedings-ecaadesigradi2019_675>.

Taha, N., Walzer, A. N., Ruangjun, J., Bürgin, T., Dörfler, K., Lloret-Fritschi, E., ... Kohler, M. (2019) Robotic AeroCrete: A novel robotic spraying and surface treatment technology for the production of slender reinforced concrete elements. In: J. P. Sousa, G. C. Henriques, & J. P. Xavier (Eds.), *Proceedings of eCAADe + SIGraDi conference: Architecture in the age of the 4th industrial revolution, Porto, Portugal*, September 11–13, 2019, 37 (3) (pp. 245–254).

Taseva, Y., Eftekhar, N. I. K., Kwon, H., Leschok, M., & Dillenburger, B. (2020). Large-scale 3d printing for functionally-graded façade. In D. Holzer, W. Nakapan, A. Globa, & I. Koh (Eds.), *RE: anthropocene, design in the age of humans; proceedings of the 25th international conference of the association for Computer-Aided Architectural Design Research in Asia (CAADRIA) 2020* (Vol. 1, pp. 183–192). Hong Kong: the Association for Computer-Aided Architectural Design Research in Asia (CAADRIA).

Tenpierik, M., Turrin, M., Wattez, Y., Cosmatu, T., & Tsafou, S. (2018). *Double Face 2.0: A lightweight translucent adaptable Trombe wall*. SPOOL, 5(2-). <https://doi.org/10.7480/spool.2018.2.2090>.

Terraperforma. (2017). *3D printed performative wall*. <https://iaac.net/project/terraperforma/>, Accessed December 2020.

Terreform ONE, & BASF. (2019). Monarch Sanctuary at Cooper Hewitt Nature Exhibition. Retrieved December 1, 2020, from <https://www.dexigner.com/news/32092>.

Thomas, D. S., & Gilbert, S. W. (2014). Costs and cost effectiveness of additive manufacturing. *A Literature Review and Discussion, 1176*.

TrigonArt. (2013). *Liebknecht portal*. Retrieved December 2, 2020, from <https://www.voxeljet.com/branchen/case-study/detailgetreue-reproduktion-eines-denkmals/>.

Vantyghem, G., Boel, V., De Corte, W., & Steeman, M. (2018). Compliance, stress-based and multi-physics topology optimization for 3d-printed concrete structures. In: T. Wangler, & R.J. Flatt (Eds), *First RILEM international conference on concrete and digital fabrication—digital concrete 2018*. DC 2018. RILEM Bookseries, 19. Springer, Cham.

Vertico. (2020). *Concrete 3D printing solutions*. <https://www.vertico.xyz> Accessed 29.10.20.

WASP, & MCA—Mario Cucinella Architects. (2019). *TECLA—Eco-housing*. Retrieved December 1, 2020, from <https://www.3dwasp.com/en/3d-printed-house-tecla/>.

WASP. (2018). *The 3D printed house Gaia*. Retrieved December 1, 2020, from <https://www.3dwasp.com/en/3d-printed-house-gaia/>.

WASP. (2020). <http://www.3dwasp.com>. Accessed 21.10.20.
Wohlers, T., Gornet, T. (2014). *History of additive manufacturing, Wohlers report.* Retrieved from: http://www.wohlersassociates.com/history2014.pdf. Accessed February 2021.
XTree. (2020). *The large-scale 3D.* <https://www.xtreee.eu> Accessed 14.11.20.
Yu, L., & Yuexiu Group. (2018). *Arachne 3D printed building façade.* Retrieved December 2, 2020, from <https://www.e-architect.com/china/arachne-3d-printed-building-façade-building-façade>.
Ziaee, M., & Crane, N. B. (2018). Binder jetting: A review of process, materials, and methods. *Additive Manufacturing, 28,* 781–801.

Mass customization as the convergent vision for the digital transformation of the manufacturing and the building industry

Gabriele Pasetti Monizza[1] and Domink T. Matt[2]
[1]Process Engineering in Constructions, Fraunhofer Italia Research, Bolzano, Italy, [2]Faculty of Science and Technology, Free University of Bozen-Bolzano, Bolzano, Italy

Abbreviations

AEC	architecture, engineering and construction
ATO	assembly-to-order
BIM	building information modelling
CD	computational design
CE	circular economy
CNC	computer numerical control
CPS	cyber-physical systems
DT	digital transformation
ETICS	External Thermal Insulation Composit System
ETO	engineer-to-order
FEM	Finite Element Method
GLT	glued-laminated timber
MGI	McKinsey Global Institute
MTO	make-to-order

18.1 Introduction

The digital transformation (DT, also known as digitalization) is changing our society and our economy. Beyond the great potential of the DT in creating new markets and new opportunities, the McKinsey Global Institute (MGI) estimates that 'Europe has a long way to go to fully tap the potential of digitalisation' (Bughin et al., 2016) even if 83% of EU citizens have access to the internet in their homes, and 76% say that they use the internet regularly. MGI highlights that only big enterprises have a higher level of digital maturity and, even among the same enterprises' size, some industrial sectors are affected by a low level of digital maturity, such as basic good manufacturing or constructions.

Both the manufacturing and the building industries are exploring potentials and criticisms of the DT by establishing and adopting new methodologies and approaches (Brunetti et al., 2020), such as advanced robotics, computational design (CD) and engineering. These methods combine computer science methods with robotics, design and engineering methods. However, potentials and criticisms look different for each industry, even whether the main goals of implementing the DT are the same: achieve quality, productivity and responsiveness in manufacturing, without increasing wastes of resources (Jack Hu, 2013).

This divergent status of potential and criticisms is mainly due to different backgrounds of each industry. They developed different methods and approaches. The manufacturing industry pursues the mass production method as the most effective manufacturing approach, since the Fordism at the beginning of the last century (Ford, 1926). This method establishes the production of large amounts of standardized products, especially on assembly lines. The building industry produces high-customized products or, in other words, buildings. Every building is almost unique, a prototype and the manufacturing approach is mainly based on craftsmanship (Motiar Rahman, 2013). Although industrial approaches from manufacturing industry have been introduced so far, such as off-site prefabrication, these approaches are not widespread adopted or integrally applied due to a lack of customization capabilities (Stoettrup Schioenning Larsen, Munch Lindhard, Ditlev Brunoe, Nielsen, & Kranker Larsen, 2019).

According to this picture, the manufacturing industry lacks customization capabilities due to mass production method, which establishes the production of large amounts of standardized products (Hart, 1995). The building industry lacks productivity in its manufacturing approach, due to high-customized products (Jensen et al., 2018). Thus they seem to aim at the mass customization method from different starting points: manufacturing industry is looking for customization; building industry is looking for mass capabilities of its manufacturing approach. In this essay the authors would discuss whether the mass customization method, as the goal of an effective DT, could reduce these lacks, being a convergent vision for both the manufacturing industry and the building industry. The authors will focus the discussion on the architecture, engineering and construction (AEC) sector, considering the manufacturing industry as part of it, for what concern industrialized products.

18.2 Digital transformation of manufacturing industry

Nowadays, the manufacturing industry is facing that the DT pushing the digital automation and the interconnection of production systems (Matt, Orzes, Pedrini, Beltrami, & Rauch, 2019). This trend is formalized within the fourth industrial revolution (also known as Industry 4.0) as expression of the DT in manufacturing (Sendler, 2013). Focus of Industry 4.0 framework is to combine production, information technology and Internet. Thus the newest information and communication technologies are combined with classical industrial processes (BBF, 2012).

This combination may enhance flexibility, changeover ability and reconfigurability relying on smart manufacturing strategies and cyber-physical systems (CPS). It ensures the future *competitiveness of the manufacturing industry by providing companies with the ability to react to rapid product changes and disturbances, efficiently and reliably* (Martínez-Olvera & Mora-Vargas, 2019). Smart manufacturing can be defined as *the dramatically intensified and pervasive application of networked information-based technologies throughout the manufacturing and supply chain enterprise* (Davis, Edgar, Porter, Bernaden, & Sarli, 2012). CPS are *the result of the combination of embedded system with cyberspace. They support real world awareness in the Internet and the access to global data and services by embedded system* (Broy, 2013).

Wang, Yang, Xie, Jin, and Deen (2017) highlight that CPS used till the design phase can enable companies to increasingly produce customized products, with shorter cycle times and lower costs than those associated with standardization and mass production. They highlight that Industry 4.0 will enable novel forms of personalization. Direct customer input to design will enable companies to increasingly produce customized products, with shorter cycle times and lower costs than those associated with standardization and mass production. The DT is affecting the customization potential of manufacturing systems, toward the utopic question: 'Could it be possible to produce a lot-size-one product with the same efficiency of mass-produced products?'.

A higher customization potential may drive the manufacturing industry toward circular economy (CE) strategies and circular value-chain systems. Several authors discuss the strong connection between customization capabilities (Hara, Sakao, & Fukushima, 2019) and circularity (Gembarski, Schoormann, Schreiber, Knackstedt, & Lachmayer, 2018) focusing on smart product and production control through smart design (Zawadzki, Żywicki, Grajewski, & Górski, 2019) as a necessary element of the smart factory of the future that is able to realize higher customization. Koller, Velte, Schötz, and Döpper (2020) introduce a rationale for using remanufacturing as a strategy to increase product individualization in combining elements and strategies from customization and remanufacturing while considering ecological and economic aspects. Remanufacturing is an industrial process that aims to *return a used product to at least it's original performance with a warranty that is equivalent or better than that of the newly manufactured product* (BSI, 2009).

This overview highlights that the DT of manufacturing industry is mainly impacting customization and reconfigurability capabilities of manufacturing systems, even toward CE strategies.

18.3 Digital transformation of building industry

Within the building industry the AEC sector may be traditionally defined an engineer-to-order (ETO) industry, according to Wortmann classification (Wortmann, 1983). This implies that every final product (building) is almost unique, a kind of prototype. Every new product must be designed and engineered time by time, according to different customers' needs, such as different contexts, preexisting building conditions, regulatory constraints and customers' customization requirements. This lack of product standardization is one of the main reasons for a lack in productivity, as it strongly increases the product variability (Poshdar, González, Raftery, & Orozco, 2014). Besides that the AEC identifies different task leaders (architects, engineers, general contractors, facility managers, etc.) for each task usually structured in a serial workflow along the value-chain system. Serial workflow reduces the efficiency of the information management by enhancing the propagation of errors (small details neglected at the beginning can trigger significant technical or operation issues in later tasks, that is, during installation and maintenance) because of a fragmented information flow among the task leaders (BSI, 2003; PMI, 2013).

In the early twentieth century, off-site prefabrication has been introduced to enhance the productivity of the building industry, applying a production process based on the assembly line, and pushing the transition from an ETO industry toward a make-to-order (MTO) or assembly-to-order (ATO) industry. MTO and ATO industries reduce the complexity of the workflow by reducing the number of tasks that have to be performed. This approach drives improvements of product quality and manufacturing efficiency. The off-site prefabrication can offer better working conditions, automation of some tasks, fewer scheduling and weather-related delays and simplified inspection processes (Huang & Krawczyk, 2006). However, the off-site prefabrication failed due to the lack of variability and an individual identified design (Kieran & Timberlake, 2004).

Off-site prefabrication has been widely applied in design and production of façade elements. Façade elements are key elements for defining several building features, such as aesthetic, control of daylight, energy demand and acoustic insulation. Considering the relevance of these elements, off-site prefabrication can strongly improve the performance (Yoon, 2019), the safety (Knyziak, 2019) and the durability (Gasparri, Lucchini, Mantegazza, & Mazzucchelli, 2015) both in new buildings and in refurbished ones (Sandberg, Orskaug, & Andersson, 2016). Several case studies focusing on off-site prefabrication of façade elements demonstrate the trend of off-site prefabrication and its production principles, the notions of open-building design and Design for X[1], as well as offering an overview of the development of automation in construction (Pan et al., 2020). The development of automation in construction may unlock the capability of handling complexity in design (Arashpour, Miletic, Williams, & Fang, 2018) toward the exploitation of DT through advanced digital technologies (Pastor, Balaguer, Rodriguez, & Ramiro, 2002).

The AEC is looking at the DT as a new opportunity to overcome the lack in productivity by pushing the collaboration in an interdisciplinary environment and by

---

[1] Design for X is a systematic approach to achieve a targeted objective. X represents targeted objectives or characteristics of product or process. It works toward improving processes in a specific field (X) like manufacturing, power, variability, cost, yield or reliability.

implementing new capabilities in prefabrication thanks to smart factory strategies and CPS. To improve the efficiency of the information management, the AEC is pushing toward the common adoption of the building information modelling (BIM), also known as building information management, as a digital information management system. The BIM is a methodology for the digital management of information along the whole life cycle of a building. Handling 3D geometry (such as technical models and blueprints) and alphanumeric data (costs, assembly, maintenance, etc.), it aims at implementing an integrated platform that spans project planning, design, construction, operations and maintenance (Sacks, Eastman, Lee, & Teicholz, 2018). It is globally recognized as a standard methodology through the ISO 19650 standard—organization and digitization of information about buildings and civil engineering works, including BIM-Information management using BIM—and the EU countries are adopting the BIM for public building commitments (European Parliament & of the Council, 2014).

Assuming that the BIM is going to establish a common digital information management system within the AEC (Schimanski, Monizza, Marcher, & Matt, 2019), data automation through CD strategies has increased in the last decades to exploit capabilities of smart factory strategies and CPS. The Design Methods (1962) conference was a pioneer event that mapped the early developments of CD in architecture. CD is a term widely adopted to describe all the disciplines and the approaches that lead the design and the engineering processes applying computer-aided methods and tools, such as parametric design, algorithmic design and generative design. The definition provided by Jabi (2013) summarizes a long debate in the scientific community since the 1940s' writings of architect Luigi Moretti (Bucci & Mulazzani, 2002) and the mathematical origins of parametric modelling (Davis, 2013). Moretti wrote extensively about *parametric architecture*, which he defines as the study of architecture systems with the goal of *defining the relationships between the dimensions dependent upon the various parameters*. Beyond stylistic implications of CD in the architecture, design and art disciplines, such as Patrick Schumacher's *Parametricist Manifesto* (Schumacher, 2008), the building industry is facing that the DT pushing a digital automation of information among design and engineering activities. This digital automation pursues new standards of efficiency, safety and quality in complex and highly customized buildings (Scheurer, 2014), aiming at handling geometry optimization, computation of quantities, production planning and machine data to facilitate the production activity and the installation process on construction sites. According to Caetano, Santos, and Leitão (2020), CD is a design process that takes advantage of computational capabilities through the following framework:

1. automating design procedures based on the following:
    a. a deduction that is applying a transformation to an element while knowing its outcome,
    b. an induction that is extrapolating the required design process to obtain a specific result and
    c. an abstraction that understands the essential design features by removing irrelevant information.

2. parallelizing design tasks and efficiently managing large amounts of information;
3. incorporating and propagating changes in a quick and flexible manner; and
4. assisting designers in form-finding processes through automated feedback, such as mapping simulation results.

This framework can be recognized among several case studies of CD strategies applied in the AEC (Albayrak Colaço & Tunçer, 2011; Yuan, Sun, & Wang, 2018; Bianconi, Filippucci, & Buffi, 2019), and among the best practices and the proofs of concept discussed later in the chapter.

CD is applied for enhancing digital fabrication strategies in the AEC, driving digital information from design and engineering toward CPS and advanced robotic systems (Gardner, Forward, Tse, & Sharma, 2020). Hack et al. (2020) discuss the application of CD for facilitating robotic in situ fabrication of nonstandard structures, such as asymmetric or doubled curved geometries.

The DT of building industry is increasing the efficiency of the information management system to avoid the propagation of errors in complex serial workflows, and it is enhancing automation capabilities to handle complexity and highly customized products (buildings) through the application of CPS and advanced robotic systems (Pasetti Monizza, Rauch, & Matt, 2017).

18.4 Mass customization as a convergence vision

The DT of manufacturing and building industries is impacting each of them in a different way: manufacturing industry is exploiting customization and reconfigurability capabilities of manufacturing systems; building industry is increasing the efficiency of the information management system and it is enhancing automation capabilities. Nevertheless, these industries are aiming at the same objective: mass customization.

The concept of mass customization is first expounded formally in the book *Future Perfect* by Stanley M. Davis in 1989 (Davis, 1989). Mass customization means manufacturing products, which have been customized for the customer, at production costs like those of mass-produced products (Kaplan & Haenlein, 2006). Production systems in a mass customization environment should be able to produce small quantities in a highly flexible way and to be rapidly reconfigurable (Qiao, Lu, & McLean, 2006). Mass customization allows customers to select attributes from a set of predefined features to design their individualized product, by which they can fulfil their specific needs and take pride in having created a unique result (Hart, 1995; Schreier, 2006; Stoetzel, 2012). To reach such a next level of changeability, it is necessary to equip manufacturing systems with cognitive capabilities to take autonomous decisions in even more complex production processes with a high product variety (Zaeh et al., 2009).

Barman and Canizares (2015) identify business processes that have applied the mass customization concept to create and deliver both tangible goods and intangible services. They describe the key elements of mass customization, such as elicitation, process flexibility and logistics. They also include a description of the four specific mass customization approaches: collaborative, adaptive, cosmetic and transparent.

According to Bock and Linner (2010), the building industry grounds its customization capabilities on information and communication technology used for forming continuous IT structures on which those information flows are then created. Customization is deeply based on the evolution and interconnection of all computer-based technologies. Huang and Krawczyk (2006) apply the mass customization method to prefabricated modular housing. They highlight that this method encourages designers to develop a series of solutions rather than single solutions for a design problem. For the technical challenge in standardizing the various building systems, it would be easier to setup a new standard system for universal and interchangeable parts.

Kasperzyk, Kim, and Brilakis (2017) present a robotic prefabrication system that employs a new concept called *re-fabrication*: the automatic disassembly of a prefabricated structure and its reconstruction according to a new design. They apply a software that employs the 3D model of a prefabricated structure as input and returns motor control command output to the hardware by two underlying algorithms:

1. A novel algorithm that automatically compares the old and new models and identifies the components which the two models do not have in common.
2. A novel algorithm that computes the optimal re-fabrication sequence.

The first algorithm enables disassembly of the original structure and its refabrication into the new design. The second algorithm transforms one model into another according to the differences identified.

Kudsk, Hvam, Thuesen, O'Brien Grønvold, and Holo Olsen (2013) carry out a case study of one of the largest construction companies in Northern Europe, according to the principles of action research. They analyse technical solutions for residential and office buildings according to a top-down approach, using a product variant master to map the relations among the identified solutions. They conclude that the principles of mass customization are best used in the building industry if used with a top-down perspective.

Mohammed Refaat Mekawy (2020) tries to identify a framework for using BIM in mass customization and prefabrication, exploiting he data-rich BIM design environment to automatically perform checks, based on stored rule sets, to determine whether the design scheme in question fulfils prefabrication prerequisites or not.

Piroozfar, Farr, Hvam, Robinson, and Shafiee (2019) set out to explore the principles of platform design, the relations between industrialization and mass customization through a serialization facilitated by BIM. Their studies rely on a given case of design, manufacturing and assembly processes of building envelopes, developing a customizable façade system to accommodate:

1. The panel components that can lodge different materials.
2. The mullions that can oblige different geometries.
3. The support structure that can accommodate a variation of different geometries and lodge components with different shapes, sizes and dimensions.

The mass customization method within its formal definition and applications is including two main aspects: efficiency of mass production systems combined with customization and reconfigurability capabilities. According to the impacts of the DT within manufacturing and building industries, the previous are enhancing customization capability and production efficiency, respectively. In other words, these industries are approaching the same objective through the DT from different starting points: manufacturing industry lacks customization and is approaching the mass customization increasing customization and reconfigurability capabilities; building industry lacks mass production efficiency and is approaching the mass customization increasing the production efficiency. According to this observation, it is possible to formalize the following hypothesis: the mass customization is the convergent vision for the DT of manufacturing and building industry.

Considering that manufacturing systems are applied for off-site prefabrication in the AEC, and manufacturing industry is part of the building industry for what concerns off-site prefabrication or industrial products, this hypothesis can be discussed referring exclusively to the building industry.

18.5 Best practices' and author's proofs of concept

Facilitating the production and the installation of special building components (such as high-customized façade elements) relies on computer-aided methods, digital fabrication and advanced manufacturing systems. The combination of these methods and systems establishes a connection between the DT applied in manufacturing industry, mainly through smart factory strategies and CPS, and in building industry, mainly through computer-aided methods and digital fabrication (Hamid, Tolba, & El Antably, 2018).

In the last years the scientific community is increasingly discussing about the combination of computer-aided methods, digital fabrication and advanced manufacturing systems (Caetano et al., 2020). In this section the authors present a short selection of best practices from the scientific community and personal research activities aiming at discussing the hypothesis stated previously.

18.5.1 The Basilica de la Sagrada Família in Barcelona

This church is universally recognized as the masterpiece of the architect Antoni Gaudí, expressing his architectural and engineering style which combines Gothic and curvilinear Art Nouveau forms. Begun in 1883, the Sagrada Família's construction progressed slowly, finishing the Nativity entrance in 1936, and was interrupted by the Spanish Civil War in the same year. Much of Gaudí's preparatory work was destroyed. It resumed to intermittent progress in the 1950s. In 1990 only 3 of the interior's 56 columns and a handful of the windows had been completed. Nevertheless, only at the end of the last century and in the early 2000s, thanks to the application of digital technologies such as computer-aided methods and computer numerical control (CNC) systems, the construction exponentially accelerated. Recently, the Sagrada Família Foundation has anticipated that the building can be completed by 2026, the centenary of Gaudí's death.

Starting from the 1980s, computers have been introduced into the design and construction process. Mark Burry from New Zealand started serving as an Executive Architect and Researcher, documenting the adoption of computer-aided methods and CNC systems (Burry, Burry, & Faulí, 2001). The introduction of digital technologies leaded the transition from craftsmanship to a computer-aided construction. Gaudí's preparatory work was mainly relying on scale plaster models both for structural and architectural prototyping. They have been applied for decades to allow workers and sculptors for understanding of Gaudí's system of proportions. Thus they were able to reproduce in real scale the complexity of the composition, such as the hyperbolic surfaces of the rose windows. Applying parametric design techniques, Burry defines a digital design system that can modify a specific surface through changing its mathematical characteristics via the computer software, thereby affecting the nature and position of the intersections of adjacent surfaces. He established a similarity between his modus operandi using powerful digital aids and the haptic iterative design modelling used by Gaudí during the predigital era (Burry et al., 2001). Thanks to the transition from design modelling toward a digital design system, a digital information system has been established. Thus the digital output of the design and engineering can be applied for rapid prototyping and digital fabrication techniques. *The particulars of this adoption and adaptation emphasise the advantages that designers have when they embrace emerging technology as closely as possible at the first opportunity by setting aside over anxiety about risk* (Burry, 2016).

Nowadays, under the direction of the Chief Architect Jordi Faulí, the construction of the Sagrada Família is completely computer aided. The construction of 10 towers (among them the 172-m Jesus tower, which would be the tallest structure in Barcelona) is completely performed relying on off-site prefabrication. These towers are assembled using prefabricated façade panels made of stone and steels that act as formwork for the concrete structure. At Galera a sprawling work site 90 minutes north of Barcelona, these panels are produced applying CNC machinery and robotic systems to handle the complexity of the overall geometry (inverted catenary) which drives to different row-by-row elements along the vertical axis. After CNC cutting, the final texture is chiselled by hand, and the interlocking row of granite stones are fixed to a steel plate and bound together into panels. This solution speeds up the rate of construction by a factor of 10 and, according to Fernando Villa—the Sagrada Familia's director of operations, *There's a symbiosis of high tech and traditional artisanship in every component.* [...] *The process of bringing them together is modelled on an automobile assembly line. Everything is done just in time* (Abend, 2018).

The combination of computer-aided methods, digital fabrication and advanced manufacturing systems leads to an outstanding rate of construction for an extremely complex building and a unique masterpiece such as the Gaudí's Sagrada Família.

18.5.2 The Ri.Fa.Re. project—energy refurbishment through timber-prefabricated façade elements

The Ri.Fa.Re. timber-based solution is the result of the Ri.Fa.Re. ('Ristrutturare con Facciate pRefabbicate'—'Energy refurbishment with prefabricated facades') project, an industrial collaborative research project funded by the Autonomous Province of Bolzano, Italy (Malacarne et al., 2016). The project has been managed by Fraunhofer Italia Research in collaboration with five local companies from five different sectors: timber buildings, windowing systems, fastenings, insulating materials and a design firm. It relied on a case study approach and it aimed at developing a standardized but customizable prefabricated solution. Applying CD, the prefabricated modules can be adapted to the specific characteristics of buildings, time by time.

The Ri.Fa.Re. timber-based solution has been designed referring to identified case study buildings and to standard production processes of cooperating companies. These companies adopt a platform frame system for new buildings. This system has been adapted for refurbishment purposes considering that wall panels will become the main insulating system of the Ri.Fa.Re. timber-based solution. Façade elements can have customizable measures: from 1.20 to 12 m of length and from 1.20 to 3.30 m of height. They can also include different external finishing, such as plaster, ventilated façade. See Fig. 18.1 for a detail of the Ri.Fa.Re. timber-based solution.

Starting from mapping the existing production processes (see Fig. 18.2) of the cooperating companies, the project identified the value adding, nonvalue adding (but necessary) and nonvalue adding (wasteful) processes and finally improved the whole chain of the processes by applying different lean techniques. According to this analysis, the simulation of production and installation processes shows that time savings range from 56% to 70% (Malacarne et al., 2016). The simulation applied the identified case study buildings, as representative for the Italian building stock.

The time savings are due to the standardization of the façade elements within the Ri.Fa.Re. timber-based solution. The standardization reduces the complexity and the variability but, at the same time, reduces the flexibility and the adaptability of the elements during installation on existing buildings. These lacks can trigger significant technical or operation issues during installation because preexisting building conditions cannot be standardized with the same approach of façade elements. Assuming that the variability of preexisting building conditions has to be solved during the design and the engineering of façade elements, these processes may become a bottleneck for the whole supply-chain system.

To speed up design and engineering processes, CD has been introduced through an algorithm that aims at driving key information from the survey activity to the machinery on the production line, assuming that it is not possible to apply faster survey techniques and assuming that actual production systems have to be applied.

The programmed algorithm automatically defines dimensions, positions and features of façade elements. This definition is programmed according to a specific list of parameters and their relationships, in detail: (1) as-is parameters, such as geometry of facades, position and dimensions of windows, position of balconies and

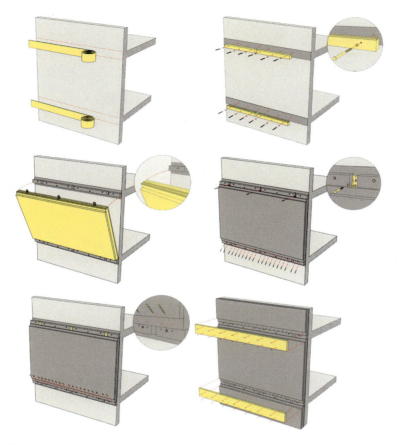

Figure 18.1 The Ri.Fa.Re. timber-based solution.

loggias and position of slabs; (2) customizable parameters, such as energy labelling requirements, selected windowing system, shading devices and external finishing.

Tolerances have been introduced according to limitations of the platform frame technology and the production system capabilities. The algorithm identifies missing coplanarities of the preexisting façade and computes different thickness of the interface layer (see Fig. 18.2). Data on thicknesses are transmitted to the production system supporting operators during the manufacturing of elements. Perpendicularity of rows and columns of façade elements relies on fastenings that are preinstalled by operators on the construction site. The algorithm calculates the placement of fastenings according to preexisting slabs. Placement is transmitted to operators on the construction site. Minimal misalignments are compensated by 2-cm expansion joints between the elements. Windows alignment relies on the same tolerances of expansion joints. Inner sealing of the new windows frame is performed by operators on the construction site during disposal of preexisting windows. The algorithm computes special elements due to geometrical discontinuities of the preexisting

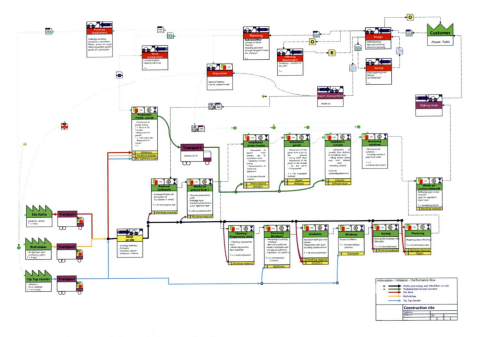

Figure 18.2 Map of the existing production processes.

façade, such as for balconies and loggias, identifying and adapting dimensions of the façade elements. These data are transmitted to the production system supporting operators during the manufacturing of elements.

Finally, the algorithm automatically recognizes portions of the preexisting façade that cannot be insulated applying the Ri.Fa.Re. timber-based solution, because of misalignments or missing coplanarities out-of-tolerance. These portions are computed as portions to be insulted applying traditional External Thermal Insulation Composit Systems (ETICS) on construction site. These data are transmitted to the operators on construction site during finishing operations at the end of façade cladding.

By this way, technicians responsible for design and engineering can provide a custom solution easily without increasing their effort on design and engineering each solution case by case.

18.5.3 Applying computational methods for enhancing glued-laminated timber capabilities

The Basilica de la Sagrada Família and Ri.Fa.Re. timber-based solution highlight that mass customization relies on the combination of computer-aided methods, digital fabrication and advanced manufacturing systems applied to single components of building elements. Thus the capacity of increasing the customization of complex building elements, such as façade elements, strongly relies on the capacity of handling the customization of single components that usually are manufactured in large lots on assembly lines.

To understand the implications of applying mass customization methods on standard assembly lines of building components, the authors developed a research activity aimed at investigating potentials and criticisms of CD techniques in mass production through a pilot case study analysis in glued-laminated timber (GLT) industry (Pasetti Monizza, Bendetti, & Matt, 2018). The analysis has been arranged through two main phases. The first phase focused on programming a parametric algorithm for GLT engineering. The second phase focused on testing impacts of implementing the algorithm in an ordinary supply-chain system of GLT.

GLT is the most ancient, engineered wood product (Laner, 2012). Since Leonardo da Vinci's studies until the official patent by Hetzer in 1905 (Rug, 2006), the main purposes of this product are to enhance mechanical performances and to increase dimension limits of the raw material (wood). To fulfil these purposes, it is manufactured by bonding together selected timber elements (laminates), reducing the natural defects of the raw material. GLT structural elements can be manufactured using a homogeneous or a combined lamination. A combined lamination allows GLT manufacturers to apply high-strength raw material only if needed. The combined lamination is regulated by the EN 14080 standard. According to this standard, combined GLT elements can be manufactured only when it is possible to identify a uniform distribution of stresses along element's cross section.

The programmed algorithm pursues two specific design intents: (1) reducing the usage of unneeded high-strength raw material; (2) improving the overall efficiency of the production system of GLT without pushing any change in the ordinary value-chain system of GLT. The programmed algorithm relies on four main steps: geometry analysis, curvature analysis, Finite Element Method (FEM) simulation and labelling for digital fabrication purposes. Through specific calculation stages, the algorithm defines the key parameters for GLT manufacturing such as position (in an X, Y, Z reference system) and strength class of laminates along the main axes of the element and along the width of the cross section; lamination thicknesses along the two main axes parallel to main curvature planes (Y- and Z-axes). The algorithm has been used for designing and engineering several GLT elements completely different one from the other, to test its robustness against multiple complex shapes. See Fig. 18.3 for an example of calculation outputs.

Testing the impacts of implementing algorithm in an ordinary supply-chain system shows that CD techniques in GLT engineering could improve the overall efficiency of the entire value-chain system by reducing the usage of unneeded high-strength raw material. According to the results collected, these savings reach up to 98% in volume. For example, in the test case of a double curved element (Pasetti Monizza et al., 2018), the EN 14080 standard establishes a homogeneous lamination, assigning C40 strength class laminates for the 100% of the cross section. On the other hand, the algorithm can identify a discrete distribution of stresses along the cross section. According to this distribution, a combined

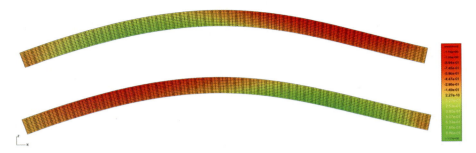

Figure 18.3 Discrete lamination of a doubled curved GLT element as calculated by the algorithm.

Table 18.1 Lamination as calculated by the algorithm.

Strength class (EN 338)	%
C14	70.19
C16	5.62
C18	7.24
C20	1.90
C22	1.81
C24	1.71
C27	5.24
C30	2.19
C35	2.76
C40	1.33

lamination can be calculated, reducing the usage of unneeded high-strength raw material (see Table. 18.1).

Nevertheless, limitations in production machineries are strongly reducing these potential benefits. The ordinary value-chain system is arranged for mass production purposes. Differently from Industry 4.0 framework, the ordinary supply-chain system of GLT does not implement any flexible reconfigurable production and logistics systems, offering interactive, collaborative decision-making mechanisms (see Fig. 18.4 for a standard layout of a GLT production system). On the other hand, saves introduced by the programmed algorithm may be achieved in a value-chain system for mass customization purposes only. The programmed algorithm may use until 10 different strength classes of raw material as discussed earlier. This variability could be reduced, but the ordinary configuration of warehouse logistics and loading truck machinery would never satisfy the needs for mass customization

Mass customization 467

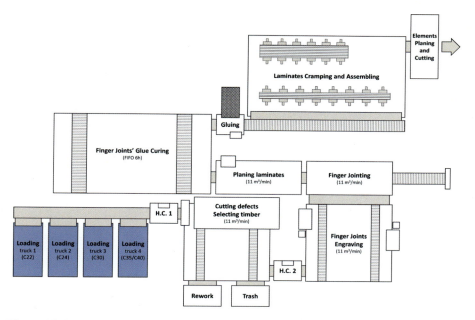

Figure 18.4 Standard layout of a GLT production system.

purposes. According to these evaluations, this research remarks that only through CPS, such as for self-organization and self-control of warehouse, intralogistics and loading trucks, it would be possible to achieve the needed variability. Increasing variability means shifting the value-chain system of GLT from mass production purposes to mass customization purposes.

Results highlight that it is possible to increase the number of product variants until every single product is different from the other. This would be possible only by reducing the changeover time in triggering the early production stages (see Fig. 18.5 for a map of standard processes). Triggering relies on the product engineering stage. It oversees providing technical features of GLT elements to drive the production process of manufacturing structural elements, according to structural engineering requirements. As measured in the case study, operators define technical features and load machinery programmes as well. They operate directly on finger jointing machine and manage the first stages of production process by arranging product's variants through loading machinery programmes. Thus any evaluation of impacts on product engineering has to include production plan and finger jointing stages as well. Product engineering and production plan stage are working in batch processing with finger jointing stage. Changeover time in finger jointing stage is spent for loading programmes into the finger jointing machine. During operation time of the machine, operators are working on programmes for the next product's variant. This workflow changes completely by introducing the algorithm along these stages. In the worst case an order includes until 50 product's variants due to the algorithm's structure. This means that every single element is different from the other. Thus

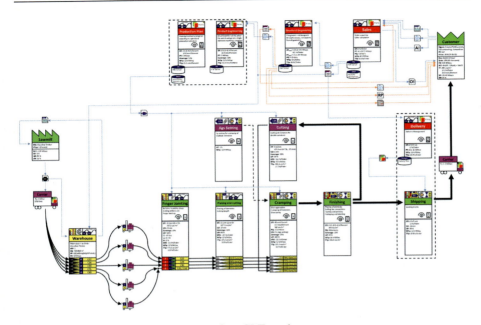

Figure 18.5 Map of the standard process in a GLT producer.

operators have to load a machinery programme per each element. The programmed algorithm takes 8 seconds to perform the entire calculation on a personal computer with a CPU-single-threaded score of 1486 (according to PassMark test). Estimating that an operator may take up to 30 seconds to load data into the algorithm, loading machinery programme has not to take longer than 48 seconds.

The programmed algorithm could produce different output formats, interacting directly with machinery as well with CPS, without any specific task performed by an operator. To enhance these capabilities of the algorithm, the ordinary supply-chain system of GLT needs an upgraded information management strategy as well as CPS embedded in the production system. According to Industry 4.0 framework, CPS have to be adopted to combine newest information and communication technologies with classical industrial processes. Besides enhancing the efficiency of the supply-chain system by exploiting benefits in mass customization, CPS introduce a more efficient information management strategy, completely paper less. Through such efficient information management strategy, benefits from the programmed algorithm would enhance not only the efficiency but also the effectiveness of the overall value-chain system.

18.6 Conclusion

The DT in the AEC drives toward a combination of computer-aided methods, digital fabrication and advanced manufacturing systems. Combining this methods and systems also means jointing the manufacturing and the building industries together. This exploits mass production capabilities of the manufacturing industry combined with the high-customized product of the building industry. The combination unlocks the capability of handling complexity and variability in an efficient way, speeding up the rate of production and assembly toward a mass customization application. In this way, mass customization has to be considered the convergent vision for the DT of manufacturing and building industry.

mass customization applications in off-site prefabrication of façade elements relies on the capacity of handling the customization of single components that usually are manufactured in large lots on assembly lines. Thus accordingly to the best practices and proofs of concept presented in the previous section, ordinary supply-chain systems of single components need an upgraded information management strategy as well as CPS embedded in the production system. CPSs have to be adopted to combine newest information and communication technologies with classical industrial processes, pursuing the Industry 4.0 framework.

Best practices and proofs of concept presented highlight that only the mutual application of the DT in manufacturing industry and building industry can exploit the benefit of mass customization applications. Completing the Basilica de la Sagrada Família means constructing one of the most complex buildings in the world. The application of computer-aided methods, digital fabrication and advanced manufacturing systems allows not only bringing to reality Gaudí's vision but also speeding up the rate of construction and efficiently managing the operations off- and on-site during manufacturing and assembling of the façade elements. Applying prefabrication and mass production approaches to energy refurbishment of facades drives toward the need of automation in design and engineering processes to avoid wastes of resources and missed savings during production and assembly processes of standardized façade elements. Finally, applying computer-aided methods (such as CD and digital fabrication) in early processes of engineering a mass-produced component requires advanced manufacturing systems to completely exploit benefits of the DT.

Although these results are promising, they cannot be extended universally. Unfortunately, every evaluation must refer to a specific production system that may differ case by case. Results highlight that benefit of mass customization applications can be achievable only in multidisciplinary environment which proactively mixes building technologies expertise and production systems engineering. Thus through their experience, the authors aim at providing a background and a framework for further initiatives in the topics, inviting referents from different discipline to cooperate side by side.

References

Abend, L., 2018. *Inside Barcelona's Unfinished Masterpiece*. <https://time.com/sagrada-familia-barcelona/>. Accessed 30.10.20.

Albayrak Colaço, C., & Tunçer, B. (2011). In *Performative architecture as a guideline for transformation of the defence line of Amsterdam, eCAADe 29—Generative and parametric design conference*. <https://www.researchgate.net/publication/321331184_Performative_architecture_as_a_guideline_for_transformation_Defense_Line_of_Amsterdam>. Accessed 30.10.20.

Arashpour, M., Miletic, M., Williams, N., & Fang, Y. (2018). Design for manufacture and assembly in off-site construction: Advanced production of modular façade systems. In *Proceedings of the 35th ISARC* (pp. 224–229), https://doi.org/10.22260/ISARC2018/0032.

Barman, S., & Canizares, A. E. (2015). A survey of mass customization in practice. *International Journal of Supply Chain Management*, 4(1), ISSN: 2050–7399.

Bianconi, F., Filippucci, M., & Buffi, A. (2019). Automated design and modeling for mass-customized housing. A web-based design space catalog for timber structures. *Automation in Construction*, 103, 13–25. Available from https://doi.org/10.1016/j.autcon.2019.03.002.

Bock, T., & Linner, T. (2010). Mass-customization in a knowledge-based construction industry for sustainable high-performance building production. In *CIB world congress, Manchester/Salford, England*.

Broy, M. (2013). Engineering cyber-physical systems: Challenges and foundations. In M. Aiguier, Y. Caseau, D. Krob, & A. Rauzy (Eds.), *Complex systems design & management*. Berlin, Heidelberg: Springer. Available from https://doi.org/10.1007/978-3-642-34404-6_1.

Brunetti, F., Matt, D. T., Bonfanti, A., De Longhi, A., Pedrini, G., & Orzes, G. (2020). Digital transformation challenges: Strategies emerging from a multi-stakeholder approach. *TQM Journal*, 32(4), 697–724. Available from https://doi.org/10.1108/TQM-12-2019-0309.

BSI—British Standards Institute (2003). Poor communication costing UK construction industry. <https://www.bsigroup.com/en-GB/about-bsi/media-centre/press-releases/2003/9/Poor-Communication-Costing-UK-Construction-Industry/>. Accessed 30.10.20.

BSI—British Standards Institution. (2009). *BS 8887-2,2009: Design for manufacture, assembly, disassembly and end-of-life processing (MADE)*. London: Terms and definitions, British Standards Institution. (2009).

Bucci, F., & Mulazzani, M. (2002). *Luigi Moretti: Works and writings*. Hudson, NY: Princeton Architectural Press, ISBN: 1568983069.

Bughin, J., Hazan, E., Labaye, E., Manyika, J., Dahlström, P., Ramaswamy, S., & Cochin de Billy, C. (2016). *Digital Europe: Realizing the continent's potential*. McKinsey Global Institute. <https://www.mckinsey.com/business-functions/mckinsey-digital/our-insights/digital-europe-realizing-the-continents-potential> Accessed 30.10.20.

Bundesministerium für Bildung und Forschung (BBF). (2012). *Zukunftsbild 'Industrie 4.0' (Vision 'Industry 4.0')*. Berlin: Bundesministerium für Bildung und Forschung (Federal Ministry of Education and Research). https://www.bmbf.de/de/zukunftsprojekt-industrie-4-0-848.html (in German, Accessed 30.10.20).

Burry, M. (2016). *Robots at the Sagrada Família Basilica: A brief history of robotised stonecutting. Robotic fabrication in architecture, art and design*. Cham: Springer. Available from https://doi.org/10.1007/978-3-319-26378-6_1.

Burry, M., Burry, J., & Faulí, J. (2001). *Sagrada Família Rosassa: Global computeraided dialogue between designer and craftsperson (overcoming differences in age, time and distance)*.

Caetano, I., Santos, L., & Leitão, A. (2020). Computational design in architecture: Defining parametric, generative, and algorithmic design. *Frontiers of Architectural Research*, *9*(2), 287−300. Available from https://doi.org/10.1016/j.foar.2019.12.008.

Davis, S. M. (1989). From 'future perfect': Mass customizing. *Planning Review*, *17*(2), 16−21.

Davis, D. (2013). Modelled on software engineering: Flexible parametric models in the practice of architecture (PhD thesis). RMIT University. Available from: <http://www.danieldavis.com>. Accessed 15.04.20.

Davis, J., Edgar, T., Porter, J., Bernaden, J., & Sarli, M. (2012). Smart manufacturing, manufacturing intelligence and demand-dynamic performance. *Computers & Chemical Engineering*, *47*, 145−156. Available from https://doi.org/10.1016/j.compchemeng.2012.06.037.

European Parliament and of the Council (2014). *Directive 2014/24/EU on public procurement*. <http://data.europa.eu/eli/dir/2014/24/oj>. Accessed 30.10.20.

Ford, H. (1926). *Today and tomorrow, doubleday*. Garden City, NY: Page and Company.

Gardner, G. E., Forward, K., Tse, K., & Sharma, K. (2020). Rapid composite formwork: An automated and customizable process for freeform concrete through computational design and robotic. In: *Second RILEM international conference on concrete and digital* fabrication. DC 2020. RILEM Bookseries, Vol 28. Springer, Cham. https://doi.org/10.1007/978-3-030-49916-7_86.

Gasparri, E., Lucchini, A., Mantegazza, G., & Mazzucchelli, E. S. (2015). Construction management for tall CLT buildings: From partial to total prefabrication of façade elements. *Wood Material Science & Engineering*, *10*(3), 256−275. Available from https://doi.org/10.1080/17480272.2015.1075589.

Gembarski, P. C., Schoormann, T., Schreiber, D., Knackstedt, R., & Lachmayer, R. (2018). Effects of mass customization on sustainability: A literature-based analysis. In S. Hankammer, K. Nielsen, F. Piller, G. Schuh, & N. Wang (Eds.), *Customization 4.0. Springer proceedings in business and economics* (pp. 285−300). Cham: Springer. Available from https://doi.org/10.1007/978-3-319-77556-2_18.

Hack, N., Dörfler, K., Nikolas Walzer, A., Wangler, T., Mata-Falcón, J., Kumar, N., ... Kohler, M. (2020). Structural stay-in-place formwork for robotic in situ fabrication of non-standard concrete structures: A real scale architectural demonstrator. *Automation in Construction*, *115*. Available from https://doi.org/10.1016/j.autcon.2020.103197.

Hamid, M., Tolba, O., & El Antably, A. (2018). BIM semantics for digital fabrication: A knowledge-based approach. *Automation in Construction*, *91*, 62−82. Available from https://doi.org/10.1016/j.autcon.2018.02.031.

Hara, T., Sakao, T., & Fukushima, R. (2019). Customization of product, service, and product/service system: What and how to design. *Mechanical Engineering Reviews*, *6*(1), 18−00184. Available from https://doi.org/10.1299/mer.18-00184.

Hart, C. W. (1995). Mass-customization: Conceptual underpinnings, opportunities and limits. *International Journal of Service Industry Management*, *6*(2), 36−45. Available from https://doi.org/10.1108/09564239510084932.

Huang, J., & Krawczyk, R. (2006). Integrating mass customization with prefabricated housing. In *The Second international conference of Arab Society for Computer Aided Architectural Design (ASCAAD)*, Sharjah, UAE.

Jabi, W. (2013). *Parametric design for architecture*. London: Laurence King, ISBN: 1780673140.

Jack Hu, S. (2013). Evolving paradigms of manufacturing: From mass production to mass customization and personalization. *Procedia CIRP*, *7*, 3−8. Available from https://doi.org/10.1016/j.procir.2013.05.002.

Jensen, K. N., Nielsen, K., Brunoe, T. D., Larsen, J. K., Hankammer, S., Nielsen, K., ... Wang, N. (2018). Productivity, challenges, and applying mass customization in the building and construction industry. In S. Hankammer, K. Nielsen, F. Piller, G. Schuh, & N. Wang (Eds.), *Customization 4.0. Springer proceedings in business and economics* (pp. 551−565). Cham: Springer. Available from https://doi.org/10.1007/978-3-319-77556-2_34.

Kaplan, A. M., & Haenlein, M. (2006). Toward a parsimonious definition of traditional and electronic mass-customization. *Journal of Product Innovation Management*, *23*(2), 168−182. Available from https://doi.org/10.1111/j.1540-5885.2006.00190.x.

Kasperzyk, C., Kim, M., & Brilakis, I. (2017). Automated re-prefabrication system for buildings using robotics. *Automation in Construction*, *83*, 184−195. Available from https://doi.org/10.1016/j.autcon.2017.08.002.

Kieran, S., & Timberlake, J. (2004). *Refabricating architecture: How manufacturing methodologies are poised to transform building construction*. New York: McGraw-Hill Companies, ISBN: 007143321X.

Knyziak, P. (2019). The impact of construction quality on the safety of prefabricated multi-family dwellings. *Engineering Failure Analysis*, *100*, 37−48. Available from https://doi.org/10.1016/j.engfailanal.2019.02.042.

Koller, J., Velte, C. J., Schötz, S., & Döpper, F. (2020). Customizing products through remanufacturing-ideation of a concept. *Procedia Manufacturing*, *43*, 598−605. Available from https://doi.org/10.1016/j.promfg.2020.02.151.

Kudsk, A., Hvam, L., Thuesen, C., O'Brien Grønvold, M., & Holo Olsen, M. (2013). Modularization in the construction industry using a top-down approach. *The Open Construction and Building Technology Journal*, *7*, 88−98. Available from https://doi.org/10.2174/1874836801307010088.

Laner, F. (2012). *Il legno. Materiale e tecnologia per progettare e costruire (The timber. Material and technology to design and buildings)*. Torino: Utet Scienze Tecniche, ISBN: 885980888X (in Italian).

Malacarne, G., Pasetti Monizza, G., Ratajczak, J., Krause, D., Benedetti, C., & Matt, D. T. (2016). Prefabricated timber façade for the energy refurbishment of the Italian building stock: The Ri.Fa.Re. project. *Energy Procedia*, *96*, 788−799. Available from https://doi.org/10.1016/j.egypro.2016.09.141.

Martínez-Olvera, C., & Mora-Vargas, J. (2019). A comprehensive framework for the analysis of industry 4.0 value domains. *Sustainability*, *11*(10), 29−60. Available from https://doi.org/10.3390/su11102960.

Matt, D. T., Orzes, G., Pedrini, G., Beltrami, M., & Rauch, E. (2019). Roadmap into a digital world: On the way to a digital world—The digital roadmap for the Tyrol-Veneto Macroregion [Roadmap in eine Digitale Welt: Auf dem Weg in eine Digitale Welt—Die Digitale Roadmap für die Makroregion Tirol-Veneto]. *ZWF Zeitschrift fuer Wirtschaftlichen Fabrikbetrieb*, *114*(9), 576−579. Available from https://doi.org/10.3139/104.112136.

Mohammed Refaat Mekawy, M. (2020). *A framework for using BIM in mass-customization and prefabrication in the AEC industry* (PhD thesis). Technische Universität München. Available from: <https://mediatum.ub.tum.de/1545783>. Accessed 15.04.20.

Motiar Rahman, M. (2013). Barriers of implementing modern methods of construction. *Journal of Management in Engineering*, *30*(1). Available from https://doi.org/10.1061/(ASCE)ME.1943-5479.0000173.

Pan, W., Iturralde, K., Bock, T., Martinez, R. G., Juez, O. M., & Finocchiaro, P. (2020). A conceptual design of an integrated façade system to reduce embodied energy in residential buildings. *Sustainability*, *12*(14), 5730. Available from https://doi.org/10.3390/su12145730.

Pasetti Monizza, G., Bendetti, C., & Matt, D. T. (2018). Parametric and generative design techniques in mass-production environments as effective enablers of Industry 4.0 approaches in the building industry. *Automation in Construction*, *92*, 270–285. Available from https://doi.org/10.1016/j.autcon.2018.02.027.

Pasetti Monizza, G., Rauch, E., & Matt, D. T. (2017). Parametric and generative design techniques for mass-customization in building industry: A case study for glued-laminated timber. *Procedia CIRP*, *60*, 392–397. Available from https://doi.org/10.1016/j.procir.2017.01.051.

Pastor, J. M., Balaguer, C., Rodriguez, F. J., & Ramiro, D. (2002). Computer-aided architectural design oriented to robotized façade panels manufacturing. *Computer-Aided Civil and Infrastructure Engineering*, *16*(3), 216–227. Available from https://doi.org/10.1111/0885-9507.00227.

Piroozfar, P., Farr, E. R. P., Hvam, L., Robinson, D., & Shafiee, S. (2019). Configuration platform for customisation of design, manufacturing and assembly processes of building façade systems: A building information modelling perspective. *Automation in Construction*, *106*. Available from https://doi.org/10.1016/j.autcon.2019.102914.

PMI—Project Management Institute (2013). *The high cost of low performance: The essential roles of communications.* <http://www.pmi.org/-/media/pmi/documents/public/pdf/learning/thought-leadership/pulse/the-essential-role-of-communications.pdf>. Accessed 30.10.20.

Poshdar, M., González, V. A., Raftery, G. M., & Orozco, F. (2014). Characterization of process variability in construction. *Journal of Construction Engineering and Management*, *140*(11). Available from https://doi.org/10.1061/(ASCE)CO.1943-7862.0000901.

Qiao, G., Lu, R. F., & McLean, C. (2006). Flexible manufacturing systems for mass customisation manufacturing. *International Journal of Mass Customisation*, *1*(2–3), 374–393. Available from https://doi.org/10.1504/IJMASSC.2006.008631.

Rug, W. (2006). 100 Jahre Hetzer-Patent. *Bautechnik*, *83*, 533–540. Available from https://doi.org/10.1002/bate.200610046.

Sacks, R., Eastman, C., Lee, G., & Teicholz, P. (2018). *BIM handbook: A guide to building information modeling for owners, designers, engineers, contractors, and facility managers.* Wiley, ISBN: 978-1-119-28753-7.

Sandberg, K., Orskaug, T., & Andersson, A. (2016). Prefabricated wood elements for sustainable renovation of residential building facades. *Energy Procedia*, *96*, 756–767. Available from https://doi.org/10.1016/j.egypro.2016.09.138.

Scheurer, F. (2014). *Materialising complexity. Theories of the digital in architecture.* New York: Routledge, Oxon, ISBN: 0415469244.

Schimanski, C. P., Monizza, G. P., Marcher, C., & Matt, D. T. (2019). Pushing digital automation of configure-to-order services in small and medium enterprises of the construction equipment industry: A design science research Approach. *Applied Sciences*, *9*(18). Available from https://doi.org/10.3390/app9183780.

Schreier, M. (2006). The value increment of mass-customized products: An empirical assessment. *Journal of Consumer Behaviour, 5*(4), 317−327. Available from https://doi.org/10.1002/cb.183.

Schumacher, P., 2008. Parametricism as style—Parametricist Manifesto. Dark Side Club1. In *11th Architecture Biennale*, Venice 2008. <http://www.patrikschumacher.com/Texts/Parametricism%20as%20Style.htm>. Accessed 30.10.20.

Sendler, U. (2013). *Industrie 4.0—Beherrschung der industriellen Komplexität mit SysLM (Industry 4.0—Mastering complexity in industry using systems lifecycle management)*. München: Springer-Verlag, ISBN: 3642369179 (in German).

Stoettrup Schioenning Larsen, M., Munch Lindhard, S., Ditlev Brunoe, T., Nielsen, K., & Kranker Larsen, J. (2019). Mass customization in the house building industry: Literature review and research directions. *Frontiers in Built Environment, 5*. Available from https://doi.org/10.3389/fbuil.2019.00115.

Stoetzel, M. (2012). Exploiting mass-customization towards open innovation. In *Proceedings 5th international conference on mass-customization and personalization in Central Europe, MCP-CE 2012*, Novi Sad, Serbia.

Wang, X., Yang, L. T., Xie, X., Jin, J., & Deen, M. J. (2017). A cloud-edge computing framework for cyber-physical-social services. *IEEE Communications Magazine, 55*(11), 80−85. Available from https://doi.org/10.1109/MCOM.2017.1700360.

Wortmann, J. C. (1983). *A classification scheme for master production schedule. Efficiency of manufacturing systems*. New York: Plenum Press. Available from https://doi.org/10.1007/978-1-4684-4475-9_10.

Yoon, J. (2019). SMP prototype design and fabrication for thermo-responsive façade elements. *Journal of Façade Design and Engineering, 7*, 41−62. Available from https://doi.org/10.7480/jfde.2019.1.2662.

Yuan, Z., Sun, C., & Wang, Y. (2018). Design for manufacture and assembly-oriented parametric design of prefabricated buildings. *Automation in Construction, 88*, 13−22. Available from https://doi.org/10.1016/j.autcon.2017.12.021.

Zaeh, M. F., Beetz, M., Shea, K., Reinhart, G., Bender, K., Lau, C., ... Herle, S. (2009). *The cognitive factory. Changeable and reconfigurable manufacturing systems* (pp. 355−371). London: Springer.

Zawadzki, P., Żywicki, K., Grajewski, D., & Górski, F. (2019). *Efficiency of automatic design in the production preparation process for an intelligent factory. Intelligent systems in production engineering and maintenance* (Vol. 52). Springer International Publishing. Available from https://doi.org/10.1007/978-3-319-97490-3_52.

Automation and robotic technologies in the construction context: research experiences in prefabricated façade modules

19

Kepa Iturralde, Wen Pan, Thomas Linner and Thomas Bock
Chair of Building Realization and Robotics, Technical University of Munich, Munich, Germany

Abbreviations

ROD	robot-oriented design
BERTIM	Building Energy Renovation Through Timber-Prefabricated Modules
STCR	single-task construction robots
CWM	curtain wall module
HEPHAESTUS	Highly automatEd PHysical Achievements and performancES using cable roboTs Unique Systems
TS	total station
IFC	industry foundation classes
MEE	modular end effector
DOF	degrees of freedom
CDPR	cable-driven parallel robot

19.1 Introduction

Robotic technology has improved the overall performance of the manufacturing industry. However, the construction industry is lagging on robotic technology adoption. In developed countries the construction industry experiences skilled labour shortages, diminishing productivity and dangerous working conditions. Consequently, since the late 1970s and early 1980s, there was a popular trend among researchers, institutes and private sectors in developing robotics applications that address certain construction tasks. In 1988 Bock proposed (Bock, 1989) the notion of robot-oriented design (ROD). The ROD concept stresses that the design of a building should not only focus on the aesthetic aspect of the building but should also consider how the building is constructed, the running costs, serviceability, maintenance, geometrical composition and life cycle management of a building. The ROD concept also emphasizes the idea that building components shall be designed to be easily handled by robots during the assembly phase. However, in a construction project plenty of stakeholders (client, developer, contractor, guilds and end users), without a previous mutual acquaintance, are involved. This is considerably different to other industries where stakeholders are more bonded (supplier, assembly and client). Moreover, the construction industry is highly fragmented. There is a lack of partnership or collaboration between various tiers of suppliers. Despite that there are other constraints to adopting robotics and automation technologies for the construction industry. This includes characteristics of the regional construction method, inflexible building regulation, improvident institutional legislations, incompatible infrastructure standards and a lack of technology awareness. The aforementioned issues have imposed challenges, such as low acceptance and high initial research costs when trying to adopt robotics technology in the construction sector.

This chapter showcases projects funded by the European Union. The BERTIM (Building Energy Renovation Through Timber-Prefabricated Modules) (BERTIM, 2020) project focused on the existing building renovation by proposing prefabricated façade modules, an automated manufacturing facility for prefabricated timber façade and automated installation procedures. The HEPHAESTUS (Highly automatEd PHysical Achievements and performancES using cable roboTs Unique Systems) project aimed to provide an automated and robotic solution for curtain wall modules (CWMs) installation (HEPHAESTUS, 2020). Both the projects demonstrate innovative approaches to building component production, building envelope installation and implementation of robotics technology in the specific construction context. The following section provides an overview of the background of construction robotics, explains the decision-making process when dealing with construction robotic development and features the key technological achievement regarding building façade installation tasks.

19.2 Context and background of construction robotics technology

Construction robot refers to robotic devices that are developed for specific construction-related activities. The introduction of the construction robot is not a novel approach as its utilization was first recognized in 1983 in Japan, gaining traction in the 1990s when close to 150 construction robots were developed and covered a huge variety of construction tasks. The term 'Single-Task Construction Robots (STCRs)' was first defined by Bock and Linner (2016), who also defined the typology, application categories and mechanical features of the STCR. An STCR usually consists of three main parts: a travel platform, a manipulator and an end effector. The STCR is normally a stand-alone system that is developed for a specific construction task, which reflects the practical need of the construction practitioner and is operated in a coordinated environment. In general, STCR can outperform human labour time and offers consistency in quality; however, if an inadequate level of automation or the incorrect function to the robot is applied, the payoffs and gains are often offset by the additional tasks required to set up, calibrate and operate the robot.

The development of automation and robotic technologies has improved the overall performance of many industry sectors, for example, the manufacturing industry has improved its productivity, quality, working environment since the introduction of highly autonomous production lines. Yet, there are fundamental differences between the construction and manufacturing industries, such as operational methods, sensor technology, human collaboration and skillsets. These differences define distinctions between industrial robots and STCR in design specification, operational experiences, the choice of mechanism and kinematic properties. Even though the robot can potentially perform many construction tasks that were once carried out by humans, there are many trade-offs when applying robotic technology in the construction context without considering the human and robot attributions. For example, robots can outperform humans in repetitive and physically demanding tasks and offer higher efficiency, endurance and accuracy. On the contrary, humans are able to surpass robots when dealing with decision-making in an unfamiliar situation. They also have the ability to learn and make adjustments according to changes in the instructions given and are able to process new information accordingly. The STCR designer needs to be aware of the optimum level of automation that they should assign to the robot and which tasks are performed more efficiently by humans and should, therefore, remain manual. The robot's design has to achieve the right balance, in terms of allocation of functions and the complexity grade to reach the desired automation level, not to create a robot that is too complex to operate. Humans and robots complement each other's performance; but robots will not remove the need for humans (Pan, 2020).

To gain a good understanding of how construction has been carried out conventionally, it is key to apply an appropriate level of automation to a specific construction task. Unlike in the manufacturing industry, where most of the tasks are performed under a controlled environment such as on the factory floor, on-site construction tasks are often carried out in an unpredictable environment that is subject to external elements such as rain, wind, snow and sunlight. Some construction tasks are repetitive and, therefore, ease the adoption of robotic technology, yet the process can be very dynamic, in which many tasks are performed by different trades in parallel. The aforementioned aspects make robotic adoption particularly challenging for the construction sector. Therefore the successful automation of an on-site construction task requires the STCR designer to have in-depth knowledge of the existing workflow and an awareness of the tools, information transfer method, collaboration required between different trades and the decision-making process associated with the task (Price & Pulliam, 1982).

Recently, a South Korean consortium developed a robot on top of a platform that helps the human operator to handle a curtain wall (Chung, Lee, Yi, & Kim, 2010). In the European context, there is an example of the installation of a façade sandwich panel with a robotic mobile crane which was prototyped by a Slovenian research institute (Činkelj, Kamnik, Čepon, Mihelj, & Munih, 2010). In façade renovation, robotic arms on top of mobile platforms have performed positioning and insulation activities. An innovative method for delivery and mounting curtain modules was developed by Brunkeberg AG (Brunkeberg, 2021). The Brunkeberg system demonstrates the technological achievement in building façade installation. The installation of curtain walls also shares the same challenges, as mentioned earlier; for instance, the harsh working environment, miscommunication between various trades, the unreliability of delivery, as well as scarcity in skilled labour and availability of cranes (Eisele & Kloft, 2003). The integrated system consists of five key elements, including vertical profiles, conveyor systems, hoisting systems and lifting systems. The main advantage of the systems is the minimalization of on-site tasks such as panel staging, material handling and lifting. This controls the installation process digitally that provides better data transparency to the stakeholders. The system was tested on a case study building in Sydney and achieved up to a 25% productivity gain and up to a 5% project cost saving. The challenges faced by the system are the limitations of existing transportation. For example, delivery lorries must be converted so that they can load and deliver oversized façade panels. Due to the loading area being close to the building and the vertical transportation require a preassembled vertical profile. Moreover, there is a risk of disturbing ongoing activities being carried out by other trades. Despite the challenges, the system has laid a foundation for automating the installation of a building façade (Friblick et al., 2009).

In the following section the BERTIM and the HEPHAESTUS projects will be presented to demonstrate the implementation of robotics and automation technologies while installing prefabricated building façade modules. The projects provide detailed analysis through designated on-site pilot testing. This features integration strategies, data acquisition methods, project planning and how the system interacts with building infrastructure and human worker in the real construction environment.

19.3 BERTIM project: optimizing façade retrofit through automation

Recently, improving the efficiency of the building renovation process has gained momentum. However, the building renovation process can be very challenging due to the unpredictable condition of the existing building. In addition, the conventional installation task is dangerous and physically demanding (Iturralde et al., 2016).

BERTIM aimed to explore the potential of using automation technology in the timber façade module prefabrication process and optimization of the existing building renovation procedure. In the BERTIM project, there are three main challenges: (1) the data flow from the measurement of the existing building to the installation process; (2) the manufacturing accuracy and the automation level of the prefabricated timber frame modules; and (3) how to automate the façade module manufacturing process. These topics are explained in the following sections.

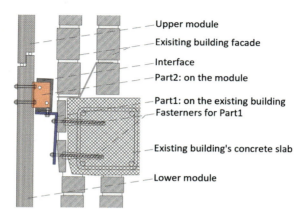

Figure 19.1 The façade interface design.
Source: *Illustration by Kepa Iturralde.*

19.3.1 Highly customized connector design

The unpredictable configuration of the existing building can present many challenges when dealing with renovation tasks, especially when a new layer of building element needs to be added directly on top of the irregular building façade. This means, when installing a prefabricated façade module, the existing façade shall require tailor-made connectors. Furthermore, both the façade module and the connectors need to be calibrated and adjusted to overcome building deviations. Usually, when installing the façade element, first, the geometry of the existing building will be collected by using a 3D laser scanner. Then the collected data will be used to produce the connectors and adjacent façade module. However, customizing connectors for each façade module, while considering each individual building surface deviation, can be a very lengthy undertaking. A highly customized connector, which acts as an interface between the existing building and the prefabricated façade module, was developed in the BERTIM project. In brief, the connectors consist of three parts, as shown in Fig. 19.1. The first part fastens with the existing wall, the second part connects to the prefabricated façade panel and the middle component is the interface between the parts. The installation position of the connectors depends on the deviation and geometry of the wall. The customized connector can adjust up to 50 mm of surface tolerance (Iturralde, Linner, Bock, 2020; Iturralde, Feucht et al., 2020).

19.3.2 Automation of manufacturing processes and the installation procedure

Currently, the manufacturing accuracy of the façade elements is dependent on the precision—routing—machining of the elements that comprise the module. In terms of the traditional manufacturing context, it was revealed that the more complex machining is, the more difficult the assembly of the elements gets. This fact impedes automation assembly. To find a new strategy the idea of routing the whole perimeter of each of the elements, to get an accurate module, must be questioned before implementing automation. Within this context, some questions aroused: can the accuracy of noncalibrated timber frame manufacturing, by using current assembly, be improved by further routing/machining the joints while avoiding unbalancing the assembly process? Can full automation be approached within the current manufacturing and assembly lines? Which are the impediments for reaching that goal within the current assembly lines? With these questions in mind, a systematic installation solution was proposed. The solution was demonstrated in a pilot project to validate its proposed logistical, data generation, manufacturing and installation procedure, see Fig. 19.2.

The overall installation procedure is described next:

- Scan the existing façade with a total station (TS). Check the alignment and the position of the installation.
- 3D print the connectors for the prefabricated façade module. The connector must then be installed on the building façade, guided by TS.
- Produce the prefabricated façade module according to the on-site data. This must then be shipped to the site.
- The final installation must then take place. The prefabricated façade module and the connectors are installed.

The BERTIM project showcased numerous methods for building renovation tasks, such as real-time building configuration, customized connectors for prefabricated module installation, automated manufacturing processes and digitalized data management tools. Those methods indicate the development of construction automation and prefabrication can potentially benefit the existing building renovation task and building life cycle energy performance. The case study at the KUBIK, at

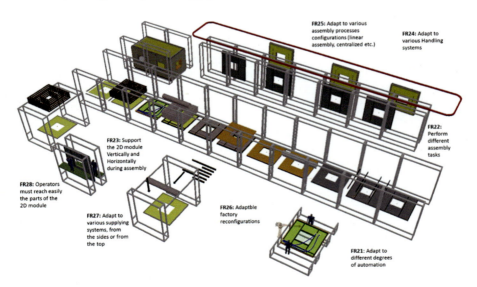

Figure 19.2 The proposed off-site manufacturing facility.
Source: Illustration by Kepa Iturralde.

Figure 19.3 (Left) Installation process of modules and (right) services embedded in the module. *Source*: Photos by Kepa Iturralde.

Tecnalia facilities in Derio (Basque Country, Spain), further validated the claims, the BERTIM solution provides a yearly heating demand reduction from 83 to 26.78 kW h/m^2 y (Garai Martinez, Benito Ayucar, & Arregi Goikolea, 2017). During the BERTIM project the following objectives were achieved:

- The proposed integrated façade system has drastically reduced on-site installation working time by 50% (Iturralde, Linner et al., 2020; Iturralde, Feucht et al., 2020).
- Integration of renewable energy services on the modules (see Fig. 19.3 right).
- Move from individual manufacturing to mass customization.
- Reduction in costs. This is related to the reduction of time of the manufacturing and installation processes.

Furthermore, the achievements of the BERTIM project ensured quality of the works, facilitated dismantling of the modules and improved on-site health and safety during manufacturing and installation.

Nevertheless, the BERTIM solution has room for improvement, especially in the field of robotics, as well as automation of the manufacturing process. Improving the installation efficiency and developing a picking place robotic end effector that can operate in a construction environment are some of the many challenges the research team will face.

19.3.3 Digital tool for renovation tasks

A collaboration platform between the potential client, designer and manufactures called RenoBIM was developed in the BERTIM project. Through technology consultation and project evaluation, the client was aware of whether the BERTIM solution is feasible. The RenoBIM relies on 3D scanners to generate a point cloud model in combination with the building information modelling data that provide comprehensive information, including module geometry, manufacturer requirements, energy consumption and cost estimation. In addition, the RenoBIM tool provides a renovation workflow by exchanging data with computer-aided design and computer-aided manufacturing software, using the industry foundation classes (IFC) format. The IFC files will be converted to intermediate data format and use

EnergyPlus to generate automated energy simulation. The RenoBIM tool provides data transparency, for stakeholders that are involved in renovation tasks, that can potentially increase the overall operational efficiency (Lasarte et al., 2017).

19.4 HEPHAESTUS project: robotic installation of unitized façade systems in tall buildings

The HEPHAESTUS project aims to develop a cost-effective, reliable, user-friendly, efficient system that can carry out the CWM installation task automatically.

The curtain wall installation requires highly skilled professionals and certified installers to work at great height that imposes a huge risk on the health and safety aspects of the operation. Even though the installation benefited from using modern construction equipment, such as cranes and forklifts, the work is still predominantly performed through manual effort. HEPHAESTUS provides an innovative concept of installing off-site manufactured CWMs by adopting a cable-driven parallel robot (CDPR). The objective of the HEPHAESTUS project is to develop a highly adaptable, self-configurable CDPR system that can install CWMs in an outdoor construction setting. This is also to increase the chance of the potential market acceptance. The proposed CDPR shall be developed based on off-the-shelf technology. This is reliable and cost-effective.

The curtain wall is commonly used for modern building construction. It is often treated as a type of cladding system. In this project the authors focused on unitized system curtain wall installation methods. The curtain wall systems are supplied by a curtain wall manufacturing company named FOCCHI Spa. The CWM installation requires high precision, skilled professionals and the entire activity has to comply with straight regulations. For example, the finished installation has to comply with the European curtain wall standard EN 13830 (Iturralde, Linner et al., 2020; Iturralde, Feucht et al., 2020). The installation operation consists of several steps, including delivery, surveying, hoisting, transferring, positioning, adjusting and sealing. The numbers of site personnel required for a specific task and time spent on each step is described in Table 19.1.

According to FOCCHI, the CWMs will be delivered by trucks and stored at the designated location. Commonly, loading platforms are used to accommodate the vertical transportation of the CWMs. The loading platforms are fabricated on-site by using steel members and installed in the building by a tower crane. Once the loading platform is fully loaded with CWMs, the installer will move the CWM storage cart to the relevant installation position. Conventionally, the installation anchor and channels are precast into the building floors. The installers will install the brackets one-by-one by using hand tools. For health and safety reason the installers are protected by safety equipment such as safety harnesses. Then the CWMs are transferred from the storage cart to the lunching table. A lunching table is a device designed for CWMs transportation within the assembly floor level. The spider crane will host the CWM from the lunching table, rotate the panel 180 degrees and

Table 19.1 Manual installation requirements based on FOCCHI Spa data.

Personnel description	Operators	Engineers	Specialist operators	Time per CWMs/ fixtures (s)
Surveyor	0	1	0	N/A
Bracketry installation	2	0	0	300
Panel preparation	2	0	0	300
Panel transportation to launching bed position	5	0	1	180
Panel on launching bed and lubrication	5	0	1	300
Panel connected to crane, lifted and rotated	5	0	1	240
Panel alignment on-site (mullions engagement)	5	0	1	120
Panel alignment to brackets	5	0	1	60
Panel leveling	2	0	0	300
Gasket placement per 5 units	1	0	0	840
Panel protection films installation per 10 units	2	0	0	960

CWM, Curtain wall module.

position the CWM in the right position. The structural engineer and surveyor shall examine the position as well as to conduct the final adjustments of the panel and all installation fixtures. Finally, the installer shall insert sealant and seal off the gaps. As described earlier, it is clear that the manual CWMs installation is physically demanding and requires high precision, see Fig. 19.4.

Moreover, manual installation activities can be augmented by automation and robotic technologies; thus HEPHAESTUS applies an interdisciplinary research approach, which involved partners from academia, industry and research institution. Besides the high degree of automation, the proposed system must also be able to adjust to various external conditions, such as rain and wind. The system shall be reconfigurable and easy to adapt to different architectural designs (Taghavi, Iturralde, & Bock, 2018). The HEPHAESTUS project proposed to use eight cables to manipulate six degrees of freedom (DOF) (Iturralde, Linner et al., 2020; Iturralde, Feucht et al., 2020). The research team (Tecnalia, TUM, CNRS-LIRMM, Fraunhofer-IPA, Cemvisa, nLink, FOCCHI and R2M solution) was inspired by alternative examples of large-scale applications in other sectors. For instance, similar approaches can be seen in contour crafting and masonry construction (Bosscher,

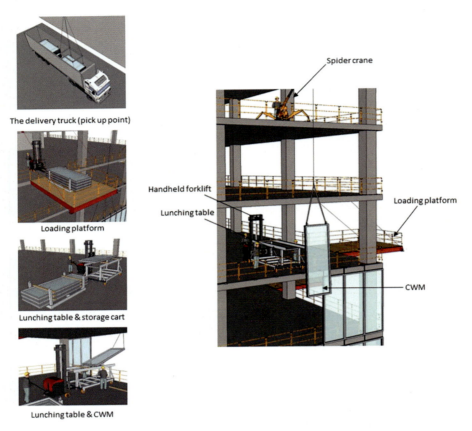

Figure 19.4 The conventional CWM installation process by FOCCHI and drawn by TUM. *CWM*, Curtain wall module. *Source: Images by Wen Pan.*

Williams, Bryson, & Castro-Lacouture, 2007). Prior to the full system development, the project team conducted a detailed task analysis, to understand the various levels of tolerances either contributed through the construction phase or from the CWM compositions and how to mitigate deviations through robotic technologies. As the study indicated, the building structure bears larger tolerances than the prefabricated CWMs; thus the placement of the CWM and the adjustment of the installation brackets require a higher level of precision. The proposed CWM installation task can be divided into two key phases:

1. The brackets need to be fixed accurately on top of the slab. For the demonstrations, holes were made in the concrete and fasteners were used. After that the CWMs have to be placed on to a picking position (on top of a rack) where the CDPR can easily reach from the ground floor.
2. The CDPR picks up the CWM and position it in the correct location and to integrates it into the preinstalled brackets.

Apart from the installation function, there are additional functions that the system should be able to perform over the building life cycle. This includes cleaning of the CWMs, replacing or upgrading the CWMs, and they must be able to change end-effectors for future functional iterations. In addition, the payload, weight and stability of the robot superstructure were taken into consideration. The aforementioned tasks have also laid the foundation for the system functional requirements engineering. Based on the notion of ROD, to optimize the system performance, the specifications of the CWM were designed to match the capability of the CDPR. All in all, the dimensions of the CWMs used in the on-site testing were 1500 mm in width, 3700 mm in height and 200 mm in depth. The CWM weighs approximately 70 kg/sqm.

One of the important aspects of functional requirement engineering is to identify nonfunctional requirements. In brief, the nonfunctional requirement focuses on the quality of the product. It can be intangible and hard to measure. In some cases the nonfunctional requirements can cast limitations over the functional requirements, which assist the design team to make concrete decisions (Capilla, Babar, & Pastor, 2012). The nonfunctional requirements of the proposed CDPR are listed next:

- Operation and control: inadequate allocation of function between operator and robot can intensify issues rather than solve them; therefore in the early stage of development, it is key to decide the appropriate control method and which task shall be assigned to the operator or the automated systems. The operation should be controlled by one operator either in close proximity of the CDPR or in the control room remotely. The control user interface must be user-friendly so that an ordinary façade installer can easily be trained. The overall performance of the operation must be more efficient than the conventional manual operation.
- Tolerance: the system shall have the ability to predict faults and adjust tolerances in advance. The fault detection function should be performed with minimal operator interventions.
- Configurability: the CDPR shall be designed as a system rather than a collection of individual components. The CDPR is required to be reconfigured regularly, such as when the system has to move from the completed building to another. When two building facades are with different dimensions, thus, the CDPR has to be able to readjust to the new façade dimension. To achieve greater efficiency, the upper supporting structures shall be able to self-climb as the building is taking shape.
- Logistics: transportation is one of the important performance indicators for the proposed CDPR, as the system will be required to move from one construction site to another. Generally, the distance between the two sites may cover a substantial distance. It is crucial to investigate the means of transportation, their size and limitations. Also consideration should be given to whether the system can be designed to be more compact or can be disassembled to smaller components for transportation.
- Health and safety: the HEPHAESTUS solution promotes a certain level of human−robot interaction in an outdoor condition, while the system is handling heavy load across the job site. Therefore a safer operational environment is required. The system needs to be developed to comply with relevant directives and regulations, which cover the areas of the European directive on machinery, the European directive and standards on electrical equipment. The European standards about automation, European ISO standards and construction work standards must also be considered.

19.4.1 Technical description of the CDPR and its components

The CDPR is the system that moves all the tools meaning the modular end effector (MEE) and the CWM along the workspace of the façade. For the HEPHAESTUS solution, the CDPR consists of several components that include two upper and four lower supporting systems, both equipped with winches and pulleys. The upper supporting structure is also referred to as the superior supporting structure, a mobile platform (that includes the MEE), picking station, control station and electrical cabinet (see Fig. 19.7). The upper supporting structure is composed of two sets of cantilever cranes, which are installed on the roof level of the building and connected to the roof structure. The lower supporting structures are anchored to the ground pit foundation. All CDPR structures are connected to the building structure to support the forces generated by the cables. The winches, motors and drivers that are connected to the supporting structures were also tested to make sure that they are compatible with the expected performance. The control room is located on the ground level that facilitates the control cabinet, power supplies, winch distribution boxes and all other communication- or control-related equipment.

As shown in Fig. 19.6, the mobile platform is constructed with structural steel bars. It is lightweight yet offers a very robust enclosure for the CDPR. As mentioned earlier, eight cables are driving the CDPR. The cables are made of nonrotation steel wire rope. Its tensile strength, yield strength, elasticity, breaking load and other key parameters were tested under laboratory condition.

19.4.2 Technical description of the MEE and its components

MEE is one of the most sophisticated subsystems within the CDPR. It is responsible for localization, setting up the anchors, placing the installation brackets, drilling, rebar scanning and CWM installation. The upper MEE frame is made of aluminium, which accommodates the robotic arm for the brackets installation, anchor magazine and sensors. Many functions provided by the MEE require the system to remain stable while operating. Hence, the allocation system has been developed to stabilize the MEE. The allocation system comprises two main subsystems. This includes two linear actuators connected with two sets of vacuum grippers. The system shall retain at least 150 kg of weight. After testing the project team noticed that the quality and evenness of the concrete surfaces has a profound implication on the performance of the vacuum grippers.

Based on the project partner's experience and recommendation (nLink), a six-DOF robotic arm was used to carry out high precision tooling, pick and place and drilling functions. The robotic arm is located inside the MEE frame, surrounded by interchangeable tools. The main functions of the robotic arm are drilling, fixing expansion bolts, rebar scanning and setting installation anchors. In general, FOCCHI has recommended the most appropriate expansion bolt system to be used in both new build and renovation projects. The project partner nLink has contributed knowledge about robotic drilling system. A custom mounting bracket was made to facilitate Hilti TE7 rotary hammer. In addition, a vacuum system was integrated with the rotary hammer to ensure that the dust particles can be vacuumed instantaneously. To position the installation bracket over the predrilled holes and insert, fastening the expansion anchor through the opening on the bracket, while keeping both the tools and the materials stable, is extremely challenging. A fully integrated anchoring tool was developed that incorporates a Hilti TE7 rotary hammer, a parallel gripper with custom anchor gripper fingers, three spring plungers, a magnetic gripper and sensors, as shown in Fig. 19.5.

After the brackets have been installed, the next task is to install the CWM. A picking station was designed to allow the CWMs to be positioned at an optimum angle that eases the on-board vacuum suction lifter to accomplish relevant pick-up tasks. Each CWM has a male and a female slot that interlocks. Once the CWM has been picked up and transferred to the installation position, the panels can be easily connected via the interlocking devices. Most of the systems described earlier are compactly integrated within the MEE frame as well as the mobile platform. The

Figure 19.5 Anchoring tool developed by nLink and TUM.
Source: Photo by Kepa Iturralde.

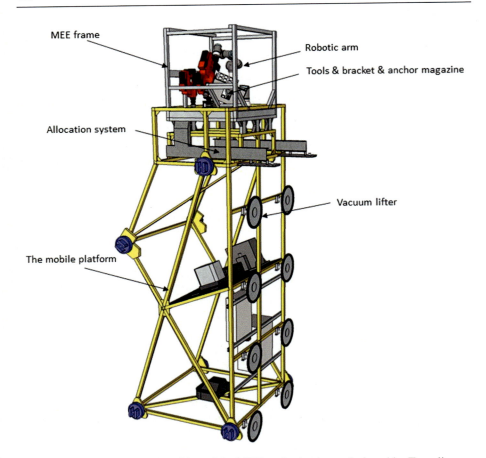

Figure 19.6 The overall composition of the MEE, and subsystems designed by Tecnalia, Cemvisa VICINAY, TUM and nLink. *MEE*, Modular end effector.
Source: Illustration by Wen Pan.

overall composition of the MEE system, vacuum lifter and other subsystems can be seen in Fig. 19.6.

19.4.3 The proposed operation sequence

This section provides an insight into the predicted operational sequence for the proposed CDPR. A computer-generated 3D building with a regular square floor plan was constructed that serves as a case study building. Additional attention is required to determine if the roof of the building is structurally sound for the additional load contributed by the CDPR. The CDPR components are delivered on-site by trucks and unloaded into the dedicated storage area. The upper supporting structure along with winches, gearboxes and pulleys is to be installed on the roof level by using the tower crane. The footings are prepared and ready to connect with the lower supporting structures. Next, the mobile platform, MEE and vacuum lifter shall be integrated. After this step the cables, controllers, sensors and other accessories will be assembled and calibrated ready for operation.

The CWMs are transported on-site by trucks and unloaded to a transfer frame by a spider crane. The transfer frame docks with the picking station and the spider crane hoists located at the side of the CWM to allow it to slowly slip into the picking station. Then the eight cables carry the MEE to the respective installation position. This keeps approximately 350 mm between the system and the building. The linear actuators and vacuum grippers will be extended out and stabilize the MEE. Then the onboard robotic arm will scan the installation area for rebar to determine the drilling positions. After the installation holes are drilled, the compacted anchoring tool will place the bracket over the holes then insert and fasten the expansion anchors. Next, the vacuum lifter picks the CWM from the picking station and transfers it to the correct position. Then the CWM is slotted into place and sealed off automatically.

Furthermore, MEE is controlled by several industrial PCs. The controllers are based on TwinCAT software featured by BECKHOFF (Beckhoff, 2021). The vacuum lifter and the onboard robotic arm are controlled through a programmable logic controller. In addition, the HEPHAESTUS system provides an interactive graphical user interface so that the operator can operate the CDPR remotely. At the time of writing the proposed CDPR is still under development. The next section will present the on-going on-site pilot testing and will evaluate the challenges imposed by a real construction scenario.

19.4.4 First and second demonstrations

The first demonstration tests were performed at Tecnalia facilities in Derio (Basque Country, Spain), see Fig. 19.7. Once all the components of the demonstrator were installed, the operation of all the components (motors, movement of the robot, positioning in relation to the steel structure, sensor, etc.) was verified. This was the first time the different elements of the robot (winches with cable pulling on the platform/base) and the higher level control of the robot that makes the coordination of the winches were put together.

The second demonstration tests were performed at Acciona demo building, in Noblejas (Toledo, Spain), see Fig. 19.8. This was a real construction scenario specifically built for the final demonstration stage in which the robot completed the installation of a set of curtain walls that covered the façade of the target construction. Based on the results of these tests, the cable robot was validated, among other performance indicators, in terms of time required to complete the operations, accuracy, efficiency and usability for workers of the constructor sector. During the whole demonstration process, special care was taken to fulfil the safety requirements and recommendations for these types of operations.

After the demonstrations held in Spain, some lessons learned were gathered. For example, the repeatability of the CDPR was quite high, at around 1 or 2 mm. On

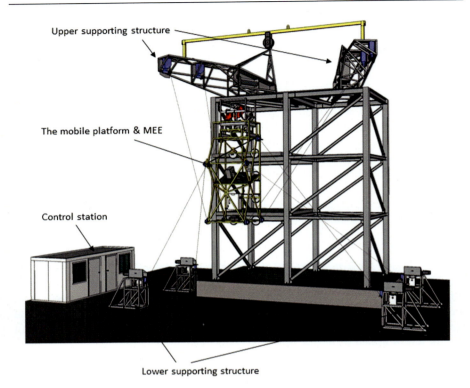

Figure 19.7 The HEPHAESTUS project pilot testing site at Tecnalia lab facilities in Derio, (Basque Country, Spain). *HEPHAESTUS*, Highly automatEd PHysical Achievements and performancES using cable roboTs Unique Systems.
Source: Illustration by Wen Pan.

the other hand, it was concluded that the use of a robotic arm for the fixing of the brackets was not an optimal solution because it involved too many tool changes and generated unstable performances. Moreover, in future instances, more flexible positioning of the cranes that host the winches and poles will be necessary. Finally, the shape of the different connectors and even the profiles of the CWM should be modified to facilitate a smooth installation process.

Nonetheless, the proposed CDPR still has many challenges, such as long installation and calibration duration, the building roof structure needs to be designed to be able to accommodate the CDPR upper supporting structure, and the building structures also need to be structurally compatible with the CDPR system. The current CDPR design is limited to buildings with a large flat open space on the roof. If the building possesses any convex, circular, concave surfaces and with any large protruding features, it will challenge the installation of the CDPR. In particular, the installation of the upper supporting structure will become very difficult. All these challenges will likely be solved in the future.

Figure 19.8 The pilot testing site of the HEPHAESTUS project at the demo building, in Acciona facilities in Noblejas (Spain). *HEPHAESTUS*, Highly automatEd PHysical Achievements and performancES using cable roboTs Unique Systems.
Source: Photos by Kepa Iturralde.

19.5 Conclusion

The chapter described the background of construction automation development and explained the method of how to adopt automation and robotic technology in the construction context. Two projects along with case studies were demonstrated. The BERTIM project presented a method to fabricate building façade elements by using robotic technology, to develop an interface that can adapt to the various building surfaces in the context of the renovation project. HEPHAESTUS focused on the development of a CDPR that integrated off-the-shelf components and was able to carry out CWM installation task in a semiautonomous manner. The demonstrations of the HEPHAESTUS project presented in this chapter are a milestone for a complete façade module installation with robotic tools in a close-to-real environment. The tasks were achieved in an automated mode and in the future, such robots might be found on real construction sites, which will improve the efficiency of such operations. Moreover, the main strength relies on a faster process when compared to manual methods, as explained in the BERTIM and HEPHAESTUS.

However, there are still many challenges and weaknesses associated with implementing automation and robotic technologies in specific construction task. In-depth pilot testing is necessary to evaluate the technology integration strategies, planning and the interaction between human and robots with the building. Besides a potential threat might be related to safety issues. The system needs to address the potential additional safety hazards and implications contributed when implementing automation and robotics. A cross-disciplinary collaboration is essential to fulfil the research gaps and address each stakeholder's interests. Nevertheless, the outcomes from the aforementioned projects will inspire new opportunities in the construction industry while they attempt to implement practical STCR applications in the future.

Acknowledgement

The research presented in this chapter was carried out in the context of the H2020 projects BERTIM and HEPHAESTUS. BERTIM and HEPHAESTUS received funding from the European Union's Horizon 2020 research and innovation programme under grant agreement no 636984 and 732513.

References

Beckhoff (2021). Available from: https://www.beckhoff.com/. Accessed 08.02.21.
BERTIM (2020). Available from: http://www.bertim.eu/index.php?option = com_content& view = featured&Itemid = 111&lang = en. Accessed 06.11.20.
Bock, T. (1989). *A study on robot-oriented construction and building system* (PhD thesis). University of Tokyo.
Bock, T., & Linner, T. (2016). *Construction robots: Elementary technologies and single-task construction robots* (Vol. 3). Cambridge: Cambridge University Press.
Bosscher, P., Williams, R. L., II, Bryson, L. S., & Castro-Lacouture, D. (2007). Cable-suspended robotic contour crafting system. *Automation in Construction, 17*(1), 45−55.
Brunkeberg (2021). Available from: http://www.brunkeberg.com/. Accessed 08.02.21.
Capilla, R., Babar, M. A., & Pastor, O. (2012). Quality requirements engineering for systems and software architecting: Methods, approaches, and tools. *Requirements Engineering, 17*(4), 255−258.
Chung, J., Lee, S. H., Yi, B. J., & Kim, W. K. (2010). Implementation of a foldable 3-DOF master device to a glass window panel fitting task. *Automation in Construction, 19*(7), 855−866.
Činkelj, J., Kamnik, R., Čepon, P., Mihelj, M., & Munih, M. (2010). Closed-loop control of hydraulic telescopic handler. *Automation in Construction, 19*(7), 954−963.
Eisele, J., & Kloft, E. (2003). *High-rise manual − typology and design, construction and technology*. Basel: Birkhäuser.
Friblick, F., Tommelein, I. D., Mueller, E., & Falk, J. H. (2009, July). Development of an integrated façade system to improve the high-rise building process. In *Proceedings of the 17th annual conference of the International Group for Lean Construction (IGLC)* (pp. 359−370).
Garai Martinez, R., Benito Ayucar, J., & Arregi Goikolea, B. (2017). Full scale experimental performance assessment of a prefabricated timber panel for the energy retrofitting of multi-rise buildings. *Energy Procedia, 122*, 3−8.
HEPHAESTUS (2020). Available from: <https://www.HEPHAESTUS-project.eu/>. Accessed 06.11.20.

Iturralde, K., Feucht, M., Hu, R., Pan, W., Schlandt, M., Linner, T., ... Elia, L. (2020). A cable driven parallel robot with a modular end effector for the installation of curtain wall modules. In *Proceedings of ISARC 2020, Kytakyushu, Japan.*

Iturralde, K., Linner, T., & Bock, T. (2016). Development of a modular and integrated product-manufacturing-installation system kit for the automation of the refurbishment process in the research project BERTIM. In Proceedings of the 33rd ISARC (pp. 1081–1089).

Iturralde, K., Linner, T., & Bock, T. (2020). Matching kit interface for building refurbishment processes with 2D modules. *Automation in Construction, 110.*

Lasarte, N., Chica, J. A., Gomis, I., Benito, J., Iturralde, K. & Bock, T. (2017). Prefabricated solutions and automated and digital tools for the optimisation of a holistic energy refurbishment process. In *Proceeding of the 8th European congress on energy efficiency and sustainability in architecture and planning and 1st international congress on advanced construction* (pp. 125–140).

Pan, W. (2020). Methodological development for exploring the potential to implement on-site robotics and automation in the context of public housing construction in Hong Kong (PhD thesis). The Technical University of Munich.

Price, H. E. & Pulliam, R. (1982). *The allocation of functions in man-machine systems.* Available from: https://ntrs.nasa.gov/archive/nasa/casi.ntrs.nasa.gov/19850070377.pdf.

Taghavi, M., Iturralde, K. & Bock, T., 2018, Cable-driven parallel robot for curtain wall modules automatic installation. In *Proceedings of the 35th international symposium on automation and robotics in construction (ISARC)* (pp. 396–403).

Life cycle assessment in façade design

20

Linda Hildebrand, Kim Tran, Ina Zirwes and Alina Kretschmer
Chair for Reuse in Architecture, Faculty of Architecture, Rheinisch-Westfälische Technische Hochschule (RWTH) Aachen, Aachen, Germany

Abbreviations

BIM	building information modelling
EC	embodied carbon
EE	embodied energy
EPD	environmental product declarations
GHG	greenhouse gas
IFC	industry foundation classes
LCA	life cycle assessment

20.1 Introduction

Environmental aspects have shaped facades in various ways over the last five decades. In the context of (Otto et al., 2020) building typologies such as the passive or active house, the building envelopes were further developed to improve energy performance, particularly solar, thermal performance and airtightness. Furthermore, it has become more common to integrate technical devices such as decentral ventilation, heating units and energy-harvesting devices (eg, photovoltaic, solar thermal panels) (Knaack, Klein, Bilow, & Auer, 2007), enhancing the performance of facades aimed at reducing the environmental impact during operation. The decreased demand for operational energy and the growing share of renewable energy lowered the amount of greenhouse gas (GHG) emissions during the operational lifetime of the building. Based on this improvement, the building fabric becomes the relevant parameter with environmental potential to unlock. While previously the focus was primarily on the operating phase, it is increasingly important to integrate the ecological dimension of material and construction into the design phase. This is done particularly by considering the energy and emissions linked to the fabrication of the building fabric and furthermore its ability for reuse and recycling.

Façade engineers can impact the extent of environmental impact linked to the building envelope in the early design phases. In the concept and schematic design phase, various approaches are useful.

Rethinking Building Skins. DOI: https://doi.org/10.1016/B978-0-12-822477-9.00026-7
© 2022 Elsevier Inc. All rights reserved.

During concept design, basic design decisions are made and design parameters are set which are relevant to the environmental impact, such as amount and type of material are uncertain and subject to change. On the other hand, the range of environmental impact is set during concept design with the choice of façade typology to be used.

The schematic design phase builds on the decisions made in earlier stages and different alternatives regarding the construction, material and product are made. Life cycle assessment (LCA) tools can help to provide recommendations for a specific design choice.

This subchapter is structured according to the approach in the concept stage followed by the schematic design phase. The last chapter reflects this process and provides an outlook.

20.1.1 Façade typology

Considering the entire building, the structure accounts for the highest amount of embodied energy (EE) and embodied carbon (EC) (for a concrete or steel structure). Significant reductions can be made by using timber where possible and reducing the weight of the construction. Generally, the EE and emissions of facades are lower, when the lower overall volume is smaller. In Hildebrand (2014), 25 office buildings were examined and roughly half of the EE and EC accounted for the structure and a third for the façade. The characteristics of environmental impact in facades are different compared to structures since a variety of materials, products and typologies can be used. Depending on the façade typology, its construction and the choice of façade materials, the EE and EC can account for the highest share (eg, with an aluminium double-skin façade or with a façade contributing to a reduction of environmental impact, eg, with punched windows in a timber wall). Due to a high variety in design parameters, the integration of environmental concerns has to be considered carefully as it has the potential to improve the overall (production-, operational- and postuse phases) assessment.

To design environmentally conscious facades, some parameters are generally valid and others need to be calculated based on the specific situation of the assignment.

The choice of façade typology is very relevant for the amount of EC (Fig. 20.1). Double facades with full glazing account for the highest amount of GHG, followed by solid walls with punched windows based on mineral materials. A lower amount of GHG is bound by curtain wall systems with a high amount of renewable materials such as wood-based products.

The façade typology predefines an impact to a certain extent. Lightweight facades, such as timber stick systems, have the lowest proportion of EE. In addition, only a few materials in this system are bonded by adhesives. Most of them are mechanically fixed, which increases the recycling potential. Solid facades, such as masonry or concrete walls, are heavier and hence are higher in EC. For lightweight enclosures, film and textile cladding offer a solution with a low EE value. Double-skin facades with a high proportion of aluminium or steel framing have a larger EE

Life cycle assessment in façade design

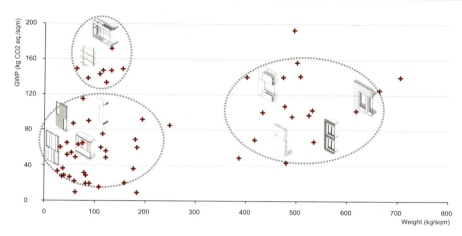

Figure 20.1 Global warming potential in different façade typologies.
Source: From Linda Hildebrand. Figure designed by the author.

impact. The different façade types and their ecological impacts are summarized in Fig. 20.1. It shows the global warming potential in relation to the weight of different façade typologies. By choosing the façade typology the bandwidth of environmental impact is predefined. A detailed case study can be found in Hildebrand (2014).

20.1.2 Design principles for facades

Considering a variety of design principles is useful throughout the duration of the design process, the most useful principles are briefly presented and categorized based on environmental potential, limitations of functionality, impact on the design, as well as built references in the hierarchy of their environmental relevance.

1. Reused façade elements

Potential	Limits
• Designing with reused building elements preserves natural resources as no resources, energy and emissions are needed as opposed to working with new products. Precast concrete panels in particular are suitable for reuse as their primary fabrication has a high environmental impact and the robustness of the construction increases the longevity of the structure (Dechantsreiter et al., 2015)	• Design process is affected and the availability of elements for reuse needs to be guaranteed in the early design stages • Reusing façade elements will determine the specific grid size. This needs to be considered when designing the envelope and any additional elements to be integrated • Warranty and guarantee of the element need to be considered. (Urszula, 2019)

Reference projects:

- Various projects reusing precast concrete elements in Germany by Prof. Angelika Mettke, TU Cottbus.
- EU Headquarters in Brussels, Belgium by Philippe Samyn and Partners (reuse of windows from different countries as symbolic unification).
- *Plattenpalast* in Berlin, Germany by wiewiorra hopp architekten and Claus Asam (reuse of precast concrete and window elements from the former Palace of the Republic of the German Democratic Republic) (Fig. 20.2).

Figure 20.2 *Plattenpalast* in Berlin by wiewiorra hopp architekten and Claus Asam. *Source:* From Chair for Reuse in Architecture.

- *Big Dig House* in Lexington, Kentucky, USA by *Single Speed Design* (reuse of steel elements for as exoskeleton).

2. Recycled content in façade products

Potential	Limits
• This strategy is relevant for stick-built facades as well as for aluminium unitized facades. The embodied emissions can be reduced significantly (up to 95%) when using profiles with recycled content (up to 75% recycled content) • A variety of cladding materials with recycled content are available	• The effort to process the recycled content needs to be lower when compared to the production of a new product • Products with recycled content are not available for all materials. Although theoretically possible, finding recycled content in float glass in practice is difficult

Life cycle assessment in façade design

Reference projects:

- Extension to *Museum Folkwang* in Essen, Germany by David Chipperfield Architects (recycled glass in the translucent façade cladding) (Fig. 20.3).

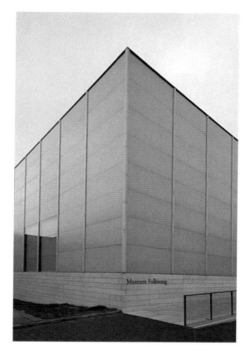

Figure 20.3 Extension to *Museum Folkwang* in Essen by David Chipperfield Architects. *Source:* From David Kasparek. This work is licensed under CC BY 2.0 license. The original version can be found here: https://www.flickr.com/photos/dave7dean/4603876035/in/photolist-7PQXow-81Q4X2−81Tdxm-81TdfS-7PQXwS-81Tdz5−81Q5Ep-81TdCf-81TdhL-81Q4UK.

- Circle house in Copenhagen, GXN, Vandkunsten, Lendager (Shingles in the façade cladding are made from recycled plastic).

3. Façade materials as a carbon storage

Potential	Limits
• Using materials that absorb GHG during their growth helps to reduce the environmental impact. Most common are wood-based products as profiles, cladding or insulation • Hemp used as an insulating brick or rammed earth has a similar effect	• Products based on renewable materials can be less robust and need to be maintained • Fire is a consideration when using wood-based cladding materials and composites

Reference projects:

- *Woodcube* in Hamburg, Germany by *Architekturagentur* (solid timber façade with wood-based insulation and timber cladding).
- *Moholt Timber Towers* in Oslo, Norway by MDH Arkitekten (high insulating façade with cross-laminated timber cladding) (Fig. 20.4).

Figure 20.4 Woodcube in Hamburg by *Architekturagentur*.
Source: From Gunnar Ries. This work is licensed under CC BY-SA 2.0 license. The original version can be found here: https://www.flickr.com/photos/gunnarries/9194049064/in/photolist-f1rUtj-f1cVbM-2ddmHV2−2ddmHRz.

4. Designing for disassembly

Potential	Limits
• The material choice and the type of joint impact the future (economic and ecological) quality of the building envelope The environmental benefit can be gained in the future when the elements are reused and recycled and thereby resources and energy are preserved	• Functional performance requirements such as air- and water tightness have to be negotiated with joints that are easy to disassemble • The environmental benefit is a potential. Frame conditions can change or information is lost so the reuse and recycling do not take place

Reference projects:

- *Urban Mining and Recycling UMAR* in Dübendorf, Switzerland by Werner Sobek with Dirk E. Hebel and Felix Heisel (The project combines different strategies described under (1) and (2). Beyond that all materials can be disassembled due to nondestructive joints).
- Product Development Test Lab in Delft, Netherlands by Tillmann Klein and Marcel Bilow (rear ventilated façade, weather barrier is added without gluing to substructure, all products can be dismantled and are ready for reuse or recycling) (Fig. 20.5).

Figure 20.5 Product development test lab in Delft by Tillmann Klein and Marcel Bilow. *Source:* From Marcel Bilow.

5. Designing with local materials

Potential	Limits
• The use of local materials helps to reduce the transport energy and emissions, thus contributing to a reduction in production energy	• The local conditions for energy supply and the type of transport need to be considered • Labour expenses can lead to higher costs when the building is located in high-wage countries

Reference projects:

- Earth ships, various locations by Biotecture (different façade materials such as rammed earth or car tires).
- Villa Welpeloo in Enschede, the Netherlands by *2012 Architecten* (locally reclaimed materials such as wood from industry products as cladding material).
- *Alnatura Campus* in Darmstadt, Germany by Haas Cook Zemerich (prefabricated rammed earth façade, regionally sourced) (Fig. 20.6).

Figure 20.6 *Alnatura Campus* in Darmstadt by Haas Cook Zemerich.
Source: From Ulrich Knaack.

6. Designing lightweight construction

Potential	Limits
• LCA is based on a structure's mass, so lightweight solutions present advantages over heavy ones. This is relevant in the design phase, as the choice for a lightweight material such as film or textile initially shapes the design and requires a different structure and organization than a conventional construction method	• Lightweight construction can be less robust that leads to short usage cycles and high exchange rate • The lack of thermal mass in lightweight construction can have a detrimental effect on the indoor climate

Reference project:

- Unilever HQ in Hamburg, Germany by *Behnisch Architekten* (film cladding as second skin) (Fig. 20.7).

Figure 20.7 Unilever HQ in Hamburg by *Behnisch Architekten*.
Source: From Marcel Bilow.

- Kukje Gallery in Seoul, South Korea by SO-IL (stainless steel mesh from welded rings covers the building and provides its cubature).

The abovementioned principles are useful for the early design phases as they can guide the façade design toward lower environmental impact. Once the façade typology and design principles have been identified, further environmental considerations are useful to decide on construction and material (Dechantsreiter et al., 2015).

20.2 Assessing environmental impact of facades in the design phase

Design decisions are subject to functional, aesthetic and financial criteria. The environmental perspective does not replace any of these, but it is an additional criterion to be integrated into the evaluation process. The consideration of environmental aspects needs to develop with a level of detail that grows during design. For an academic discussion a quantification of an alternate façade typology can be done. It requires a high level of detail and is work-intensive that translates to higher costs in the context of a project. Even more relevant, assessment in the concept phase (when the typology is discussed) needs assumptions (such as material and construction details) that develop in the design process. With wrong assumptions the assessment becomes obsolete. Once the typology is defined, an assessment is useful to demonstrate the opportunities and limitations of a specific design alternative. For this the application of LCA is useful.

To understand the method of LCA, this subchapter provides a brief overview of its development and integration in the building sector and the façade industry. This is followed by a brief summary of the LCA method, tools and databases.

20.2.1 Origin of life cycle assessment and its implementation in façade industry

The first ideas about what is understood as LCA today were mentioned in a simplified way by biologist and economist P. Geddes in 1884 (Frischknecht, 2009). In his text, he developed a method to monitor energy and material flows (Geddes, 1884). The first LCA for building materials was assessed based on the calculation of energy generation flows due to the use of energy for any production. LCA evolved over the last 50 years at different sites. Research activity took place in several institutes, motivated by the desire to reduce waste (glass bottles vs cans, cloth vs disposable diapers) and the efficiency enhancement of energy generation due to the oil crisis in the 1970s. According to Kümmel Fink, the *Eidgenössische Materialprüfanstalt* in St. Gallen is the institute that coined the term LCA in 1978 (Kümmel, 2000). In 1995 another Swiss, Daniel Spreng, introduced the term Grey Energy (Spreng, 1995) referring to the quantification of the primary energy used for harvesting resources, transport, processing, maintaining and deconstructing a product or energy used for a service.

LCA is a method to quantify all input and output flows related to an assessed item, service or product of any discipline, based on researched information and estimations. It has a descriptive or comparing nature that quantifies the flows and is used as an instrument to screen flows and identify optimization potential. While the basic structure and content are regulated, different systematic approaches can be integrated (Beuth, 2006).

The application of LCA in the building industry was initiated when green building certificates became (more) common at the beginning of this century. While in Great Britain BREEAM was already successful in the 1990s, awareness and a broader application of the certificates grew at the beginning of the 1990s in other Western European countries such as Switzerland, the Netherlands and Germany. LCA played an indirect role in the British Certificate. Environmental performance of materials was rated by letters from A to F, which was reflected in credits as part of the evaluation scheme. In the German-speaking countries, an LCA database was growing and made available to the public in 2009 by the German Government. This database contained a variety of building materials and as it was compiled by one company, the criteria for the data were similar and therefore comparable to a certain extent. It included general data on product level based on literature review and environmental product declaration (EPD) according to DIN EN 14025 (Nagus, 2006) that were anonymized. Open databases for LCA are hard to find as the calculation requires detailed information on, that is, energy sources or upstream processes and is therefore work- and cost-intensive. Furthermore, some companies prefer to keep this information confidential. One way to access environmental data for specific products is through platforms for EPD, which provide information also for façade and cladding materials. LCA in façade design is growing in relevance that can be observed by the increasing amount of certificates and information supplied by façade suppliers that provide environmental data. Studies comparing different façade alternatives regarding the environmental impact over different life cycle stages can be found published in scientific and other publications (Guardigli, Monari, & Bragadin, 2011; Kellenberger & Althaus, 2009; Union, 2014−2017). However, the integration of LCA in façade design (as well as for other building elements) is limited to a small number of projects. Open access databases help to integrate the EE and emissions into the façade planning process. Still, the data refer to a national context with specific energy and transport infrastructure. Consequently, this leads to imprecise results when LCA data from foreign databases (with a certain safety addition) are used. EPDs contributed to the increased consideration of environmental impact, especially in Western Europe. Generally speaking, the availability of EPDs contributed to the increased consideration of environmental impact, especially in Western Europe. A reason for this might be the product-oriented approach of the façade industry in Europe. On the contrary, in the United States, EPDs are rare. One reason for that is that facades are typically bespoke and very much tailored to a specific project.

20.2.2 The method of life cycle assessment

Multiple guidelines and building standards that regulate the assessment are available. Most relevant is DIN EN 14040 (Beuth, 2006) as it regulates the procedure of the assessment. A comprehensive description relevant to facades can be found in Bach, Mohtashami & Hildebrand (2019) or Wadel & Gerardo (2013). The most relevant steps of the assessment are itemized next. The first step is known as Goal and Scope that in the context of façade design is driven by two motivations:

- comparing alternate façade approaches to optimize the environmental impact of the design (occurs during the design).
- assessing the EC and EE for a comprehensive assessment (including all building elements and the operational energy) for communication purposes like a green building certificate (occurs after completion).

As part of the system boundaries, the life cycle phases that are considered in the assessment need to be specified. DIN 15804 (DIN, 2012) differentiates 17 life cycle phases in the building context that are grouped into four categories (Fig. 20.8):

- product stage A1−A3
- construction process A4−A5
- usage stage B
- end of life C1−C4

For a building LCA the façade would be considered in the production and end-of-life phase. Focusing on the façade, the building operation phase can include 'B2 maintenance', 'B3 repair', 'B4 replacement', 'B5 refurbishment', and 'B7 water use'. Since no energy is used to operate the façade, B6 is not relevant. The operational phase is typically not considered in the LCA for facades and currently, no scientific studies are available.

Primarily, the production phase (with life cycle phases A1−A3) is relevant when assessing the façade.

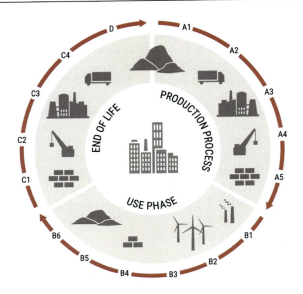

```
A1 Raw material supply      B1 Use                    C1 Deconstruction | Demolition
A2 Transport                B2 Maintenance            C2 Transport
A3 Manufacturing            B3 Repair                 C3 Waste processing
A4 Transport                B4 Replacement            C4 Disposal
A5 Construction process     B5 Refurbishment          D  Reuse | Recycling potential
                            B6 Operational energy use
                            B7 Operational water use
```

Figure 20.8 Life cycles phases of a building.
Source: From Linda Hildebrand. Figure designed by the author.

The phase after the operation phase, the so-called end of life, lays in the not foreseeable future (most likely 30–100 years ahead) and an LCA expert approaches these by using scenarios. End-of-life scenarios are associated on a material level based on the average values. For most mineral materials, this is 'processing of building rubble' as it is the most common scenario. Looking at the EE and carbon, this end of life accounts for less than 10% compared to the building production. A prefabricated building element that is suitable for reuse will also be considered recycling with this flow as the association is on material not on the construction level. This misguidance also works the other way around; credits are associated with most metals for which the environmental footprint can be reduced when recycling material is integrated into the production, even when these are bonded to other materials and corrupt the recycling process (eg, insulated sandwich panels made from polyurethane foam bonded to a metal sheet). This leads to misguided results as not only the material is relevant for the postuse but also the type of joints and the context of the construction to assess the actual environmental impact (Hartwig & Hildebrand, 2017).

To address this, different solutions can be offered. An experienced LCA expert with knowledge of the construction can assess each element individually and calculate the resources and emissions linked to the end-of-life phase in detail. This requires LCA and engineering skills and certain time capacities as it is more labor-intensive than working with scenarios. An alternative is described in Heesbeen Zabek, and Hildebrand (2021); Hildebrand, Kosanovic, Cukovic-Ignjatovic, Bach, and Radivojevic (2017); Hildebrand, Schwan, Vollpracht, Brell-Cokcan, and Zabek (2017) that introduce the postuse classes: 'Reuse', 'Pure Recycling', 'Mixed Recycling', and 'Critical'. Based on this evaluation, the engineer can optimize the design with more certainty. A similar approach can be found in the Recycling Atlas (Hillebrandt, Riegler-Floors, Rosen, & Seggewies, 2018) in which a 'closed-loop potential' is calculated based on a similar systematic. This discussion gains relevance in the context of the Circular Economy as discussed by Heesbeen et al. (2021).

In contrast to the beverage industry, where LCA has its roots and the consideration of postuse scenarios is clear, LCA in the building context is more complex because the number of materials or products fixed together is higher and components are used for a longer duration. When LCA is applied in architecture, these particular qualities need to be considered.

20.2.3 Categories for life cycle assessment tools in façade design

To find a suitable tool to support LCA, the scope of the software needs to be specified. Based on the categories for LCA tools described in Bach et al. (2018), seven categories can be used to compare LCA software, which are (1) origin, (2) required user knowledge, (3) data source, (4) entry format, (5) optimization, (6) default settings, (7) and life cycle phases.

1. Origin
    The origin of the software tools depends on the country of origin, how current the data are (year of publication, updates) and the developer background (research institution or company). This can affect the information about the accessibility, scope and specific purpose of the programme. The national background is also relevant, as there are differences in the national databases to which the programmes are linked.
2. Required user knowledge
    Tools are available for different levers of user's knowledge. They mainly differ in the editability of the individual parameters. The more these settings can be adjusted manually, the more precise the results can be generated. However, this requires a certain amount of expertise. Depending on their conception and editability, the software can be categorized as 'no previous knowledge', 'basic knowledge', and 'expert knowledge'.
3. Data source
    LCA software are linked to various databases, which are either predefined or changeable. Basically, databases are placed in a national context, which, for instance, feed in the CO_2 equivalents due to the transport effort of a country. The data sources can be divided into the category-specific (or product) data and generic data. While the primary data refer to a specific product, the generic data are based on average values generated by averaging values from research and literature.

4. Entry format

In general, LCA is based on mass- and volume-related data. The input of this data can be determined either geometrically via 3D-geometric data or through spreadsheets. Some tools offer standard settings for this, others must be entered individually. The calculation of the environmental impact is based on the entered building mass and volume linked with the values from the stored databases.

5. Optimization

After LCA is completed, an optimization can be implemented, ideally. Here, the results can either be analysed manually to make changes in the design or the LCA is reassessed based on an algorithm. The façade element that is based on environmental data should be integrated into the design. This could refer to the type of material, for example, an optimization in respect to environmental criteria of insulation material could favour a wood-based product over a synthetic product. Functional criteria, like construction thickness or cost, are taken into consideration as part of the algorithm.

6. Default settings

Default settings provide a basic structure to facilitate the applicability of the LCA. The more default settings are predefined, the faster and easier it is to make initial statements. To achieve the most precise result, as many parameters as possible should be adjusted manually.

7. Life cycle phases

As stated earlier, 17 life cycle phases are distinguished in the building context. (A specification for facades is not available.) There are only a few LCA tools that are capable of considering all phases. All LCA programmes include the production phase (phase A1−A3). A few programmes cover the usage phase of the building, and others include the exchange rate of elements due to maintenance. More detailed programmes additionally integrate the end-of-life flows 'C1 disposal' or 'D recycling potential'.

20.2.4 Introduction to life cycle assessment tools in façade design

When comparing LCA software it needs to be stated that the application in the building industry is still limited, particularly for façade assessment. The obstacles include added complexity to the design process and higher costs (Meex, Hollberg, Knapen, Hildebrand, & Verbeeck, 2018). However, some tools are available, and their number is growing. Tools specially designed for façade designers and engineers are not common, so either a façade engineering company develops their own calculation tool on top of other functionalities or uses existing tools that can be applied to façade design. This chapter refers to the latter.

In the following the most established LCA software are introduced. Table 20.1 gives an overview of the 11 analysed tools and compares their scope in the 7 categories considered previously. Fig. 20.9 provides an overview of the applications based on the evaluation of the categories. The first software was the German *Ganzheitliche Bilanzierung* (German Holistic Assessment) GaBi and the Dutch SimaPro. Both tools and the subsequent software Umberto combine LCA modelling and reporting, content databases with intuitive data collection. During this period, they have become expert tools for LCA to affect assessment of detailed

Table 20.1 Software programmes for life cycle assessment.

Software	Data source	Required user's knowledge	Access	Entry format	Level	Default settings and adaptability	Life cycle phases
Athena (impact estimator for buildings)	Athena database details	Basic (tutorials)	Free (registration needed)	Spreadsheet	Material, component	Default setting partly adaptable	A1–A3D, A4–A5, B1–B7, C1–C2, C3, C4, D
Bauteileditor eLCA	Ökobau.dat	No prior knowledge	Free (registration needed)	Values entered in default spreadsheet	Component	Default setting not adaptable	A1–A3, C3, C4, D
BEES	Environ. Perform. Score, Econ. Perform. Score	Basic, expert	Free (no registration)	Spreadsheet	Component	Default setting partly adaptable	A1–A3
CAALA	Ökobau.dat	No prior knowledge	Beta version is free	3D geometrical	Building	Default setting partly adaptable	A1–A3, C3, C4, D
Ganzheitliche Bilanzierung	GaBi Database, Ökobau.dat	Expert (tutorials and manuals)	Conditional access (30 days trial), paid access (license needed)	Spreadsheet with graphical elements	Material, component, add-on 'Built-it'	No default setting	A1–A3, A4–A5, C3, C4, D

(*Continued*)

Table 20.1 (Continued)

Software	Data source	Required user's knowledge	Access	Entry format	Level	Default settings and adaptability	Life cycle phases
Legep	Ecoinvent, Ökobau.dat	No prior knowledge, basic, expert	Paid access (license needed)	Spreadsheet	Component, building	Default setting adaptable, default setting partly adaptable	A1–A3, C3, C4, D
Umberto	GaBi Database, Ecoinvent	Expert (no tutorials or manuals)	Limited access (14 days free trial), paid access (license needed)	Spreadsheet	Material, component	Default setting adaptable	A1–A3, C3, C4, D
Simapro	Various	Basic, expert	Conditional access, paid access (license needed)	Values are entered in default spreadsheet	Component, building	Default setting adaptable	A1–A3, C3, C4, D
Tally	–	–	Conditional access	3D geometrical	Building	Default setting partly adaptable	A1–A3, C3, C4, D
360 optimi	Various, for example Ökobau.dat, EPD	No prior knowledge	Free access (registration part access), conditional access	3D geometrical	Building	Default setting partly adaptable	A1–A3, C3, C4, D

EPD, environmental product declarations.
Source: Reprinted from Bach and Hildebrand (2018). This work is licensed under a CC by 4.0 license. The original version can be found in: Bach, R. & Hildebrand, L. (2018). Review on tools for assessment of buildings' environmental performance. In S. Kosanovic, T. Klein, T. Konstaniou, A. Fikfak, & L. Hildebrand (Eds.), *Sustainable and resilient building design—Approaches, methods and tools.* Delft: BK Books.

Life cycle assessment in façade design

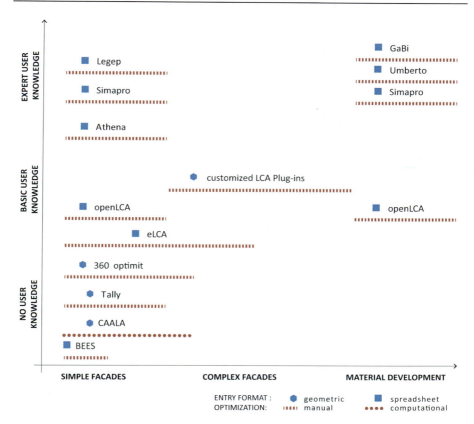

Figure 20.9 LCA software programmes in façade design. *LCA*, life cycle assessment. *Source:* From Linda Hildebrand. Reprinted and modified from Bach, Mohtashami, and Hildebrand (2018). This work is licensed under a CC by 4.0 license. The original version can be found in: Bach, R., Mohtashami, N., & Hildebrand, L. (2018). Comparative overview on LCA software programmes for application in the façade design process. *Journal of Façade Design and Engineering, 7.*

environmental product information (Betz, 2012; Goedkoop, De Schryver, Oele, Durksz, & De Roest, 2010; Prox, 2016).

The German software Legep was developed to support during the phases of design, construction and evaluation of new buildings and to compare alternatives. For example, a company would use it to analyse their building product or specialized architects evaluate their building design when most details are developed. At the same time the Canadian software Athena was developed to cover all planning and life cycle phases (Trusty & Meil, 1999). A large number of required parameters (eg, durability of elements, time span or reference unit) as well as the spreadsheet input restrict the application mainly to experienced users (Bach et al., 2018).

The growing interest in LCA data has simplified the application: GaBi developed the 'Build-it' interface, which provides iterative optimization steps for building elements (according to DIN 276). Several tools augment the input format of the spreadsheet with graphical elements. Furthermore, freely accessible software such as openLCA, which uses partly free databases, were established. Default settings for the input mask were offered to simplify and accelerate calculations, for example, eLCA and BEES, to strengthen applications as a basis for decision-making. The free access increased the motivation to integrate LCA into the German market (BBSR, 2016).

A 3D digital model provides the link to semantic information by building information modelling (BIM) with which the bill of material can automatically be exported. Software such as Tally, 360 Optimi and CAALA established the connection between 3D data and various LCA databases. They provide industry foundation classes interfaces so that a 3D model can be inserted and LCA values could be extracted. A growing number of add-ons for BIM-based design programmes such as Revit (Autodesk) or Rhino (McNeel) are available to optimize design decisions with LCA data (Bach et al., 2018).

The comparison of the different LCA tools also illustrates a range of different software to suit almost every level of user experience. While Gabi, Simapro, Umberto and Legep require a high level of user knowledge, CAALA, 360 Optimi and BEES provide a simplified application through default settings. Similarly, eLCA offers more customizable options but is still easy to use. eLCA provides a variety of facades that can be adapted to a specific design, and it is open access. The programmes differ in their application objectives: Tally and CAALA offer functional optimization by supplying geometrical data, while 360 Optimi is recommended for comparing results from different data sources. CAALA is designed to show optimization potential on the building level (when the best solution of operational and EE and emissions is found by its algorithm) but could be also used for facades. However, it is to be noted that the application of BIM software for façade design is not always practical, as the level of detail contained in full-scale drawings would require processing power exceeding typical desktop or laptop computers to be able to represent an entire façade. In addition, since most tools are spreadsheet-based transferring previously created, 3D geometries from the design phase into the software can be challenging. In addition, the use of 3D geometries allows an easier recalculation of the environmental impact.

20.2.5 *Brief summary of life cycle assessment databases*

Databases contain information on input and output flows at the material or component level, which are processed by software tools. Most software products support the import of all types of databases and provide an export in spreadsheet format. While single data can be opened with common tools like a web browser, PDF reader or Excel, the database (the multitude of data) requires software products. National databases have been published for greater clarity. For instance, Ökobau.dat, which was derived from GaBi data, was sold to the German government and later published as a freely accessible dataset containing general and specific information from EPDs. Today, many databases are available. The decision to use a particular database depends on the national context, as the data refer to the specific values of a country (Bach & Hildebrand, 2018).

20.3 Integrating life cycle assessment in the design process of facades

Designing a façade that includes environmental aspects starts with a client that is demanding this particular quality for their building. (This can change when LCAs for facades become mandatory.) Architects and engineers need to sensitize the client to the relevance of this subject. Either an expert for LCA and circularity is included in the team of specialists and planners or the façade engineer has skills in this discipline. In both cases the façade engineer needs to be aware of the relevance of the façade typology regarding the environmental range that is predefined by its choice. Furthermore, it can help to clarify the design strategy as it can impact the design process due to necessary additional steps (such as extra research). Once the framework for the design is set, it is useful to conduct an LCA that could be done manually. A variety of LCA software and databases are available, as well as scientific studies.

The application in façade design is not yet common but is growing. In façade design, LCA tools are either used by industry, architects or façade engineers. For the façade industry, it is useful to either work with a customized LCA tool that is integrated into other functions like costs or structural calculations or to use software that focuses on the production process like openLCA, GaBi or Legep. This is also true for façade engineers as the level of detail is similar. The difference relates to optimization; while the industry can improve the energy source for the production or use more efficient machines, the façade engineer can impact the geometry (by reducing the volume) and the material. The exact calculation and therefore the use of software make sense when different design options are compared against each other, and most details are developed. Façade engineers can compare material alternatives to understand benefits and challenges in the context of their functionality. Environmental performance is hereby negotiated with more traditional performance criteria like fire, thermal and acoustics. In the technical design phase where the materials are defined and the details developed, LCA can be used to compare different products. The (renewable) energy source of the façade material production can contribute to a significant reduction of environmental impact as well as transport distances and the type of transport (Cortes Vargas, Hildebrand, Rammig, & Zani, 2021). The geographical preferences (system-/product-based approach in Europe, bespoke approach in the United Kingdom and the United States) are reflected in the data availability and need to be integrated into the type of assessment. This also counts for green building certificates (such as BREAAM, LEEDS, GreenStar, NABERS) that have specific requirements for the assessment of EC and

EE. LCA can initiate close cooperation with industry and suppliers to make comparable data available for different markets.

In academia the discussion on LCA in buildings and especially facades started at the beginning of this century. The main fields of interest can be grouped in:

1. comparison of construction and material alternatives (sometimes in cooperation with industry research) (Gustavsson, Pingoud, & Sathre, 2006; Kim & University of, 2009; Wadel & Gerardo, 2013; Krug & Ullrich, 2010; Müller et al., 2007),
2. integration into design phase (including tools, legal framework for closed-loop approach or a methodology for the end-of-life phase) (Azcarate-Aguerre, Den Heijer, & Klein, 2017; Hollberg, 2016; Meex et al., 2018) and
3. discussing indicators to quantify environmental impact (Ismaeel, 2018; Nissinen, Suikkanen, & Salo, 2019; Passer et al., 2015).

Answers to the open question in designing the environmental impact of facades need to balance accuracy and applicability at the same time. Knowledge on low-carbon solutions is available but often not implemented. The first two fields have a proximity to design practice and are therefore very relevant to architects and engineers. The later includes the political discussion and the tendencies to provide a broader variety of indicators that would lead to more accurate results but at the same time to increased complexity.

More research is needed to overcome the long usage cycles in the façade industry (belonging to the second field). New liability concepts (eg, via a third party or within a company) need to be developed with industry partners that include financial incentives for all stakeholders.

20.4 Conclusion

The growing trend to integrate LCA in façade design is helping to unlock the environmental potential bound in the material and construction. To be efficient, knowledge of the interdependencies of typology and EE/EC is key. LCA tools provide quantified arguments for choice of material and construction and, later in the process, also for the specific façade product.

The integration of LCA in façade design needs to be more widespread. For this, (reliable) environmental data need to be accessible for different national contexts.

Either façade engineers extend their competencies in this field or an LCA/circularity expert is included in the project team.

A full assessment needs to include the postuse phase according to the particularity of the building sector. The ability for reuse and recycling has to be reflected as a parameter in the design process. The available methods in the postuse phase need to be translated into software to evaluate the design regarding circularity.

By this, an incentive can be created to initiate new use models based on the information from LCA and circularity. These models could, for example, take up leasing concepts, whereby manufacturers take back their products after the usage and thus bear the responsibility for postuse.

Life cycle assessment in façade design 515

In conclusion the LCA method represents a major step toward revolutionizing environmental design in facades as it helps to quantify and compare different alternatives objectively. The façade industry as well as designers and engineers can obtain a holistic perspective with LCA, including the postuse scenarios, so that a circular building construction is reached (Anderson and Shiers, 2009).

References

Anderson, J., & Shiers, D. (2009). 28 Jan *The green guide to specification: Breeam specification*. Wiley-Blackwell.

Azcarate-Aguerre, J. F., Den Heijer, A., & Klein, T. (2017). Integrated facades as a product-service system: Business process innovation to accelerate integral product implementation. *Journal of Façade Design and Engineering*, 6, 41–56.

Bach, R., & Hildebrand, L. (2018). Review on tools for assessment of buildings' environmental performance. In S. Kosanovic, T. Klein, T. Konstaniou, A. Fikfak, & L. Hildebrand (Eds.), *Sustainable and resilient building design—Approaches, methods and tools*. Delft: BK Books.

Bach, R., Mohtashami, N., & Hildebrand, L. (2018). Comparative overview on LCA software programs for application in the façade design process. *Journal of Façade Design and Engineering*, 7.

BBSR (2016). Ökologische baustoffwahl—Aspekte zur komplexen planungsaufgabe 'schadstoffarmes bauen'. In BBSR (Ed.), *Zukunft bauen, forschung für die praxis bd. 4*. Berlin.

Betz, M. (2012). *GaBi 5*. Stuttgart: Pe International. Available from http://www.gabi-software.com/deutsch/software/gabi-software/gabi-5/.

Beuth. (2006). *Environmental management—Life cycle assessment—Principles and framework (ISO 14040:2006); German and English version EN ISO 14040:2006*. Berlin: Beuth.

Cortes Vargas, T. C., Hildebrand, L., Rammig, L., & Zani, A. (2021). Carbon conscious! The impact of embodied emissions on design decisions for building envelopes. *Journal of Façade Design and Engineering*.

Dechantsreiter, U., Horst, P., Mettke, A., Asmus, S., Schmidt, S., Knappe, F., ... Lau, J. J. (2015). *Instrumente zur wiederverwendung von bauteilen und hochwertigen verwertung von baustoffen dessau roßlau: Umweltbundes amt für mensch und umwelt*.

DIN, N.B.I. (2012). *Nachhaltigkeit von bauwerken—Umweltproduktdeklarationen. Grundregeln für die produktkategorie bauprodukte; Deutsche fassung EN 15804:2012*. Berlin: DIN Deutsches Institut für Normung e. V.

Frischknecht, R. (2009). *Umweltverträgliche technologien: Analyse und beurteilung. Teil 2: Ökobilanzen*. Zürich.

Geddes, P. (1884). An analysis of the principles of economics. In *Proceedings of the Royal Society of Edinburgh* (Vol. XII; pp. 943–980).

Goedkoop, M., De Schryver, A., Oele, M., Durksz, S., & De Roest, D. (2010). In P. E. Consultants, (Ed.), *SimaPro7 introduction into LCA*. Amersfoort: PRé Consultants bv.

Guardigli, L., Monari, F., & Bragadin, M. A. (2011). Assessing environmental impact of green buildings through LCA methods: A comparison between reinforced concrete and wood structures in the European context. *Procedia Engineering*, 21, 1199–1206.

Gustavsson, L., Pingoud, K., & Sathre, R. (2006). Carbon dioxide balance of wood substitution: Comparing concrete- and wood-framed buildings. *Mitigation and Adaptation Strategies for Global Change, 11*, 667−691.

Hartwig, J., & Hildebrand, L. (2017) Nutzung von ökobilanzen im planungsprozess zur kurz- und mittelfristigen realisierung positiver umwelteffekte in. In: IBO (Ed.), *BauZ wiener kongress für zukunftsfähiges bau.* Vienna, Austria.

Heesbeen, C., Zabek, M., & Hildebrand, L. (2021). A definition of essential characteristics for a method to measure circularity potential in architectural design. *Journal of Façade Design and Engineering.*

Hildebrand, L., & Hollberg, A. (2018). Methodology for assessing environmental quality of materials and construction. In S. Kosanovic, T. Klein, T. Konstaniou, A. Fikfak, & L. Hildebrand (Eds.), *Sustainable and resilient building design − Approaches, methods and tools.* Delft: BK Books.

Hildebrand, L. (2014). *Strategic investment of embodied energy during the architectural planning process.*

Hildebrand, L., Kosanovic, S., Cukovic-Ignjatovic, N., Bach, R. & Radivojevic, A. (2017). *Methods and tools for integrated design.*

Hildebrand, L., Schwan, P., Vollpracht, A., Brell-Cokcan, S., & Zabek, M. (2017). Methodology to evaluate the building construction regarding the suitability for further application. *Creativity game, 5.*

Hillebrandt, A., Riegler-Floors, P., Rosen, A., & Seggewies, J.-K. (2018). *Recycling atlas: Gebäude als materialressource.* München: Edition Detail.

Hollberg, A. (2016). Parametric life cycle assessment (PhD). Bauhaus Universität Weimar.

Ismaeel, W. S. E. (2018). *Midpoint and endpoint impact categories in green building rating systems.*

Kellenberger, D., & Althaus, H.-J. (2009). Relevance of simplifications in LCA of building components. *Building and Environment, 44*, 818−825.

Kim, K.-H., & University of, M. (2009). *Structural evaluation and life cycle assessment of a transparent composite façade system using biofiber composites and recyclable polymers.*

Knaack, U., Klein, T., Bilow, M., & Auer, T. (2007). *Fassaden: Prinzipien der konstruktion.* Basel: Birkhäuser.

Krug, R., & Ullrich, B. (2010). *Nachhaltigkeitsstudie-ökobilanzen von fassadenkonstruktionen mit naturstein und glas.* Würzburg: Deutscher Naturwerkstein Verband.

Kümmel, J. (2000). Life cycle assessment of construction materials in recycling structural light weight aggregate concrete (PhD). Universität Stuttgart.

Meex, E., Hollberg, A., Knapen, E., Hildebrand, L., & Verbeeck, G. (2018). Requirements for applying LCA-based environmental impact assessment tools in the early stages of building design. *Building and Environment, 133*, 228−236.

Müller, M., Schlüter, C., Fehrmann, R., Voss, K., Morhenne, J., Engelmann, P., ... Kaiser, C. (2007). *Ökologische/ökonomische bewertung zweier fassadenkonzepte—Glasfassade vs kunststofffassade zur sanierung eines verwaltungsgebäudes der 1960-er jahre.* Remscheid: Deutschen Bundesstiftung Umwelt.

Nagus. (2006). *Environmental labels and declarations—Type III environmental declarations— Principles and procedures (ISO 14025:2006).* Berlin: Normenausschuss Grundlagen des Umweltschutzes (NAGUS) im DIN, *ISO/TC 207/SC 3 'Environmental Labelling'.*

Nissinen, K., Suikkanen, J., & Salo, H. (2019). *Product environmental information and product policies: How product environmental footprint (PEF) changes the situation?.*

Otto, I. M., Donges, J. F., Cremades, R., Bhowmik, A., Hewitt, R. J., Lucht, W., ... Schellnhuber, H. J. (2020). Social tipping dynamics for stabilizing Earth's climate by 2050. *Proceedings of the National Academy of Sciences of the United States of America, 117*, 2354−2365.

Passer, A., Lasvaux, S., Allacker, K., de Lathauwer, D., Spirinckx, C., Wittstock, B., ... Wallbaum, H. (2015). Environmental product declarations entering the building sector: critical reflections based on 5 to 10 years experience in different European countries. *The International Journal of Life Cycle Assessment, 20*, 1199−1212.

Prox, M. (2016). *Know the flow-sharing experience for sustainability & productivity, factor for success of LCA: Data, data and again data* [Online]. Hamburg. Available: https://www.ifu.com/knowtheflow/2016/erfolgsfaktoren-fur-okobilanzen-life-cycle-assessment-lca-daten-daten-und-nochmals-daten/ [Accessed].

Spreng, D. (1995). *Graue energie-energiebilanzen von energysytemen.* B.G. Teubner Stuttgart.: Zürich, Hochschulverlag AG an der ETH Zürich.

Trusty, W. & Meil, J. (1999). Building life cycle assessment: residential case study. In *Proceedings: Mainstreaming green: Sustainable design for buildings and communities.* Chattanooga, TN.

Union, E. (2014−2017). European platform on life cycle assessment *[Online]*. Brussels: European Union Available. Available from http://eplca.jrc.ec.europa.eu.

Urszula, K. (2019). Circular design: reused materials and the future reuse of building elements in architecture. *Process, Challenges and Case Studies, 225.*

Wadel Raina, G. F., Alonso, P., Zamora i Mestre, J.-L. s., & Garrido Torres, P. (2013). Simplified LCA in skin design: The FB 720 case. *International Journal of Sustainable Building Technology and Urban Development, 4*(1), 68−81.

Circular economy in facades

21

Mikkel K. Kragh[1] and Nebojša Jakica[2]
[1]Department of Civil and Architectural Engineering, Aarhus University, Aarhus, Denmark,
[2]Department of Technology and Innovation, University of Southern Denmark (SDU), Odense, Denmark

Abbreviations

3R	reduce, reuse, recycle
AI	artificial intelligence
BAMB	buildings as material banks
BIM	building information modelling
BIPV	building-integrated photovoltaics
BM	business model
C&D	construction and demolition
C2C	Cradle-to-cradle
CE	circular economy
CEBMs	circular economy business models
CLC	closed-loop cycles
DfD	design for disassembly
GDP	global domestic product
IEQ	indoor environmental quality
IGU	insulating glazing units
LCA	life cycle assessment
MEP	mechanical, electrical and plumbing
RFID	radio-frequency identification
UMAR	urban mining and recycling
XR	extended reality

21.1 Introduction

Increased human activity and rapidly rising global population put enormous pressure on energy supply, which results in increasing consumption of fossil fuels and urgency to transition to renewable sources. However, carbon emissions are just part of the challenges we face in an era of climate change and the biodiversity crisis. Within the broader sustainable development agenda, there are many actions and approaches that have the potential to create a perfect storm for a rethink of design and construction processes. The global challenges require new mindsets

Rethinking Building Skins. DOI: https://doi.org/10.1016/B978-0-12-822477-9.00016-4
© 2022 Elsevier Inc. All rights reserved.

and—fortunately—present opportunities for professionals, businesses and society. *Circular economy* (CE) is set to fundamentally alter how we consider resources and is likely to require radical rethinking across disciplines.

The concept of circularity and interdependence between the environment and the economic system has been known for years (Pearce & Turner, 1990). The importance for the construction industry to adopt CE principles may be seen through the statistics. The construction sector is one of the largest industrial sectors, 9% of the *global domestic product* (GDP) of the *European Union* (European Commission, 2016) and 6% of global GDP (Deloitte, 2020). It uses 30%–50% of the raw materials and produces about 40% of waste to landfill (Walker, Grant, & McAlister, 2007) while generating close to 40% of all greenhouse gas emissions in Europe (European Committee for Standardization (CEN), 2017) and globally (UN Environment & International Energy Agency, 2017). Within the construction sector, six areas of CE are identified: development of CE, reuse of materials, material stocks, CE in the built environment, *life cycle assessment* (LCA) analysis and material passport (Benachio, Freitas, & Tavares, 2020). Above all, government policies, such as legislation and tax incentives, are crucial in the transition to a CE (Munaro, Tavares, & Bragança, 2020; Stanojev & Gustafsson, 2019).

Building skins constitute an intricate part of architecture, which affects—and is affected by—a range of disciplines and many different parts of the value chain. Moreover, facades have a relatively limited service life compared with the main structure of a building. This not only means that the impact of the building skin will be significant in the short term, but it also means that the impact of the façade includes maintenance and possible replacement over the service life of a given building. Therefore it is not only challenging but also potentially very important to resolve CE in facades with a view to transitioning from the take−make−waste mindset to one of reduce−reuse−recycle.

21.2 Reduce, reuse, recycle

Over the last decades, economic and ecological crises significantly influenced architecture that began to embrace sustainability and ecology as key aspects of design. In combination with the economy, circular principles signify a reduction in wasteful deployment of resources through efficient design and implementation of products and processes for improved resource-efficiency in a closed-loop material flow (Jawahir & Bradley, 2016). The environmental impact has been commonly shown by means of LCA (Guinée et al., 2011; The European Committee for Standardization (CEN), 2010), tracking carbon emissions and embodied carbon through four stages of a life cycle (premanufacturing, manufacturing, use and postuse). Numerous research studies have focused on the assessment of environmental impact through LCA (Bolin & Smith, 2010; Jönsson, Björklund, & Tillman, 1998). Contrary to the predominant, linear *take−make−waste* model (De Wolf, Hoxha, & Fivet, 2020), the concept of CE has been in general subsequently associated with the 3R: *reduce, reuse, recycle* (Kirchherr, Reike, & Hekkert, 2017).

The building envelope is estimated to represent the largest single contribution to the total embodied carbon in buildings (Cole & Kernan, 1996). The embodied carbon is a measure of the carbon emissions associated with the production and installation of a building envelope. Even in the extended version of 3R, the concept of *reduce* is integral to all stages of the life cycle, whereas the first necessary step in the postuse stage is *recover* from which other four innovation-based Rs originate (*reuse, recycle, redesign* and *remanufacture*) (Jawahir & Bradley, 2016). Although the method is widely adopted, LCA tools still require development. For several reasons the tools still do not fully suffice in the design stage: data intensiveness, lack of available data at the design stage, and to some extent, the lack of knowledge among decision-makers as regards how to carry out and interpret LCAs (Malabi Eberhardt, Rønholt, Birkved, & Birgisdottir, 2021).

Although *reduce* and *recycling* have been the primary focus of the debate on sustainability over the last century, in a building context, designing and building for *Reuse* of components might actually have a more profound effect on CE. A case study found that *Reuse* offsets greenhouse gas emissions by 88% and, therefore, environmental benefits of reuse practices could far surpass recycling, as shown on the *legacy living lab* modular building (Minunno, O'Grady, Morrison, & Gruner, 2020). Contrary to *Reduce* and LCA, where the focus is to reduce negative environmental impact, the concept of *Reuse*, however, proposes a positive vision of a future approach described through the concept of *cradle-to-cradle* (C2C) certified product standard (MBDC LLC, 2014) and a concept of *closed-loop cycles* (CLC) in the case of multiple C2C cycles. Therefore the C2C approach considers not just the service life of buildings and their components like LCA, but the whole C2C or CLCs that include *Reuse* as an alternative to *Recycling* that prolongs building component service life (Bjørn & Hauschild, 2017). Both LCA and C2C can be used to cross-check results and offer different insights and perspectives, as shown in a study of reuse of exterior wall framing systems (Cruz Rios, Grau, & Chong, 2019). The study tested the benefits of reusing a given material with high embodied energy compared with a single-use alternative. The analyses showed that process-based LCA and hybrid LCA could generate conflicting results in a C2C LCA. Furthermore, it was found that the *Reuse* benefits depend on multiple reuses, aggressive reuse rates (in excess of 70%), short transportation distances and the embodied energy of thermal insulation materials. The study suggests that materials with high embodied carbon and water consumption should be designed for future reuse. In the case of the material steel, for example, this would require the design of bolted connections as opposed to welded ones.

Today, the importance of *Reuse* is emphasized through policy-making (Ellen MacArthur Foundation, 2015; Ministry of Environment & Food and Ministry of Industry, 2018), industry (Lendager Group, 2019; Vandkunsten Architects, 2016; 3XN, 2016) and research (Malmqvist et al., 2018; Nußholz, Rasmussen, Whalen, & Plepys, 2020). Furthermore, a number of services, products and processes have been developed to further the development of CE (European Commission, 2020, 2021).

However, despite all of the benefits of CLC and C2C, the industry still largely relies on recycling as the dominant waste management option. Recycling induces complete remanufacturing of the material, requiring energy (eg, steel melting) or downcycling of materials (eg, concrete crushing). *Reuse* on the other hand entails a minimized amount of transformation by using the component again with its original features (De Wolf et al., 2020).

The adoption of the 3R design approach is currently limited as there are barriers for the implementation in the façade industry. The barriers are mainly due to limited return on investment, the lack of infrastructure and inappropriate design of buildings for material recovery (Adams, Osmani, Thorpe, & Thornback, 2017; Hart, Adams, Giesekam, Tingley, & Pomponi, 2019). Moreover, 3R for toxic waste management is still underregulated, and further attention will be necessary to ensure the same treatment and the quality and safety protocols for such materials. Specification of recyclable materials, their origin and exposure to potentially hazardous substances and agents should be introduced and regulated. Therefore while the adoption of CE in the façade industry is an urgent matter for environmental, social and economic reasons, much remains to be done (De Wolf et al., 2020). Hopefully, the necessary adoption may happen through increased integrated efforts across sectors, governance levels and stakeholders, pushed by the climate change agenda.

21.3 Design for disassembly and maintenance

Increasing concerns over intensive consumption of resources and low recycling rates have put pressure on all facets of the façade industry. One of the fundamental principles of the CE is *design for disassembly* (DfD). This chapter is discussing specifically the application of DfD in the field of facades. The concept implies the design of facades that facilitate future changes, maintenance and disassembly and guarantees multiple life options of facades and their systems, components and materials. It advocates throughout the design process the consideration of end-of-life options (demolition) that ensures minimized waste. The nature of the DfD process is essentially alternating between construction and deconstruction, making possible reversible design and construction. The philosophy of reversible construction may significantly impact on both design processes and *business models* (BMs) of construction. The circularity of the reversible façade design process implies transformability at both technical and material levels. Such transformation opens up opportunities for new value propositions throughout the façade life cycle. Design goals and actions could be harmonized as proposed by Durmisevic (2018, 2019a,b), where an integrated design protocol integrates aspects of transformation and building-level reversibility with technical aspects related to the recovery and high reuse potential of products. In a business-as-usual scenario, the added value of disassembly is not easily assessed and monetized, which, in turn, poses both challenges and opportunities in a possible green construction transition. The building skin is an architectural element that offers excellent potential for disassembly, provided that the design processes are developed to include end-of-life.

Technical aspects determine the transformability of façade technology. These aspects range from periodic maintenance to full disassembly. The following aspects are particularly important: structural integrity, design of connections, accessibility, safety, *mechanical, electrical and plumbing* (MEP) systems integration, standardization and interchangeability. The structural integrity of the façade during maintenance, repair and 'reskinning' determines the potential for technical transformation prior to, and after, the end-of-life. Therefore masonry facades have a longer lifespan, but low transformability, compared with curtain walls with a shorter lifespan, but high transformability. For load-bearing masonry facades, interventions may include structural enhancements, humidity control and vapour barrier integration, resurfacing on both interior and exterior for weather and impact protection, and most commonly the replacement of *insulating glazing units* (IGU), glass coatings and frames for thermal performance enhancement. The end-of-life of solid masonry facades usually coincides with the end-of-life of the building, since a masonry wall façade is usually an integral part of the building's structure. In the case of a rainscreen façade, thermal insulation and protective layers may be replaced and upgraded without interventions to the structural façade layer which, as a consequence, extends the lifespan of the building. The differences in potential for disassembly may occur in the technology that is used to support rainscreen cladding in cold and warm facades. Warm facades are insulated, but not ventilated, and usually have lower disassembly potential since the protection layer is frequently glued or directly attached to the insulation layer. Cold facades are ventilated and usually consist of components that are supported by brackets and detached from the insulation layer. Therefore the disassembly process for cold rainscreen facades is highly transformable as it represents a reversible construction process.

On the other hand, the transformability of curtain walling varies greatly depending on the specifics of the assembly and can be from partial to complete, as it may be advantageous to transform parts of the assembly or the entire system. Furthermore, since curtain wall facades are not a part of the structural building system, their transformability is independent from that of the building. Maintenance mostly targets thermal performance enhancements, including infiltration, gasket replacement, glass delamination and glass breakage. Similarly to the construction process, accessibility and safety are essential aspects of maintenance and disassembly of stick systems and unitized curtain wall facades.

Naturally, the viability of DfD strongly depends on the connection details. The DfD potential is higher when connection details rely on mechanical fixings, such as bolts, screws and nails, rather than chemical bonding, sealing and gluing. Structural silicone adhesives, bonding, glass lamination, ceramic frit coating and sealing of IGUs limit their disassembly and recyclability potential. Connections that cannot be disassembled such as structural curtain walling should preferably be avoided and replaced with solutions that allow for maintenance and servicing to extend the desired performance and service life. Solutions for IGUs may include refilling the cavity with gases, while structural silicones may be replaced with hidden or minimal mechanical fasteners. When considering MEP systems integrated in facades, the complexity of installations and potential technological obsolescence present challenges to long-term maintenance. Electrochromic glazing and *building-integrated photovoltaics* (BIPVs), despite their benefits in terms of overall building energy balance, are challenging to maintain, replace, disassemble and recycle. DfD should focus on more flexible, standardized and interchangeable technical solutions that take into account multiple extensions of their life span through maintenance, zero-waste disassembly and potential adaptive reuse. The whole life performance should be considered and the energy performance in use should be taken into account when considering design options.

Material aspects of DfD rely on the precautionary principle, as suggested by the Hippocratic oath, of 'first, do no harm'. Eliminating materials that emit volatile organic compounds—for example, from paints and varnishes, the toxicity of impregnating agents, materials that release toxic fumes during a fire, asbestos-based materials and radioactive materials—is one of the keys in sustainable constructions (Pacheco-Torgal & Jalali, 2011). In general, the façade industry should rethink the criteria for choosing materials that go beyond aesthetics, thermal performance and cost and rather focus more on material durability, low maintenance and DfD to minimize hidden operational and environmental costs. An Autopsy—Façade teardown at Frener & Reifer case study (Arup, Systems, & WithinLab, 2015) demonstrates one such integrated approach that revolves around DfD and recyclability. Furthermore, it emphasizes the required additional effort of integrating stakeholders through events such as workshops and factory visits to facilitate informed decision-making.

21.4 Material passports, material banks

Deconstruction can permit an appropriate recovery of components and materials for either reuse or recycling in the most cost-effective manner and with a substantial reduction of *embodied greenhouse* gas emissions (Malabi Eberhardt, Rønholt, Birkved, & Birgisdottir, 2021). For this reason, in the framework of DfD, the potential reusability of materials and components shall be planned globally, in order for all of the phases of collection, assessment, management, reassembly, or recycling of materials and components to realize cost-effectiveness, material-effectiveness and a high degree of reusability during the construction and disassembly phases.

An independent agent or contractor acting as a materials and component bank could handle all of the businesses involved in the construction industry, as an integrated part of a global strategy for CE. The agent could organize the transfer of materials and components extracted from demolished or deconstructed structures to a new structure and help manage the dismantling and postdismantling processes, as a database/management agency of materials and components that promotes their direct reuse in the next cycle of construction activities (Cai & Waldmann, 2019).

Nevertheless, creating this building database requires an in-depth exploration of the makeup of building systems, as not all components can be easily dismantled and reused. This depends on their degree of interdependency, as the more a component is independent, the easier will be its alteration or replacement without affecting other components.

Building skins play an important role as mediators between the inside and the outside environment, controlling the flow of heat, light, noise, information and other media. Once they enter a material bank, when appropriately designed for DfD and as independent elements, they could be technologically improved and adapted. This would potentially make them suited for either new constructions or for existing buildings in need of retrofitting, thanks to their high level of construction adaptability. In any case, in DfD it is of utmost importance to obtain detailed knowledge about buildings' materials and components to programme possible, alternative future scenarios. The availability of certified information about components and materials is a precondition for future maintenance, next reuse and recycling. For this reason the concept of the material passport becomes relevant.

As the material passport concept is relatively new, there are only a few studies on the topic so far. One study defines the material passport as a design optimization tool and an inventory of all materials embedded in a building, that displays its recycling potential and environmental impact (Honic, Kovacic, & Rechberger, 2019).

Building information modelling (BIM) is a technology capable of handling design decisions and information during the whole building life cycle that can help turn buildings into banks of materials and communicate component information. It can bring together the entire supply chain thanks to interoperability and enable the end customer to know what is in the building and what the building and its components have been used for (Ellen Macarthur Foundation & Arup, 2019), potentially contributing to waste minimization (Akinade et al., 2015).

Some studies combined BIM technology and *radio-frequency identification* (RFID) (Naranje & Swarnalatha, 2019; Xue, Chen, Lu, Niu, & Huang, 2018) to create a cyber-physical system, forming a bridge between the physical and the virtual tool of DfD. RFID is a system that uses radio waves to read and capture data, and it has been widely used already in different sectors of industry (Teknologisk-Institut, 2016). In the framework of DfD, its adoption could be useful in resource management, logistics, tracking processes and safety, as it would carry and store materials' and components' ID.

In this regard a recent study by Xue, Chen, Lu, Niu, & Huang, 2018 theorized a common operative framework for the construction sector that harnessed the power of using BIM and RFID combined. Their application could be even more impactful when dealing with adaptive facades. Real-time information regarding changes in their environments and climatic loads and past events could be stored to improve their future performance or maintain serviceability. In this way, manufacturers could gain real-time feedback, adapting dynamic settings of facades systems and components.

However, the façade industry still has a long way to go and, in any case, the transition toward circularity still requires a deeper understanding of material flows and stock (Honic et al., 2019). Nevertheless, research goes ahead: several models have recently been experimented to estimate the existing or future quantity of materials reused on buildings at their end-of-life at a city scale (Gepts, Meex, Nuyts,

Knapen, & Verbeeck, 2019; Oezdemir, Krause, & Hafner, 2017). Even if the methods are different, both groups agreed on the feasibility to create a city-scale material stock from the existing data, even though they still need more research on material properties to reliably estimate which ones could be reused, entering the material banks of cities.

Currently, policy-making and research are headed in this direction. The BAMB project (*buildings as material banks*) attempts to create ways to understand and increase the value and reuse of building materials so that they can finally sustain their value, avoiding waste of resources. In this regard the city of Reburg was introduced as a future circular city concept—the world's most circular city (Reburg—BAMB, 2020).

In Switzerland the project by Experimental Unit NEST [UMAR—Urban Mining & Recycling (UMAR), 2018] emphasized the possibility for the building to be at the same time a material laboratory, temporary material storage and a public repository of information. Heisel and Rau-Oberhuber (2020) had analysed this project with Madaster (European Commission, 2021) and demonstrated at the same time not just its circularity (96%), but also the reliability of the Madaster Circularity Indicator as a design tool and the capability of materials passports to document material stocks and flows within a circular built environment. Madaster is the register for materials and products. In the online platform, buildings are registered including the materials and products that are used. However, so far within the Madaster database, there is no other circular case study with a comparable level of detail to that of UMAR. This shall boost the development of more detailed information regarding buildings of/for the future and their further analysis in terms of circularity.

21.5 Circular business models

Exploiting competitive advantage produced through value creation represents the core of every business. A BM defines the architecture of the value creation, delivery and capture mechanisms it employs. Furthermore, a BM is conceptual rather than merely financial (Teece, 2010). Fundamentally, value creation consists of delivering value to customers, convincing customers to pay for value and finally turning those payments into profit. In market economies with free consumer choice and diversity of consumers, products, and companies, BMs constantly evolve to satisfy market demand. Particular industries share the same BMs over the decades until new BMs challenge the old ones and force the industries to adapt. Traditionally, buildings have been understood as static objects that undergo specific and closed processes with individual time schedules of interaction by diverse stakeholders. Until the introduction of modular prefabricated construction, the façade industry relied on on-site construction with associated BMs dependent on local labour and technology. Later on, unitized curtain wall systems offered a new BM with a value proposition, including quality, safety, installation speed and outsourcing to locations with lower costs of labour. Today, the magnitude of the climate and biodiversity crises pushes the industry to go beyond technical solutions for products and establish sustainable business and economic incentives for present and future stakeholders to manage them in circular ways (Witjes & Lozano, 2016).

Most façade BMs originate from a physical end product BM, invented centuries ago, where costs are calculated based on a presumption that an end façade product value reaches its peak when installed. Durability and operational costs are usually neglected in favour of lower capital cost of installed façade systems. However, new high-tech adaptive facades exploit advantages in operational costs, increased human comfort and building energy performance and even energy generation from BIPVs. Such advanced façade systems emphasize value creation over the product life cycle. Value gains through energy generation by BIPV might lead to *zero-energy buildings* where operational energy costs are entirely offset by the energy that is generated on-site. In some cases, depending on the energy policies, *plus-energy buildings* value creation over the building life cycle might even lead to offsetting the capital cost and making green investment financially very attractive. Apart from the electricity revenues, green identity has also been recognized as a crucial value increasing aspects that allow the selling and renting of green buildings at a premium (Macé et al., 2018).

When considering a CE perspective and a CLC basis, *circular economy business models* (CEBMs) are still in their infancy. The shift toward a CE is a complex process that cannot be realized by product and process innovation alone but rather by profoundly altering the logic of value creation underlying current production and consumption systems (Roome & Louche, 2016). In the realm of circularity the concept of value is understood more broadly to encompass a wider range of stakeholders, such as value chain partners, the environment and society (Bocken, Short, Rana, & Evans, 2014). Hence, defining CEBMs depends on the value chain structure, identifying material reuse processes, transportation distances, site conditions and quantities of materials.

One research project suggests that the most common and successful BMs for capturing value in a CE are resale-, internalization- and performance-based (Hopkinson, De Angelis, & Zils, 2020). The performance-based model is particularly interesting as it captures value through pay-per-use revenue systems, with customers benefitting from reduced upfront costs and manufacturers from control over the product, components and materials during the usage cycle (Hopkinson, Chen, Zhou, Wang, & Lam, 2018). Similarly, the product service system exploits leasing and sharing (as opposed to purchasing), aiming to slow down the general rate of resource consumption and to create the highest possible use of value for the longest possible time while consuming as few material resources and energy as possible (Mont, 2002). Therefore CEBMs tend to shift from selling volumes to selling functions.

Although other industries effectively use this model (Han, Heshmati, & Rashidghalam, 2020), the façade industry still lags in implementation. Nevertheless, the façade as a service model has recently shown its practical application to a real project. Starting from September 2016, Delft University installed a pilot project temporarily replacing a section of the façade of one of its campus buildings to develop a circular BM based on the use of multifunctional facades as performance-delivering tools (Azcarate-Aguerre, Andaloro, & Klein, 2018, 2021). In this scenario the client is no longer the owner of physical building components but instead leases them from a service provider who takes responsibility for the product to perform over its entire operative life. This system has plenty of positive side effects, but it still asks for a complete rethinking of the entire social, economic and legal background on which our engineering and construction projects are based.

Hence, the CE transition demands innovation far beyond the scope of technological advancement and a revision of concepts such as utilitarian value, ownership and bankability. Nevertheless, projects as the one described earlier attract a growing interest of academic and professional partners and display how these barriers are not insurmountable, as they can involve at the same time investors, bankers, contract managers, legal advisors and property developers.

21.6 Circular design processes

The circularity of the design process (as opposed to the linear one) is probably one of the most overlooked aspects of CE. The combination of material flow, information flow and value creation requires new processes where C2C thinking is at the core and forms the basis of design decisions. It is found that poor design decisions are responsible for up to one-third of the *construction and demolition* (C&D) waste (Osmani, Glass, & Price, 2008).

Looking back in history, in a vernacular construction process, knowledge creation and information sharing happened spontaneously and had been accumulating over extended periods. Scarcity of resources in the past, and still this day in developing countries, has imposed various limits that have led to symbiosis and dependence on local builders' skills, local sourcing of materials and available technologies. UNESCO world heritage sites like Chan Chan in Peru, Old Towns of Djenné in Mali, Uruk in Iraq and Old Walled City of Shibam in Yemen demonstrate vernacular brilliance harmonizing buildings with their respective environments and climates with a minimal environmental impact. Therefore the nature of the vernacular construction processes could be seen as intrinsically circular.

The disruption in circular practices started to happen with an influx of technological innovations such as internal combustion engines, artificial lighting and mechanical ventilation. Early enthusiasm, fuelled with rapid technological development and an apparent abundance of natural resources, even resulted in showcase buildings that were utterly airtight, mechanically ventilated and artificially lit, and therefore potentially wholly isolated from the urban context and climate (Leslie, Panchaseelan, Barron, & Orlando, 2018). Concerns about climate and energy shortages occasionally happened after *World War II*, as the *Organization of the Petroleum Exporting Countries* energy crisis of 1970s, but they did not last long enough to shift the design processes toward sustainability completely. However, the

Olgyay brothers' work (Olgyay & Olgyay, 1957) followed by Givoni (1992) provided rules-of-thumb for bioclimatic architecture. This design process may be considered circular due to the reusability of successful design strategies that minimize environmental impact.

With the rapid advances of information and communication technologies starting from mid-1990s, many industries that adopted these technologies have seen a boost in productivity, while the construction industry has stagnated and, over some periods, even recorded negative growth rates (Barbosa et al., 2017; Fulford & Standing, 2014). Ever since, the construction industry has seen unprecedented challenges to digitalize and improve efficiency and its usage of resources. In its first phase, rapid digitalization focused on digitizing analogue processes and collaboration in a linear fashion. Furthermore, the early digital design process was oriented toward the end product or façade system, without the capacity to develop a digital circular design model, based on data, that keeps track of information flow during design, construction, occupancy, demolition and recycling. Consequently, the absence of information storing and sharing to provide feedback on successful practices has resulted in a linear digital design process with very limited learning and improvement potential.

Over time, digitalization has reached the point that the world's most valuable resource is no longer oil but data (The Economist, 2017). As a result, this has increased the capacity for gathering and reusing information among stakeholders over the whole façade life cycle in a C2C manner. Information exchange is essential to design end-of-life options successfully and integrate reused and recycled materials and products. At a 'meta level' the reuse of design information and successful design processes may enhance the performance and quality of building skins performance. To achieve high performance and low cost, façade products require integration of manufacturability and supply chain knowledge earlier than usual in the design process (Montali, Overend, Pelken, & Sauchelli, 2018).

Moreover, façade designers and engineers can benefit from more control over the product's manufacturability, performance and architectural intent in less time by using product-oriented knowledge bases and digital tools (Day, Gasparri, & Aitchison, 2019; Jakica & Zanelli, 2015; Montali, Sauchelli, Jin, & Overend, 2019). Consequently, the development of a façade circular design framework such as Product Model for automatic rule checking and knowledge reuse is needed to shift façade products from *engineer-to-order* to *make-to-order* types (Montali, Overend, Pelken, & Sauchelli, 2017). However, to include multiobjective optimization and mass customization, the circular design process has to upgrade knowledge-based tools to include machine learning and *artificial intelligence* (AI) algorithms that are more powerful in dealing with complex façade design and performance requirements.

One of the main challenges ahead lies in the standardization and integration of all of the data from different façade performances, stakeholders' input and external and internal loads. One study (Dokter, Thuvander, & Rahe, 2021) showed that industrial designers more intensely focus on CEBMs, while architects focus more on reusing materials at a building level. The study advocates establishing dedicated CE research and design teams within companies to facilitate knowledge-sharing and circular strategies and methods. Long-term client relationships and designers' engagement with the designed artifacts' life cycles are also identified as crucial for learning and improvement. To address this, Leising, Quist, and Bocken (2018) developed a conceptual framework to study supply chain collaboration in circular buildings and created an empirically based tool to enhance collaboration for CE in the building sector.

However, like construction, the façade domain still does not have methods and industry-specific theoretical frameworks to complement robust tools that can address the complexity of the circular design of façade systems and components. The façade industry still has a long way to go before digital façade tools are eventually capable of handling big data, including generation, storage and analyses, being the synthetic data created by performance simulations or data derived from real-time performance monitoring of occupants and building operation.

21.7 Circular Construction 4.0

The so-called Fourth Industrial Revolution—Industry 4.0—or Construction 4.0—is gaining pace and offers new tools and methods that are likely to impact the feasibility of CE and its viability in the field of facades. In the context of industry 4.0, the digitalization of facades represents only a stepping stone for the implementation of all technologies associated with the fourth industrial revolution. BIM has been introduced years ago, yet only recently BIM software has reached a state where the same file formats are used and exchanged during the design and construction phases, keeping track of changes and performing intelligent in-depth model checking clash detection. Notwithstanding this, the use of BIM for the management of a building's end-of-life still remains relatively rare (Charef & Emmitt, 2020) performed an analysis of the BIM uses that may foster a CE approach in construction. The research identified 35 uses, categorized in existing and new BIM uses, that significantly impact all aspects of a project. More specifically, they identified eight existing BIM uses to be used for C&D waste management and minimization: (1) phase planning, (2) quantity take-off, (3) design review, (4) 3D coordination, (5) site utilization planning, (6) 3D control and planning, (7) digital fabrication and (8) construction system design. Additionally, they identified seven new BIM uses: (1) *digital model for end-of-life*, (2) material passport development, (3) project database, (4) data checking, (5) circularity assessment (8D), (6) materials recovery processes and (7) materials bank. The study also pointed out that further research is required to better understand the occupancy phase and occupants' and facilities managers' role.

In this regard, studies using immersive *extended reality* (XR) environments could shorten the postoccupancy evaluation time by bringing it into a design process as a preoccupancy evaluation, including all stakeholders' active involvement. In this immersive environment, stakeholders would be more aware of the benefits of CE and engaged with the facades leading to a better understanding of the role of facades in CE transition. Moreover, these devices' ability to capture spatial−temporal occupant behaviour represents an unprecedented opportunity to learn more about personalized and collective occupant preferences and their relationship with adaptive facades. Occupant−façade interactions are often disruptive and a source of dissatisfaction because of conflicts, for example, energy efficiency and *indoor environmental quality* (IEQ) (Luna-Navarro et al., 2020). Understanding and identifying Occupant−façade interaction strategies would greatly influence the reduction of operational carbon and energy consumption, while increasing occupant productivity, well-being and satisfaction (Luna-Navarro, Allen, Meizoso, & Overend, 2019). However, concerns about privacy, collection, use of and access to personal data have led to the *general data protection regulation* (European Commission, 2016), which significantly hindered the implementation of technologies such as XR that rely on massive data collection, even when collected anonymously.

As a result of a circular design loop, a circular information flow reduces errors and costly project modifications at the later design stages and facilitates DfD. Furthermore, *computational design* within the BIM software represents a powerful tool, especially in the façade domain, where the *meta*-model of a unitized façade system may be effectively designed through iterations while automatically being populated on a façade surface. Mutual linking of multiple performances allows for finding an optimum solution range for a given set of criteria in a unique parametric framework (Jakica & Kragh, 2020), where the CE may be prioritized. Different geometry and *digital manufacturing* optimizations may be performed to increase material efficiency further, structural strength (Arup, Systems, & WithinLab, 2015) or thermal insulation properties, while minimizing material usage and associated embodied carbon. Capabilities of *distributed autonomous collaborative robots* and *drones* for on-site *additive manufacturing* have been growing and have finally reached the tipping point, where the building regulations can be met (PERI, 2020). Large-scale and high-volume *3D concrete printing* applications could reduce unnecessary construction waste of moulds and scaffolding with associated logistics and workforce costs (Katzer & Szatkiewicz, 2019). Furthermore, the freedom and flexibility of 3D printing, including material choice and distribution, make this process extremely versatile, resilient for stricter future requirements and open for improvements (Buswell, Leal de Silva, Jones, & Dirrenberger, 2018).

However, the real potential lies in asynchronous development and integration of digital and physical processes, environments and tools. In particular, *circular Construction 4.0* will depart from static and linear behaviour of predicted standard-based performance, toward smart circular process loops of real-time sensing, processing and adapting, where all actions are stored, analysed and adjusted to increase efficiency with every iteration and provide personalized responses. Moreover, sensing processes in performance testing represent a key driver for building smart readiness (Markoska, Jakica, Lazarova-Molnar, & Kragh, 2019; Markoska, Sethuvenkatraman, Jakica, & Lazarova-Molnar, 2021). Sensing can be enabled through embedded electronics within adaptive façade elements or *Internet-of-things* performance sensors positioned in interiors or exteriors, including local weather stations recording climate parameters, sensors monitoring IEQ and software applications on mobile devices tracking occupant requirements and satisfaction through a *human-centred design*. *Building operating systems* will have to be developed to handle complex scenarios of adaptive facades (Attia, Lioure, & Declaude, 2020) through processing and storing *Big Data*, either through decentralized in-house building servers or through centralized *cloud computing* (Toth et al., 2011) and *data lakes* (Marinakis, 2020). Opportunities include the automation of urban mining and repurposing of elements, components and materials (Koutamanis, van Reijn, & van Bueren, 2018). *Blockchain technologies* and RFID are also promising technologies to provide data access and quality throughout the life cycles of façade assets and materials (Van Gassel & Jansen, 2008). The ability to demonstrate operational effectiveness of specific façade systems, learn from errors and upscale while offering adequate product declarations and warranties is likely to prove critical to the uptake of novel circular models.

A jump from individual toward multiagent façade systems response and collective behaviour tracking across climates worldwide will increase the robustness and efficiency of *machine learning* and AI algorithms. Furthermore, algorithms may be reused on synthetic data created by performance simulations in *digital twin* environments during the design process to augment decision-making (Płoszaj-Mazurek, Ryńska, & Grochulska-Salak, 2020). Appending planned and actual value pairs into the training database in synthetic and ground truth data lakes will significantly increase accuracy, decrease discrepancies, reducing the performance gap between planned and actual performance. Therefore technology facilitates access to information and makes possible design decisions based on evidence as opposed to qualitative and emotional stimuli.

Centralized databases with a constant circular feedback loop will offer direct field updates to the database and over-the-air updates for improved façade response. Collective building smartness will, thus, continuously rise and improve resilience to future increased demands. Further *Construction 4.0* infrastructure for resilience may include *drones* and *swarm robots* for façade installation (Taghavi, Heredia, Iturralde, Halvorsen, & Bock, 2018) and inspection during construction and operation, commissioning, maintenance and cleaning (Fawzy, El Sherif, & Khamis, 2019). Identifying weak spots and potential construction issues and monitoring failures at inaccessible or high-risk areas may drastically reduce safety risks and associated costs and, therefore, increase the frequency of checks to extend the life cycle of facades (Thornton Tomasetti CORE Studio, 2020). Furthermore, timely signals from embedded façade monitoring devices on potential future failures may prevent malfunctioning and costly repairs.

21.8 Conclusion

CE is confidently expanding to every aspect of our society. The façade domain has also been impacted and challenged to transcend toward circular models of thinking, designing, acting, collaborating, making, disposing, reducing, reusing and recycling. This chapter showed multiple facets of CE in facades, analysing and envisioning a future where industrial and social activity is in harmony with the environment and where CE practices drive economic growth and value creation. The chapter started with 3R concepts and continued with a DfD and maintenance section that emphasized the importance of widening and extending the design perspective to include closed-loop cycles. Furthermore, Section 21.4 demonstrated the need for tracking, collection, storing and accessing the information on building materials. The section on CEBMs brought a business perspective of CE in facades and presented innovative BMs that may enable a large-scale CE transition in the façade sector. Lastly, Section 21.7 offered an overview of advanced technologies and their potential application in the future development of CE. A future where there are no concepts of 'waste' but only resources and assets, where products have multiple lives and where stakeholders compete in the circular value creation. This is the future that is not just plausible but plain necessary.

The main barrier for a wider CE adoption in the façade sector is the transition toward resource-conscious design. Decision-making across the value chain requires the structured and well-documented coordinated effort of all parties involved. This barrier's roots can be found in the lack of knowledge, tools, frameworks and methods, timely and holistic information for decision-making, resource-conscious demand from the market and in the lack of adoption of the so-called Industry 4.0 technologies, among many others.

As seen in other industries, information technologies have boosted productivity and may represent a framework for knowledge generation, sharing, collaboration and innovation to boost productivity. The ability to repeat and upscale while offering adequate product declarations is likely to prove critical to the general uptake of novel circular models. Industry 4.0—or Construction 4.0—is gaining pace and offers a novel palette of opportunities through new tools and methods that are likely to impact on the feasibility and viability of the CE in facades. The façade is an architectural element that offers excellent potential to revolutionize our perception of the built environment and what it can become. It is a visual and ubiquitous expression of our life, culture, goals and ideals, and as such, it may become one of the critical factors that will facilitate the green transition and secure a truly circular future.

References

3XN. (2016). *Building a circular future—3rd Ed.*

Adams, K. T., Osmani, M., Thorpe, T., & Thornback, J. (2017). *Circular economy in construction: Current awareness, challenges and enablers. Proceedings of institution of civil engineers: Waste and resource management* (pp. 15–24). ICE Publishing. Available from https://doi.org/10.1680/jwarm.16.00011.

Akinade, O. O., Oyedele, L. O., Bilal, M., Ajayi, S. O., Owolabi, H. A., Alaka, H. A., & Bello, S. A. (2015). *Waste minimisation through deconstruction: A BIM based*

Deconstructability Assessment Score (BIM-DAS). Resources, conservation and recycling (Vol. 105, pp. 167–176). Elsevier B.V. Available from http://doi.org/10.1016/j.resconrec.2015.10.018.

Arup, CDRM, 3D Systems, WithinLab & EOS. (2015). *Additive manufacturing in construction & engineering*. Available at: https://www.arup.com/projects/additive-manufacturing. Accessed: 02.12.20.

Attia, S., Lioure, R., & Declaude, Q. (2020). Future trends and main concepts of adaptive façade systems. *Energy Science & Engineering, 8*(9), 3255–3272. Available from https://doi.org/10.1002/ese3.725.

Azcarate-Aguerre, J. F., Andaloro, A., & Klein, T. (2018). Façade leasing: Drivers and barriers to the delivery of integrated Facades-as-a-Service. *Real Estate Research Quarterly, 17*(3), 11–22.

Azcarate-Aguerre, J. F., Andaloro, A., & Klein, T. (2021). Facades-as-a-Service: A business and supply-chain model for the implementation of a circular façade economy. In E. Gasparri, et al. (Eds.), *Rethinking building skins—Transformative technologies and research trajectories*. Elsevier.

Barbosa, F., Woetzel, J., Mischke, J., Ribeirinho, M.J., Sridhar, M., Parsons, M., Bertram, N., & Brown, S. (2017). *Reinventing construction: A route to higher productivity*. Available at: http://www.mckinsey.com/mgi. Accessed: 20.02.21.

Benachio, G. L. F., Freitas, MdoC. D., & Tavares, S. F. (2020). Circular economy in the construction industry: A systematic literature review. *Journal of Cleaner Production, 260*, 121046. Available from https://doi.org/10.1016/j.jclepro.2020.121046.

Bjørn, A., & Hauschild, M. Z. (2017). *Cradle to cradle and LCA. Life cycle assessment: Theory and practice* (pp. 605–631). Springer International Publishing. Available from https://doi.org/10.1007/978-3-319-56475-3_25.

Bolin, C. A. & Smith, S. T. (2010). '*Life cycle assessment of borate-treated lumber with comparison to galvanized steel framing*'. https://doi.org/10.1016/j.jclepro.2010.12.005.

Bocken, N. M. P., Short, S. W., Rana, P., & Evans, S. (2014). A literature and practice review to develop sustainable business model archetypes. *Journal of Cleaner Production, 65*, 42–56. Available from https://doi.org/10.1016/j.jclepro.2013.11.039.

Buswell, R. A., Leal de Silva, W. R., Jones, S. Z., & Dirrenberger, J. (2018). 3D printing using concrete extrusion: A roadmap for research. *Cement and Concrete Research, 112* (October 2017), 37–49. Available from https://doi.org/10.1016/j.cemconres.2018.05.006.

Cai, G., & Waldmann, D. (2019). A material and component bank to facilitate material recycling and component reuse for a sustainable construction: Concept and preliminary study. *Clean Technologies and Environmental Policy, 21*(10), 2015–2032. Available from https://doi.org/10.1007/s10098-019-01758-1.

Charef, R., & Emmitt, S. (2020). Uses of building information modelling for overcoming barriers to a circular economy. *Journal of Cleaner Production, 285*, 124854. Available from https://doi.org/10.1016/j.jclepro.2020.124854.

Cole, R. J., & Kernan, P. C. (1996). Life-cycle energy use in office buildings. *Building and Environment, 31*(4), 307–317.

Cruz Rios, F., Grau, D., & Chong, W. K. (2019). Reusing exterior wall framing systems: A cradle-to-cradle comparative life cycle assessment. *Waste Management, 94*, 120–135. Available from https://doi.org/10.1016/j.wasman.2019.05.040.

Day, G., Gasparri, E., & Aitchison, M. (2019). Knowledge-based design in industrialised house building: A case-study for prefabricated timber walls. *Lecture notes in civil engineering* (pp. 989–1016). Springer. Available from https://doi.org/10.1007/978-3-030-03676-8_40.

De Wolf, C., Hoxha, E., & Fivet, C. (2020). Comparison of environmental assessment methods when reusing building components: A case study. *Sustainable cities and society*, *61*(May 2019), 102322. Available from https://doi.org/10.1016/j.scs.2020.102322.

Deloitte (2020). 'GPoC 2019 global powers of construction', *Deloitte*, p. 54. Available from: https://www2.deloitte.com/content/dam/Deloitte/at/Documents/real-estate/2017-global-powers-of-construction.pdf.

Dokter, G., Thuvander, L., & Rahe, U. (2021). How circular is current design practice? Investigating perspectives across industrial design and architecture in the transition towards a circular economy. *Sustainable Production and Consumption*, *26*, 692–708. Available from https://doi.org/10.1016/j.spc.2020.12.032.

Durmisevic, E. (2018). *BAMB: WP3 reversible building design reversible building design guidelines*.

Durmisevic, E. (2019a). *Circular economy in construction design strategies for reversible buildings*.

Durmisevic, E. (2019b). *Explorations for reversible buildings*.

Ellen MacArthur Foundation (2015). *Potential for Denmark as a circular economy a case study from: Delivering the circular economy—A toolkit for policy makers*. Available from: https://www.ellenmacarthurfoundation.org/books-and-reports. Accessed: 27.01.21.

Ellen Macarthur Foundation & Arup (2019). *Urban buildings systems summary, circular economy in cities*.

European Commission (2016). *The European construction sector: A global partner* (p. 16) European Commission.

European Commission (2020). *REBRICK H2020 project, market uptake of an automated technology for reusing old bricks*. Available from: https://ec.europa.eu/environment/eco-innovation/projects/en/projects/rebrick. Accessed: 27.01.21.

European Commission (2021). *Madaster platform*. Available from: https://www.madaster.com/en. Accessed: 27.01.21.

European Committee for Standardization (CEN) (2017). *Business plan CEN/TC 442 building information modelling*.

Fawzy, H., El Sherif, H. & Khamis, A. (2019). 'Robotic façade cleaning system for high-rise building', in: *Proceedings—ICCES 2019: 2019 14th international conference on computer engineering and systems* (pp. 282–287). Institute of Electrical and Electronics Engineers Inc. https://doi.org/10.1109/ICCES48960.2019.9068112.

Fulford, R., & Standing, C. (2014). Construction industry productivity and the potential for collaborative practice. *International Journal of Project Management*, *32*(2), 315–326. Available from https://doi.org/10.1016/j.ijproman.2013.05.007.

Givoni, B. (1992). *Comfort, climate analysis and building design guidelines, energy and buildings*.

Gepts, B., Meex, E., Nuyts, E., Knapen, E., & Verbeeck, G. (2019). Existing databases as means to explore the potential of the building stock as material bank. *IOP Conference Series: Earth and Environmental Science*, *225*(1). Available from https://doi.org/10.1088/1755-1315/225/1/012002.

Guinée, J. B., Heijungs, R., Huppes, G., Zamagni, A., Masoni, P., Buonamici, R., ... Rydberg, T. (2011). Life cycle assessment: Past, present, and future. *Environmental Science and Technology*, *45*(1), 90–96.

Han, J., Heshmati, A., & Rashidghalam, M. (2020). Circular economy business models with a focus on servitization. *Sustainability (Switzerland)*, *12*(21), 1−17. Available from https://doi.org/10.3390/su12218799.

Hart, J., Adams, K., Giesekam, J., Tingley, D. D., & Pomponi, F. (2019). *Barriers and drivers in a circular economy: The case of the built environment. Procedia CIRP* (pp. 619−624). Elsevier B.V. Available from https://doi.org/10.1016/j.procir.2018.12.015.

Heisel, F., & Rau-Oberhuber, S. (2020). Calculation and evaluation of circularity indicators for the built environment using the case studies of UMAR and Madaster. *Journal of Cleaner Production*, *243*, 118482. Available from https://doi.org/10.1016/j.jclepro.2019.118482.

Honic, M., Kovacic, I., & Rechberger, H. (2019). Improving the recycling potential of buildings through Material Passports (MP): An Austrian case study. *Journal of Cleaner Production*, *217*, 787−797. Available from https://doi.org/10.1016/j.jclepro.2019.01.212.

Hopkinson, P., Chen, H. M., Zhou, K., Wang, Y., & Lam, D. (2018). Recovery and reuse of structural products from end-of-life buildings. *Proceedings of the Institution of Civil Engineers: Engineering Sustainability*, *172*(3), 119−128. Available from https://doi.org/10.1680/jensu.18.00007.

Hopkinson, P., De Angelis, R., & Zils, M. (2020). Systemic building blocks for creating and capturing value from circular economy. *Resources, Conservation & Recycling*, *155*. Available from https://doi.org/10.1016/j.resconrec.2019.104672.

Jakica, N., & Kragh, M. K. (2020). Assessing self-shading benefits of twisting towers. *Journal of Façade Design and Engineering*, *8*(1), 115−130. Available from https://doi.org/10.7480/jfde.2020.1.5043.

Jakica, N., & Zanelli, A. (2015). *Knowledge based expert system tool for optimization of the complex glass BIPV system panel layout on the cable net structural skin. Energy Procedia* (pp. 2226−2231). Elsevier Ltd. Available from https://doi.org/10.1016/j.egypro.2015.11.339.

Jawahir, I. S., & Bradley, R. (2016). *Technological elements of circular economy and the principles of 6R-based closed-loop material flow in sustainable manufacturing. Procedia CIRP* (pp. 103−108). Elsevier B.V. Available from https://doi.org/10.1016/j.procir.2016.01.067.

Jönsson, Å., Björklund, T., & Tillman, A. M. (1998). LCA of concrete and steel building frames. *International Journal of Life Cycle Assessment*, *3*(4), 216−224. Available from https://doi.org/10.1007/BF02977572.

Katzer, J., & Szatkiewicz, T. (2019). Properties of concrete elements with 3-D printed formworks which substitute steel reinforcement. *Construction and Building Materials*, *210*, 157−161. Available from https://doi.org/10.1016/j.conbuildmat.2019.03.204.

Kirchherr, J., Reike, D., & Hekkert, M. (2017). Conceptualizing the circular economy: An analysis of 114 definitions. *Resources, Conservation and Recycling*, *127*(April), 221−232. Available from https://doi.org/10.1016/j.resconrec.2017.09.005.

Koutamanis, A., van Reijn, B., & van Bueren, E. (2018). Urban mining and buildings: A review of possibilities and limitations. *Resources, Conservation and Recycling*, *138*, 32−39. Available from https://doi.org/10.1016/j.resconrec.2018.06.024.

Leising, E., Quist, J., & Bocken, N. (2018). Circular Economy in the building sector: Three cases and a collaboration tool. *Journal of Cleaner Production*, *176*, 976−989. Available from https://doi.org/10.1016/j.jclepro.2017.12.010.

Lendager Group (2019). *A changemaker's guide to the future*. Available from: https://issuu.com/lendagertcw/docs/achangemakersguidetothefuture_2.udg. Accessed: 27.01.21.

Leslie, T., Panchaseelan, S., Barron, S., & Orlando, P. (2018). Deep space, thin walls: Environmental and material precursors to the Postwar Skyscraper. *Journal of the Society of Architectural Historians*, *77*(1), 77−96. Available from https://doi.org/10.1525/jsah.2018.77.1.77.

Luna-Navarro, A., Loonen, R., Juaristi, M., Monge-Barrio, A., Attia, S., & Overend, M. (2020). Occupant-Façade interaction: A review and classification scheme. *Building and Environment*, *177*(April). Available from https://doi.org/10.1016/j.buildenv.2020.106880.

Macé, P., Larsson, D., Benson, J., Woess-Gallasch, S., Kappel, K., Frederiksen, K., ... Warneryd, M. (2018). Transition towards sound BIPV business models Report IEA-PVPS T15−03: 2018. Olivier Jung.

Luna-Navarro, A., Allen, M., Meizoso, M., & Overend, M. (2019). BIT-Building Impulse Toolkit: A novel digital toolkit for productive, healthy and resource efficient buildings. *Journal of Physics: Conference Series*, *1343*.

Malabi Eberhardt, L. C., Rønholt, J., Birkved, M., & Birgisdottir, H. (2021). Circular economy potential within the building stock—Mapping the embodied greenhouse gas emissions of four Danish examples. *Journal of Building Engineering*, *33*(September 2020). Available from https://doi.org/10.1016/j.jobe.2020.101845.

Malmqvist, T., Nehasilova, M., Moncaster, A., Birgisdottir, H., Rasmussen, F. N., Wiberg, A. H., & Potting, J. (2018). Design and construction strategies for reducing embodied impacts from buildings—Case study analysis. *Energy & Buildings*, *166*, 35−47. Available from https://doi.org/10.1016/j.enbuild.2018.01.033.

Marinakis, V. (2020). Big data for energy management and energy-efficient buildings. *Energies*, *13*(7). Available from https://doi.org/10.3390/en13071555.

MBDC LLC (2014). *Version 3.1 cradle to cradle certified product standard*.

Ministry of Environment and Food and Ministry of Industry, B. and F. A. (2018). *Strategy for circular economy: More value and better environment through design, consumption, and recycling*.

Markoska, E., Jakica, N., Lazarova-Molnar, S., & Kragh, M. K. (2019). *Assessment of building intelligence requirements for real time performance testing in smart buildings*, in: 2019 4th International conference on smart and sustainable technologies, SpliTech. Institute of Electrical and Electronics Engineers Inc. Available from https://doi.org/10.23919/SpliTech.2019.8783002.

Markoska, E., Sethuvenkatraman, S., Jakica, N., & Lazarova-Molnar, S. (2021). Are we ready to evaluate the smart readiness of Australian buildings? In J. Littlewood, R. J. Howlett, & L. C. Jain (Eds.), *Sustainability in energy and buildings 2020. Smart innovation, systems and technologies* (vol 203, pp. 549−559). Singapore: Springer. Available from https://doi.org/10.1007/978−981-15−8783-2_46.

Minunno, R., O'Grady, T., Morrison, G. M., & Gruner, R. L. (2020). Exploring environmental benefits of reuse and recycle practices: A circular economy case study of a modular building. *Resources, Conservation and Recycling*, *160*(May), 104855. Available from https://doi.org/10.1016/j.resconrec.2020.104855.

Mont, O. K. (2002). Clarifying the concept of product-service system. *Journal of Cleaner Production*, *10*. Available from https://www.cleanerproduction.net.

Montali, J., Overend, M., Pelken, P. M., & Sauchelli, M. (2017). Towards facades as Make-To-Order products—The role of knowledge-based-engineering to support design.

Journal of Façade Design and Engineering, 5(2), 101−112. Available from https://doi.org/10.7480/jfde.2017.2.1744.

Montali, J., Overend, M., Pelken, P. M., & Sauchelli, M. (2018). Knowledge-based engineering in the design for manufacture of prefabricated facades: Current gaps and future trends. *Architectural Engineering and Design Management, 14*(1−2), 78−94. Available from https://doi.org/10.1080/17452007.2017.1364216.

Montali, J., Sauchelli, M., Jin, Q., & Overend, M. (2019). Knowledge-rich optimisation of prefabricated facades to support conceptual design. *Automation in Construction, 97*, 192−204. Available from https://doi.org/10.1016/j.autcon.2018.11.002.

Munaro, M. R., Tavares, S. F., & Bragança, L. (2020). Towards circular and more sustainable buildings: A systematic literature review on the circular economy in the built environment. *Journal of Cleaner Production, 260*. Available from https://doi.org/10.1016/j.jclepro.2020.121134.

Naranje, V., & Swarnalatha, R. (2019). Design of tracking system for prefabricated building components using RFID technology and CAD model. *Procedia Manufacturing, 32*, 928−935. Available from https://doi.org/10.1016/j.promfg.2019.02.305.

Nußholz, J. L. K., Rasmussen, F. N., Whalen, K., & Plepys, A. (2020). Material reuse in buildings: Implications of a circular business model for sustainable value creation. *Journal of Cleaner Production, 245*, 118546. Available from https://doi.org/10.1016/j.jclepro.2019.118546.

Oezdemir, O., Krause, K., & Hafner, A. (2017). Creating a resource cadaster—A case study of a district in the Rhine-Ruhr metropolitan area. *Buildings, 7*(2). Available from https://doi.org/10.3390/buildings7020045.

Olgyay, A., & Olgyay, V. (1957). *Solar control and shading devices*. Princeton: Princeton University Press.

Osmani, M., Glass, J., & Price, A. D. F. (2008). Architects' perspectives on construction waste reduction by design. *Waste Management, 28*(7), 1147−1158. Available from https://doi.org/10.1016/j.wasman.2007.05.011.

Pacheco-Torgal, F., & Jalali, S. (2011). Toxicity of building materials: A key issue in sustainable construction. *International Journal of Sustainable Engineering, 4*(3), 281−287. Available from https://doi.org/10.1080/19397038.2011.569583.

Pearce, D. W., & Turner, R. K. (1990). *Economics of natural resources and the environment*. Herts: Harvester Wheatsheaf.

PERI (2020). *3D construction printing*. Available from: https://www.peri.com/en/business-segments/3d-construction-printing.html. Accessed: 20.02.21.

Płoszaj-Mazurek, M., Ryńska, E., & Grochulska-Salak, M. (2020). Methods to optimize carbon footprint of buildings in regenerative architectural design with the use of machine learning, convolutional neural network, and parametric design. *Energies, 13*(20). Available from https://doi.org/10.3390/en13205289.

Reburg—BAMB (2020). Available from: https://www.bamb2020.eu/future/reburg/. Accessed: 29.01.21.

Roome, N., & Louche, C. (2016). Journeying toward business models for sustainability: A conceptual model found inside the black box of organisational transformation. *Organization and Environment, 29*. Available from https://doi.org/10.1177/1086026615595084.

Stanojev, J., & Gustafsson, K. (2019). 'Circular economy concepts for cultural heritage adaptive reuse implemented through smart specialisations strategies'. In G. Getzinger, & M. Jahrbacher (Eds.), *Proceedings of the STS conference Graz 2019* (pp. 415−436).

Graz: Verlag der Technischen Universität Graz. Available from https://diglib.tugraz.at/ download.php?id = 5df338c7b72f4&location = browse.
Taghavi, M., Heredia, H., Iturralde, K., Halvorsen, H., & Bock, T. (2018). Development of a modular end effector for the installation of curtain walls with cable-robots. *Journal of Façade Design and Engineering*, *6*(2), 1−8. Available from https://doi.org/10.7480/jfde.2018.2.2067.
Teece, D. J. (2010). Business models, business strategy and innovation. *Long Range Planning*, *43*, 172−194. Available from https://doi.org/10.1016/j.lrp.2009.07.003.
Teknologisk-Institut (2016). '*Review of the current state of RFID*', pp. 16−34.
The Economist (2017). *Regulating the internet giants—The world's most valuable resource is no longer oil, but data*. Available from: https://www.economist.com/leaders/2017/05/06/the-worlds-most-valuable-resource-is-no-longer-oil-but-data. Accessed: 20.02.21.
The European Committee for Standardization (CEN) (2010). *LCA European Standards, building level (EN 15978), product level (EN 15804)*. Available from: http://amanac.eu/wiki/european-standards/. Accessed: 27.01.21.
Thornton Tomasetti CORE Studio (2020). *T2D2©: Thornton Tomasetti Damage Detector*. Available from https://t2d2.ai/. Accessed: 20.02.21.
Toth, B., Salim, F., Burry, J., Frazer, J., Drogemuller, R., & Burry, M. (2011). Energy-oriented design tools for collaboration in the cloud. *International Journal of Architectural Computing*, *9*(4), 339−359. Available from https://doi.org/10.1260/1478-0771.9.4.339.
UMAR—Urban Mining and Recycling (UMAR) (2018). Available from: http://nest-umar.net/portfolio/umar/. Accessed: 29.01.21.
UN Environment and International Energy Agency (2017). *Towards a zero-emission, efficient, and resilient buildings and construction sector. Global status report 2017*.
Van Gassel, F. & Jansen, G. (2008). 'A simulation tool for radio frequency identification construction supply chains', in: ISARC 2008—Proceedings from the 25th international symposium on automation and robotics in construction, (June) (pp. 64−68). https://doi.org/10.3846/isarc.20080626.64.
Vandkunsten Architects (2016). *Nordic built component reuse*.
Walker, A., Grant, T. & McAlister, S. (2007). '*The environmental impact of building materials*', (May), p. 12.
Witjes, S., & Lozano, R. (2016). Towards a more circular economy: Proposing a framework linking sustainable public procurement and sustainable business models. *Resources, Conservation and Recycling*, *112*, 37−44. Available from https://doi.org/10.1016/j.resconrec.2016.04.015.
Xue, F., Chen, K., Lu, W., Niu, Y., & Huang. (2018). Linking radio-frequency identification to building information modeling: Status quo, development trajectory and guidelines for practitioners. *Automation in Construction*, *93*(April), 241−251. Available from https://doi.org/10.1016/j.autcon.2018.05.023.

Facades-as-a-Service: a business and supply-chain model for the implementation of a circular façade economy

22

Juan F. Azcarate-Aguerre[1], Annalisa Andaloro[2] and Tillmann Klein[1]
[1]Department of Architectural Engineering + Technology, Chair Building Product Innovation, Faculty of Architecture and the Built Environment, Delft University of Technology, Delft, The Netherlands, [2]Energy Efficient Buildings Group, Institute for Renewable Energy, Eurac Research, Bolzano, Italy

Abbreviations

CBM	circular business models
FaaS	Facades-as-a-Service
KPI	key performance indicators
PSS	product-service systems
TCO	total cost of ownership

22.1 Introduction

The transition toward an energy-efficient and circular built environment demands the creation of new business models, financing strategies and organizational structures. The slow rate of improvement of our building stock, in terms of both energy performance and resource stewardship, must be accelerated through the use of innovative and long-term collaboration models that align incentives and create shared value across the construction supply chain.

As designers and engineers of buildings and building systems, our approach to the circular economy transition is frequently as an engineering challenge: Design for adaptability and disassembly, standardization and reusability, design based on reused and resourced materials and components, are all highly valuable engineering strategies. These strategies 'enable' the technical circularity of building components and materials, can potentially extend their service lives and increase their chances of being given a second, third or indefinite number of technical cycles.

Designing an integral circular economy for building envelopes, however, requires us to go far beyond the design and engineering of circularity-enabled components. It requires us to analyse the entire life cycle of our products, as well as the stakeholders involved throughout these life cycles and the economic incentives

Rethinking Building Skins. DOI: https://doi.org/10.1016/B978-0-12-822477-9.00005-X
© 2022 Elsevier Inc. All rights reserved.

under which they operate (European Commission, 2020). A 'circularity-enabled' product can be managed and decommissioned in a linear way, rendering its circular design intent unsuccessful, while a linearly designed product might be treated and reprocessed in a circular way (Fig. 22.1). The challenge is to not only technically enable our products to remain efficiently and effectively part of a closed industrial system, but to also establish lasting business and economic incentives for present and future stakeholders to manage these products in circular ways (Witjes & Lozano, 2016).

Facades-as-a-Service (FaaS) is a business-centred organizational model aiming to approach the circular building envelope question from the opposite direction: rather than designing and engineering circularity-enabled products to operate in a predominantly linear economic system, it aims to create a 'circularity-enhancing' business environment. Such an environment would foster a circular management of materials and components, providing financial incentives for product suppliers, installers and managers to design and engineer products and processes to fit their new circular business priorities.

The present chapter describes the latest developments in the near-future implementation of FaaS models. It starts by establishing general circular economy and circular business model theory relevant to the FaaS model. It then describes the multidisciplinary stakeholders involved in the process of building and managing the building envelope – from its commissioning and financing through its maintenance, upgrading and decommissioning – and how these roles must change to enable a FaaS model. Lastly, it discusses the circular economy potential of the FaaS model, and remaining open questions that need to be addressed.

Figure 22.1 Certain electric junction box models (left) are designed to be adaptable, modular and reusable. In practice, however, they are often discarded due to their low economic value. Ornamental cement tiles (right), on the other hand, were designed with no specific circular intent in mind and installed with mortars and glues. The existence of a strong second-hand market for such tiles leads to them being carefully removed, cleaned and reused, a difficult and expensive process justified by the fact that such 'vintage' tiles are frequently valued more highly than their new equivalent.
Source: (Left) Photograph by Azcarate-Aguerre, J.F., Modular electric sockets, 2021, Rotterdam, NL. (Right) Photograph by Azcarate-Aguerre, J.F., Second-hand cement tiles, 2019, Eemnes, NL.

22.2 Circular business models and product-service systems

The FaaS model builds on academic and practical examples of circular business models (CBM) and product-service systems (PSS). We must, therefore, start by describing these two concepts.

CBM are those that focus on the delivery of a value proposition based on regenerative supply chains, reverse logistic channels and extended material value preservation and recovery. They can be categorized into three subtypes, determined by the extent to which they influence or achieve the circularity of resource flows (Bocken, De Pauw, Bakker, & van der Grinten, 2016; Geissdoerfer, Morioka, de Carvalho, & Evans, 2018) (Fig. 22.2):

- *Narrowing loops* focuses on reducing the volume of resources required to fulfil a certain functional requirement. For example, optimizing a structure to reduce the volume of steel required. Such models already operate in a linear economy and have limited effect on the circularity of resources as they focus on decreasing volume rather than setting up regenerative flows.

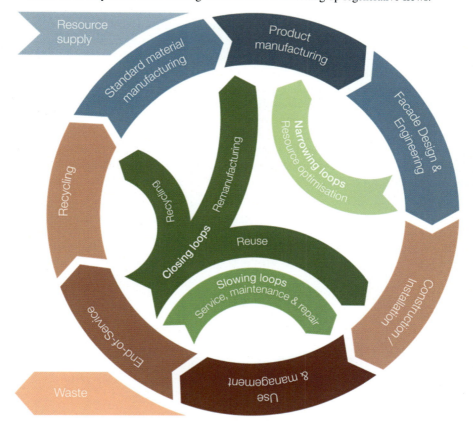

Figure 22.2 Narrowing, slowing and closing loops in the field of façade design and engineering. *Source:* Diagram by Klein (2021).

- *Slowing loops* aims to extend the service life of products to slow down the general rate of resource consumption. PSS fit within this category as they incentivize the production of more durable, higher quality products, the preservation and regeneration of residual value through extended maintenance and servicing, and, thus, the general dematerialization of economic transactions. PSS create a business environment that facilitates the closing of loops but do not necessarily lead to fully closed and circular systems if the products are eventually discarded or downcycled due to a lack of economically feasible regenerative alternatives.
- *Closing loops* relates to the preservation of resources within a closed regenerative system through reuse, remanufacturing and recycling activities. A PSS circular business model combined with an effective reverse logistics chain and remanufacturing process can effectively close loops. One of the key overarching challenges to the implementation of 'Closing loops' CBM is the extremely long time frame within which buildings operate and the fact that it might take decades for us to confirm whether or not today's circular plans translate into truly circular results 15 or 25 years down the road.

Within this context, PSS models are a broad category within CBM which aim to gradually shift the value proposition behind business transactions from the transfer of material products to the delivery of performance services. The linear economic system revolves around the transfer of legal and economic ownership of products between parties. These products are used, over a determined length of time, to deliver certain utilitarian results. When the product is no longer capable of delivering these results, or of doing so in an efficient way, they are discarded; most often through low-level recycling or landfilling.

PSS are categorized according to the extent to which transactional value is focused on performance rather than product-delivery. The basic classifications for models currently available on the market are as follows (Cong, Chen, Zheng, Li, & Wang, 2020; da Costa Fernandes, Pigosso, McAloone, & Rozenfeld, 2020; Pergande et al., 2012; Tukker, 2004) (Fig. 22.3):

- *Product-oriented PSS* models deal with tangible products, ownership of which is transferred to the consumer (client), while additional services are offered by the service provider, for example, maintenance contracts.
- *Use-oriented PSS* models also deal with tangible products, the ownership of which is retained by the service provider who sells product functionalities to the client, for

Figure 22.3 Broad categorization of product-service systems.
Source: Diagram by Azcarate-Aguerre (2021).

example, car or other equipment leasing contracts. However, in this case the product can also be an intangible asset, for example, content streaming platforms such as Spotify or Netflix. These models are also referred to as access-oriented, as they provide access to customers or end users to the product or service they require, without conveying ownership of the delivering product on to them.
- *Result-oriented PSS* models emphasize the value of the delivered performance over the tangible assets used to deliver such performance. As in the use-oriented PSS model, the provider retains ownership of the product and then sells the final performance to the client/end user while retaining technical responsibility and economic incentives over how efficiently this performance is delivered. An example of such a result-oriented model would be a scenario in which a building owner hires a certain indoor comfort – based on indicators such as temperature, humidity rate and air quality – for a fixed price, and regardless of how much it costs the provider to install and maintain the equipment needed to deliver this performance.

Applied to the façade industry, Product-as-a-Service models promote long-term relationships between the façade service provider and its supply chain on the one hand, and the final client on the other (Leising, Quist, & Bocken, 2016). It does so by focusing on value creation through ongoing service delivery and shared performance objectives, rather than the traditional procurement and sales contracting mechanism. This responds to a twofold aim: (1) shifting the guarantee over performance to the service provider who has more extensive technical expertise over its products, whereas (2) incentivizing a reductive use of materials and other finite resources in the delivery of these performance values (Baines & Lightfoot, 2013; Vezzoli et al., 2017).

The 'FaaS' provider will retain responsibility and control over the product during its entire life cycle, by retaining ownership of the equipment and by guaranteeing the façade's performance with provision of continuous maintenance and monitoring services. So, the client (ie, the building owner/operator) pays for the service delivered to the building user and does not need to take direct responsibility over façade management and maintenance. The FaaS model hereby described, therefore, lies between a use-oriented and a performance-oriented model. The drivers and barriers described in this chapter will have a determinant effect on the extent to which an integral performance contract can be provided in the near future (Azcarate-Aguerre et al., 2018).

22.3 A cross-value chain perspective: life cycle engineering for facades and façade performance contracting

A systemic shift such as the one proposed by the FaaS model must be founded on a deep understanding of the diverse stakeholders involved in a building's life cycle, and the economic incentives and motivations under which they operate.

22.3.1 The life cycle of a linear façade

Traditionally, buildings have been understood as static objects that undergo a series of specific and delimited processes with clear time demarcations. The main stages of construction, operation and decommissioning/demolition closely mirror the take, make, dispose process underlying the linear economy (Rampersad, Prins, Heurkens, & Ploeger, 2016). Stakeholders related to the building project have, therefore, been categorized according to their roles along this linear process, in both time and space. The following breakdown will focus on the façade industry, but close similarities can be applied to other building disciplines (Fig. 22.4).

Before any specific project begins, *product and system developers* are the earliest contributors to a building project's future circularity. Product and system developers design and manufacture the standard building blocks out of which buildings will eventually be assembled. Apart from heavily customized, high-profile projects, *architects and engineers* generally depend on these basic components in the planning and design process of a building. Framing systems, glazing units, hinges and connectors and sun-shading systems are among the most easily recognizable building blocks in the façade engineering process, and they are generally standardized by the developers and manufacturers behind these systems (Klein, 2013; Leising, Klein, Geldermans, & Azcárate-Aguerre, 2016).

Once the design of the façade has been established, a *façade fabricator* will assemble all the prescribed building blocks into a unitary or curtain-wall façade system. Component and service integration will vary from one project to the next, but often sun-shading and other building service components will be assembled and delivered by separate parties from those that deliver the basic façade elements. All these parties will work under the supervision of a *general contractor or another similar central project management coordinator*, working for the *project's commissioner, or building developer*. The extent to which these systems are integrated in

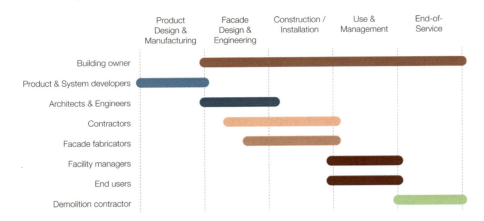

Figure 22.4 Stakeholder involvement over the life cycle of a linear façade.
Source: Diagram by Klein (2021).

both construction and operation will depend on the success of the planning and coordination process. In general, these parties are only responsible for delivering and installing the specified components and systems and hold a limited technical guarantee over their performance once the building is operational.

As the project's construction phase is finalized, the project developer will frequently transfer ownership of the building and all its physical systems to a *building owner*, private or institutional, which will either use it for its own operational functions or will own it as an investment object and rent it to external end users. While building owners tend to be the longest standing stakeholder tied to a building over its life cycle, buildings can be traded surprisingly often, leading to a general change of management team with each new owner. During its operation, and depending on the size of the project, building owners will engage *facility management and building maintenance teams* in the ongoing servicing of the digital and physical systems responsible for keeping the building effectively operational. In some cases, these teams will prioritize preventive maintenance, such as frequent cleaning of the façade components and replacement of key elements. In other cases, only reactive maintenance will be performed, by servicing or entirely replacing systems when they are close to failure (preventive maintenance), or once they have already failed (reactive maintenance). Preventive maintenance is key to extending product durability and performance and is, therefore, a crucial 'Slowing loops' strategy. Unfortunately, reactive maintenance still dominates the built environment.

Over its life cycle, buildings will frequently undergo numerous renovation and refurbishment cycles, each one normally performed by a different general contractor or coordinator, and with systems delivered by different fabricators and suppliers according to the specifications of different architecture and engineering consultants. This lack of continuity results in every cycle being an essentially new project with limited knowledge transferred and 'lessons learnt'. Lastly, when the building's operational life is deemed to be no longer economically or functionally extendable through renovation or refurbishment, a decommissioning process will begin. *Demolition contractors* are still the most mainstream alternative around the world, with buildings often being shredded into piles of recyclable, down-cyclable or only landfillable materials in an effort to make the process as quick and cheap as possible. *Building disassemblers*, often themselves previous demolition contractors, are slowly emerging as a higher value decommissioning alternative, balancing a slower and more costly disassembling process against the recovery of higher residual value from entirely reusable or remanufacturable components, or from at least higher grade recyclable materials.

The clearly defined moments at which diverse stakeholders interact with, and are responsible for the building, result in a heavily linear process in which incentives for long-term collaboration and shared responsibility are minimized or eliminated. Setting up an alternative model such as FaaS, therefore, requires overcoming cultural, technical and contractual barriers related to these traditional roles. In the authors' opinion the most promising way of overcoming these barriers is by developing and highlighting the economic value proposition of alternative models, not only in terms of financial gains but also in terms of soft values and managerial streamlining benefits.

22.3.2 Design and engineering of integral façade products and strategic life cycle planning for PSS and CE

Within the linear stakeholder process described earlier, specific moments of intervention can be recognized as moments of intense resource investment. Construction and renovation processes require vast volumes of materials, energy, labour and capital. The effectiveness of the facility management and building maintenance process bridging these key moments will largely determine the durability of these investments, and the likelihood that components will have to be replaced or upgraded sooner than expected. Furthermore, the financial cycles resultant from these resource investment moments limit the systemic flexibility of the building to face – in both a technically and economically feasible way – commercial changes in market demand and user requirements. Changes in occupancy and lettability, building safety, technical performance or other commercially relevant factors can render the property less attractive as an operating facility or an investment object, hence contributing to its technical decline.

Products-as-a-Service rely largely on product integration and system demarcation. The better integrated a product is and, therefore, the more clearly its performance delivery can be demarcated from that of other components or disciplines, the more likely it is to offer a significant and specific value proposition to the building owner and end user groups. The first step in the design and engineering of integrated façade products is, therefore, a focus on vertical supply chain integration which eliminates boundaries or demarcation lines between sub-suppliers, providing a single point of contact for the building owner; recognizes a single responsible party as system integrator and PSS provider responsible for a clearly-demarcated technical discipline, and; clearly defines specific technical KPI's (key performance indicators) according to which the value proposition will be evaluated.

From an initial product development perspective, integration of FaaS-ready technical components requires a holistic understanding of the (expanding) range of functional requirements which the building envelope can fulfil. Not only are unitary and curtain wall facades highly influential to the aesthetic appearance, thermal comfort and energy performance of a building, they are also increasingly relevant in the provision of decentralized, envelope-integrated functions such as solar energy generation, thermal and ventilation management, smart monitoring and response capability, among many others (Mach, Grobbauer, Streicher, & Müller, 2015). In some buildings, such systems could entirely replace centralized building services, while in others they will need to continue working in combination with these services. As a steppingstone, fully integrated facades that deliver complete building service functionality are most efficiently integrated into a FaaS product proposition, as they offer a clearer performance demarcation leading to a more specific and measurable value proposition backed by tangible performance indicators.

Moving on to the operational phase, the FaaS model requires the integration of technical solutions, which do not necessarily benefit the end user through enhanced comfort and energy performance, but which rather fulfil new functions necessary to the new PSS activities of other stakeholders (Ardolino, Saccani, Gaiardelli, &

Facades-as-a-Service

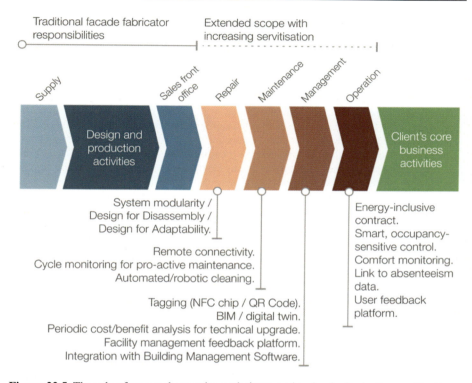

Figure 22.5 The role of new and emerging techniques and technologies in the servitization path of the façade industry (inspired by Baines & Lightfoot, 2013).
Source: Diagram by Azcarate-Aguerre (2020), inspired by Baines, T., & Lightfoot, H. (2013). Made to serve: How manufacturers can compete through servitization and product service systems: John Wiley & Sons.

Rapaccini, 2016). As façade fabricators shift their operations and processes to deliver not only the façade but also extended life cycle services such as monitoring, maintenance, upgrading and replacement, they require access to information which has until now been mostly only relevant to facility managers and building owners, or which has not been relevant at all to any stakeholder in a linear economy (Fig. 22.5). These technologies are both physical and digital and can be roughly categorized into two subgroups.

Performance monitoring and management technologies

Performance monitoring and management technologies are those that control, optimize and report system operation with the goal of continuously fulfilling the expected KPIs. Indoor and outdoor climate sensors connected to specific and well-tuned response algorithms can extract the maximum value from decentralized façade-integrated technologies by responding quickly to changes in temperature, humidity, air quality, etc. brought about by changes in external weather conditions or internal occupancy and use trends.

Energy consumption metering can, meanwhile, provide specific feedback and live recommendations to end users on the most optimal use of façade-integrated systems for maximum comfort and/or minimum energy consumption. Occupancy monitoring can guarantee that energy-saving modes are activated when no end users are present in a delimited area, while personalized tracking technologies can be tied to user-specific indoor comfort parameters.

The decentralized nature of façade-integrated technologies can also result in faster and more effective localized response to these changing conditions than when relying on large and centralized building systems. Lastly, functional-cycle metering allows service providers of keep track of their components' service-life trajectory to schedule effective preventive maintenance and replacement of key components. Electromechanically moving parts such as engines and actuators, air filters, batteries and such sensitive and shorter-lived elements can continuously be monitored or periodically inspected to forecast possible failure or performance decrease.

Component tracking and material circularity technologies

Component tracking and material circularity technologies have the primary goal of enhancing the residual value of components once they (or the elements they are assembled from) reach the end of their service lives. Relying on a combination of physical products, such as near-field communication-chips, quick-response-codes or other tagging technology, and digital platforms, such as digital twins and remotely accessible and updatable bills of quantities, these components facilitate the tracking of a component's life cycle, and the transfer of information between stakeholders (Wang & Wang, 2019).

CBM largely rely on the capturing of the residual value of components at their end-of-service. Maximum value can be achieved when components are reused or remanufactured rather than recycled, as the energy and resources invested in manufacturing them is preserved. High-value relife options, however, rely largely on the confidence of future customers, end users and financiers that the preowned product has been properly maintained and remanufactured, and that it is, therefore, capable of fulfiling its performance requirements over the next expected service cycle. By closely tracking façade information such as installation date, components type, embodied materials, maintenance and replacement schedules, use profile, among other data, façade providers can reliably evaluate the state of a façade element and determine which activities – if any – are necessary to extend the service life of the assembly for a second service cycle.

22.3.3 From purchasing bricks to hiring key performance indicators: the demand pull perspective

Having established the shifting role of technology and supply chain organization behind the FaaS service provider's expanding operations, we will now explore systemic changes that determine the near-future implementation opportunity of FaaS models. While some of these changes are of broad economic and legal nature, and do not originate within the client's scope of action, we here refer to them as demand-pull mechanisms. These mechanisms have a significant influence on enabling customer choice toward a PSS model, once this model has been developed and is being offered by a FaaS supplier.

Conflicts and limitations of product-based commissioning

Currently, the basis for real estate transactions at all levels, from the sale of components from suppliers to commissioners, to the transfer of entire buildings between investors, lies in the transfer of legal ownership and technical responsibility (and liability). Project developers and building commissioners hire planning specialists to detail the new building or renovation project, purchase the standard and customized products from which the building will be composed, and hire the services of contractors and subcontractors to assemble, deliver and install these products on site.

Apart from some technical performance guarantees (which are quite limited in time and typology), the transfer of ownership taking place in these transactions also conveys a transfer of technical responsibility from the supplier to the customer. This results in a series of overarching conflicts of interest between parties:

- *Commissioners focus on technical solutions rather than functional requirements*, by leading procurement processes based on prescribed initial product characteristics rather than ongoing performance indicators. Suppliers, subcontractors and general contractors then compete with each other to deliver these products at the lowest possible price, rather than offering life cycle alternatives prioritizing lowest total cost of ownership (TCO).
- *The building owner is left in charge of the technical maintenance* of components and systems which it did not engineer or fabricate, and which it only requires to deliver necessary performances to the building's end users.
- Lastly, *the tracking of materials and components is lost*, together with legal ownership, by the parties with the most technical knowledge of them, and who would be most qualified to extract maximum residual value from them.

As buildings and building envelopes become ever more technically complex, the role of the building owner as technically responsible party is made increasingly difficult, while its financial and legal position is made increasingly risky by the growing chance of technical failure (leading to a drop in technical building performance) or TCO miscalculation (leading to a decline in business case performance). Long and expensive legal battles thus continuously emerge from the miscommunication and misalignment of incentives in the current stakeholder culture.

The path toward performance-based commissioning

Standardized processes and policy mechanisms are required to translate innovative initiatives, such as the research and pilot projects on which this chapter is based (Fig. 22.6), into commercially viable mainstream solutions (Ing Economics Department, 2015). These can be categorized as follows:

- *Procurement standards* focus on the shift from specifying technical solutions to specifying performance requirements. Increasing freedom can, therefore, be provided to specialized suppliers to offer the most effective technical solution to provide the desired functional

Figure 22.6 FaaS pilot project at the building of the Faculty of Civil Engineering and Geosciences, TU Delft. The project included the renovation of 2600 m^2 of façade area, and the preparation of legal, financial and managerial processes to enable the implementation of a FaaS model. Digital twin technology reports data related to occupant comfort and technical condition of components. *FaaS*, Facades-as-a-Service.
Source: Photograph and rendering by Azcarate-Aguerre (2019).

outcome. Competition between suppliers could, therefore, shift from providing the product with the lowest initial cost, to deliver the desired performance over a determined period of time for the lowest total cost. Technical guarantees would become obsolete, as the commissioned performance needs to be provided on an ongoing basis, regardless of the condition or quality of the technical means used to deliver it. The balancing between quality and cost, complexity and robustness, durability and serviceability will be entirely up to the performance supplier as ultimate decision-maker.

- *Legal standards* are set in place to enforce the implementation of procurement standards and facilitate conflict resolution between parties when needed. Legal standards of both a policy and regulatory nature are built on the linear principles of ownership transfer and technical demarcation. Property ownership laws conceptualize buildings as entities made up of physical technical components. The preservation of value of such buildings, therefore, lies in the collective ownership of all crucial building components by the same party (building owner), to prevent the possibility of various partial owners of components reclaiming their building parts and dismantling the building unit in the process. Customized contracts have been developed to allow long-term collaboration between parties based on performance indicators, with or without transfer of legal ownership of the products used to deliver these performances. This legal customization results in higher setup costs for such contracts, as well as a higher perceived risk from investors and financiers due to their non-standard nature. Mainstream implementation of such collaboration models will not be possible as long as procurement standards, and their underlying legal standards, are not accessible to a wider market sector.

- *Economic and financial standards* impact the bankability of a building or building renovation project. Currently, the bankable unit of measure is the complete functional building or compartmentalized building space (as in the case of an office or apartment unit). This is tied to the same building-conceptualization logic previously described in the legal standards section. Real estate finance, the most common form of which is the property mortgage, is based on and backed by the resale value of the property when transferred and/or the potential cash-flow generated by the property when rented or otherwise exploited. Such cash flow—based models could easily be translated to specific building components, basing the value of such components on the ongoing service fee paid by the building owner or end user for the performances delivered by the service provider's product-service combination. Finance, however, is a risk-based discipline in which innovation needs to be gradually tested and backed by historical performance. Once standard procurement and legal models are better understood, and the intrinsic value of components and materials as independent units potentially disconnected from the building is more widely accepted, bankability of individual technical solutions to performance requirements will become a more likely proposition in the built environment.

22.4 Propositions

22.4.1 Lessons and following steps

The later points are the cross-validated results of the two research projects which have acted as foundation for this chapter. These results articulate how an integrated façade PSS value chain should be organized in terms of skills integration and partnerships (Fig. 22.7). They are the key evidence supporting the near-future implementation of FaaS models and at the same time point to open fields of research:

- *Commissioner (client):* Under the hypothesis that a third-party subject can manage full Façade-as-a-Service operation along its life cycle, this would ensure stability to the whole process (partially releasing the contractor from responsibility and dealing with the risk of contractor bankruptcy). This scheme is advantageous for the prospective client, who

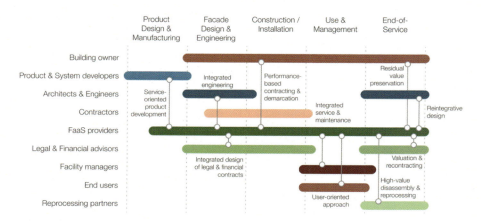

Figure 22.7 Enhanced stakeholder involvement, collaboration and integration over the life cycle of a circular Façade-as-a-Service.
Source: Diagram by Klein and Azcarate-Aguerre (2021).

would be willing to pay a premium for guaranteed ease of mind during the project life cycle. A contract, including full maintenance for 10/20 or more years (whole service life at best), could be a starting point for tenders, ensuring the client full operation of the building envelope at all stages.

- *Investment capabilities and bankability:* These are to be embedded in the entity running the business (eg, an energy service company). Otherwise, a solid partnership with an investor pool or credit institutions should be put in place, to ensure that the PSS model has sufficient financial support to bear exposure during the project life cycle (Överholm, 2017). There is an option for the façade service provider to remain the unique façade owner, having the adequate competences for ensuring a long-term service of the product and delivering performances with the minimum use of energy and material resources. However, this implies the PSS provider to bear the whole financial exposure for the entire duration of the PSS contract, which may not be feasible without the presence of an external investor.
- *Legal advice* should be sought at all stages to manage procurement details. This skill could either be integrated in the PSS business or outsourced to specialized legal consultants. The development of standardized contracts will eventually reduce the investment, in both time and money, required by currently customized contracts. Regional regulation across Europe will still demand specific solutions to fit local building law standards.
- *Rounded technical skills* must be well integrated by the FaaS provider, as it must be able to understand information exchanged and verify technical compliance at all stages. Such skills could as well be outsourced to a technical consultancy, but only if specific workflows are put in place to avoid bottlenecks in the information flow between the long-term collaborating and invested parties.
- *Engagement of material suppliers*, transformation industry and façade manufacturers to ensure quality of delivery and retention of tangible and intangible value over time. Those actors could benefit from entering this integrated value chain, as it would improve the financial stability of their business in the face of economic and real estate cycles.
- *Facility managers* should become key partners in the FaaS model, participating also in the system definition phase. They can then contribute to the optimization of life cycle costs of the PSS offering with respect to predictive maintenance.
- *Engagement of recycling and dismantling companies* should be increased in the future, to transform them into resource prosumers. Their involvement in the system definition phase is crucial to boost circularity practices in terms of resource recovery and reuse by identifying feasible relife options (Tebbatt Adams, Osmani, Thorpe, & Thornback, 2017).
- *Digital tools* can support long-term asset management records, as well as the setting up of technically and economically feasible reverse logistics chains to resupply future operations. Converging information packages on a central hub (accessible to all invested parties) can maximize circularity throughout all project phases and enhance the social value of the building as a future material bank.

22.4.2 Discussion: the potential benefits of, and barriers to, PSS and FaaS in delivering a more circular façade economy

The opportunities and barriers described in this chapter will resonate in the exploration of any product-service system model for the construction industry, and in fact to all types of CBM. The circular economy transition demands innovation far beyond the scope of technological advancement. It requires that we rethink the entire social, economic and legal background on which our engineering and construction projects take place, and that we revise preconceptions on topics such as utilitarian value, ownership and bankability, which have in many cases remained fundamentally unchanged for hundreds or even thousands of years.

The authors of this chapter believe that the restructuring of economic incentives and forms of value creation, based on shared benefits and long-term collaboration (Fig. 22.8), will play a key role in (1) the acceleration of investment in the (energy) renovation and upgrade of buildings and building envelopes; (2) the closing of resource gaps through local remanufacturing and redistribution channels, reducing transportation costs and emissions and cascading legacy equipment; and (3) optimizing and streamlining the façade construction value chain from its current structure as a zero-sum game, to one based on strategic partnerships and shared goals.

Barriers to implementation are diverse and deeply grounded but are not insurmountable. The ongoing projects on which this chapter is based continue to attract the growing interest of academic and professional partners, as well as significant investment of resources. Commercial pilot projects implementing the FaaS model are currently under development in the Netherlands and involve investors, bankers, contracts managers, legal advisors, property developers and – of course – FaaS providers.

Strengths	Weaknesses
- Well-integrated linear logistics process as foundation. - Total Value of Ownership approach. - Unique market reputation. - Experience as product system integrator. - Willingness to share incentives.	- Lack of knowledge/experience delivering integrated services. - Access to (long-term) finance. - Management of performance-based liabilities. - Need for corporate and technical reorganisation.
Opportunities	**Threats**
- New and more stable market, forms of value & revenue flows. - Material security through sustainable high-value recovery. - Access to performance data. - Fewer technical demarcations & potential conflict boundaries.	- Lack of standard procurement models and mechanisms. - Short-termism of investors/building owners. - Fragmented construction industry. - (Linear) fiscal and legal frameworks. - Lack of a (mainstream) market.

Figure 22.8 Noncomprehensive strength, weakness, opportunity and threat (SWOT) analysis for a traditional façade fabricator aiming to deliver Facades-as-a-Service.
Source: Diagram by Azcarate-Aguerre (2021).

22.5 Conclusion

This chapter has described the theoretical and practical foundations for the near-future implementation of a FaaS model for building envelope procurement. It has described how such a model could support the energy efficiency and material circularity transitions in the built environment. The changing roles of stakeholders, as well as the incentives these parties would have to assume new scopes and responsibilities have been discussed, and a summary of the present economic and industrial landscape has been presented.

The chapter, as well as the research work on which it is based, highlights the relevance of close and constant contact between academy and industry, and the role that practicing researchers must assume as mediators and advocates for collaborative models such as the one presented. FaaS (and PSS models in general) require a wide systemic perspective that looks beyond the present interests and modes of operation of individual stakeholders and by doing so aims to reorganize entire project consortia into more ecologically and economically sustainable structures. Achieving this goal requires a gradual adjustment of macroeconomic and legal policy, as well as of microeconomic decision-making mechanisms of individual stakeholders.

The multidisciplinary outline hereby described is the basis for future research work (some of it ongoing) across a number of fields: from the design and engineering of technically and economically recoverable products, through financial tools to support clients', suppliers' and financiers' decision-making processes and long-term balance-sheet analysis, and up to comprehensive recommendations for policymakers and industry organizations aiming to standardize and facilitate the implementation of PSS and FaaS models. Radical systemic change is considered an unlikely proposition. This chapter argues that the strength of CBM and PSS is that they can gradually find their way into our existing legal and economic system, benefiting current stakeholders, and while doing so lead to a more sustainable way of building, retrofitting and managing our built environment.

Acknowledgements

This chapter collects lessons learnt from two parallel research projects:

1. Since 2015 TU Delft's Faculty of Architecture and the Built Environment has worked on the Façade Leasing research project (a.k.a. Facades-as-a-Service). Largely funded by EIT Climate-KIC, and with significant additional investment – both financial and in-kind – from a range of industry partners and key stakeholders, the project has explored the range of questions, from the technological to the managerial, legal, and financial, which have been discussed in this chapter. Two pilot projects have been planned, engineered and built in the process, and a spin-off commercial pilot project is currently in late development phases.
2. Eurac and Arup are working in partnership for the development of an Arup-funded research work on the topic of 'Envelope for Service' (E4S). The research aims at developing an 'envelope for service' business model, providing a proof of concept for the

commercialization of a full-service façade package. E4S is a 'performance procurement'—integrated business model coupling technical complexity and multidisciplinary performance assessment, including a thorough façade value chain analysis. While the research is still ongoing, outcomes of the first phase have contributed to provide further insights and argument to this chapter. Authors kindly acknowledge Matteo Orlandi and Giulia Santoro from Arup Italy, Conor Cooney and Darren Walsh from Arup Cork.

References

Ardolino, M., Saccani, N., Gaiardelli, P., & Rapaccini, M. (2016). Exploring the key enabling role of digital technologies for PSS offerings. In *Paper presented at the 8th CIRP IPSS CONFERENCE-Product-Service Systems across Life Cycle, 2016*.

Azcarate-Aguerre, J. F., Klein, T., Den Heijer, A. C., Vrijhoef, R., Ploeger, H. D., & Prins, M. D. I. (2018). Façade leasing: Drivers and barriers to the delivery of integrated Facades-as-a-Service. *Real Estate Research Quarterly, 17*(3).

Baines, T., & Lightfoot, H. (2013). *Made to Serve: How manufacturers can compete through servitization and product service systems*. John Wiley & Sons.

Bocken, N. M., De Pauw, I., Bakker, C., & van der Grinten, B. (2016). Product design and business model strategies for a circular economy. *Journal of Industrial and Production Engineering, 33*(5), 308–320.

Cong, J.-C., Chen, C.-H., Zheng, P., Li, X., & Wang, Z. (2020). A holistic relook at engineering design methodologies for smart product-service systems development. *Journal of Cleaner Production, 272*, 122737.

da Costa Fernandes, S., Pigosso, D. C., McAloone, T. C., & Rozenfeld, H. (2020). Towards product-service system oriented to circular economy: A systematic review of value proposition design approaches. *Journal of Cleaner Production, 257*, 120507.

European Commission. (2020). *Circular economy—Principles for buildings design*.

Geissdoerfer, M., Morioka, S. N., de Carvalho, M. M., & Evans, S. (2018). Business models and supply chains for the circular economy. *Journal of Cleaner Production, 190*, 712–721. Available from https://doi.org/10.1016/j.jclepro.2018.04.159.

ING Economics Department. (2015). *Rethinking finance in a circular economy: Financial implications of circular business models*. Retrieved from Amsterdam: https://www.ing.nl/media/ING_EZB_Financing-the-Circular-Economy_tcm162-84762.pdf.

Klein, T. (2013). *Integral façade construction. Towards a new product architecture for curtain walls* (Doctoral dissertation). Delft: Delft University of Technology: TU Delft.

Leising, E. J., Quist, J., & Bocken, N. M. P. (2016). *Circular supply chain collaboration in the built environment* (MSc thesis). Delft: TU Delft.

Leising, R. M., Klein, T., Geldermans, R. J., & Azcárate-Aguerre, J. F. (2016). *Steel curtain walls for reuse* (MSc thesis). Delft: TU Delft.

Mach, T., Grobbauer, M., Streicher, W., & Müller, M. J. (2015). *MPPF—The multifunctional plug & play approach in façade technology* (p. 339) Technischen Universität Graz.

Överholm, H. (2017). Alliance formation by intermediary ventures in the solar service industry: Implications for product–service systems research. *Journal of Cleaner Production, 140*, 288–298. Available from https://doi.org/10.1016/j.jclepro.2015.07.061.

Pergande, B., Nobre, P. L., Nakanishi, A. C., Zancul, E. S., Loss, L., & Horta, L. C. (2012). Product-service system types and implementation approach. *Leveraging Technology for a Sustainable World* (pp. 43–48). Springer.

Rampersad, R., Prins, M., Heurkens, E. W. T. M., & Ploeger, H. D. (2016). *Financiële business cases voor circulaire vastgoedontwikkeling voor beleggende vastgoedontwikkelaars* (MSc thesis). Delft: TU Delft.

Tebbatt Adams, K., Osmani, M., Thorpe, T., & Thornback, J. (2017). Circular economy in construction: Current awareness, challenges and enablers. *Proceedings of the Institution of Civil Engineers—Waste and Resource Management*, *170*(1), 15−24. Available from https://doi.org/10.1680/jwarm.16.00011.

Tukker, A. (2004). Eight types of product−service system: Eight ways to sustainability? Experiences from SusProNet. *Business strategy and the environment*, *13*(4), 246−260.

Vezzoli, C., Kohtala, C., Srinivasan, A., Xin, L., Fusakul, M., Sateesh, D., & Diehl, J. (2017). *Product-service system design for sustainability*. Routledge.

Wang, X. V., & Wang, L. (2019). Digital twin-based WEEE recycling, recovery and remanufacturing in the background of Industry 4.0. *International Journal of Production Research*, *57*(12), 3892−3902. Available from https://doi.org/10.1080/00207543.2018.1497819.

Witjes, S., & Lozano, R. (2016). Towards a more circular economy: Proposing a framework linking sustainable public procurement and sustainable business models. *Resources, Conservation and Recycling*, *112*, 37−44. Available from https://doi.org/10.1016/j.resconrec.2016.04.015.

Afterword

When you are asked to write an afterword, you know that the next generation of researchers, designers and builders are ready to move the subject forward. And yes, when, around the turn of the millennium, we started to rigorously establish the topic of façade technology in research and teaching, there were only a few who saw this as becoming a discipline and a profession in its own right. Individual books such as the 'intelligent glass facades' by Andrea Compagno or the 'Façade Manual' by Thomas Herzog recognized the potential, but the market was still focused on purely production-oriented or technological solutions to the issues of air tightness, rain tightness and structural integrity of the building envelope. Today, it has become very clear that the building façade takes on much more than purely envelope-related functions. Rather, it contributes significantly to the overall performance of the building. This performance concerns both the conductional energy and the consideration of the energy used for the production of the façade (embodied energy) in relation to the consumption of energy. It also concerns, of course, the well-being of the users of the building.

Thus the time has arrived for a book that reflects the current state of knowledge in façade design, engineering and research. In addition to providing a valid collection of state-of-the-art knowledge, this book facilitates further development at an international level and aims to drive the interaction between research, design and industry.

Energy generation by the building envelope/façade will contribute to the provision of renewable energy, and this generation needs to be aligned with the perfect functionality of the façade to optimize the overall energetic performance of the building. This means that the hitherto unresolved issue of the complete integration of photovoltaics cells into the façade (BIPV) is an essential area that needs to be addressed. Yes, some good examples can be seen, but there still seems to be a large undefined obstacle to overcome to realize the enormous potential that the technology offers and to greatly increase the limited realization of BIPV. In this context, solar thermal collector solutions may also offer great potential for nontransparent surfaces!

Nevertheless, energy is not going to be the one and only permanent problem: yes, we will have to continue to save energy, no question, but if, as planned by the EU in 2030/2050, energy is produced entirely from renewable sources, energy consumption will no longer be the central problem. The question of material consumption will then become much more important, as it is undeniable that nonbio-based materials are limited. In this regard the three topics that will take on a central role are material harvesting, recycling of material/components and efficiency of material use.

On the subject of material harvesting, it is logical that bio-based materials have an advantage over mineral or metallic materials, which are more costly to procure and are finite. Accordingly, we see intensive research and development in this area. For the area of recycling, two different approaches are of interest: on the one hand, materials that can be recycled well such as glass, metal, polymers and paper, and on the other, the idea of dismantling entire structures and repurposing them with new functions. For the area efficiency of material use, new manufacturing technologies such as digital process control, digital subtractive manufacturing and digital additive manufacturing are exciting avenues. These manufacturing technologies are complemented by digital planning tools that optimize the performance within design, manufacturing, construction and use.

When considering digital design tools, the possibility opens up to include user satisfaction of a construction in a more complex planning process. It is then no longer exclusively about minimizing energy for a building but rather about maximizing user satisfaction and health, which must be ensured by the building and the façade.

A next level of consideration regarding the building envelope is the urban context: it is evident that in the next decades, half of the world's population will be concentrated in cities. The resulting problems of urban climate, that is, overheating, acoustic stress and energy consumption are obvious.

The building envelope can make a significant positive contribution in all of the aforementioned areas and is, therefore, a central component in terms of energy optimization, energy recovery, reduction of acoustic stress and reduction of overheating. It is exciting to observe how green facades, for example, can make excellent contributions in these areas; no wonder that many researchers are actively on the move here.

And of course, following the developments in research, engineering and industrial production must adapt these topics and make them marketable. This not only requires direct operations but also new business models such as leasing concepts for materials and components; exciting developments that can be found in this publication.

Naturally, we also know that this is not the end of the story. Development never ends—it always leans on the existing developed knowledge and is followed by the next solution. Themes and foci might adapt to future problems and change the position of the involved players—but it will not end. So, this is what this book is about: identifying themes and expressing the state of the art of knowledge while also pushing for the next ideas and steps to be taken.

Ultimately, this development is about people who are skilled and well educated and motivated to push the façade further: The Next Generation.

Ulrich Knaack[1,2]

[1]Institute of Structural Mechanics and Design, Chair of Façade Structures, Department of Civil and Environmental Engineering, Technical University of Darmstadt, Darmstadt, Germany, [2]Department of Architectural Engineering + Technology, Chair Design of Construction, Faculty of Architecture and the Built Environment, Delft University of Technology, Delft, The Netherlands

Index

Note: Page numbers followed by "*f*" and "*t*" refer to figures and tables, respectively.

A

Actual and committed costs, 407
   vs. time in product, 408*f*
Adaptability, defined, 185
Adaptation, 23–24
Adaptive building envelopes, 156
   challenges and requirements for automation of, 159–162, 161*f*
   control characteristics of, 164*t*
   current examples from real-life implementation and research activities, 167–174
      chromogenic technologies and solar shading, 167–170, 169*f*
      double-skin façades, 171–174
      operable windows, 170–171, 172*f*
   elements, characteristics, and logics for embedding intelligence in, 162–167, 166*t*
   purpose of automation in, 156–159
Adaptive capacity, 364–366, 365*f*
Adaptive resilience, 372–373
Adaptive Solar Façade, 191*f*
Additive fabrication, 100
Additive Manufacturing ISO/ASTM 52900, 428, 428*t*
Advanced building envelope materials, systems/operation, inverse design for
   building envelope research, 380–387
      building technology level, 383–384, 384*f*, 385*f*, 387*f*
      façade system level, 385–387
      material level, 380–383, 381*f*, 383*f*
   demonstration of, 389–399, 391*f*
      façade design elements, 392–395
      façade design option 1, 395–397
      façade design option 2, 397–398
      performance goals and indicators, 391–392
      simulation workflow, 392, 393*f*
   methods and digital tools for, 387–389, 388*f*
Advanced Fenestration Systems (AFS)
   advanced integrated façade/fenestration systems, 136
      airflow control in DSFs, 136–138
      double skin façades, 135–136, 135*f*
      fluid-flow façades, 139–140
      solar energy conversion in DSFs, 138–139
   characteristics, limitations and research trends for, 133*t*, 143*t*
   components, 121–135
      advanced transparent technologies for solar energy conversion, 131–135
      aerogel glazing, 125–126, 125*f*, 126*f*
      complex solar shading systems, 129–131
      glazing integrated phase change materials, 122–124, 123*f*
      smart glazing, 121–122, 122*f*
      vacuum insulated glazing, 127–129, 128*f*
   electricity generator in, 135
   overview of, 120*f*
   performance evaluation to support design and operation decision making, 141–143
Advanced integrated façade/fenestration systems (AIFS), 136
   airflow control in DSFs, 136–138
   double skin façades, 135–136, 135*f*
   fluid-flow façades, 139–140
   solar energy conversion in DSFs, 138–139

Advanced robotic systems, 458
Advanced transparent technologies, for solar energy conversion, 131–135
Aerogel, 125
Aerogel glazing, 125–126, 125f, 126f
Algorithm, 466t
Amorphous silicon PV solar cells (aSi), 209
Angular-selective retroreflective (AS-RR) materials, 70–71, 70f
Application programming interface (API), 409
Approved Document B (ADB), 353–354
Architectural styles, 4–5
Architecture, engineering, and construction (AEC), 409, 455–458
Artificial intelligence (AI), 529, 532
Artificial neural networks (ANNs), 172–174
AskNature, 188–189
Assembly-to-order (ATO), 456
Atocha Station Memorial Madrid, 108
Australian Building Code, 31
Australian National Construction Code (NCC), 235
Australian National University (ANU), 239–240
Autoreactive façade concept, 138f

B

Barcelona, Basilica de la Sagrada Família in, 460–461, 464–465
Baseline dynamic solar shading strategies, 394, 395t
Bending strength, 360–362
BERTIM project, 478–482
Bespoke Carousel System (BSC), 413
Bessemer process, 4
Big Data, 531–532
Bioinspiration, 184
Biology, 182–184
Biomimetic adaptive building skins (Bio-ABS), 181
  application of, 187–192
  bio-driven design concepts in building design, 182–184, 184f
  case study, 190–192
  definition of, 182
  designing, 185–187
    natural design principles, 185–187
    processes and levels, 185, 186f
  examples of, 182, 183f
  performance evaluation and benefits of, 192–194, 194t
  challenges and opportunities in, 195
  significance of, 182
Biomimetic design processes, defined, 188
Biomimetics, 181–182, 184
Biomimicry, 184
Biornametics, 184
Blinkenlight project, 314f
Blockchain technologies, 531–532
Bottom-up approach, 185, 188
Brick approach at TU Delft, 108–110, 110f
BrightFarm Systems, 293–294
Brock Commons façade system, 236–238, 239f, 240f, 241f, 242f, 243f, 244f, 245f
Brussels Expo, 314–315
Building Area Method, 419
Building automation, 160–161, 165
Building disassemblers, 547
Building envelope, 155, 521, 541–542, 548, 551, 553–554
  fire safety strategy
    assumptions embedded, 347
    for high-rise buildings, 344–347
    fundamental performance principles for, 348–353
    compartmentalization, 349
    fire ingress, 348–349
    fire spread, 350–353
    performance principles, 348
    incompatibilities, 355–356
    major misconceptions, 353–355
Building façade, 43, 291, 318
Building industry, 508
Building information modeling (BIM), 42, 222–223, 242, 251, 406, 456–458, 525, 530–531
Building-integrated agriculture (BIA), 288, 295
Building-integrated photovoltaics (BIPV), 201–202, 523–524, 527
Building-integrated solar thermal (BIST) collectors, 201–202
Building, life cycles phases of, 506f
Building operating systems, 531–532
Building owner, 547–548, 551
Building performance, 182

Index

Building performance simulation (BPS), 378–379, 382–384, 389
Building regulations, 346, 348–349, 353–354
Building "skins". *See* Building façades
BuiltHEAT façade, 270*f*
"Burn-out," design for, 347
Business models (BMs), 522, 526–528

C

Cable-driven parallel robot (CDPR), 482, 484–486, 488–490
Carbon storage, façade materials, 499
Casa de Musica, Porto, OMA/ABT, 103–105
Centre for Window and Cladding Technology (CWCT), 27
Circular business models (CBM), 526–528, 543–545, 550, 554–555
Circular Construction 4.0, 530–532
Circular design processes, 528–530
Circular economy (CE), 20, 22–23, 507, 519–522, 524, 527–531, 541–542, 554–555
Circular economy business models (CEBMs), 527
Circularity, 513
"Circularity-enabled" product, 541–542
"Circularity-enhancing" business environment, 542
Cladding, 437–439, 438*f*, 438*t*
Climate adaptive building shell (CABS), 386, 387*f*
Climate crisis, 359, 366, 368–369
Closed Cavity Façades (CCF), 136–138
Closed-loop cycles (CLC), 521–522, 527
Closing loops, 543*f*, 544
Cloud computing, 531–532
Comfort, 23–24
Common misconception, 349
Compartmentalization, 349, 350*f*
Competency, 341–343
Complex Fenestration Systems (CFS), 129
Complex solar shading systems, 129–131
Component tracking/material circularity technologies, 550
Computational design (CD), 454, 457–458, 462, 465, 531

Computer numerical control (CNC), 461
Concept, best practices/author's proofs of, 460–468
Construction and demolition (C&D), 528, 530
Construction context, automation and robotic technologies in
  BERTIM project, 478–482
    highly customized connector design, 479
    manufacturing processes/installation procedure, automation of, 479–481
    renovation tasks, digital tool for, 481–482
  construction robotics technology, context and background of, 476–478
  HEPHAESTUS project, 482–490
    CDPR/components, technical description of, 486
    first and second demonstrations, 489–490
    MEE/components, technical description of, 486–488
    proposed operation sequence, 488–489
Construction decarbonization, 234–235
Construction outcomes, 341
Construction robotics technology, 476–478
Construction 3D printing (C3DP), 425–432, 426*f*, 428*t*
Context, 403–406
Controlled-environment agriculture (CEA), 290
Convergence vision, mass customization, 458–460
Corning Gorilla Glass (GG), 111
Cosimulation, 384, 389
Covid-19 pandemic, 24, 325
Cradle-to-cradle (C2C), 521–522, 528–529
Cross-contamination, 48
Cross-laminated timber (CLT), 234–235
Crystal Houses (Amsterdam), 108, 109*f*
Crystalline solar cells (cSi), 208–209
Curtain wall, 360–361, 364–370, 365*f*, 441, 441*f*, 442*t*, 482
Curtain wall modules (CWMs), 475–476, 482–489, 484*f*
Cyber-physical systems (CPS), 454–455, 460, 466–468

D

Daily light integral (DLI), 291, 297–298
Daramu House, in Sydney, 233f
Daylighting performance, 392
Deep renovation, defined, 259
Degrees of freedom (DOF), 483–485
Demolition contractors, 547
Design decisions, 503
Design for disassembly (DfD), 522–525, 531
Design, resilience by
 ages/resilient future, designing for, 367–370, 369f
 bending strength, 360–362
  durability, stresses link resilience to, 361–362
  resilience/building skin, 360–361
  shocks and stresses, 361
 embodied resilience, 362–367
  adaptive capacity, 364–366
  forces of obsolescence, 362–363, 363t
  forces of obsolescence, adaptability as a vaccine for, 366–367
  maintenance/renovation, planned cycles of, 363–364
 strategies to enhance resilience, 370–373
  adaptive resilience, 372–373
  resilience planning, 370–371
  thermal resilience, 371–372
Design-for-compliancy approach, 20
Designing and evaluating interactive media façades, 326–328, 328f
Designing lightweight construction, 502
Digital fabrication, 458, 460–461, 464–465
Digital placemaking, 326
Digital transformation (DT), 10, 12, 453–458, 460
 building industry, 455–458
 manufacturing industry, 454–455
Directed energy deposition, 432
Disassembly and maintenance, design for, 522–524
Disassembly, designing for, 500
Distributed intelligence, 163
Double evacuated glazing, 127f
Double-skin façade (DSF), 158–159, 291, 295–296
Drones and swarm robots, 532
Dry joint, 93, 93f
Durability and adaptability, 362–367
Durability, stresses link resilience to, 361–362
Durable façade, 23–24
Dye-based thermochromics, 71–72, 73f
Dye-sensitized solar cells (DSSC), 131–132, 212, 213f
Dynamic building skins, 314, 319
Dynamic solar shading control strategy, 395, 395t
Dynamo, 192–193

E

Economic and financial standards, 552–553
Eidgenössische Materialprüfanstalt, 504
Electric junction box models, 542f
Electrochromic glazing, 523–524
Electrosynbionics, 184
Embodied carbon (EC), 496–497
Embodied energy (EE), 21–22, 496–497
Embodied greenhouse gas emissions, 524
Embodied resilience, 362–367
Emergent techniques, 432
End-effector, 485
Energiesprong, 267
Energy efficiency, 530–531
EnergyPlus, 192–193
Engineer-to-order (ETO), 455–456
 types, 529
Environmental performance, 181, 193, 195–196
Environmental product declaration (EPD), 504–505
Environmental regulation in buildings, 181
ESP-r, 192–193
European Union (EU), 257
Experimental Unit NEST, 526
Explanation provided earlier, 351
Explore Industrial Park, 413
Extend service life, maintenance/renovation as a strategy to, 363–364
Extended reality (XR), 530–531
External fire spread, 353–354
External fires, 348–349
External Wall Construction, 353–354

F

Fabric formwork, 437–439
Façade design, 404f, 405, 405f

Index

life cycle assessment in
  categories for life cycle assessment tools in, 507–508
  design principles for, 497–503
  façade industry, life cycle assessment/implementation in, 504–505
  façade typology, 496–497
  introduction to life cycle assessment tools in, 508–512
  life cycle assessment databases, brief summary of, 512–513
  method of life cycle assessment, 505–507
life cycle assessment tools in
  categories for, 507–508
  introduction to, 508–512
  LCA software programs in, 511*f*
Façade engineer. *See* Façade expert
Façade expert, role of, 6, 6*f*
Façade farming, 286–288, 293
"Façade first" approach, 55
Façade industry, 2–3, 12
  life cycle assessment and its implementation in, 504–505
Façade innovation, industry perspective by design, 28–30
  drivers, challenges, and opportunities, 18–30
  industry insights, 30–56
  method, 16–18
  organizational innovation, 25–28
  sustainability, 19–25
    adaptation, 23–24
    circular economy, 22–23
    comfort, 23–24
    durability, 23–24
    embodied energy, 21–22
    frameworks, 24–25
    operational energy, 21–22
    policies, 24–25
    resilience, 23–24
Façade interface design, 479*f*
Façade process, 11
  construction practices, 11
  design methods, 11
  fabrication processes, 11
  life cycle approaches, 11
Façade product, 10–11
  circular façades, 11

  eco-active façades, 10
  inter-active façades, 10–11
  re-active façades, 10
  recycled content in, 498
Façade typology, 496–497, 497*f*
Façades
  circular economy in
    circular business models, 526–528
    circular construction 4.0, 530–532
    circular design processes, 528–530
    disassembly and maintenance, design for, 522–524
    material passports/material banks, 524–526
    reduce, reuse, recycle, 520–522
  design principles for, 497–503, 498*f*, 499*f*, 500*f*, 501*f*, 502*f*, 503*f*
  environmental impact, 496–497, 503–513
  farming on, 290–296
    façades and technologies, vertical farming on, 291–293, 292*f*
    vertical farming implementation on, 293–296, 295*f*
    vertical farming, 290–291
Façades today, 2–8
  design and construction, 5–8
    actors' early-stage involvement in, 7, 7*f*
    barriers to construction innovation, 8*f*
  technological advancements, 3–5
Façades tomorrow, 8–11
  rethinking building skins, 10–11, 12*f*
  transformational change, 8–10
Façades, vertical farming on
  façades, farming on, 290–296
    façades and technologies, vertical farming on, 291–293, 292*f*
    vertical farming implementation on, 293–296, 295*f*
    vertical farming, 290–291
  productive façades, 296–303
    design criteria and optimization, 296–298, 299*f*
    optimal productive façade prototypes, 298–300, 300*t*, 301*f*
    system's social acceptance, 303, 304*f*
    tropical technologies laboratory, test setup and experimental results, 301–303, 302*t*

Façades, vertical farming on (*Continued*)
  urban agriculture, 287–288
  vertical greenery systems, 289–290
    benefits and drawbacks, 289–290
    definition and classification, 289
Façades-as-a-Service (FaaS)
  circular business models and product-service systems, 543–545
  cross-value chain perspective, 545–553
    hiring key performance indicators, from purchasing bricks to, 550–553
    linear façade, life cycle of, 546–547
    PSS and CE, Design and engineering of integral façade products and strategic life cycle planning for, 548–550, 549*f*
  propositions, 553–555
    lessons and following steps, 553–554, 553*f*, 555*f*
Facility management and building maintenance teams, 547
Fenner Hall (Canberra), 236–237, 239–240, 239*f*, 247*f*, 248*f*, 249*f*, 250*f*
Fire ingress, 348–349
Fire performance, 341, 343, 355
Fire resistance, 345
Fire safety strategy, 341, 347–348
Fire spread, 350–353, 351*f*, 353*f*
Flame spread, 350, 352*f*
Flectofin, 182
Fluid-Flow Glazing (FFG), 139–140, 141*f*
FOCCHI Spa, 482–483, 483*t*, 484*f*, 486–487
Food security, 286–288, 303
Forces of obsolescence, 362–363, 363*t*
  adaptability as a vaccine for, 366–367
Form-finding, 380, 399
Fourth Industrial Revolution (Industry 4.0), 2, 530
Freeform architecture, 37
Fused deposition modeling (FDM), 100, 428–429, 429*f*, 437

G

General data protection regulation, 530–531
Glass, 89
  alternative transparency concepts with, 101–103, 103*f*
  Vakko Headquarters (Istanbul), 103, 104*f*
  approach to transparent connections, 93–101
    dry joint, 93, 93*f*
    glass corners, 95–98, 96*f*
    glass–glass connections, 95, 95*f*
    glass–metal connections, 94, 94*f*, 95*f*
    heat-bonded glass connections, 98–100, 100*f*
    silicone adhesive–bonded connections, 93–94
    3D-printed glass connections, 100–101, 102*f*
  Atocha Station Memorial Madrid, 108
  vs. building materials, 89
  Casa de Musica, Porto, OMA/ABT, 103–105
  connections, 90–92, 92*f*
  Crystal Houses (Amsterdam), 108, 109*f*
  further development of brick approach at TU Delft, 108–110, 110*f*
  K11 Musea, Hong Kong, SOIL/EOC, 105, 106*f*
  as solid material, 107–108
  state of the art, 90
  thin glass, 110–112, 112*f*, 113*f*
  timeline, in architecture, 91*f*
  transparency property, 89
  Vidre slide, Duesseldorf, 105–107, 107*f*
Glass corners, 95–98, 96*f*
Glass–glass connections, 95, 95*f*
Glass–metal connections, 94, 94*f*, 95*f*
Glass 3D printing (G3DP), 432
Glazing integrated phase change materials, 122–124, 123*f*
Global domestic product (GDP), 520
Glued-laminated timber (GLT), 464–468
Glue-laminated timber (Glulam), 234–235
Graphical user interface (GUI), 409
Grasshopper, 192–193
Great Recession, 45–46
Greenhouse gas (GHG) emissions, 231–232, 495–496
The Grenfell Tower fire, 341
Gymnasium Façade, 182

H

Heat bonding process, 98

Index

Heat-bonded glass connections, 98–100, 100*f*
HEPHAESTUS project, 482–490, 490*f*, 491*f*
Heterogeneity, defined, 187
Hierarchical edge bundling, 416*f*
Hierarchy, defined, 186
Highly customized connector design, 479
Highly reflective (HR) materials, for UHI mitigation, 67–71
High-rise buildings, fire safety strategy for, 344–347
Horizontal overhang, vertical blinds with, 397–398, 397*f*
Horizontal shading angle (HSA), 392–397, 394*f*, 394*t*
Horticulture, 291, 303
Human-centered design, 531–532

I

IES VE, 192–193
In The Air Tonight project, 320–323, 321*f*
Incompatibilities, 355–356
Indoor environmental quality (IEQ), 118, 530–531
Industrialization, 258, 260
Industrialized construction, 260
Industrialized renovation, of building envelope
   degrees of, 260–261, 262*t*
   importance of building envelope for deep renovation, 259–260
   outlook for future, 272–277
      adaptability and circularity, 275–276
      process optimization, 276
      renovation as product, 276–277
      renovation market, 276
   and state of art, 261–272
      design and construction principles of, 265–272, 273*t*
      industrialized renovation process, 261–264, 263*f*
Industry 4.0, 454–455, 466–468
Industry experts, list of, 17*t*
Industry Foundation Class (IFC), 406, 481–482
Information and communication technology (ICT), 163
Innovation. *See also* Façade innovation
   definition of, 15, 34
Installation, 476–490, 481*f*, 483*t*, 484*f*
Insulating glazing units (IGU), 93, 523–524
Integrated design approach, 20
Interaction concepts, technologies/applications, 323–326, 324*t*
Interaction design, 313–314, 326–327
Interactive architecture, urban screens/media façades, 314–318, 317*f*
Interactive media façade design, 328–331, 330*t*, 331*f*
Interactive media façades—research prototypes, application areas/future directions
   designing and evaluating interactive media façades, approaches for, 326–328
   interaction concepts, technologies/applications, 323–326
      digital placemaking, 326
      participatory civics, 326
      playful encounters, 325
      raising awareness, 325
   interactive architecture, urban screens/media façades, 314–318
   interactive media façade design, framework for, 328–331
   practice and research, interactive media façades in, 318–322
International Energy Agency, 231–232
Internet of things (IoT), 163
Inverse design, 379–380
   building envelope research, 380–387
   demonstration of, 389–399, 391*f*
   direct design approach, comparison to, 379*f*
   methods and digital tools, 387–389, 388*f*
ISO 19650 standard, 456–457

K

K11 Musea, Hong Kong, SOIL/EOC, 105, 106*f*
Knowledge base, 409–410, 414–417
Knowledge-based engineering (KBE) in façades, 412–419
   application, constructing, 408–412, 410*f*, 411*f*, 411*t*, 412*f*
   context, 403–406

Knowledge-based engineering (KBE) in façades (*Continued*)
  example, 412–419, 413*f*, 414*f*, 415*f*, 416*f*, 418*f*, 420*f*
  knowledge base, 414–417
  knowledge capture, 414
  to optimize performance and cost, 406–408
  product model—digital tool implementation, 417–419
  results, 419–421
  UML modeling, 417
Knowledge capture, 409, 414
Knowledge Nurture for Optimal Multidisciplinary Analysis and Design (KNOMAD), 408–409
Knowledge Wheel, 418–419

L
Legal standards, 552
Lendlease, 27, 50–52
Life cycle assessment (LCA), 21, 142, 496, 503–514, 520–521
  databases, 512–513
  method of, 505–507
  software programs for, 509*t*
Light-emitting diodes (LEDs), 313–316, 318–322
Linear façade, life cycle of, 546–547, 546*f*
Local materials, designing with, 501
Longer service life, 363–364
Luminescence solar concentrator (LSC), 132, 209–211, 211*f*

M
Machine learning, 532
Major misconceptions, 353–355
Make-to-order (MTO), 456, 529
Manual heat bonding process, 98
Manufacturing processes/installation procedure, automation of, 479–481
Mass timber technologies, 234–240
  envelope prefabrication in timber buildings, 235–240
Material passports/material banks, 524–526
McKinsey Global Institute (MGI), 453
The MegaFaces media façade, 316*f*, 319–320

Mechanical, electrical, and plumbing (MEP), 523–524
Media architecture, 323, 329, 332
Media Architecture Awards, 318–319
Media surfaces, 318–319
Methodologies and tools Oriented to Knowledge-based engineering Applications (MOKA), 408–409, 418–419
Modular end effector (MEE), 486–489, 488*f*
Modular façade (MF), 268–269
Module, installation process of modules and services embedded in, 481*f*
Mud-spraying drones, 432
Multibiomechanism approach, 189, 190*f*
Multicriteria decision making (MCDM), 296–298, 389
Multifunctionality, defined, 186
Multistorey building, 232, 234–235, 241–242

N
Narrowing loops, 543, 543*f*
National University of Singapore (NUS), 296, 297*f*
Natural design principles, Bio-ABS, 185–187
Next generation of building envelopes, 40
Nondye thermochromics, 71
Nutrient film technique (NFT), 291–293, 295–296

O
Off-site construction, 276
Off-site manufacturing facility, 480*f*
"One-off" nature, of construction projects, 7–8
Open Studio, 192–193
Operational energy, 21–22
Optimal control, 159, 389–391, 399
Optimization, 378–380, 383–387, 389–390, 398
  algorithms, 389
Organic Solar Cells (OSC), 131–132
Organization of the Petroleum Exporting Countries, 528–529

P
Paris Agreement, 19

Index

Participatory civics, 326
Particle-bed 3D printing and binder jetting, 430, 431f
Path toward performance based commissioning, 551−553, 552f
Penalty weights, 395t
Performance monitoring/management technologies, 549−550
Performance principles, 348
Perovskite, 131−132
Phase change materials (PCMs), 64−65, 122−124
Photovoltaic (PV) technologies, 131−132
Photovoltaic (PV)-thermal collectors, 201−202
Playful encounters, 325
Plug-and-play and hackable systems, 370
Plus-energy buildings, 527
Policy-making and research, 526
Practice and research, innovations in interactive media façades in, 318−322, 319f, 321f
Preassembled configurations, 271−272, 272f
Precast concrete cladding (PCC), 406
Prefabricated rainscreen façades, 270−271
Prefabricated sandwich panels, 266−267, 266f, 267f
Prefabrication, 478, 480−481
  of timber envelopes, 235−240, 237f, 238f
Process innovation, 5−8, 12f
Procurement standards, 551−552
Product and system developers, 546
Product-based commissioning, conflicts and limitations of, 551
Product innovation, 3−5, 12f
Product model (PM), 406−412, 414−415, 417−418
Product model—digital tool implementation, 417−419
Productive façade (PF), 293, 296−303, 298f
Product-service systems (PSS), 543−545, 548−550
  broad categorization of, 544f
  product-oriented, 544
  result-oriented, 545
  use-oriented, 544−545
Proposed and experimental farming façades, 294f
Proposed operation sequence, 488−489

Q

Quantum dots, thermal quenching/antiquenching effects in, 72f
Quasiunitized façade system, 238

R

Radiative coatings, 75−78, 76f, 77f
Radiative cooling, 75−77, 76f
Radio-frequency identification (RFID), 525, 531−532
Raising awareness, 325, 326f
Reduce, reuse, recycle, 520−522
Remanufacturing, 544, 555
Renewable Energy Sources (RES), 118
Renovation tasks, digital tool for, 481−482
Renovation, 364
ResearchGate, 202
Resilience, 23−24
  building skin, 360−361
  planning, 370−371
Retroreflective (RR) materials, 69
Reused façade elements, 497
Reverse logistics, 544, 554
Reversible construction, 522−523
Ri.Fa.Re. project, 462−465
Robotic technology, 475−476
Robotic Wood Printing, 432
Robot-oriented design (ROD), 475−476
RU2114_BracketSelection rule, 415−417

S

ScienceDirect, 202
Scopus, 202
Semifluid material extrusion (SME), 429−430, 430f
Semi-transparent BIPV technologies, example of, 132f
Sentry glass (SG)-laminated corner, 97, 97f
Shading, 440−441, 440f, 440t
Shocks/stresses, 361
Shotcrete 3D printing, 432
Silicone adhesive−bonded connections, 93−94
Simulation workflow, 392, 393f
Singapore Construction Industry Transformation Map, 406
Single-Task Construction Robots (STCRs), 476−477
SkecthUp, 192−193

Skin systems, additive manufacturing in construction 3D printing, AM techniques for, 427–432
  directed energy deposition, 432
  emergent techniques, 432
  fused deposition modeling, 428–429
  particle-bed 3d printing and binder jetting, 430
  semifluid material extrusion, 429–430
  wire and arc additive manufacturing, 430–431
 projects for, 432–441
  cladding, 437–439
  curtain walls, 441
  shading, 440–441
  walls and envelopes, 433–437
Skyscrapers, 55
Slowing loops, 543f, 544, 547
Smart glazing, 121–122, 122f
Smart Window concept, 211–212, 212f
Social objectives, 341
Solar buildings, 202
Solar control in architectural glazing, 120f
Solar photovoltaic façades
 archetypes of innovation, 207–218, 208f
  innovation in opaque PV façade, 216–218, 216f
  innovation in transparent PV façades, 207–216, 208f, 210f
 challenges and needs, 203–207, 205t
  availability of products, 207
  durability and maintenance, 206
  energy integration, 206–207
  façade esthetics and technical complexity, 206
  performance levels, 204–205
 digitalization of, 222–223
 integrating with building energy system, 218–222, 219f
  energy-matching approach, 219–220, 220f
  renewable electricity distribution system, 220–222, 221f
Solar Thermal Blinds (STB), 138–139
Solar window block, 218f
Solid material, glass, 107–108
Stakeholders, 541–542, 545–546, 548
Standard temperature *vs.* time curve, 349

Static solar shading devices, 392–394, 394f
"Stay-put" strategy, 346–347, 352
Stick curtain wall systems, 4–5
Supercool materials, 63–64
Suspended particle device (SPD), 139–140
Sustainability, 368–369
 in construction, 9
Sustainable Development Goals, 19

T
Tall buildings, and façades, 232–234
Technology readiness level (TRL), 433, 441
Temperature-triggered Air Flow(er), 182
TetraBIN project, 322, 322f
Thermal resilience, 371–372
Thermochromics, 71–75, 72f
 mortar, reversibly, 73f
 window, 74f
Thin glass, 110–112, 112f, 113f
3D-printed glass connections, 100–101, 102f
3R design approach, 522
Timber-frame panels, 267, 268f
Top-down approach, 185, 188
Total Façade Management Systems, 42
Toxic waste management, 3R for, 522
Transparency
 in façade, 4
 glass, 89
Transparent connections, 93–101
 dry joint, 93, 93f
 glass corners, 95–98, 96f
 glass–glass connections, 95, 95f
 glass–metal connections, 94, 94f, 95f
 heat-bonded glass connections, 98–100, 100f
 silicone adhesive–bonded connections, 93–94
 3D-printed glass connections, 100–101, 102f
Transparent structural silicone adhesive (TSSA), 93
Triple Helix model, 28
Tropical Technologies Laboratory (T^2 Lab), 296, 297f, 299, 301–302, 302f
Trump Tower fire, 346, 346f
TRYSNS, 192–193

Index

U

Unified modeling language (UML) modeling, 410–412, 417, 417f
Unitized curtain walls (UCWs), 233, 365f
  benefits of, 233–234
Unitized glazed systems, 4–5
Unitized timber envelopes, 241–252, 251f
  design integration, 251–252
  design methodology, 242
  design process, 242–250, 252f
    frame, 245–250, 249t
    rainscreen, 250
    wall, 243–245
University of British Columbia (UBC), 237
Urban agriculture, 287–288
Urban heat island (UHI) mitigation, 61–62, 62f, 65f, 66t
  highly reflective materials, 67–71
  radiative coatings, 75–78, 76f, 77f
  thermochromics, 71–75, 72f
Urban mining, 531–532
Urban overheating effect, 61–62
Urban resilience, 303, 306
Urban screens, 315
Urbanization, in architecture, 232
U-values and window technologies, development of, 119f

V

Vacuum insulated glazing (VIG), 127–129, 128f
Vakko Headquarters (Istanbul), 103, 104f
Venetian blinds systems, 129–130
Ventilated façade, 270
Vertical farming (VF), 286–288, 290–298, 303
Vertical fins, roller shade with, 395–397, 396f
Vertical greenery systems (VGS), 288–290, 289f
Vertical shading angle (VSA), 392–394, 394f, 394t
Vidre slide, Duesselorf, 105–107, 107f
View fraction (VF), 392
VIKOR optimization method, 298

W

Walls and envelopes, 433–437, 434t, 435t, 436f
Window-to-wall ratio (WWR), 385–386, 385f
Wire and arc additive manufacturing (WAAM), 430–431, 431f
World War II, 528–529

Z

Zero-energy buildings, 527

Printed in the United States
by Baker & Taylor Publisher Services